AF356378

Advances in Developmental Biology
in Tree Fruit and Nut Crops

Advances in Developmental Biology in Tree Fruit and Nut Crops

Guest Editors

Zhaohe Yuan

Bo Li

Yujie Zhao

Basel • Beijing • Wuhan • Barcelona • Belgrade • Novi Sad • Cluj • Manchester

Guest Editors

Zhaohe Yuan
College of Forestry
Nanjing Forestry University
Nanjing
China

Bo Li
Shandong Academy of Grape
Shandong Academy of
Agricultural Science
Jinan
China

Yujie Zhao
College of Horticulture
Henan Agricultural
University
Zhengzhou
China

Editorial Office
MDPI AG
Grosspeteranlage 5
4052 Basel, Switzerland

This is a reprint of the Special Issue, published open access by the journal *Horticulturae* (ISSN 2311-7524), freely accessible at: https://www.mdpi.com/journal/horticulturae/special_issues/2YE02K1N11.

For citation purposes, cite each article independently as indicated on the article page online and as indicated below:

Lastname, A.A.; Lastname, B.B. Article Title. *Journal Name* **Year**, *Volume Number*, Page Range.

ISBN 978-3-7258-3799-1 (Hbk)
ISBN 978-3-7258-3800-4 (PDF)
https://doi.org/10.3390/books978-3-7258-3800-4

Contents

About the Editors

Zhaohe Yuan

Zhaohe Yuan, Ph.D. and professor, in College of Forestry and Grassland, Nanjing Forestry University. He was the former chairman of the Working Group Pomegranate and Minor Mediterranean Fruits of the International Society for Horticultural Science (ISHS) from Sept. 2013 to Feb. 2022 and is the Honorary Chairman of the Pomegranate Section of the Chinese Society for Horticultural Science. He mainly engages in research on tree fruit germplasm resources and genetic breeding, molecular biology, and genomics, especially in the evaluation and innovative utilization of pomegranate and other tree fruit germplasm resources. From January 1994 to December 1996, he successively worked as a visiting scholar at the Irrigated Agricultural Research and Extension Center, Washington State University, and the Department of Horticulture, University of Maryland in the United States. He has presided over more than 30 national and provincial research projects, including the National Natural Science Foundation of China, the National Key Basic Research Program, the National Science and Technology Support Program, and the Science and Technology Development Plan. He has published more than 230 articles in *Plant Biotechnology Journal*, *Plant Physiology*, *Food Chemistry*, *International Journal of Molecular Sciences*, *Journal of Genetics and Genomics*, and other publications. He obtained more than 50 patents in China and abroad, and edited more than 10 books, including *China Fruit Science and Practice—Pomegranate*, *Acta Horticulturae 1089*, *Modern Fruit Cultivation in China*, *Integrative Management of Pomegranate Diseases and Pests in China*, and *Pomegranate Culture, Art and Functional Utilization*. He serves as a reviewer for journals such as *Plant Biotechnology Journal*, *The Plant Journal*, *Journal of Integrative Agriculture*, *Forestry Science*, and *Journal of Fruit Science*, as well as being an expert reviewer for the National Natural Science Foundation of China. His research results have been widely applied in the national tree fruit industry and provided technological support for China's pomegranate industry upgrading.

Bo Li

Bo Li, Ph.D. from Shanghai Jiao Tong University, is a Research Professor and Director of the Shandong Academy of Grape. He concurrently serves as the Head of the Jinan Comprehensive Experimental Station of the China Agriculture Research System, Vice President of the Grape Branch of the Chinese Society of Agronomy, and a member of the Shandong Province Forestry and Grass Variety Review Committee. He is also a Master's supervisor at Shandong Agricultural University, Qingdao Agricultural University, and Yangzhou University. His primary research areas include fruit tree breeding and efficient cultivation. He has approved five new fruit tree varieties, including grapes, cherries, and apples, and registered two new grape varieties. He pioneered a precise parent selection technique for grape breeding and a greenhouse seedling tray propagation technology. Li also established a high-efficiency cultivation system for grapes on saline-alkali soils based on "root zone restriction," featuring "alternate irrigation with brackish and freshwater" and supported by "efficient canopy management". This system has been widely demonstrated and promoted in the Yellow River Delta area. In the past five years, he has led more than 10 national and provincial research projects, including the National Key Research and Development Program of China, the China Agriculture Research System, the Shandong Major Science and Technology Innovation Project, and the Shandong Agricultural Seed Project. He has been granted 10 national invention patents, developed two provincial and municipal standards, and five group standards, and recognized six agricultural technology innovations promoted by Shandong Province. Li has authored two books and published over 100 papers in renowned international and domestic journals such as *Food Research International, Frontiers in Plant Science,* and *Acta Horticulturae Sinica.* He has received more than ten awards, including the Shandong Science and Technology Progress Award, the Higher Education Scientific Research Outstanding Achievement Award, and various provincial-level honors.

Yujie Zhao

Yujie Zhao, Ph.D., College of Horticulture, Henan Agricultural University. She is mainly engaged in fruit tree molecular biology and genetic breeding, as well as fruit tree development in biology-related research. From April 2022 to April 2023, she studied as a Joint PhD student at Chiba University. She has published 7 SCI papers and 3 Chinese papers as the first author. She presided over the project of the Science and Technology Department of Henan Province: Creation of new germplasm of apple suitable for laborsaving cultivation (242102110284). She participated in the publication of two monographs, 'Integrative Management of Pomegranate Diseases and Pests in China' and 'Pomegranate Culture, Art and Functional Utilization'. Research interests: (1) Apple cross-breeding and molecular genetic breeding, (2) study on callus stem cell induction and its developmental mechanism in apple leaves, and (3) fruit tree developmental biology such as pomegranate ovule development.

 *horticulturae*

Editorial

Advances in Developmental Biology in Tree Fruit and Nut Crops

Zhaohe Yuan [1,*], **Bo Li** [2] **and Yujie Zhao** [3]

[1] College of Forestry and Grassland, Nanjing Forestry University, Nanjing 210037, China
[2] Shandong Academy of Grape, No. 3666, Erhuandong Rd., Jinan 250199, China
[3] College of Horticulture, Henan Agricultural University, No. 63 Agricultural Road, Zhengzhou 450002, China
* Correspondence: zhyuan88@hotmail.com

1. Introduction

Tree fruit and nut crops constitute a vital component of global agriculture, not only by virtue of their nutritional and economic importance, but also because of the complexity of their developmental processes. These crops often exhibit intricate regulatory networks that govern flowering, fruit set, fruit expansion, ripening, and senescence, all of which are significantly influenced by environmental conditions, genetic makeup, and horticultural practices [1]. Over the last two decades, extensive research has revealed that the developmental biology of tree fruit and nut crops is shaped by a multilayered interplay of hormonal signaling, transcriptional regulation, epigenetic modification, and metabolic reprogramming [2]. Advances in molecular tools and high-throughput "omics" approaches have accelerated our ability to decipher these complex networks, thus opening new frontiers in fruit quality improvement, stress tolerance, and sustainable orchard management [3].

Despite these remarkable strides, many gaps remain. One such gap concerns the interplay between hormone homeostasis and the transcription factors controlling fruit development. Classical phytohormones—such as auxins, gibberellins, cytokinins, abscisic acid, and ethylene—are well established as central regulators of fruit development and ripening [4]. However, new classes of regulators, including microRNAs and long noncoding RNAs, have been identified as having profound influences on gene expression and metabolic pathways during the progression of fruit growth and senescence [5]. The role of these non-coding RNAs in perennial fruit crops is less well characterized than in annual species such as tomato or Arabidopsis, indicating a pressing need for more comprehensive functional studies in tree fruit and nut crops [6].

In addition to hormonal and transcriptional regulation, fruit development is strongly impacted by carbohydrate metabolism, cell wall remodeling, and secondary metabolite biosynthesis. Carbohydrate availability in the developing fruit depends on source–sink relationships, which are dynamic throughout the growing season. Sink strength is modulated by the activity of sugar transporters, invertases, and other enzymes that allocate carbon to the developing fruit [7]. Meanwhile, the biosynthesis of pigments and secondary metabolites—such as anthocyanins, flavonols, and terpenoids—plays a key role in fruit quality, influencing color, flavor, and nutritional value [8]. These traits are crucial for consumer acceptance and market competitiveness. Modern breeding programs increasingly target these quality attributes, seeking to integrate molecular markers for sugar, acid, and secondary metabolite content into conventional breeding pipelines [9]. However, the genetic and biochemical networks underlying these traits can vary significantly across different tree fruit and nut species, complicating efforts in translating knowledge from model systems to diverse germplasm resources [10].

Received: 5 March 2025
Accepted: 10 March 2025
Published: 17 March 2025

Citation: Yuan, Z.; Li, B.; Zhao, Y. Advances in Developmental Biology in Tree Fruit and Nut Crops. *Horticulturae* 2025, 11, 327. https://doi.org/10.3390/horticulturae11030327

Climate change further complicates our understanding of developmental biology in perennial fruit and nut crops. Rising temperatures, altered rainfall patterns, and the increased frequency of extreme weather events challenge the stability of orchard systems worldwide [1]. Changes in winter chill accumulation, for instance, can disrupt dormancy release and flowering times, leading to mismatches between pollinators and bloom periods, ultimately affecting fruit set. Concurrently, abiotic stresses such as drought and salinity can drastically reduce photosynthetic efficiency, alter hormonal homeostasis, and accelerate or delay ripening [2]. Consequently, the need to develop climate-resilient cultivars and adaptive orchard management practices has never been more urgent. Researchers are thus intensifying their efforts to identify stress-responsive genes, investigate their regulatory mechanisms, and incorporate this knowledge into breeding programs and horticultural protocols [3].

In this rapidly evolving landscape, the application of cutting-edge technologies—ranging from CRISPR/Cas9-based gene editing to single-cell transcriptomics—has opened up unprecedented opportunities for dissecting and manipulating key developmental pathways [4]. Tree fruit and nut crops have historically lagged behind annuals in adopting such technologies, partly due to their extended juvenile phases, large genome sizes, and complex polyploidy in certain genera [5]. Yet, recent successes in the transformation and editing of fruit tree species suggest that the gap is closing. For instance, breakthroughs in Agrobacterium-mediated transformation protocols and the development of genotype-independent transformation systems have significantly improved the feasibility of functional genomics studies in fruit crops [6].

Furthermore, the field is witnessing an expansion of integrated multi-omics approaches. By combining genomics, transcriptomics, proteomics, metabolomics, and epigenomics, researchers are painting an increasingly comprehensive picture of how multiple layers of regulation converge to shape fruit and nut development [7]. In particular, integrative omics has proven instrumental in pinpointing candidate genes and pathways linked to traits of agronomic interest, such as fruit size, flavor, texture, and stress tolerance [8]. Once identified, these candidate genes can be validated through functional assays, thus expediting the process of varietal improvement.

However, the ultimate success of such endeavors hinges on effective translational pipelines. Breeders, growers, and policymakers must collaborate to ensure that the knowledge gleaned from molecular and physiological studies translates into practical innovations that benefit the entire horticultural value chain [9]. From advanced phenotyping platforms that enable high-throughput trait evaluation to orchard-level management strategies that optimize resource use, the potential for transformative impact is immense. Equally important is the need to address sustainability and environmental stewardship. Future orchard systems must be designed to reduce inputs such as water, fertilizers, and pesticides while maintaining or enhancing yield and quality [10]. By leveraging insights into the developmental biology of tree fruit and nut crops, researchers and stakeholders can chart a path toward more resilient and sustainable horticultural systems worldwide.

This Special Issue, entitled "Advances in Developmental Biology in Tree Fruit and Nut Crops", sought to address some of these critical gaps by bringing together a diverse array of studies. While we will not cite these articles directly here, the collective body of work provides new insights into fruit development from the perspectives of molecular genetics, physiology, and applied horticulture. In the following sections, we will first provide an overview of the published articles, summarizing how they contribute to the understanding of developmental biology in perennial fruit systems. We will then conclude with a broader discussion on the future directions and research opportunities that have emerged from this compilation of studies, underscoring the importance of interdisciplinary approaches and global collaboration.

2. Overview of Published Articles

In total, this Special Issue comprises 15 papers covering a wide range of tree fruit and nut species, including blueberry, peach, grape, pomegranate, pear, breadfruit, sweet cherry, almond, avocado, strawberry, *Torreya grandis*, and olive. These contributions span multiple thematic areas, ranging from transcriptomic and metabolomic analyses to phylogenetic studies, genetic characterization, and investigations into the hormonal regulation of fruit development. Such diversity reflects the multifaceted nature of developmental biology in perennial crops and highlights the breadth of cutting-edge research being conducted in this arena.

One unifying theme among several papers is the focus on the molecular underpinnings of fruit quality traits. This emphasis aligns with a longstanding interest in identifying and characterizing the genes involved in biosynthetic pathways for pigments, sugars, acids, and flavor volatiles. By integrating transcriptomic data with metabolite profiling, these studies shed light on the key regulatory nodes that govern fruit color, texture, and taste. Another cluster of articles delves into the responses of fruit trees to abiotic stressors, such as drought and salinity, offering novel insights into how environmental conditions can perturb developmental programs and how plants adapt on both physiological and molecular levels.

Several contributions also explore the role of hormonal regulation in fruit set, enlargement, and ripening, examining how exogenous or endogenous shifts in hormones like abscisic acid, gibberellic acid, or ethylene can trigger distinct developmental outcomes. Notably, these studies highlight the context dependency of hormonal signaling, emphasizing that cultivar-specific genetic backgrounds and orchard practices can modulate the plant's responsiveness to hormonal cues.

In addition, this Special Issue includes research that expands our knowledge on the genetic diversity of underutilized or regionally important crops such as breadfruit and Torreya grandis. By leveraging molecular markers or chloroplast genome profiling, these studies underline the need to conserve and characterize genetic resources for future breeding and germplasm enhancement. Likewise, work on the self-incompatibility systems in almond germplasm illustrates the intricate reproductive biology of nut crops, with direct implications for orchard pollination strategies.

A number of articles in this collection also adopt integrative omics frameworks, pairing transcriptomics with metabolomics or combining genetic and phenotypic data to build comprehensive models of fruit development. These integrative approaches not only deepen our fundamental understanding, but also pave the way for translational applications in breeding and crop management. From a technical standpoint, such studies serve as exemplars of how high-throughput sequencing, bioinformatics, and advanced analytical chemistry can be harnessed to elucidate complex biological processes.

Lastly, several papers explore the evolutionary and phylogenetic relationships among cultivars or species, offering clues about how developmental pathways have diverged or been conserved across lineages. Such evolutionary perspectives can inform breeding strategies by pinpointing ancestral alleles or pathways that may be manipulated for improved performance under contemporary agricultural conditions.

Collectively, the contributions to this Special Issue provide a rich tapestry of the current state of developmental biology in tree fruit and nut crops. They highlight emerging trends, such as the growing reliance on integrative omics and the heightened interest in stress adaptation, while reaffirming the central importance of hormone signaling and genetic diversity in shaping fruit traits. Together, these articles reinforce the notion that developmental biology is not a static discipline, but rather a rapidly evolving field that sits at the nexus of molecular genetics, physiology, ecology, and agronomy.

3. Summary and Future Outlook

The body of work featured in this Special Issue underscores the extraordinary progress that has been made in understanding the developmental biology of tree fruit and nut crops. Yet, the journey is far from complete. Looking ahead, we anticipate several key directions that will define the research agenda in this domain. These directions will likely encompass advanced genetic and genomic tools, integrative approaches to studying stress responses, and a renewed emphasis on translating fundamental discoveries into orchard-level solutions. By capitalizing on emerging methodologies, researchers and practitioners can tackle the pressing challenges of climate change, resource limitation, and evolving consumer preferences.

One promising avenue for future research is the continued expansion of CRISPR/Cas-based genome editing. Already, CRISPR/Cas9 and its related systems have demonstrated considerable potential in modifying genes related to fruit ripening, disease resistance, and stress tolerance. Despite the challenges posed by long juvenile phases and complex reproductive cycles, several fruit crops have been successfully edited, signaling a paradigm shift in functional genomics. As the off-target effects are minimized and transformation efficiencies improve, these technologies will become increasingly feasible for perennial species. In particular, the refinement of base editing and prime editing techniques offers the potential to introduce precise nucleotide changes without inducing double-stranded breaks, thus further accelerating crop improvement.

In parallel to this, advanced phenotyping tools and horticultural precision practices are poised to play a greater role in bridging the gap between laboratory research and orchard management. Automated sensor networks, drone-based imaging, and high-throughput phenotyping platforms enable the real-time monitoring of fruit development, allowing for the early detection of stress or disease and the fine-tuning of irrigation, fertilization, and pruning regimes. Coupled with integrative omics, these phenotyping strategies can help unravel the complex interactions between genotype, environment, and management practices, thereby offering actionable insights for growers. Such data-rich approaches will also facilitate the development of predictive models for yield forecasting, fruit quality, and resource allocation.

From a breeding perspective, genome-wide association studies (GWAS) and quantitative trait locus (QTL) mapping will continue to be indispensable for identifying the genetic loci that govern key developmental traits. As the cost of next-generation sequencing continues to decline, we can expect an influx of high-quality reference genomes and pan-genomes for a wider array of tree fruit and nut species. These genomic resources, in turn, will serve as the foundation for marker-assisted selection, genomic selection, and the eventual incorporation of genome editing technologies. In addition, epigenetic modifications—such as DNA methylation and histone modifications—are emerging as important layers of regulation that can influence traits across multiple generations. Future research should prioritize understanding how epigenetic states are established, maintained, and reset in perennial crops, and how they might be harnessed for stable trait improvement.

Equally critical is the need to deepen our understanding of how tree fruit and nut crops will respond to a rapidly changing climate. As global temperatures rise and weather patterns become more erratic, traits such as heat tolerance, water-use efficiency, and resilience to biotic and abiotic stress will become even more pivotal. Researchers must continue to dissect the molecular and physiological basis of stress responses, integrating knowledge on hormone signaling, transcriptional networks, and metabolic shifts. This work will be essential for breeding or engineering cultivars to maintain high yields and quality under suboptimal or fluctuating environmental conditions.

Furthermore, the conservation and utilization of genetic diversity will be paramount for ensuring long-term sustainability. Many fruit and nut species have wild relatives or landraces that harbor alleles conferring resistance to pests, diseases, or environmental extremes. Efforts to collect, characterize, and conserve these germplasm resources must be intensified. Genebanks, botanical gardens, and in situ conservation sites will play pivotal roles, as will collaborations with local and Indigenous communities that have historically stewarded diverse cultivars. The continued development of genomic tools will make it easier to screen these resources for beneficial traits, thereby broadening the genetic base available to breeders.

On a more holistic level, interdisciplinary collaboration will be essential for advancing developmental biology in horticulture. Plant physiologists, molecular biologists, geneticists, breeders, pathologists, entomologists, soil scientists, and socio-economists must work together to address the complex challenges facing orchard systems. This includes not only the biological dimensions of crop development, but also the socio-economic contexts that shape how research is funded, how new varieties are released, and how orchard practices are adopted. Partnerships with industry stakeholders can facilitate the rapid deployment of new technologies and ensure that scientific advances align with market demands and regulatory frameworks.

Finally, education and outreach are indispensable for translating scientific breakthroughs into tangible benefits for growers and consumers. Extension programs, workshops, and digital platforms can help disseminate new knowledge and best practices, ensuring that smallholder farmers and large-scale producers alike can capitalize on the latest developments. Public engagement is also critical for addressing concerns related to genetically modified or gene-edited crops, and for fostering a broader appreciation of the complexities involved in horticultural innovation.

In conclusion, the future of developmental biology in tree fruit and nut crops is exceedingly bright. The studies featured in this Special Issue have drawn attention to the sophistication of the current research and the breadth of challenges that remain. By embracing new technologies, forging interdisciplinary collaborations, and committing to sustainable and inclusive approaches, we can ensure that the insights gained in the laboratory translate into resilient orchard systems and high-quality produce for generations to come.

On behalf of the Editorial Board, we extend our heartfelt thanks to all the authors, reviewers, and editorial staff whose dedication and expertise made this Special Issue possible. We look forward to witnessing the continued evolution of this field and the transformative impact that it will have on horticultural science and global food security.

Acknowledgments: The author thanks all the contributors and reviewers for their valuable contributions and support from the section editors of this Special Issue.

Conflicts of Interest: The author declares no conflicts of interest.

List of Contributions

1.   Ma, B.; Song, Y.; Feng, X.; Guo, Q.; Zhou, L.; Zhang, X.; Zhang, C. Exogenous Abscisic Acid Regulates Anthocyanin Biosynthesis and Gene Expression in Blueberry Leaves. *Horticulturae* **2024**, *10*, 192. https://doi.org/10.3390/horticulturae10020192.

2.   Daley, O.O.; Alleyne, A.T.; Roberts-Nkrumah, L.B.; Motilal, L.A. Microsatellite Sequence Polymorphisms Reveals Substantial Diversity in Caribbean Breadfruit

[*Artocarpus altilis* (Parkinson) Fosberg] Germplasm. *Horticulturae* **2024**, *10*, 253. https://doi.org/10.3390/horticulturae10030253.

3. Guo, Y.; Song, C.; Gao, F.; Zhi, Y.; Zheng, X.; Wang, X.; Zhang, H.; Hou, N.; Cheng, J.; Wang, W.; et al. Exploring the PpEXPs Family in Peach: Insights into Their Role in Fruit Texture Development through Identification and Transcriptional Analysis. *Horticulturae* **2024**, *10*, 332. https://doi.org/10.3390/horticulturae10040332.

4. Li, X.; Cai, Z.; Liu, X.; Wu, Y.; Han, Z.; Yang, G.; Li, S.; Xie, Z.; Liu, L.; Li, B. Effects of Gibberellic Acid on Soluble Sugar Content, Organic Acid Composition, Endogenous Hormone Levels, and Carbon Sink Strength in Shine Muscat Grapes during Berry Development Stage. *Horticulturae* **2024**, *10*, 346. https://doi.org/10.3390/horticulturae10040346.

5. Wan, R.; Yang, Z.; Liu, J.; Zhang, M.; Jiao, J.; Wang, M.; Zhang, K.; Hao, P.; Liu, Y.; Bai, T.; et al. Identification of Laccase Genes in Grapevine and Their Roles in Response to *Botrytis cinerea*. *Horticulturae* **2024**, *10*, 376. https://doi.org/10.3390/horticulturae10040376.

6. Yin, K.; Cheng, F.; Ren, H.; Huang, J.; Zhao, X.; Yuan, Z. Insights into the PYR/PYL/RCAR Gene Family in Pomegranates (*Punica granatum* L.): A Genome-Wide Study on Identification, Evolution, and Expression Analysis. *Horticulturae* **2024**, *10*, 502. https://doi.org/10.3390/horticulturae10050502.

7. Xu, P.; Wang, L.; Wang, X.; Xu, Y.; Ablitip, Y.; Guo, C.; Ayup, M. Identification of S-RNase Genotypes of 65 Almond [*Prunus dulcis* (Mill.) D.A. Webb] Germplasm Resources and Close Relatives. *Horticulturae* **2024**, *10*, 545. https://doi.org/10.3390/horticulturae10060545.

8. Hapuarachchi, N.S.; Kämper, W.; Hosseini Bai, S.; Ogbourne, S.M.; Nichols, J.; Wallace, H.M.; Trueman, S.J. Selective Retention of Cross-Fertilised Fruitlets during Premature Fruit Drop of Hass Avocado. *Horticulturae* **2024**, *10*, 591. https://doi.org/10.3390/horticulturae10060591.

9. Jiang, L.; Song, R.; Wang, X.; Wang, J.; Wu, C. Transcriptomic and Metabolomic Analyses Provide New Insights into the Response of Strawberry (*Fragaria × ananassa* Duch.) to Drought Stress. *Horticulturae* **2024**, *10*, 734. https://doi.org/10.3390/horticulturae10070734.

10. Jiao, H.; Chen, Q.; Xiong, C.; Wang, H.; Ran, K.; Dong, R.; Dong, X.; Guan, Q.; Wei, S. Chloroplast Genome Profiling and Phylogenetic Insights of the "Qixiadaxiangshui" Pear (*Pyrus bretschneideri* Rehd.1). *Horticulturae* **2024**, *10*, 744. https://doi.org/10.3390/horticulturae10070744.

11. Ma, B.; Song, Y.; Feng, X.; Guo, P.; Zhou, L.; Jia, S.; Guo, Q.; Zhang, C. Integrated Metabolome and Transcriptome Analyses Reveal the Mechanisms Regulating Flavonoid Biosynthesis in Blueberry Leaves under Salt Stress. *Horticulturae* **2024**, *10*, 1084. https://doi.org/10.3390/horticulturae10101084.

12. Chen, Y.; Wang, L.; Zhang, J.; Chen, Y.; Jin, S. Altered Photoprotective Mechanisms and Pigment Synthesis in Torreya grandis with Leaf Color Mutations: An Integrated Transcriptome and Photosynthesis Analysis. *Horticulturae* **2024**, *10*, 1211. https://doi.org/10.3390/horticulturae10111211.

13. Fu, Q.; Xu, D.; Hou, S.; Gao, R.; Zhou, J.; Chen, C.; Zhu, S.; Wei, G.; Sun, Y. Comprehensive Identification and Expression Profiling of Lipoxygenase Genes in Sweet Cherry During Fruit Development. *Horticulturae* **2024**, *10*, 1361. https://doi.org/10.3390/horticulturae10121361.

14. Cuevas, J. Why Olive Produces Many More Flowers than Fruit—A Critical Analysis. *Horticulturae* **2025**, *11*, 26. https://doi.org/10.3390/horticulturae11010026.

15. Li, H.; Chen, L.; Liu, R.; Lu, Z. Role of Endogenous Hormones on Seed Hardness in Pomegranate Fruit Development. *Horticulturae* **2025**, *11*, 38. https://doi.org/10.3390/horticulturae11010038.

References

1. Busatto, N.; Herrera, R. Fruit Development and Ripening—A Molecular and Physiological View Modulating and Enhancing Fruit Quality. *J. Plant Growth Regul.* **2025**, *44*, 1069–1071. [CrossRef]
2. Giovannoni, J.J. Molecular Biology of Fruit Maturation and Ripening. *Annu. Rev. Plant Biol.* **2001**, *52*, 725–749. [CrossRef] [PubMed]
3. Gillaspy, G.; Ben-David, H.; Gruissem, W. Fruits: A Developmental Perspective. *Plant Cell* **1993**, *5*, 1439–1451. [CrossRef]
4. Seymour, G.B.; Poole, M.; Manning, K.; King, G.J. Genetics and Epigenetics of Fruit Development and Ripening. *Curr. Opin. Plant Biol.* **2008**, *11*, 58–63. [CrossRef] [PubMed]
5. Zhu, G.; Wang, S.; Huang, Z.; Zhang, S.; Liao, Q.; Zhang, C.; Lin, T.; Qin, M.; Peng, M.; Yang, C. Rewiring of the Fruit Metabolome in Tomato Breeding. *Cell* **2018**, *172*, 249–261. [CrossRef] [PubMed]
6. Dalla Costa, L.; Piazza, S.; Pompili, V.; Salvagnin, U.; Cestaro, A.; Moffa, L.; Vittani, L.; Moser, C. Strategies to produce T-DNA free CRISPRed fruit trees via *Agrobacterium tumefaciens* stable gene transfer. *Sci Rep* **2020**, *10*, 20155. [CrossRef] [PubMed]
7. Lu, Y.; Yang, W.; Huang, D.; Liang, L.; Xu, C. Advances in understanding of the mechanisms regulating anthocyanin accumulation in peach. *Sci. Hortic.* **2025**, *342*, 114022. [CrossRef]
8. Yu, A.; Zou, H.; Li, P.; Yao, X.; Zhou, Z.; Gu, X.; Sun, R.; Liu, A. Genomic characterization of the NAC transcription factors, directed at understanding their functions involved in endocarp lignification of iron walnut (*Juglans sigillata* Dode). *Front. Genet.* **2023**, *14*, 1168142. [CrossRef] [PubMed]
9. Wang, L. Physiological and molecular responses to drought stress in rubber tree (Hevea brasiliensis Muell. Arg.). *Plant Physiol. Biochem.* **2014**, *83*, 243–249. [CrossRef] [PubMed]
10. Baima, L.; Nari, L.; Nari, D.; Bossolasco, A.; Blanc, S.; Brun, F. Sustainability analysis of apple orchards: Integrating environmental and economic perspectives. *Heliyon* **2024**, *24*, e38397. [CrossRef] [PubMed]

 *horticulturae*

Article

Exogenous Abscisic Acid Regulates Anthocyanin Biosynthesis and Gene Expression in Blueberry Leaves

Bin Ma, Yan Song, Xinghua Feng, Qingxun Guo, Lianxia Zhou, Xinsheng Zhang and Chunyu Zhang *

College of Plant Science, Jilin University, Changchun 130062, China
* Correspondence: cy_zhang@jlu.edu.cn

Abstract: Blueberry (*Vaccinium corymbosum*) leaves have a positive influence on health because of their phenolic contents, including anthocyanins. Phytohormone abscisic acid (ABA) promotes anthocyanin accumulation, but the underlying mechanisms are unclear in blueberry leaves. In this study, we found that exogenous ABA promotes anthocyanin accumulation in blueberry leaves and we explored the global molecular events involved in these physiological changes by treating in vitro-grown blueberry seedlings with ABA and performing transcriptome deep sequencing (RNA-seq). We identified 6390 differentially expressed genes (DEGs), with 2893 DEGs at 6 h and 4789 at 12 h of ABA treatment compared to the control. Kyoto Encyclopedia of Genes and Genomes (KEGG) pathways related to plant hormone signal transduction and phenylpropanoid and flavonoid biosynthesis were significantly enriched at both stages of the ABA treatment. Analysis of DEGs in plant hormone signal transduction pathways revealed that exogenous ABA affected the expression of genes from other plant hormone signaling pathways, especially brassinosteroid, auxin, and gibberellin signaling. To elucidate the mechanism driving anthocyanin biosynthesis in blueberry in response to ABA treatment, we screened anthocyanin biosynthesis structural genes (ASG) from the phenylpropanoid and flavonoid biosynthetic pathways, MYB transcription factor genes from R2R3-MYB subgroups 5, 6, and 7 and ABRE-binding factor (ABF) genes from the ABA signal transduction pathway. Pearson's correlation coefficient (*r*) analysis indicated that the *ABFs*, *MYBs*, and structural genes form a network to regulate ABA-induced anthocyanin biosynthesis and *MYBA1* is likely to play an important role in this regulatory network. These findings lay the foundation for improving anthocyanin biosynthesis in blueberry leaves.

Keywords: blueberry; RNA sequencing; ABA; anthocyanin; MYB transcription factor; plant hormone

Citation: Ma, B.; Song, Y.; Feng, X.; Guo, Q.; Zhou, L.; Zhang, X.; Zhang, C. Exogenous Abscisic Acid Regulates Anthocyanin Biosynthesis and Gene Expression in Blueberry Leaves. *Horticulturae* **2024**, *10*, 192. https://doi.org/10.3390/horticulturae10020192

Academic Editor: Sherif M. Sherif

Received: 19 January 2024
Revised: 16 February 2024
Accepted: 17 February 2024
Published: 19 February 2024

1. Introduction

Plant hormones (phytohormones) play crucial roles in plant growth and development [1–3]. The roles of the major plant hormones abscisic acid (ABA), gibberellins (GAs), ethylene, auxin, salicylic acid (SA), brassinosteroids (BRs), jasmonates (JAs), and cytokinins (CKs) have been well characterized and their signal transduction pathways elucidated [4,5]. ABA, a stress response phytohormone, plays a role in plant defense against stress conditions and stimulates fruit coloration by accelerating the accumulation of anthocyanins [6–8]. The ABA signal transduction pathway includes pyrabactin resistance 1 (PYR)/PYR1-like (PYL)/regulatory component of ABA receptor (RCAR), type 2C protein phosphatase (PP2C), and sucrose non-fermenting 1-related protein kinase 2 (SnRK2) family members. These proteins function via a double negative regulatory system in which ABA-bound PYR/PYL/RCAR inhibits PP2C activity, releasing SnRK2 from repression [9,10]. In the absence of ABA, PP2C inactivates SnRK2. In the presence of ABA, SnRK2 phosphorylates basic leucine zipper (bZIP)-type ABA-responsive element (ABRE)-binding factors (ABFs) to regulate the expression of downstream target genes [11–13]. Most ABA-induced genes contain conserved cis-acting ABA response elements (ABREs) in their promoter regions. ABF transcription factors regulate downstream genes by binding to these elements [14,15].

ABA regulates multiple aspects of plant physiology via its signal transduction pathway, but how ABA regulates anthocyanin biosynthesis via the signal transduction pathway in blueberry leaves is unclear.

Anthocyanins are visible flavonoid pigments that produce the red and blue colors of many ripe fruits to attract seed dispersers [16,17]. Anthocyanins are biosynthesized via the general phenylpropanoid pathway, with key steps catalyzed by phenylalanine ammonia-lyase (PAL), cinnamate 4-hydroxylase (C4H), and 4-coumarate CoA ligase (4CL), followed by the flavonoid pathway catalyzed by chalcone synthase (CHS), chalcone isomerase (CHI), and flavanone 3-hydroxylase (F3H). Finally, dihydroflavonol 4-reductase (DFR), leucoanthocyanidin dioxygenase (LDOX), UDP glucose-flavonoid 3-O-glucosyltransferase (UFGT), and UDP-glycosyltransferase 75 (UGT75) successively catalyze anthocyanin biosynthesis. However, for blueberry leaves, the link between ABF transcription factors and anthocyanin biosynthesis genes remains unknown in the ABA-induced anthocyanin synthesis pathway.

The transcription of anthocyanin biosynthesis structural genes (ASGs) is directly controlled by the MBW complex, consisting of R2R3-MYB transcription factors, basic helix–loop–helix (bHLH) transcription factors and WD40 proteins [18–20]. Within this complex, R2R3-MYBs are generally the key factors in determining the spatial and temporal occurrence of anthocyanins [21,22]. R2R3-MYB transcription factors are divided into 22 subgroups and subgroups 5, 6, and 7 play regulatory roles in flavonoid biosynthesis, including anthocyanins, proanthocyanins, and flavonols [21,23,24]. For instance, blueberry VcMYBA (subgroup 6) and Arabidopsis AtMYB113 and AtMYB114 (subgroup 6) control anthocyanin biosynthesis, AtMYB12 (subgroup 7) control flavonol biosynthesis, and At-MYB123/TT2 (subgroup 5) control proanthocyanidin biosynthesis [19,21,25,26]. Therefore, we hypothesized that ABF transcription factor, MYB transcription factor and anthocyanin biosynthesis structural genes may form a regulatory network to regulate ABA-induced anthocyanin synthesis.

Blueberry (*Vaccinium corymbosum*) is a popular fruit worldwide, in part due to the health benefits associated with its rich flavonoid compounds, particularly antho-cyanins [27,28]. The fruits of the Vaccinium species are the main commercial products, whereas the leaves have been used mainly for medicinal purposes [29–32]. In the wild lowbush blueberry (*V. angustifolium*) and lingonberry (*V. vitis-idaea*), the total antioxi-dant capacities in terms of radical scavenging activity and reducing power were much higher in the leaves of both plants as compared to their fruits, which were in correlation with phenolic contents including total flavonoids, anthocyanins, and tannins [33]. The total flavonoids from blueberry leaves had both hypolipidemic and antioxidant effects, which could help the cure and prevention of hyperlipidemia diseases [34]. Thus, it is very important to study the mechanism of ABA induced anthocyanin biosynthesis in blueberry leaves.

In this study, to obtain differential expression genes (DEGs) from blueberry leaves in response to ABA treatment, we performed transcriptome deep sequencing (RNA-seq) analysis by treating in vitro-grown blueberry seedlings with ABA and annotated DEGs from KEGG pathways related to plant hormone signal transduction. To elucidate the mech-anism driving anthocyanin biosynthesis in blueberry in response to ABA treatment, we assembled a regulatory network of ABFs–MYB–ASGs by gene co-expression analysis. Our findings increase understanding of ABA-induced anthocyanin biosynthesis and provide a basis for improving the value of blueberry leaves via ABA treatment.

2. Materials and Methods

2.1. Plant Materials and ABA Treatments

In vitro-grown seedlings were obtained from the blueberry (*Vaccinium corymbosum*) cultivar 'Northland'. The materials were cultured on modified woody plant medium (WPM) containing Murashige and Skoog vitamins under a 16-h-light/8-h-dark photoperiod of 2000 Lux from cool white fluorescent tubes at 25 °C. The seedlings were cultured on WPM containing 1.0 mg/L trans-Zeatin for 5 weeks, and then, seedlings with six to ten blades were transferred to the medium containing 100 µM ABA for 0, 6 or 12 h for RNA-seq and RT-qPCR analysis and measurement of the anthocyanin contents [35,36].

2.2. Transcriptome Sequencing

Total RNA was extracted from seedlings after 0, 6, or 12 h of ABA treatment using an RNAprep Pure Plant Kit (Tiangen, Beijing, China), and RNA concentration and purity were measured using a NanoDrop 2000 (Thermo Fisher Scientific, Wilmington, DE, USA). Sequencing libraries were generated using a Hieff NGS Ultima Dual-mode mRNA Library Prep Kit for Illumina (Yeasen Biotechnology (Shanghai) Co., Ltd., Shanghai, China). The libraries were sequenced on an Illumina NovaSeq platform to generate 150-bp paired-end reads. The raw reads were processed with a bioinformatic pipeline tool, BMKCloud (www.biocloud.net (accessed on 1 July 2023)). Clean data (clean reads) were obtained by removing reads containing adapters, reads containing ploy-N, and low-quality reads from raw data. The Q30, GC content, and sequence duplication level of the clean data were calculated. All downstream analyses were based on clean, high-quality data. The clean reads were mapped to the reference *Vaccinium corymbosum* cv. Draper V1.0 genome sequence (https://www.vaccinium.org/genomes (accessed on 1 August 2023)) using Hisat2 tools software version 2.0.4.

2.3. Differential Expression Analysis and Functional Annotation

Gene expression levels were estimated based on fragments per kilobase of transcript per million fragments mapped (FPKM). DEGs between two samples were identified using a false discovery rate (FDR) < 0.01 and absolute $\log_2$ (fold-change) $\geq$ 1 by DESeq2 (version 1.30.1). Gene function was annotated based on Clusters of Orthologous Groups (COG) (http://www.ncbi.nlm.nih.gov/COG (accessed on 20 August 2023)), Gene Ontology (GO) (https://geneontology.org/ (accessed on 20 August 2023)), Kyoto Encyclopedia of Genes and Genomes (KEGG) (https://www.genome.jp/kegg/ (accessed on 20 August 2023)), Eukaryotic Orthologous Groups (KOG) (ftp://ftp.ncbi.nih.gov/pub/COG/KOG (accessed on 25 August 2023)), Protein family (Pfam) (https://pfam.sanger.ac.uk/ (accessed on 25 August 2023)), a manually annotated and reviewed protein sequence database (Swiss-Prot) (https://www.uniprot.org/ (accessed on 25 August 2023)), the evolutionary genealogy of genes using the Non-supervised Orthologous Groups (eggNOG) (http://eggnog-mapper.embl.de (accessed on 30 August 2023)), and NCBI non-redundant protein sequence (NR) databases.

KEGG pathway enrichment analysis of the DEGs was performed using the KOBAS database and clusterProfiler software (version 4.4.4) [37]. Heatmaps of the expression of DEGs in the plant hormone signal transduction pathways and the anthocyanin biosynthetic pathway under ABA treatment were assembled using $\log_2$(FPKM) values with Tbtools (v1.098761) software [38]. The regulatory network of the anthocyanin biosynthetic pathway was constructed based on Pearson's correlation coefficients (r) between R2R3-MYB transcription factor genes and anthocyanin biosynthesis/ABF genes using IBM SPSS software (version 19.0).

2.4. Analysis of Abscisic Acid Responsiveness Element in the Promoter

The upstream sequences of screened MYB and anthocyanin biosynthesis genes were downloaded from the GDV. About 2000-bp upstream sequences were submitted to the online program PlantCARE (https://bioinformatics.psb.ugent.be/webtools/plantcare/html/ (accessed on 1 November 2023)) to predict abscisic acid responsiveness element (ABRE) or MYB binding sites.

2.5. Reverse-Transcription Quantitative PCR

Total RNA was extracted from each sample using an RNA Extraction Kit (Sangon Biotech, Shanghai, China), and first-strand cDNAs were synthesized using PrimeScriptTM RT Master Mix (TaKaRa, Kusatsu, Japan). RT-qPCR analysis was performed using an ABI 7900HT real-time PCR system. Nine ABA signal transduction pathway genes (*ABF2d*, *ABF4b*, *PP2C-24a*, *PP2C-37a*, *PP2C-7a*, *PP2C-51b*, *PYL1*, *SnRK2A*, and *PYL4a*) were selected to validate the RNA-seq data. Glyceraldehyde-3-phosphate dehydrogenase housekeeping (*GAPDH*; GenBank accession no. AY123769) was used as the reference gene, and the relative expression levels of each gene were calculated using the $2^{-\Delta\Delta Ct}$ method. Primer sequences for RT-qPCR are shown in Table S1.

2.6. Measurement of Anthocyanin Contents

Total anthocyanin contents were measured as described by Yang et al. [39]. Each sample was ground to a powder, and 1 g was extracted in 6 mL of 1% (*v*/*v*) HCl in methanol and incubated at 4 °C for 16 h. After the mixtures were centrifuged ($8000\times g$, 10 min, 4 °C), 4 mL of the supernatant was diluted with 4 mL ddH$_2$O, and the absorbance was measured at 530 and 650 nm. Total anthocyanin content was calculated using the equation $A_{530} - 0.25 \times A_{650}$ [40].

2.7. Cluster Analysis and Sequence Alignment of MYB Transcription Factors

A phylogenetic tree was constructed through amino sequencing of 23 differentially expressed R2R3-MYB genes from blueberry and 124 R2R3-MYB proteins from Arabidopsis using the neighbor-joining method with MEGA X software (https://www.megasoftware.net (accessed on 30 October 2023)). The phylogenetic tree was divided into different subgroups based on MYB transcription factors from Arabidopsis [24,41]. The MYBs from blueberry were named based on cluster and functional annotations from SwissProt databases. Sequence alignment and phylogenetic analysis between MYBA1, MYBA2 and VcMYBA (GenBank accession no. MH105054) were performed using DNAMAN software, Version 8.192.

2.8. Statistical Analysis

The gene expression levels or anthocyanin contents were carried out using three independent biological replicates, and three technical replicates were performed for each biological replicate. One-way analysis of variance (ANOVA) was used to assess the differences in gene expression levels or anthocyanin contents during ABA treatment, and Tukey's test was used to identify significant differences at *p*-value $\leq$ 0.05 using SPSS 19.0 software.

3. Results

3.1. ABA Treatment Promotes Anthocyanin Accumulation

To characterize the effect of exogenous ABA on anthocyanin accumulation in blueberry leaves, we treated in vitro-grown blueberry seedlings with exogenous ABA and without ABA for 0, 6, or 12 h (Figure 1a). The tip of the leaves turned red after 6 h of ABA treatment. We measured the anthocyanin contents of in vitro-grown blueberry seedlings under ABA treatment (Figure 1b). The anthocyanin contents significantly increased after 6 h and 12 h of ABA treatment compared to the 0 h control. At the same time, without ABA treatment, the content of anthocyanin did not change at 0 h, 6 h or 12 h. Thus, exogenous ABA promotes anthocyanin accumulation in vitro-grown blueberry seedlings.

 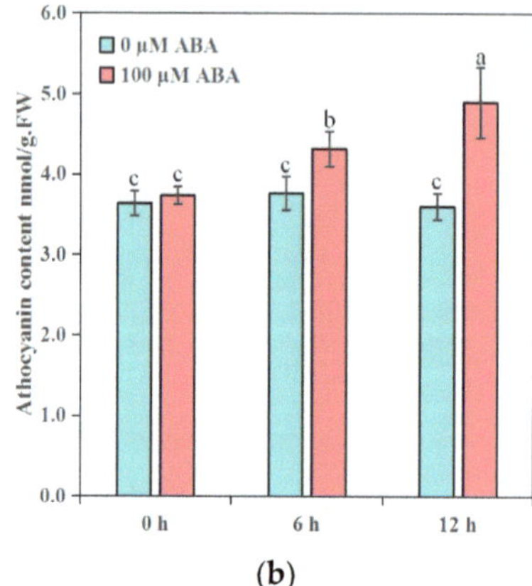

Figure 1. ABA induces anthocyanin accumulation in in vitro-grown blueberry seedlings. Representative phenotypes (**a**) and anthocyanin contents (**b**) of in vitro-grown blueberry seedlings under 100 µM ABA treatment and without ABA treatment for 0 h, 6 h and 12 h. The top and bottom plants showed 100 µM ABA treatment and without ABA treatment, respectively. Values are means ± standard error (SE) of three independent biological replicates; different letters represent significant differences at $p < 0.05$ among samples by one-way ANOVA and Tukey's test. Bar, 1 cm.

3.2. Transcriptome Sequencing and Analysis of DEGs

To elucidate the molecular changes underlying the response of blueberry to exogenous ABA, we performed RNA-seq analysis using in vitro-grown blueberry seedlings under ABA treatment for 6 or 12 h, using untreated seedlings as the 0 h control. Table 1 summarizes the nine samples used for RNA-seq. More than 19.3 million clean reads were obtained, generating more than 5.78 Gb of clean bases for each sample. We mapped the clean reads to the blueberry reference genome at a rate of approximately 89.9% to 91.2% per sample. The percentage of Q30 bases was 90.45% or above, and the GC content ranged from 45.7% to 46.55%.

Table 1. Summary of sequencing data.

Sample	Clean Reads	Clean Bases	GC Content	$\geq$Q30	Mapped Reads
0 h rep. 1	1.93×10^7	5.78×10^9	45.9%	92.6%	91.2%
0 h rep. 2	2.14×10^7	6.42×10^9	45.7%	92.3%	91.05%
0 h rep. 3	2.22×10^7	6.65×10^9	45.7%	92.1%	90.6%
6 h rep. 1	2.15×10^7	6.42×10^9	45.75%	92.8%	91.1%
6 h rep. 2	1.97×10^7	5.84×10^9	45.7%	91.0%	90.3%
6 h rep. 3	2.18×10^7	6.53×10^9	45.7%	91.4%	90.3%
12 h rep. 1	2.09×10^7	6.27×10^9	46.55%	92.1%	90.9%
12 h rep. 2	1.94×10^7	5.80×10^9	46.2%	90.45%	89.9%
12 h rep. 3	2.13×10^7	6.37×10^9	45.8%	91.4%	90.9%

Genes with FDR < 0.01 and absolute log₂ (fold-change) $\geq$ 1 between control samples (0 h) and samples under ABA treatment (6 and 12 h) were considered to be DEGs. We identified 2893 DEGs in the 0 h vs. 6 h comparison, consisting of 1765 upregulated and 1128 downregulated DEGs (Figure 2a). For the 0 h vs. 12 h comparison, the number of DEGs increased to 4789, with more downregulated DEGs (2984) than upregulated DEGs (1805) (Figure 2b). As shown in the Venn diagram in Figure 2c, across the two comparisons, we detected 6390 DEGs, with 1292 DEGs shared between 0 h vs. 6 h and 0 h vs. 12 h.

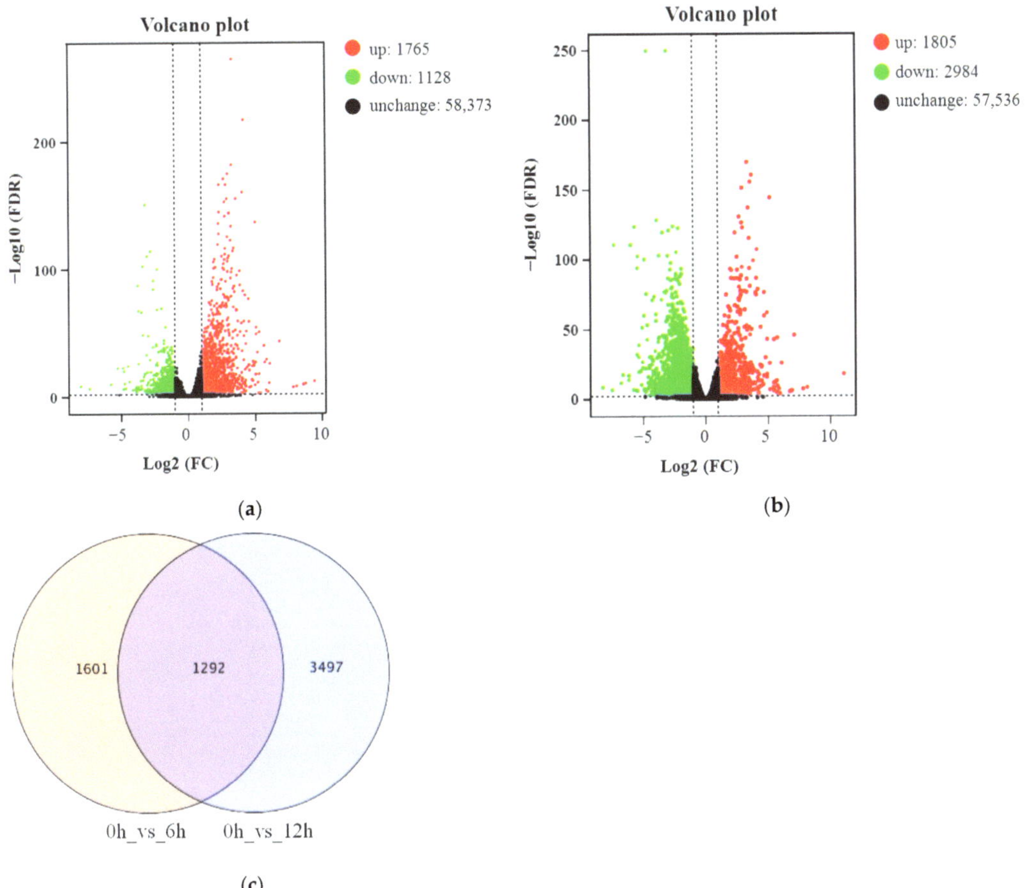

(a)

(b)

(c)

Figure 2. Differentially expressed genes in response to ABA treatment in blueberry, as revealed by transcriptome deep sequencing (RNA-seq). Volcano plots showing all the genes at 6 h (**a**) or 12 h (**b**) compared to 0 h under ABA treatment. Each point represents a single gene. The red and green dots indicate upregulated and downregulated differentially expressed genes (DEGs), respectively, and the black dots indicate genes with unchanged expression. (**c**) Venn diagram showing the number of DEGs across pairwise comparisons. Pink, DEGs for 0 h vs. 6 h; light blue, DEGs for 0 h vs. 12 h; purple, DEGs for 0 h vs. 6 h and 0 h vs. 12h.

3.3. Functional Annotation and KEGG Pathway Analysis of DEGs under ABA Treatment

We annotated the functions of the DEGs under ABA treatment using the COG, GO, KEGG, KOG, NR, Pfam, Swiss-Prot, and eggNOG databases (Table S2). The total number of annotated DEGs increased from 2730 (94.37%) for the 0 h vs. 6 h comparison to 4623 (96.53%) for the 0 h vs. 12 h comparison; we annotated 1952 and 3453 of these DEGs using the KEGG database for each comparison, respectively.

To identify the biological functions of the DEGs in blueberry, we subjected all DEGs to KEGG pathway enrichment analysis and KEGG functional classification (Figure 3). Analysis of the top 20 enriched KEGG pathways showed that the plant hormone signal transduction pathway (ko04075), phenylpropanoid biosynthetic pathway (ko00940), and flavonoid biosynthetic pathway (ko00941) are significantly enriched at all stages (Figure 3a–c). At 6 and 12 h into ABA treatment, 291, 197, and 85 DEGs were enriched in the plant hormone signal transduction pathway, phenylpropanoid biosynthetic pathway, and flavonoid biosynthetic pathway, respectively (Figure 3c). KEGG functional classification showed that the DEGs are involved in cellular processes, environmental information processing, genetic information processing, metabolism, and organismal systems (Figure 3d–f).

The KEGG pathway with the greatest number of DEGs was the plant hormone signal transduction pathway for all stages, with more upregulated DEGs than downregulated DEGs at 6 h and more downregulated than upregulated DEGs at 12 h. Among metabolic pathways, those with the greatest number of DEGs were the phenylpropanoid biosynthetic pathway (197, 7.9%), the starch and sucrose metabolism pathway (158, 6.3%), and the flavonoid biosynthetic pathway (85, 3.4%) (Figure 3f). Thus, DEGs from the plant hormone signal transduction pathway, the phenylpropanoid biosynthetic pathway, and the flavonoid biosynthetic pathway play important roles in the response of blueberry to exogenous ABA.

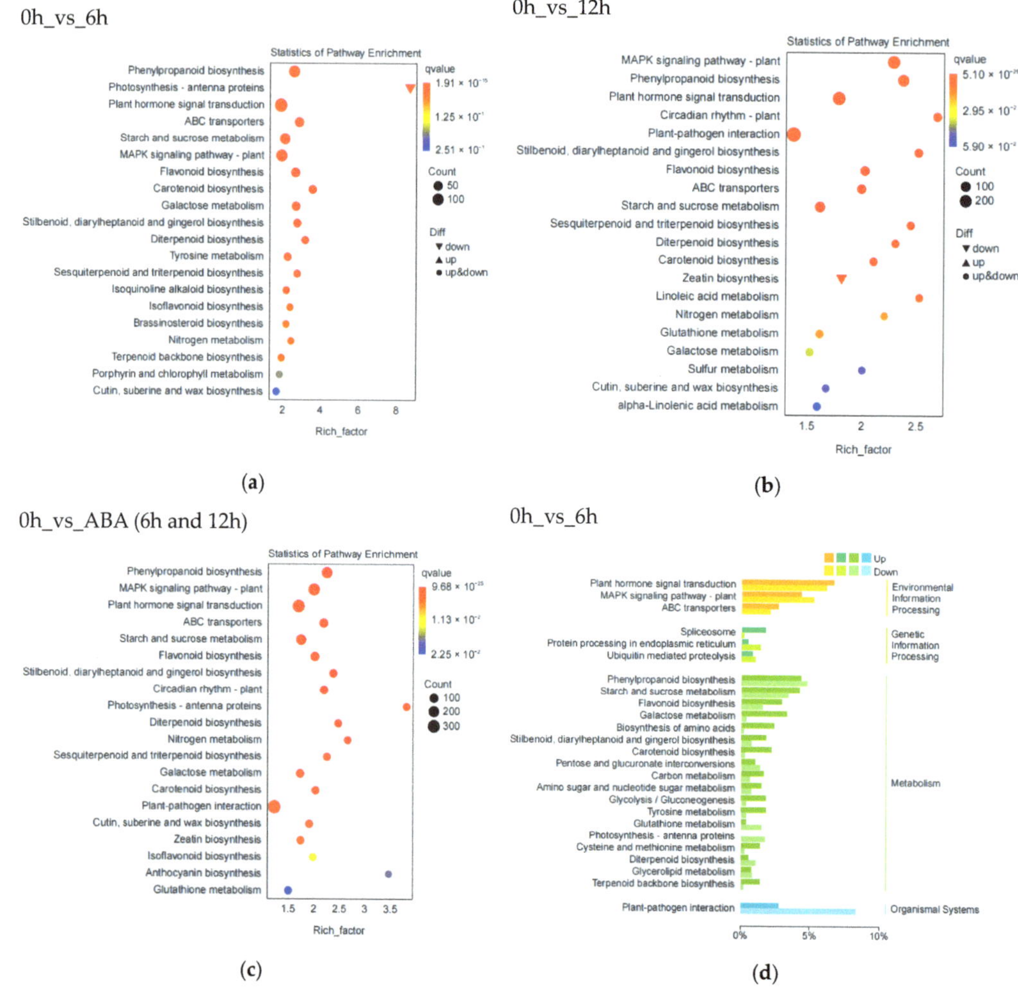

Figure 3. *Cont.*

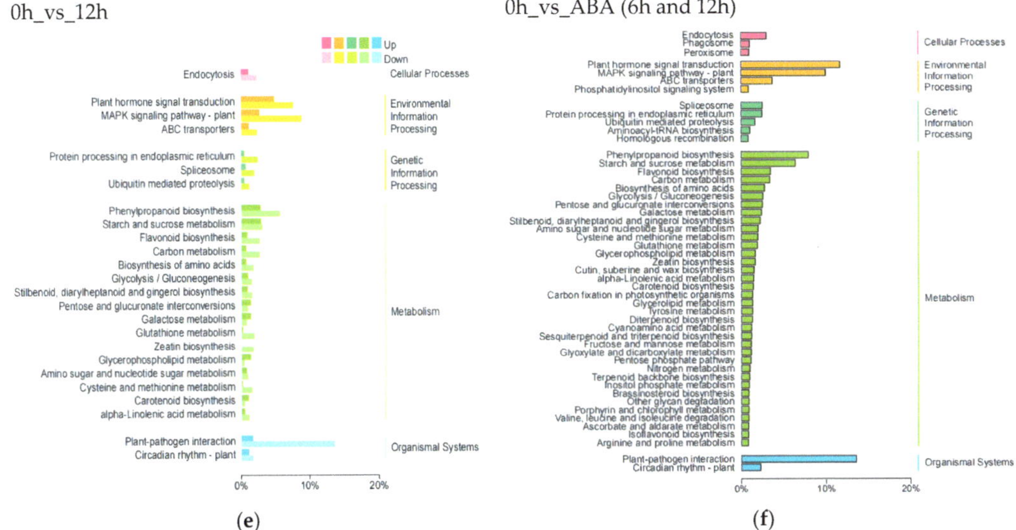

Figure 3. KEGG pathway enrichment analysis and functional classification of differentially expressed genes in blueberry under ABA treatment. Analysis of the top 20 significantly enriched DEGs after 6 h (**a**), 12 h (**b**), or 6 and 12 h (**c**) of ABA treatment compared to 0 h. The number of DEGs is indicated by the size of the circle. The solid upright and inverted triangles represent upregulated and downregulated DEGs, respectively. KEGG functional classification of DEGs after 6 h (**d**), 12 h (**e**), or 6 and 12 h (**f**) of ABA treatment compared to 0 h.

3.4. Exogenous ABA Induces the Expression of Plant Hormone Signal Transduction Pathway Genes

To elucidate the functions of DEGs from the plant hormone signal transduction pathway, we identified 209 DEGs (Table S3). These DEGs were mapped to the ABA, GA, JA, auxin, SA, BR, ethylene, and CK biosynthetic pathways (Figure 4). Of these DEGs, 101, 35, 26, and 18 were mapped to the BR, ABA, auxin, and GA biosynthetic pathways, respectively. In the ABA signaling pathway, eight PYR/PYL family members were downregulated and one was upregulated. All PP2C gene family members were upregulated at 6 and 12 h of ABA treatment relative to the 0 h control. Of the five *SnRK2* family members, one was downregulated and the remaining four were upregulated. All the ABFs were upregulated under ABA treatment. In the BR signal transduction pathway, 50 *brassinosteroid insensitive 1* (*BRI1*) genes and 34 *BRI1 associated protein kinase 1* (*BAK1*) genes were differentially expressed, including 9 that were upregulated and 75 that were downregulated. In the auxin signal transduction pathway, six *AUX/IAA* genes were upregulated and two were downregulated. In the GA signal transduction pathway, 17 genes encoding DELLA family members were identified, with 6 being upregulated and 11 being downregulated. These results indicate that ABA treatment not only induced the expression of ABA signal transduction pathway genes, but it also affected the expression of other plant hormone signal transduction pathway genes, especially the BR, auxin, and GA signal transduction pathways.

We selected nine genes from the ABA signal transduction pathway to validate the accuracy and reliability of the RNA-seq data by RT-qPCR (Figure S1a). Pearson's correlation coefficient (*r*) analysis showed that $\log_2 (2^{-\Delta\Delta Ct})$ values from RT-qPCR and $\log_2$ (FC) from RNA-seq are significantly correlated for the 0 h vs. 6 h (0.943) and 0 h vs. 12 h (0.872) comparisons (Figure S1b). These results indicate that the RNA-seq data generated in this study are accurate and reliable.

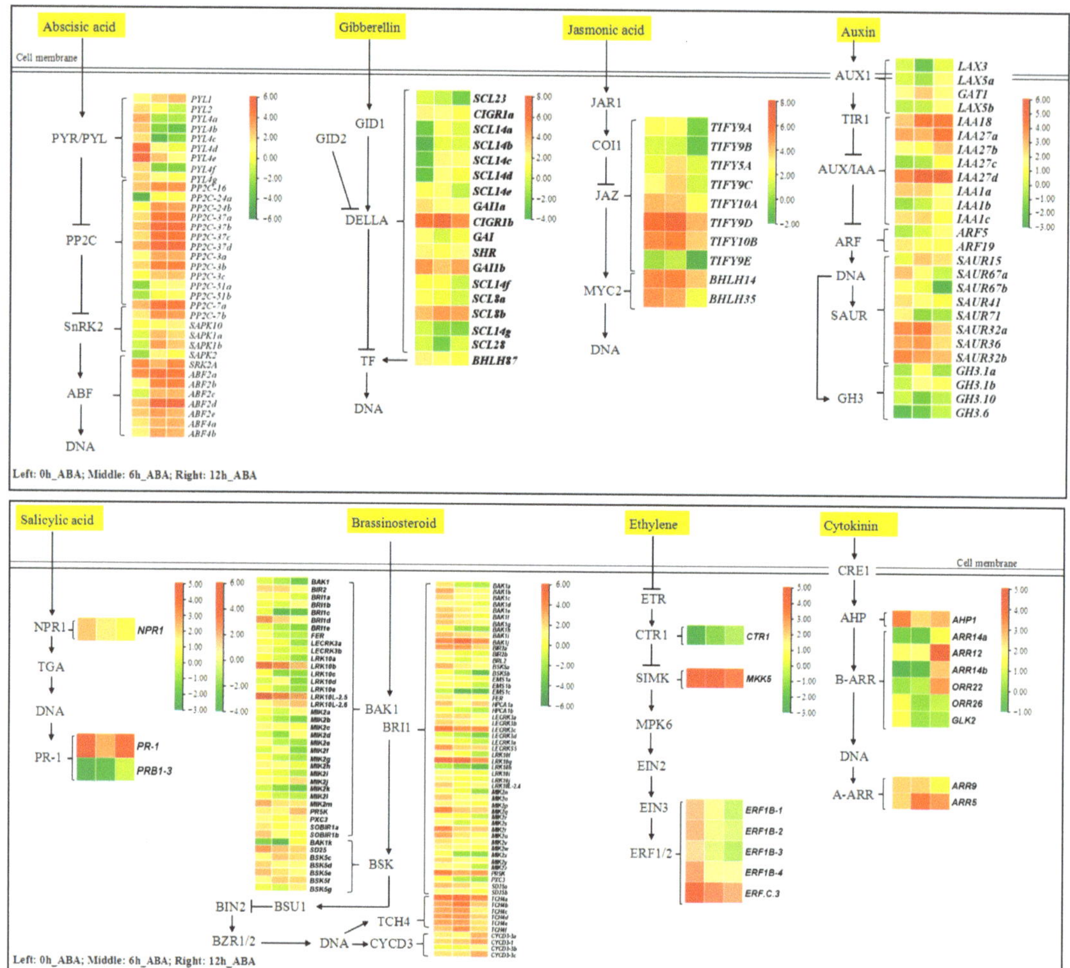

Figure 4. Differentially expressed genes of plant hormone signal transduction KEGG pathways in blueberry under ABA treatment. Green indicates low expression, and red indicates high expression, based on log$_2$ (FPKM) values.

3.5. Exogenous ABA Treatment Regulates the Expression of R2R3-MYB Transcription Factor Genes

To elucidate the functions of transcription factors in response to ABA treatment, we analyzed the differentially expressed R2R3-MYB transcription factor genes under ABA treatment according to Pfam, Swiss-Prot, and NR functional annotations. After removing the repeat MYB transcription factor genes based on sequence analysis, we identified 23 R2R3-MYB subfamily genes, of which 18 were upregulated and 5 were downregulated under ABA treatment (Table S4). To predict the potential functions of the proteins encoded by these R2R3-MYB genes, we reconstructed a phylogenetic tree based on their deduced protein sequences and 124 R2R3-MYB proteins from Arabidopsis (Figure 5). The blueberry R2R3-MYB transcription factors belong to 12 subgroups, which were previously defined for Arabidopsis R2R3-MYB proteins. MYBA1 and MYBA2 were clustered in subgroup 6; their encoding genes were upregulated under exogenous ABA treatment. Sequence alignment and phylogenetic analysis showed that MYBA1 shared 97.64% amino sequence identity with VcMYBA (GenBank accession no. MH105054) (Figure S2). Thus, MYBA1 is probably homologous to VcMYBA, which belongs to subgroup 6 and induces anthocyanin accumulation by heterologous expression in *Nicotiana benthaminana* [21]. MYB5, MYB11, and MYB12 belong to subgroup 7, and their encoding genes were upregulated under ABA treatment. MYB23 and MYB123 clustered in subgroup 5, and their encoding genes were repressed in response to exogenous ABA treatment. In general, R2R3-MYB transcription factors of subgroups 5, 6, and 7 regulated flavonoid biosynthesis [41]. Thus, these MYB transcription factors may be regulated ABA-induced anthocyanin accumulation.

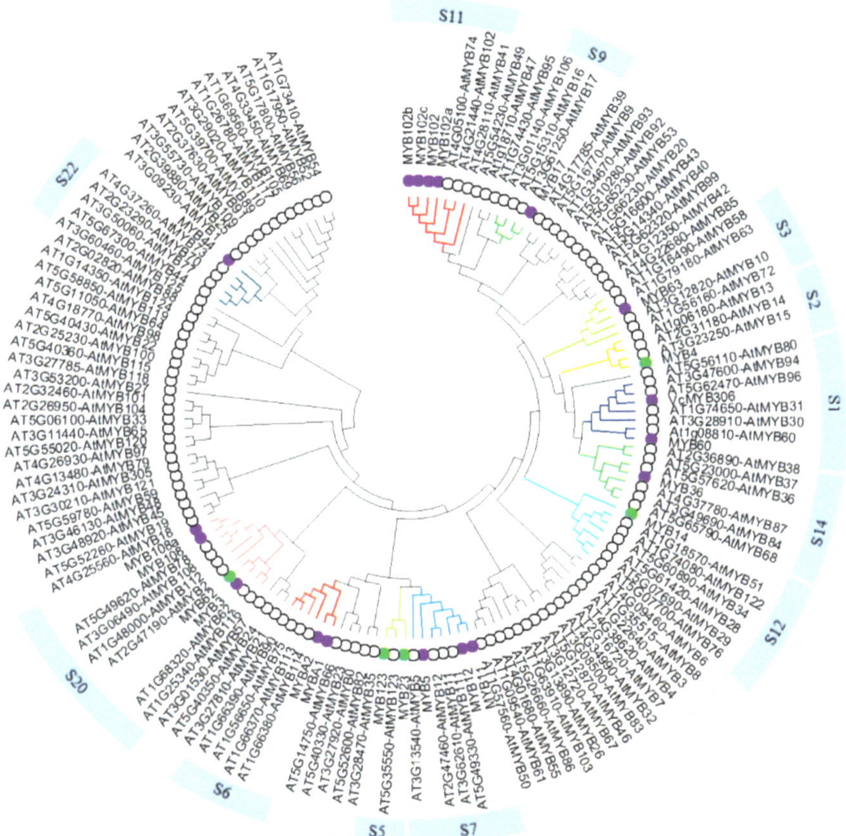

Figure 5. Phylogenetic analysis of R2R3-MYB proteins from blueberry and Arabidopsis using MEGA X with the neighbor-joining method. Purple and green dots represent upregulated and downregulated differentially expressed R2R3-MYB transcription factor genes from blueberry under ABA treatment, respectively. Black circles represent R2R3-MYB transcription factors from Arabidopsis. S1, S2, S3, S5, S6, S7, S9, S11, S12, S14, S20, and S22 represent subgroups of blueberry R2R3-MYB transcription factors referring to the subgroups of *Arabidopsis* R2R3-MYB proteins [24].

3.6. Exogenous ABA Induces the Expression of Anthocyanin Pathway Genes via MYB or ABF Transcription Factors

To clarify the regulatory network of ABA-induced anthocyanin biosynthesis, we screened the DEGs involved in anthocyanin biosynthesis from the phenylpropanoid and flavonoid biosynthetic pathway based on KEGG annotation. After removing repeat sequences, we retained seven DEGs (Table S5). Of these, *C4H*, *CHI*, *CHI3*, *VcUFGT*, and

UGT75C1 were upregulated and *4CL7* and *LDOX* were downregulated in response to ABA treatment (Figure 6 and Table S5). We collected the 2000 bp upstream sequences from the ATG start codons of the differentially expressed MYB and anthocyanin biosynthesis genes and looked for ABREs or MYB binding sites using the PlantCARE online tool (Figure S3). The promoter sequences of all the *MYBs* contain ABREs, including *MYB23*, which contains six ABREs. Among the anthocyanin biosynthesis genes, the *C4H*, *4CL7*, *CHI*, and *LDOX* promoters contained one to four ABREs. However, the promoter sequences of anthocyanin biosynthesis genes contain three to nine MYB binding sites. In general, bZIP-type AREB/ABF transcription factors target ABREs and MYB transcription factors target MYB binding sites in the promoter regions of genes to regulate their expression.

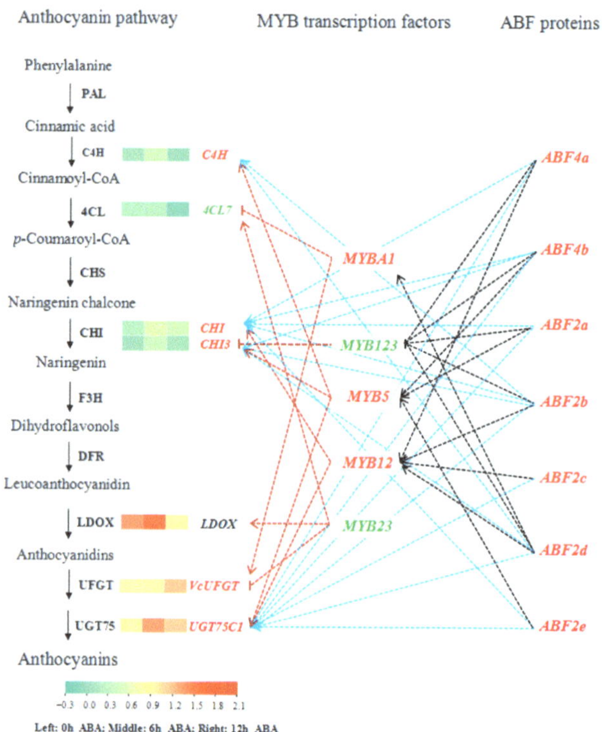

Figure 6. Proposed model of the ABF-regulated anthocyanin biosynthetic pathway under ABA treatment. Green squares indicate low expression and red squares indicate high expression, based on $\log_2$ (FPKM). Solid lines indicate the known biosynthetic pathway; dotted lines indicate the putative signaling pathway based on co-expression analysis. Arrows indicate positive correlation; short lines indicate negative correlation. Red and green text indicates upregulated and downregulated DEGs, respectively.

To investigate the functions of MYB transcription factors from subgroups 5, 6, and 7 and ABF transcription factors in the anthocyanin biosynthetic pathway, we calculated Pearson's correlation coefficients (*r*) between *ABFs* and *MYBs*, between *ABFs* and anthocyanin biosynthetic genes, and between *MYBs* and anthocyanin biosynthetic genes according to FPKM values under ABA treatment (Table S6). The expression of *MYB12*, *MYB23*, *MYB5*, and *MYBA1* was positively correlated with that of most anthocyanin metabolism genes. However, the expression of some *MYBs* was negatively correlated with that of anthocyanin pathway genes, such as *MYB123* and *CHI3*, *MYB23* and *VcUFGT*, and *MYBA1* and *4CL7*. The expression of *ABFs* was positively correlated with that of most anthocyanin metabolism genes. Correlation analysis between *ABFs* and *MYBs* showed that the expression of most *ABFs* is positively correlated with that of *MYBs*, and the expression of most *ABFs* and *MYBs* was positively correlated with that of anthocyanin metabolism genes. Based on the results of correlation analysis, we constructed a proposed regulatory network of ABA-induced anthocyanin biosynthetic pathway genes by MYB and ABF transcription factors (Figure 6).

4. Discussion

4.1. Exogenous ABA Promoter Anthocyanin Accumulation of Blueberry Leaves

Blueberries are the richest sources of polyphenolic antioxidants, especially anthocyanins, among all fruits and vegetables [42]. Blueberry leaves, which have been used mainly for medicinal purposes, are considered essentially as waste or by-products. Various health benefits were reported for leaf extracts from the *Vaccinium* species [29–32]. A large number of studies have shown that exogenous ABA stimulates anthocyanin accumulation of blueberry fruits [7,43,44]. In this study, we also found that exogenous ABA also promotes anthocyanin biosynthesis in high blueberry leaves. Therefore, exogenous ABA treatment can improve the utilization value of blueberry leaves.

4.2. Exogenous ABA Regulated the Expression of ABA Signal Transduction Pathway Genes

The plant hormone ABA mediates various physiological responses in plants via the PYR/PYL/RCAR, PP2C, and SnRK2 family members; these three families constitute the core components of the ABA signal transduction pathway [5,9,10]. ABA treatment regulates the expression of the genes encoding these family members. For example, in the primary roots of maize (*Zea mays*) seedlings, ABA treatment upregulated *ZmPYL1–3* genes (encoding homologs of dimeric-type Arabidopsis ABA receptors) but downregulated *ZmPYL4–10* (encoding homologs of monomeric-type Arabidopsis ABA receptors). However, the opposite trend was observed in the leaves of maize plants growing in medium containing ABA. At the same time, three PP2C and three SnRK2 family genes were upregulated in primary roots in response to ABA treatment [45]. In this study, among the PYR/PYL/RCAR, PP2C, and SnRK2 family members, PYL1 was upregulated and *PYL4a–g* was downregulated in in vitro-grown blueberry seedlings in response to ABA treatment (*PYL1* and *PYL4* were named based on their homologs in *Arabidopsis*). In addition, the genes encoding all PP2C and most SnRK2 family members were upregulated in in vitro-grown blueberry seedlings under ABA treatment (Table S3 and Figure 4). These results indicate that the primary roots of maize seedlings and in vitro-grown blueberry seedlings employ similar mechanisms in response to exogenous ABA treatment.

4.3. Exogenous ABA Regulated the Expression of Other Plant Hormone Signal Transduction Pathway Genes

Exogenous ABA treatment not only induces the expression of ABA signal transduction pathway genes but also affects the expression of genes in other plant hormone signal transduction pathways, highlighting the complex crosstalk between ABA and other plant hormones in regulating plant defense responses and anthocyanin biosynthesis [1,46,47]. Transcriptome analysis in mangrove (*Kandelia obovata*) showed that plant hormone signal transduction pathways were significantly enriched in DEGs related to response to ABA treatment during vivipary [48]. Our data indicate that signal transduction pathways of other plant hormones are also significantly enriched during exogenous ABA treatment in in vitro-grown blueberry seedlings, particularly BR, followed by auxin and GA signaling (Table S3 and Figure 4).

BRs are steroidal plant hormones that are essential for plant growth and development [49,50]. BRs have significant effects on plant secondary metabolism, particularly anthocyanin biosynthesis [51,52]. However, several reports indicate that BR treatment inhibited red-flesh coloration and repressed high light-induced anthocyanin accumulation in red-fleshed apple (*Malus domestica*) seedlings [53,54]. In this study, ABA treatment downregulated most *BRI1* and *BAK1* family genes. These family members form surface receptor kinase complexes that are responsible for perceiving BRs [55]. Thus, exogenous ABA may inhibit endogenous BR biosynthesis to repress plant growth and promote flavonoid biosynthesis in blueberry.

Auxin, the first plant hormone discovered, promotes plant growth and regulates secondary metabolism [56,57]. Auxin affects the levels of other phytohormones; in turn, other plant hormones affect the level of auxin. In sweet cherry (*Prunus avium*), 1-naphthaleneacetic acid (NAA) treatment altered ethylene production, which induced fruit ripening and enhanced anthocyanin production [58]. However, in apple calli, NAA and 2,4-dichlorophenoxyacetic acid (2,4-D) treatment inhibited anthocyanin accumulation [59,60]. Aux/IAAs repress auxin responses and play central roles in the auxin signal transduction pathway [61,62]. Our transcriptome analysis showed that most Aux/IAA genes were upregulated under exogenous ABA treatment in in vitro-grown blueberry seedlings. Thus, exogenous ABA also increased endogenous auxin contents in blueberry.

GA promotes plant growth and cell division and regulates anthocyanin biosynthesis [2,63]. DELLA proteins are repressors of GA signaling that modulate all aspects of GA-induced growth and development [4,64]. ABA attenuates GA biosynthesis and increases DELLA gene expression, leading to a reduction in bioactive GA levels [2]. DELLAs play a major role in mediating the crosstalk among plant hormonal signals [65]. Our transcriptome analysis showed that 17 DELLA genes were regulated by ABA treatment in in vitro-grown blueberry seedlings, indicating that DELLAs modulate the crosstalk between the ABA and GA signaling pathways.

4.4. The regulatory Network of ABA-Induced Blueberry Anthocyanin Accumulation

ABA plays important roles in promoting anthocyanin production and fruit ripening [37,66]. In general, ABA induces gene expression by binding to conserved cis-acting ABREs in the promoters of their target genes. ABF transcription factors regulate the ABRE-mediated transcription of downstream target genes [12,14,67]. R2R3-MYB transcription factors from subgroups 5, 6, and 7 regulate flavonoid biosynthesis via the transcriptional regulation of flavonoid structural genes [21,23,68]. In Arabidopsis, MYB49 interacts with ABSCISIC ACID-INSENSITIVE5 (ABI5) to regulate ABA-repressed cadmium accumulation [69]. Here, we showed that exogenous ABA promoted anthocyanin accumulation in in vitro-grown blueberry seedlings (Figure 1). The promoter sequences of all R2R3-MYB genes from subgroups 5, 6, and 7 and the ASGs (*C4H*, *4CL7*, *CHI*, and *LDOX*) contain one or more ABREs, and all the ASGs contain more than three MYB binding sites (Figure S3). The expression of ABF genes was upregulated under ABA treatment and also positively or negatively correlated with that of MYBs and ASGs. At the same time, MYBs also positively or negatively correlated with ASGs in expression levels under ABA treatment. These findings indicate that ABFs, MYBs, and ASGs form a regulatory network to co-regulate ABA-induced anthocyanin biosynthesis.

Most studies showed that R2R3-MYB genes from subgroup 6 control anthocyanin biosynthesis [19,21]. In this study, the expression of *MYBA1* and *MYBA2* genes (subgroup 6) were upregulated under exogenous ABA treatment (Figure 5). In which, the *MYBA2* gene expression was upregulated, was positively correlated with the expression of ASGs under UV-B radiation and played an important role in the UV-B-induced anthocyanin biosynthetic pathway [70]. *MYBA1* has been reported as *VcMYB437* and the expression of *VcMYB437* was higher in the skin of ripe blueberry fruit than in the pulp, implying the potential roles in anthocyanin biosynthesis [71]. VcMYBA (probably homologous to MYBA1) promoted anthocyanin accumulation by heterologous expression in *Nicotiana benthaminana* [21]. Our study found that *MYBA1* expression was significantly correlated with that of *4CL7* and *VcUFGT*, At the same time, *4CL7* and *VcUFGT* contain multiple MYB binding sites in their promoter sequences (Figure 6 and Figure S3). However, the expression of *VcMYBA2* was not significantly correlated with that of ASGs. Thus, MYBA1 is likely to play a core role in the network of ABA-induced anthocyanin biosynthesis; however, its authentic roles still need to be documented in future studies.

5. Conclusions

Exogenous ABA promotes anthocyanin accumulation in blueberry leaves and mainly affects gene expression from plant hormone signal transduction and phenylpropanoid and flavonoid biosynthetic pathways. Pearson's correlation coefficient (r) analysis indicated that ASGs from the phenylpropanoid and flavonoid biosynthetic pathways, MYB transcription factor genes from R2R3-MYB subgroups 5, 6, and 7 and ABF genes from the ABA signal transduction pathway form an ABFs–MYB–ASGs regulatory network to co-regulate ABA-induced anthocyanin biosynthesis. The MYBA1 transcription factor is likely to play an important role in the ABA-induced anthocyanin biosynthesis pathway.

Supplementary Materials: The following supporting information can be downloaded at: https://www.mdpi.com/article/10.3390/horticulturae10020192/s1, Figure S1: Validation of transcript changes by RT-qPCR. (a) Expression of ABA signal transduction pathway genes under ABA treatment. (b) The figure is based on $\log_2(2^{-\Delta\Delta Ct})$ data from RT-qPCR and $\log_2$(fold-change) data from RNA-seq. Pearson's correlation coefficients (r) between $\log_2(2^{-\Delta\Delta Ct})$ and $\log_2$(fold-change) are shown; Figure S2: Sequence alignment (a) and phylogenetic analysis (b) of MYBA1 (VaccDscaff1486-snap-gene-0.3), MYBA2 (VaccDscaff13-augustus-gene-105.26) and VcMYBA (MH105054) from blueberry (*Vaccinium corymbosum*); Figure S3: Schematic diagram of the predicted regulatory cis-elements involved in the abscisic acid responsiveness in the upstream sequences of screened MYB transcription and structure genes involved in anthocyanin biosynthesis in blueberry. The black line indicates the promoter length of genes. The blue and red boxes represent cis-acting elements involved in the abscisic acid responsiveness and MYB binding sites, respectively. '+' Sense strand; '−' Antisense strand; Table S1: Primers used in this study; Table S2: Statistics of functional annotation of differential expressed genes (DEGs); Table S3: The differential expressed genes (DEGs) from plant hormone signal transduction pathway under ABA stress; Table S4: Differentially expression R2R3-MYB transcription factors during ABA treatment; Table S5: The differential expressed genes (DEGs) from flavonoid biosynthetic pathway under ABA stress; Table S6: Pearson's correlation coefficients (r) in the expression levels between MYB transcription factors and anthocyanin biosynthetic pathway gene or genes from ABA responsive element binding factor.

Author Contributions: Conceptualization, C.Z. and B.M.; methodology, Y.S.; software, B.M. and Y.S.; validation, Y.S., B.M. and X.F.; formal analysis, B.M. and X.F.; investigation, Q.G.; resources, X.Z.; data curation, L.Z.; writing—original draft preparation, B.M.; writing—review and editing, C.Z.; visualization, L.Z.; supervision, L.Z.; project administration, C.Z. and Q.G.; funding acquisition, C.Z. All authors have read and agreed to the published version of the manuscript.

Funding: This research was funded by the Natural Science Foundation of Jilin Province, China (Grant no. 20210101006JC).

Data Availability Statement: Data are contained within the article and Supplementary Materials. We have uploaded the RNA-Seq data generated in this study to BioProject in the NCBI repository with the accession number of PRJNA997066.

Conflicts of Interest: The authors declare no conflicts of interest.

References

1. Verma, V.; Ravindran, P.; Kumar, P.P. Plant hormone-mediated regulation of stress responses. *BMC Plant Biol.* **2016**, *16*, 86. [CrossRef]
2. Loreti, E.; Povero, G.; Novi, G.; Solfanelli, C.; Alpi, A.; Perata, P. Gibberellins, jasmonate and abscisic acid modulate the sucrose-induced expression of anthocyanin biosynthetic genes in *Arabidopsis*. *New Phytol.* **2008**, *179*, 1004–1016. [CrossRef]
3. Chen, K.; Li, G.; Bressan, R.A.; Song, C.; Zhu, J.; Zhao, Y. Abscisic acid dynamics, signaling, and functions in plants. *J. Integr. Plant Biol.* **2020**, *62*, 25–54. [CrossRef]
4. Davière, J.; Achard, P. Gibberellin signaling in plants. *Development* **2013**, *140*, 1147–1151. [CrossRef] [PubMed]
5. Umezawa, T.; Nakashima, K.; Miyakawa, T.; Kuromori, T.; Tanokura, M.; Shinozaki, K.; Yamaguchi-Shinozaki, K. Molecular basis of the core regulatory network in ABA responses: Sensing, signaling and transport. *Plant Cell Physiol.* **2010**, *51*, 1821–1839. [CrossRef] [PubMed]
6. Kretzschmar, A.A.; Lerin, S.; Fagherazzi, A.; Mario, A.E.; Bastos, F.E.A.; Allebrandt, R.; Rufato, L. Application of abscisic acid increases the colour of "Rubi" grape berries in Southern Brazil. *Acta Hortic.* **2016**, *1115*, 231–236. [CrossRef]

7. Oh, H.D.; Yu, D.J.; Chung, S.W.; Chea, S.; Lee, H.J. Abscisic acid stimulates anthocyanin accumulation in 'Jersey' highbush blueberry fruits during ripening. *Food Chem.* **2018**, *244*, 403–407. [CrossRef] [PubMed]
8. Kumar, L.; Guy, H.; Hadas, S.; Reut, P.; Moshe, A.F. Anthocyanin accumulation is initiated by abscisic acid to enhance fruit color during fig (*Ficus carica* L.) ripening. *J. Plant Physiol.* **2020**, *251*, 153192. [CrossRef]
9. Gonzalez-Guzman, M.; Pizzio, G.A.; Antoni, R.; Vera-Sirera, F.; Merilo, E.; Bassel, G.W.; Fernández, M.A.; Holdsworth, M.J.; Perez-Amador, M.A. Arabidopsis PYR/PYL/RCAR receptors play a major role in quantitative regulation of stomatal aperture and transcriptional response to abscisic Acid. *Plant Cell.* **2012**, *24*, 2483–2496. [CrossRef] [PubMed]
10. Boursiac, Y.; Léran, S.; Corratgé-Faillie, C.; Gojon, A.; Krouk, G.; Lacombe, B. ABA transport and transporters. *Trends Plant Sci.* **2013**, *18*, 325–333. [CrossRef] [PubMed]
11. Cutler, S.R.; Rodriguez, P.L.; Finkelstein, R.R.; Abrams, S.R. Abscisic acid: Emergence of a core signaling network. *Annu. Rev. Plant Biol.* **2010**, *61*, 651–679. [CrossRef] [PubMed]
12. Fujita, Y.; Yoshida, T.; Yamaguchi-Shinozaki, K. Pivotal role of the AREB/ABF-SnRK2 pathway in ABRE-mediated transcription in response to osmotic stress in plants. *Physiol. Plant.* **2012**, *147*, 15–27. [CrossRef] [PubMed]
13. Weiner, J.J.; Peterson, F.C.; Volkman, B.F.; Cutler, S.R. Structural and functional insights into core ABA signaling. *Curr. Opin. Plant Biol.* **2010**, *13*, 495–502. [CrossRef] [PubMed]
14. Narusaka, Y.; Nakashima, K.; Shinwari, Z.K.; Sakuma, Y.; Furihata, T.; Abe, H.; Narusaka, M.; Shinozaki, K.; Yanmaguchi-Shinozaki, K. Interaction between two cis-acting elements, ABRE and DRE, in ABA-dependent expression of *Arabidopsis rd29A* gene in response to dehydration and high-salinity stresses. *Plant J.* **2003**, *34*, 137–148. [CrossRef] [PubMed]
15. Fujita, Y.; Fujita, M.; Satoh, R.; Maruyama, K.; Parvez, M.M.; Seki, M.; Hiratsu, K.; Ohme-Takagi, M.; Shinozaki, K.; Yamaguchi-Shinozaki, K. AREB1 is a transcription activator of novel ABRE-dependent ABA signaling that enhances drought stress tolerance in *Arabidopsis*. *Plant Cell.* **2005**, *17*, 3470–3488. [CrossRef] [PubMed]
16. Guo, X.; Shakeel, M.; Wang, D.; Qu, C.; Yang, S.; Ahmad, S.; Song, Z. Metabolome and transcriptome profiling unveil the mechanisms of light-induced anthocyanin synthesis in rabbiteye blueberry (*Vaccinium ashei*: Reade). *BMC Plant Biol.* **2022**, *22*, 223. [CrossRef] [PubMed]
17. Kuhn, N.; Ponce, C.; Arellano, M.; Time, A.; Multari, S.; Martens, S.; Carrera, E.; Sagredo, B.; Donoso, J.M.; Meisel, L.A. ABA influences color initiation timing in *P. avium* L. fruits by sequentially modulating the transcript levels of ABA and anthocyanin-related genes. *Tree Genet. Genomes* **2021**, *17*, 20. [CrossRef]
18. Gonzalez, A. Pigment loss in response to the environment: A new role for the WD/bHLH/MYB anthocyanin regulatory complex. *New Phytol.* **2009**, *182*, 1–3. [CrossRef]
19. Gonzalez, A.; Zhao, M.; Leavitt, J.M.; Lloyd, A.M. Regulation of the anthocyanin biosynthetic pathway by the TTG1/bHLH/Myb transcriptional complex in *Arabidopsis* seedlings. *Plant J.* **2008**, *53*, 814–827. [CrossRef]
20. An, X.H.; Tian, Y.; Chen, K.Q.; Wang, X.F.; Hao, Y.J. The apple WD40 protein MdTTG1 interacts with bHLH but not MYB protein to regulate anthocyanin accumulation. *J. Plant Physiol.* **2012**, *7*, 710–717. [CrossRef]
21. Plunkett, B.J.; Espley, R.V.; Dare, A.P.; Warren, B.A.W.; Grierson, E.R.P.; Cordiner, S.; Turner, J.L.; Allan, A.C.; Albert, N.W.; Davies, K.M.; et al. MYBA from blueberry (*Vaccinium* Section *Cyanococcus*) is a subgroup 6 type R2R3MYB transcription factor that activates anthocyanin production. *Front. Plant Sci.* **2018**, *9*, 1300. [CrossRef] [PubMed]
22. Cao, Y.; Li, K.; Li, Y.; Zhao, X.; Wang, L. MYB transcription factors as regulators of secondary metabolism in plants. *Biology* **2020**, *9*, 61. [CrossRef]
23. Zhang, Y.; Wang, K.; Albert, N.W.; Elborough, C.; Espley, R.V.; Andre, C.M.; Fang, Z. Identification of a strong anthocyanin activator, VbMYBA, from berries of *Vaccinium bracteatum* Thunb. *Front. Plant Sci.* **2021**, *12*, 697212. [CrossRef] [PubMed]
24. Kranz, H.D.; Denekamp, M.; Greco, R.; Jin, H.; Leyva, A.; Meissner, R.C.; Petroni, K.; Urzainqui, A.; Bevan, M.; Martin, C.; et al. Towards functional characterisation of the members of the R2R3-MYB gene family from *Arabidopsis thaliana*. *Plant J.* **1998**, *16*, 263–276. [CrossRef]
25. Stracke, R.; Ishihara, H.; Huep, G.; Barsch, A.; Mehrtens, F.; Niehaus, K.; Weisshaar, B. Differential regulation of closely related R2R3-MYB transcription factors controls flavonol accumulation in different parts of the *Arabidopsis thaliana* seedling. *Plant J.* **2007**, *50*, 660–677. [CrossRef] [PubMed]
26. Baudry, A.; Heim, M.A.; Dubreucq, B.; Caboche, M.; Weisshaar, B.; Lepiniec, L. TT2, TT8, and TTG1 synergistically specify the expression of BANYULS and proanthocyanidin biosynthesis in *Arabidopsis thaliana*. *Plant J.* **2004**, *39*, 366–380. [CrossRef]
27. Norberto, S.; Silva, S.; Meireles, M.; Faria, A.; Pintado, M.; Calhau, C. Blueberry anthocyanins in health promotion: A metabolic overview. *J. Funct. Foods* **2013**, *5*, 1518–1528. [CrossRef]
28. Ribera, A.E.; Reyes-Díaz, M.; Alberdi, M.; Zuniga, G.E.; Mora, M.L. Antioxidant compounds in skin and pulp of fruits change among genotypes and maturity stages in highbush blueberry (*Vaccinium corymbosum* L.) grown in southern Chile. *J. Soil Sci. Plant Nutr.* **2010**, *10*, 509–536. [CrossRef]
29. Tanaka, W.; Yokoyama, D.; Matsuura, Y.; Nozaki, M.; Hirozawa, N.; Kunitake, H.; Sakono, M.; Sakakibara, H. Subchronic toxicity evaluation of leaves from rabbiteye blueberry (*Vaccinium virgatum* Aiton) in rats. *Toxicol. Rep.* **2019**, *6*, 272–278. [CrossRef]
30. Inoue, N.; Nagao, K.; Nomura, S.; Shirouchi, B.; Inafuku, M.; Hirabaru, H.; Nakahara, N.; Nishizono, S.; Tanaka, T.; Yanagita, T. Effect of Vaccinium ashei reade leaf extracts on lipid metabolism in obese OLETF rats. *Biosci. Biotechnol. Biochem.* **2011**, *75*, 2304–2308. [CrossRef]

31. Yuji, K.; Sakaida, H.; Kai, T.; Fukuda, N.; Yukizaki, C.; Sakai, M.; Tsubouchi, H.; Kataoka, H. Effect of dietary blueberry (*Vaccinium ashei* Reade) leaves on serum and hepatic lipid levels in rats. *J. Oleo Sci.* **2013**, *62*, 89–96. [CrossRef] [PubMed]

32. Cezarotto, V.S.; Franceschi, E.P.; Stein, A.C.; Emanuelli, T.; Maurer, L.H.; Sari, M.H.M.; Ferreira, L.M.; Cruz, L. Nanoencapsulation of Vaccinium ashei leaf extract in Eudragit® RS100-based nanoparticles increases its in vitro antioxidant and in vivo antidepressant-like actions. *Pharmaceuticals* **2023**, *16*, 84. [CrossRef] [PubMed]

33. Venskutonis, P.R.; Barnackas, Š.; Kazernavičiūtė, R.; Maždžierienė, R.; Pukalskas, A.; Šipailienė, A.; Labokas, J.; Ložienė, K.; Abrutienė, G. Variations in antioxidant capacity and phenolics in leaf extracts isolated by different polarity solvents from seven blueberry (*Vaccinium* L.) genotypes at three phenological stages. *Acta Physiol. Plant.* **2016**, *38*, 33. [CrossRef]

34. Li, Y.C.; Li, B.X.; Geng, L.J. Hypolipidemic and antioxidant effects of total flavonoids from blueberry leaves. *Eur. Food Res. Technol.* **2011**, *233*, 897–903. [CrossRef]

35. Choi, H.; Hong, J.; Ha, J.; Kang, J.; Kim, S.Y. ABFs, a family of ABA-responseive element binding factors. *J. Biol. Chem.* **2000**, *275*, 1723–1730. [CrossRef] [PubMed]

36. Ji, L.; Wang, J.; Ye, M.; Li, Y.; Guo, B.; Chen, Z.; Li, H.; An, X. Identification and characterization of the Populus AREB/ABF subfamily. *J. Integr. Plant Biol.* **2013**, *55*, 177–186. [CrossRef]

37. Mao, X.; Cai, T.; Olyarchuk, J.G.; Wei, L. Automated genome annotation and pathway identification using the KEGG Orthology (KO) as a controlled vocabulary. *Bioinformatics* **2005**, *21*, 3787–3793. [CrossRef]

38. Chen, C.; Chen, H.; Zhang, Y.; Thomas, H.R.; Frank, M.H.; He, Y.; Xia, R. TBtools: An integrative toolkit developed for interactive analyses of big biological data. *Mol. Plant* **2020**, *13*, 1194–1202. [CrossRef]

39. Yang, B.; Song, Y.; Li, Y.; Wang, X.; Guo, Q.; Zhou, L.; Zhang, Y.; Zhang, C. Key genes for phenylpropanoid metabolite biosynthesis during half-highbush blueberry (*Vaccinium angustifolium* × *Vaccinium corymbosum*) fruit development. *J. Berry Res.* **2022**, *12*, 297–311. [CrossRef]

40. Rabino, I.; Mancinelli, A.L. Light, temperature, and anthocyanin production. *Plant Physiol.* **1986**, *81*, 922–924. [CrossRef]

41. Dubos, C.; Stracke, R.; Grotewold, E.; Weisshaar, B.; Martin, C.; Lepiniec, L. MYB transcription factors in *Arabidopsis*. *Trends Plant Sci.* **2010**, *15*, 573–581. [CrossRef]

42. Häkkinen, S.H.; Törrönen, A.R. Content of flavonols and selected phenolic acids in strawberries and *Vaccinium* species: Influence of cultivar, cultivation site and technique. *Food Res. Int.* **2000**, *33*, 517–524. [CrossRef]

43. Karppinen, K.; Tegelberg, P.; Haggman, H.; Jaakola, L. Abscisic acid regulates anthocyanin biosynthesis and gene expression associated with cell wall modification in ripening bilberry (*Vaccinium myrtillus* L.) fruits. *Front. Plant Sci.* **2018**, *9*, 1259. [CrossRef]

44. Chung, S.W.; Yu, D.J.; Oh, H.D.; Ahn, J.H.; Huh, J.H.; Lee, H.J. Transcriptional regulation of abscisic acid biosynthesis and signal transduction, and anthocyanin biosynthesis in 'Bluecrop' highbush blueberry fruit during ripening. *PLoS ONE* **2019**, *14*, e0220015. [CrossRef]

45. Fan, W.; Zhao, M.; Li, S.; Bai, X.; Li, J.; Meng, H.; Mu, Z. Contrasting transcriptional responses of PYR1/RCAR ABA receptors to ABA or dehydration stress between maize seedling leaves and roots. *BMC Plant Biol.* **2016**, *16*, 99. [CrossRef]

46. Chen, J.; Mao, L.; Lu, W.; Ying, T.; Luo, Z. Transcriptome profiling of postharvest strawberry fruit in response to exogenous auxin and abscisic acid. *Planta* **2016**, *243*, 183–197. [CrossRef]

47. Liao, X.; Li, M.; Liu, B.; Yan, M.; Yu, X.; Zi, H.; Liu, R.; Yamamuro, C. Interlinked regulatory loops of ABA catabolism and biosynthesis coordinate fruit growth and ripening in woodland strawberry. *Proc. Natl. Acad. Sci. USA* **2018**, *115*, E11542–E11550. [CrossRef]

48. Hong, L.; Su, W.; Zhang, Y.; Ye, C.; Shen, Y.; Li, Q.Q. Transcriptome profiling during mangrove viviparity in response to abscisic acid. *Sci. Rep.* **2018**, *8*, 770. [CrossRef]

49. Nolan, T.M.; Vukašinović, N.; Liu, D.; Russinova, E.; Yin, Y.H. Brassinosteroids: Multidimensional regulators of plant growth, development, and stress responses. *Plant Cell.* **2020**, *32*, 295–318. [CrossRef]

50. Planas-Riverola, A.; Gupta, A.; Betegón-Putze, I.; Bosch, N.; Ibañes, M.; Caño-Delgado, A.I. Brassinosteroid signaling in plant development and adaptation to stress. *Development* **2019**, *146*, dev151894. [CrossRef]

51. Vergara, A.; Torrealba, M.; Alcalde, J.A.; Pérez-Donoso, A.G. Commercial brassinosteroid increases the concentration of anthocyanin in red tablegrape cultivars (*Vitis vinifera* L.). *Aust. J. Grape Wine R.* **2020**, *26*, 427–433. [CrossRef]

52. Yuan, L.B.; Peng, Z.H.; Zhi, T.T.; Zho, Z.; Liu, Y.; Zhu, Q.; Xiong, X.Y.; Ren, C.M. Brassinosteroid enhances cytokinin-induced anthocyanin biosynthesis in *Arabidopsis* seedlings. *Biol. Plant.* **2015**, *59*, 99–105. [CrossRef]

53. Wang, Y.; Mao, Z.; Jiang, H.; Zhang, Z.; Wang, N.; Chen, X. Brassinolide inhibits flavonoid biosynthesis and red-flesh coloration via the MdBEH2.2-MdMYB60 complex in apple. *J. Exp. Bot.* **2021**, *72*, 6382–6399. [CrossRef]

54. Wang, Y.; Zhu, Y.; Jiang, H.; Mao, Z.; Zhang, J.; Fang, H.; Liu, W.; Zhang, Z.; Chen, X.; Wang, N. The regulatory module MdBZR1-MdCOL6 mediates brassinosteroid- and light-regulated anthocyanin synthesis in apple. *New Phytol.* **2023**, *238*, 1516–1533. [CrossRef] [PubMed]

55. Nam, K.H.; Li, J. BRI1/BAK1, a receptor kinase pair mediating brassinosteroid signaling. *Cell* **2002**, *110*, 203–212. [CrossRef] [PubMed]

56. Rahman, A.; Amakawa, T.; Goto, N.; Tsurumi, S. Auxin is a positive regulator for ethylene-mediated response in the growth of *Arabidopsis* roots. *Plant Cell Physiol.* **2001**, *42*, 301–307. [CrossRef]

57. Cecchetti, V.; Altamura, M.M.; Falasca, G.; Costantino, P.; Cardarelli, M. Auxin regulates *Arabidopsis* anther dehiscence, pollen maturation, and filament elongation. *Plant Cell.* **2008**, *20*, 1760–1774. [CrossRef]

58. Clayton-Cuch, D.; Yu, L.; Shirley, N.; Bradley, D.; Bulone, V.; Bőttcher, C. Auxin treatment enhances anthocyanin production in the non-climacteric sweet cherry (*Prunus avium* L.). *Int. J. Mol. Sci.* **2021**, *22*, 10760. [CrossRef] [PubMed]

59. Ji, X.; Zhang, R.; Wang, N.; Yang, L.; Chen, X. Transcriptome profiling reveals auxin suppressed anthocyanin biosynthesis in red-fleshed apple callus (*Malus sieversii* f. *niedzwetzkyana*). *Plant Cell Tiss. Organ Cult.* **2015**, *123*, 389–404. [CrossRef]

60. Ji, X.; Wang, Y.; Zhang, R.; Wu, S.; An, M.; Li, M.; Wang, C.; Chen, X.; Zhang, Y.; Chen, X. Effect of auxin, cytokinin and nitrogen on anthocyanin biosynthesis in callus cultures of red-flfleshed apple (*Malus sieversii* f. *niedzwetzkyana*). *Plant Cell Tiss. Organ Cult.* **2015**, *120*, 325–337. [CrossRef]

61. Wang, Y.; Wang, N.; Xu, H.; Jiang, S.; Fang, H.; Su, M.; Zhang, Z.; Chen, X. Auxin regulates anthocyanin biosynthesis through the Aux/IAA-ARF signaling pathway in apple. *Hortic. Res.* **2018**, *5*, 59. [CrossRef] [PubMed]

62. Reed, J.W. Roles and activities of Aux/IAA proteins in Arabidopsis. *Trends Plant Sci.* **2001**, *6*, 420–425. [CrossRef] [PubMed]

63. Aya, K.; Ueguchi-Tanaka, M.; Kondo, M.; Hamada, K.; Yano, K.; Nishimura, M.; Matsuoka, M. Gibberellin modulates anther development in Rice via the transcriptional regulation of GAMYB. *Plant Cell* **2009**, *21*, 1453–1472. [CrossRef] [PubMed]

64. Zhang, Y.; Liu, Z.; Liu, J.; Lin, S.; Wang, J.; Lin, W.; Xu, W. GA-DELLA pathway is involved in regulation of nitrogen deficiency-induced anthocyanin accumulation. *Plant Cell Rep.* **2017**, *36*, 557–569. [CrossRef]

65. Weiss, D.; Ori, N. Mechanisms of cross talk between gibberellin and other hormones. *Plant Physiol.* **2007**, *144*, 1240–1246. [CrossRef]

66. Wang, L.; Yang, S.; Ni, J.; Teng, Y.; Bai, S. Advances of anthocyanin synthesis regulated by plant growth regulators in fruit trees. *Sci. Hortic.* **2023**, *307*, 111476. [CrossRef]

67. Yoshida, T.; Fujita, Y.; Sayama, H.; Kidokoro, S.; Maruyama, K.; Mizoi, J.; Shinozaki, K.; Yamaguchi-Shinozaki, K. AREB1, AREB2, and ABF3 are master transcription factors that cooperatively regulate ABRE-dependent ABA signaling involved in drought stress tolerance and require ABA for full activation. *Plant J.* **2010**, *61*, 672–685. [CrossRef]

68. Karppinen, K.; Lafferty, D.J.; Albert, N.W.; Mikkola, N.; McGhie, T.; Allan, A.C.; Afzal, B.M.; Häggman, H.; Espley, R.V.; Jaakola, L. MYBA and MYBPA transcription factors co-regulate anthocyanin biosynthesis in blue-coloured berries. *New Phytol.* **2021**, *232*, 1350–1367. [CrossRef]

69. Zhang, P.; Wang, R.; Ju, Q.; Li, W.; Tran, L.P.; Xu, J. The R2R3-MYB transcription factor MYB49 regulates cadmium accumulation. *Plant Physiol.* **2019**, *180*, 529–542. [CrossRef]

70. Song, Y.; Ma, B.; Guo, Q.; Zhou, L.; Zhou, X.; Ming, Z.; You, H.; Zhang, C. MYB pathways that regulate UV-B-induced anthocyanin biosynthesis in blueberry (*Vaccinium corymbosum*). *Front. Plant Sci.* **2023**, *14*, 1125382. [CrossRef]

71. Wang, H.; Zhai, L.; Wang, S.; Zheng, B.; Hu, H.; Li, X.; Bian, S. Identification of R2R3-MYB family in blueberry and its potential involvement of anthocyanin biosynthesis in fruits. *BMC Genom.* **2023**, *24*, 505. [CrossRef] [PubMed]

horticulturae

Article

Microsatellite Sequence Polymorphisms Reveals Substantial Diversity in Caribbean Breadfruit [*Artocarpus altilis* (Parkinson) Fosberg] Germplasm

Oral O. Daley [1,*], Angela T. Alleyne [2], Laura B. Roberts-Nkrumah [1] and Lambert A. Motilal [3]

[1] Department of Food Production, Faculty of Food and Agriculture, The University of the West Indies, St. Augustine 330912, Trinidad and Tobago; laurarobertsnkrumah@gmail.com
[2] Department of Chemical and Biological Sciences, Faculty of Science and Technology, The University of the West Indies, Cavehill BB11000, Barbados; angela.alleyne@cavehill.uwi.edu
[3] Cocoa Research Centre, The University of the West Indies, St. Augustine 330912, Trinidad and Tobago; lambert.motilal@sta.uwi.edu
* Correspondence: oral.daley@sta.uwi.edu

Abstract: Breadfruit [*Artocarpus altilis* (Parkinson) Fosberg] is recognized as a tropical fruit tree crop with great potential to contribute to food and nutrition security in the Caribbean and other tropical regions. However, the genetic diversity and germplasm identification in the Caribbean and elsewhere are poorly understood and documented. This hampers the effective conservation and use of the genetic resources of this tree crop for commercial activities. This study assessed the genetic identity, diversity, ancestry, and phylogeny of breadfruit germplasm existing in the Caribbean and several newly introduced accessions using 117 SNPs from 10 SSR amplicon sequences. The results showed that there was high and comparable genetic diversity in the breadfruit germplasm in the Caribbean, and the newly introduced breadfruit accessions were based on nucleotide diversity (π_T) 0.197 vs. 0.209, respectively, and nucleotide polymorphism (θ_W) 0.312 vs. 0.297, respectively. Furthermore, the existing Caribbean breadfruit accessions and the newly introduced breadfruit accessions were statistically genetically undifferentiated from each other ($p < 0.05$). Ancestry and phylogeny analysis corroborated the genetic relatedness of these two groups, with accessions of these groups being present in both main germplasm clusters. This suggests that the existing Caribbean breadfruit germplasm harbors a higher level of genetic diversity than expected.

Keywords: crop germplasm; genetic diversity; nucleotide diversity; nucleotide polymorphism; SSR markers; underutilized crop

Citation: Daley, O.O.; Alleyne, A.T.; Roberts-Nkrumah, L.B.; Motilal, L.A. Microsatellite Sequence Polymorphisms Reveals Substantial Diversity in Caribbean Breadfruit [*Artocarpus altilis* (Parkinson) Fosberg] Germplasm. *Horticulturae* **2024**, *10*, 253. https://doi.org/10.3390/horticulturae10030253

Academic Editors: Zhaohe Yuan, Bo Li and Yujie Zhao

Received: 31 January 2024
Revised: 25 February 2024
Accepted: 29 February 2024
Published: 6 March 2024

1. Introduction

Breadfruit [*Artocarpus altilis* (Parkinson) Fosberg] belongs to the family Moraceae and was domesticated in the Pacific, where it has been a traditional staple on many islands in the region [1]. Starting in the 16th century, small numbers of breadfruit cultivars were introduced worldwide because of the low maintenance requirement of the crop while being a nutritious food source [2–4]. It is currently cultivated in over 90 countries spread across the continents of Africa, Asia, Australia, North America, and South America [5]. Breadfruit is recognized as a tree crop with great potential to contribute to food and nutrition security and to alleviate hunger in many countries in these regions [6,7]. Hence, commercial breadfruit production systems are being encouraged. Yet, there is a paucity of reliable knowledge of the genetic diversity and cultivar identification of breadfruit in many of these countries, which could compromise commercial breadfruit production activities.

Breadfruit (*A. altilis*) belongs to a complex that includes its two closest relatives, the breadnut, *A. camansi* (Blanco), native to New Guinea in Western Melanesia, and dugdug, *A. mariannensis* (Trécul), native to the Mariana Islands in Micronesia, as well as interspecific hybrids of *A. altilis* × *A. mariannensis* [1]. *A. camansi*, a fertile diploid (2n = 2x = 56), is considered the wild ancestor of *A. altilis*, which, however, varies in ploidy and seededness, from fertile and sterile diploids to seedless triploids (2n = 3x = 84) [8]. These seedless triploids are most common in Eastern Polynesia and were distributed worldwide. *A. mariannensis* is also a seeded diploid (2n = 2x = 56), which, through hybridization with *A. altilis*, has also produced diploid and triploid hybrids that vary in seededness [8]. Seededness is related to pollen viability [9]. Breadfruit is mainly cross-pollinated because, typically, most male inflorescences appear before the female inflorescence [9,10]. Wind pollination apparently predominates because, while insects, mainly bees, may actively collect pollen from the male flowers, they are much less attracted to the female flowers [10]. Pollen viability in the seedless breadfruit is as low as 6%, and the fruits are parthenocarpic [9]. Apart from the variation in seededness, the tree, leaf, and fruit morphology of breadfruit cultivars may differ according to age and environmental conditions [11,12]. Seedless breadfruit cultivars are propagated vegetatively only, whereas the seeded types are propagated by seed and vegetative methods.

Traditionally, diversity studies in breadfruit have employed morphological traits that are time-consuming to use, and show environmental plasticity [11,13]. Furthermore, in breadfruit-growing areas, it is common to find morphologically distinct cultivars having the same name or a single cultivar having multiple names in one or more locations [14,15]. These conditions make it difficult to rely only on morphological characterization to understand the diversity and range of cultivars within the species. The use of molecular or DNA markers has helped to simplify the estimation of plant genetic diversity in several species and offers increased reliability over morphological techniques [16]. In contrast to morphological techniques, DNA-based methods are independent of environmental factors and are usually highly polymorphic for each locus [17,18]. Several DNA marker techniques are available and are important tools for diversity studies in plant germplasm, and some have been applied to breadfruit. Analysis of Restriction Fragment Length Polymorphism (RFLP) in chloroplast DNA (cpDNA) of breadfruit, jackfruit, and nine related species showed 30 mutation sites on eight endonucleases, while 12 other endonucleases were monomorphic but were unable to distinguish among four breadfruit genotypes [19]. Using sequence data from both plastid (trnL intron and trnL-F spacer) and nuclear (internal transcribed spacers 1 and 2, ITS) genome sequences, Zerega, et al. [1] confirmed the close relationship between *A. altilis* and *A. camansi*, as well as between *A. altilis* and *A. mariannensis*, and showed that all three species had a monophyletic lineage. Furthermore, Amplified Fragment Length Polymorphism (AFLP) analyses revealed that not only were *A. camansi* and *A. mariannensis* breadfruit's closest relatives, but they were progenitor species [8]. AFLP marker analyses also revealed a high degree of genetic variation among six breadfruit populations in the Western Ghats of India [20]. Witherup, et al. [21] isolated 25 nuclear microsatellite loci from enriched genomic libraries of breadfruit, which were all polymorphic in at least four Artocarpus species. Zerega, et al. [22] used 19 of the 25 markers developed by Witherup, et al. [21] and characterized the diversity among 349 individual trees, which included three *Artocarpus* species. Fifteen chloroplast microsatellite loci were also identified in chloroplast sequences from four Artocarpus transcriptome assemblies [23]. The use of nuclear and microsatellite markers to examine 423 individual breadfruit tree from Oceania, the Caribbean, India, and the Seychelles revealed that there was a range in the level of genetic diversity across regions, and the diversity in some areas was greater than expected [24]. A further 50 new microsatellite loci were characterized in *Artocarpus altilis* (Moraceae) and two congeners to increase the number of available markers for genotyping breadfruit cultivars using next-generation sequencing [25]. Next-generation sequencing (NGS) was also used along with phylogenetic reconstruction of breadfruit lineage to attempt a match of breadfruit cultivars in the Caribbean with existing Polynesian types [26].

The studies using simple sequence repeats (SSRs) markers relied on the detection of variation in SSRs based on scoring SSRs alleles as the length of polymorphisms, with differences in amplicon size taken to represent differences in the repeat number in the SSRs [21–25]. However, the amplicon size also includes the length of the flanking regions, which may contain additional information such as single nucleotide polymorphisms (SNPs) and insertion/deletions [27]. Therefore, substantial polymorphic data are neglected when SSRs are described through amplicon size alone, and this type of result is prone to size homoplasy [28]. Furthermore, Barthe, et al. [27] reported higher levels of genetic diversity for amplicon sequence variation than for amplicon size variation. Additionally, the use of marker sequences offers the opportunity to analyze evolutionary events based on the presence of mutations and the rate at which they occur within the sequence [27]. Therefore, the objectives of this study were to assess the genetic diversity of breadfruit germplasm using SSRs amplicon sequences and to investigate the genetic relatedness and structure of the breadfruit germplasm, with various cultivar names, existing in the Caribbean since the 18th century, hereafter referred to as existing Caribbean accessions (ECA) and newly introduced accessions (NIA), consisting of breadfruit germplasm introduced to the Caribbean by the Ministry of Food Production, Trinidad and Tobago, in 1989 and The University of the West Indies (UWI) during the 1990s [29,30].

2. Materials and Methods

2.1. Plant Materials

Leaf tissue samples from 153 individual breadfruit trees were collected and placed in separate labeled plastic zipper bags containing silica gel at a ratio of 10 parts silica gel to 1 part leaf sample. The bagged samples were stored in iceboxes and transported to the laboratory at The University of the West Indies (UWI), St. Augustine, Trinidad and Tobago. These samples were collected in Jamaica (25 samples), Trinidad and Tobago (21 samples), St. Vincent and the Grenadines (19 samples), and St. Kitts and Nevis (2 samples) (Table 1). The remaining samples came from the UWI, St. Augustine campus, breadfruit germplasm collection (Table 1).

Table 1. Accession data for breadfruit (*Artocarpus altilis*), breadfruit hybrid (*A. altilis* × *A. mariannensis*), and breadnut (*A. camansi*) samples used in the study.

No.	Sample ID	Cultivar Name	Taxon *	Ploidy *	Sample Collection Site	Accession Grouping
1	12A	Huehue	Aa × Am	3n	UWI	NIA
2	13A	Hope Marble	Aa	3n	UWI	ECA
3	15A	Ulu'ea	Aa	2n	UWI	NIA
4	18A	Meitehid	Aa	3n	UWI	NIA
5	19A	White	Aa	3n	UWI	ECA
6	1A	Yellow	Aa	3n	UWI	ECA
7	41A	Macca	Aa	3n	UWI	ECA
8	42B	Yellow Heart	Aa	3n	UWI	ECA
9	43B	Yellow Heart	Aa	3n	UWI	ECA
10	44A	Aveloloa	Aa	3n	UWI	NIA
11	45A	Creole	Aa	3n	UWI	ECA
12	47A	White	Aa	3n	UWI	ECA
13	48B	White	Aa	3n	UWI	ECA
14	49B	White	Aa	3n	UWI	ECA
15	50B	Porohiti	Aa	3n	UWI	NIA
16	51A	Porohiti	Aa	3n	UWI	NIA
17	522B	Captain Bligh	Aa	3n	SVG	ECA
18	523B	Unidentfied 1	Aa	3n	SVG	ECA
19	524A	Creole	Aa	3n	SVG	ECA
20	52B	Toneno	Aa	3n	UWI	NIA
21	532B	Sally Young	Aa	3n	SVG	ECA
22	534A	White	Aa	3n	SVG	ECA
23	537A	Hog Pen	Aa	3n	SVG	ECA
24	539B	Dessert	Aa	3n	SVG	ECA
25	53B	Roiha'a	Aa	3n	UWI	NIA

Table 1. *Cont.*

No.	Sample ID	Cultivar Name	Taxon *	Ploidy *	Sample Collection Site	Accession Grouping
26	544B	Kashee Bread	Aa	3n	SVG	ECA
27	545A	Hope Marble	Aa	3n	SVG	ECA
28	546A	Lawyer Caine	Aa	3n	SVG	ECA
29	548A	Dessert	Aa	3n	SVG	ECA
30	54A	Tapeha'a	Aa	3n	UWI	NIA
31	552B	Sally Young	Aa	3n	SVG	ECA
32	553A	White	Aa	3n	SVG	ECA
33	555B	Waterloo/Cotton	Aa	3n	SVG	ECA
34	556A	Soursop	Aa	3n	SVG	ECA
35	558A	Liberal	Aa	3n	SVG	ECA
36	559A	Yellow Heart	Aa	3n	JAM	ECA
37	55A	Piipiia	Aa × Am	3n	UWI	NIA
38	561A	Timor	Aa	3n	JAM	ECA
38	562B	Yellow Heart	Aa	3n	JAM	ECA
40	563A	Couscous	Aa	3n	JAM	ECA
41	565A	Timor	Aa	3n	JAM	ECA
42	567B	Yellow Heart	Aa	3n	JAM	ECA
43	569B	Yellow Heart	Aa	3n	JAM	ECA
44	56A	Meinpadahk	Aa × Am	3n	UWI	NIA
45	571A	Timor	Aa	3n	JAM	ECA
46	572A	Couscous	Aa	3n	JAM	ECA
47	573B	White Heart	Aa	3n	JAM	ECA
48	574B	Macca	Aa	3n	JAM	ECA
49	575B	Brambram	Aa	3n	JAM	ECA
50	57B	Momolega	Aa	2n	UWI	NIA
51	581B	Yellow Heart	Aa	3n	JAM	ECA
52	58B	Unidentified 2	Aa	3n	UWI	NIA
53	590A	White Heart	Aa	3n	JAM	ECA
54	591A	Monkey Breadfruit	Aa	3n	JAM	ECA
55	592A	Monkey Breadfruit	Aa	3n	JAM	ECA
56	596A	Banjam	Aa	3n	JAM	ECA
57	59A	Pua'a	Aa	3n	UWI	NIA
58	5B	Meitehid	Aa	3n	UWI	NIA
59	601A	Portland Breadfruit	Aa	3n	JAM	ECA
60	602A	Man Bread	Aa	3n	JAM	ECA
61	603A	Ma'afala	Aa	2n	JAM	NIA
62	60A	Unidentified 3	Aa	-	UWI	-
63	61B	Mahani	Aa	3n	UWI	NIA
64	62A	Afara	Aa	3n	UWI	NIA
65	63B	Fafai	Aa	3n	UWI	NIA
66	64B	Yellow	Aa	3n	UWI	NIA
67	65B	Otea	Aa	3n	UWI	NIA
68	66B	Puou	Aa	2n	UWI	NIA
69	69A	Breadnut/ Chataigne	Ac	2n	UWI	ECA
70	730A	Masunwa	Aa	2n	TRI	NIA
71	731A	Ma'afala	Aa	2n	TRI	NIA
72	741B	Yellow	Aa	3n	TRI	ECA
73	742A	Ma'afala	Aa	2n	TRI	NIA
74	743A	White	Aa	3n	TRI	ECA
75	764C	Unidentified 4	Aa	3n	SKN	ECA
76	780C	Unidentified 5	Aa	3n	SKN	ECA
77	782A	Unidentified 6	Aa	3n	TOB	ECA
78	783B	Butter Breadfruit	Aa	3n	TOB	ECA
79	788B	Choufchouf	Aa	3n	TOB	ECA
80	790C	Unidentified 7	Aa	3n	TOB	ECA
81	798A	Ma'afala	Aa	2n	TRI	NIA
82	7A	Timor	Aa	3n	UWI	ECA
83	800 B	White	Aa	3n	TOB	ECA
84	805 A	Pu'upu'u	Aa	3n	TOB	NIA

Table 1. *Cont.*

No.	Sample ID	Cultivar Name	Taxon *	Ploidy *	Sample Collection Site	Accession Grouping
85	806 A	Meitehid	Aa	3n	TOB	ECA
86	808 A	Timor	Aa	3n	TOB	ECA
87	821 B	White	Aa	3n	TOB	ECA
88	827 B	Ma'afala	Aa	2n	TOB	ECA
89	828 A	Local Yellow	Aa	3n	TOB	ECA
90	833A	Local Yellow	Aa	3n	TRI	ECA
91	835A	Ma'afala	Aa	2n	TRI	NIA
92	9B	Cassava	Aa	3n	UWI	ECA
93	BF12	Unidentfied 8	Aa	-	UWI	NIA
94	NO.17 B	Pu'upu'u	Aa	3n	UWI	NIA
95	SV4A	Cocobread	Aa	3n	UWI	ECA

Taxon (Aa = *Artocarpus altilis*; Ac = *Artocarpus camansi*; Am = *Artocarpus mariannensis*; Aa × Am = *A. altilis* × *A. mariannensis* hybrid); Ploidy (3n = Triploid; 2n = diploid; - = unknown); * Taxon and ploidy are based on previous publications [22,24,26]; Collection site (UWI = The University of the West Indies, St. Augustine Campus Breadfruit Germplasm Collection; TRI = Trinidad; TOB = Tobago; JAM = Jamaica; SVG = St. Vincent and the Grenadines; SKN = St. Kitts and Nevis); Accession grouping (ECA = Existing Caribbean Breadfruit Accessions; NIA = Newly Introduced Breadfruit Accessions).

2.2. DNA Isolation and Purification

Total genomic DNA was extracted from the leaf tissue following the Wizard® Genomic DNA Purification Kit following the manufacturer's recommended protocol (Promega Corporation, Madison, WI, USA). Approximately 0.4 g of leaf tissue was ground in liquid nitrogen using a mortar and pestle. The ground tissue was transferred to a 1.5 mL microcentrifuge tube, treated with 600 μL of nuclei lysis solution, and incubated at 65 °C for 15 min. Samples were retrieved, and 3 μL of RNase solution (4 mg/mL) was added to each sample, mixed by inversion, and incubated at 37 °C for 15 min. The samples were then cooled at ambient room temperature for 5 min, treated with 200 μL of protein precipitation solution, vortexed, and then centrifuged at 13,000× *g* for three minutes. The supernatant of each sample was pipette-transferred to a clean, labeled microcentrifuge tube containing 600 μL of room-temperature isopropanol, mixed by inversion, and then centrifuged at 13,000× *g* for one minute. The supernatant was decanted, and 600 μL of room temperature 70% ethanol was added to each sample. The samples were mixed by inversion and centrifuged at 13,000 rpm for one minute. The ethanol was then aspirated, and the DNA pellet was allowed to air dry at ambient temperature for 15 min, after which 100 μL of DNA rehydration solution was added. Immediately after rehydration, the DNA concentration and quality of the samples were measured and evaluated using a Nanodrop spectrophotometer 2000. All DNA samples were diluted using DNA rehydration solution to 25 ng/μL and stored at −80 °C until use.

2.3. PCR Amplification and Sequencing

Polymerase chain reaction was performed in a 50 μL reaction volume containing 25 μL PCR master mix (Promega Corporation), 0.5 μL CXR dye, 2 μL of 10 μM forward primer, 2 μL of 10 μM reverse primer, 12 μL sample DNA (25 ng/μL), and 8.5 μL sterile distilled water. Twenty-three primers (Witherup et al., 2013) synthesized by Integrated DNA Technologies (Integrated DNA Technologies, Coralville, IA, USA) were screened. The PCR reactions were performed in a 96-well microtiter plate using an Applied Biosystems 7300 Real-Time PCR System (Thermo Fisher Scientific Corporation, Waltham, MA, USA). After initial screening of 23 primers, 10 SSR markers with sequences previously described by Witherup, et al. [21] were selected for further use (Table 2).

The PCR reaction was optimized and carried out under the following conditions: initial denaturation at 95 °C for 15 min; 40 cycles of 94 °C for 30 s; 55 °C for 90 s; and 72 °C for 60 s; and a final extension at 60 °C for 30 min. Amplification was confirmed by electrophoresis in a 1.5% agarose gel, followed by staining with ethidium bromide, then visualized under UV light.

Table 2. Characteristics of microsatellite loci used to amplify breadfruit (*Artocarpus altilis*), breadfruit hybrid (*A. altilis* × *A. mariannensis*), and breadnut (*A. camansi*) samples used in the study.

Locus	Primer Sequence (5′–3′)	Repeat Motif [a]
MAA40	F: AGCATTTCAGGTTGGTGAC R: TTGTTCTGTTTGCCTCATC	$(TG)_{16}$
MAA54a	F: AACCTCCAAACACTAGGACAAC R: AGCTACTTCCAAAACGTGACA	$(CA)_5,(AT)_4$
MAA71	F: TTCCTATTTCTTGCAGATTCTC R: AGTGGTGGTAAGATTCAAAGTG	$(CT)_{11}(CA)_{19}$
MAA85	F: TCAGGGTGTAGCGAAGACA R: AGGGCTCCTTTGATGGAA	$(CA)_{11}$
MAA96	F: GGACCTCAAGGATGTGATCTC R: ACACGGTCTTCTTTGGATAGC	$(CA)_{14}(TA)_7(TG)_3(GT)$
MAA140	F: CCATCCCCCATCTTTCCT R: TCCTCGTTTGCCACAGTG	$(CT)_{25}$
* MAA178a	F: GATGGAGACACTTTGAACTAGC R: CACCAGGGTTTAAGATGAAAC	$(GT)_3,(GT)_6,(GT)_3,(GA)_3,(GA)_{10}$
* MAA178b	F: GATGGAGACACTTTGAACTAGC R: CACCAGGGTTTAAGATGAAAC	$(GT)_3,(GT)_3,(GA)_3,(GA)_{11}$
MAA182	F: TACTGGGTCTGAAAAGATGTCT R: CGTTTGCGTTTGGATAAAT	$(CT)_{19}$
MAA251	F: ATCGTCTTTGTCACCACCAC R: ATAGCCGAGTAACTGGATGGA	$(ATC)_{10}$

[a] Comma indicates the presence of nonrepeating nucleotides between repeats. * Primers amplified two separate loci. Source: Witherup, et al. [21].

Ninety-five of the initial 153 samples collected were selected for sequencing. This included one breadnut (*A. camansi*), 91 *A. altilis*, and three *A. altilis* × *A. mariannensis* hybrids (Table 1). This selection was based on the quality of amplicons produced over all ten primer pairs and a deliberate attempt to represent as many cultivar names as possible, leaf and fruit morphological variations (Figures 1 and 2), and collection sites. This resulted in multiple samples for some cultivars and single samples for other cultivars. Amplified amplicons were sent to Macrogen Inc. (Seoul, Republic of Korea) for sequencing using the Sanger method.

Figure 1. Some of the leaf morphological variations observed among cultivars in the UWI breadfruit germplasm collection. (**A**) 'Timor'; (**B**) 'Kashee Bread'; (**C**) 'Cassava'; (**D**) 'Meitehid'; (**E**) 'Puou' and (**F**) 'Yellow'. Source: Daley, et al. [11].

Figure 2. Some fruit morphological variations observed among cultivars in the UWI breadfruit germplasm collection. Smooth-skinned cultivars: [(**A**) 'Yellow'; (**B**) 'Afara'; (**C**) 'Timor']; Rough skinned cultivars: [(**D**) 'Kashee Bread'; (**E**) 'Toneno' and (**F**) 'Piipiia']. Source: Daley, et al. [11].

2.4. Data and Statistical Analysis

2.4.1. Sequence Editing and Alignment

Base calling, sequence editing, trimming, and multiple sequence alignment were accomplished using Geneious version 9.1.3 [31].

2.4.2. Polymorphism and Diversity Analysis

The level of genetic variation at the nucleotide level assessed as nucleotide diversity (π_T) and nucleotide polymorphism (θ_W) as well as SNP (parsimony informative sites), Tajima's D test, Fu's F test, and Harpending index, number of haplotypes, haplotype diversity (h), minimum number of recombination events (R_m), estimate of population recombination, and Wall's B statistic were calculated and analyzed for each locus using the software package DnaSP version 5 [32]. This software was also used to estimate the extent of geographic structure at individual loci among the breadfruit samples using the S_{nm} method [33] as well as estimate the population subdivision (F_{ST}) between the ECA and NIA groups.

2.4.3. Linkage Disequilibrium

The decay of linkage disequilibrium (LD) between parsimony informative sites within loci was estimated as r^2 using DnaSP 5, following the methods of Remington, et al. [34].

2.4.4. Population Structure and Phylogenetic Analysis

SequenceMatrix [35] was used to concatenate all the sequences for each of the 95 accessions. The population structure was evaluated with the software BAPS: Bayesian Analysis of Population Structure [36,37] using an admixture model with no linkage. A phylogeny of the combined sequences was then constructed using the Neighbor Joining algorithm of PAUP Ver. 4.0b10 with a bootstrapping of 1000 iterations [38].

3. Results

3.1. Sequence Analysis

Amplicons from 10 SSR loci were sequenced in each of 91 *A. altilis*, three *A. altilis* × *A. mariannensis*, and one *A. camansi* sample. Sequence length varied from 150 to 370 bp and included indels. A total of 2560 bps of sequences were aligned over the 10 loci per individual sample, and close to 240,640 bps of sequence data were generated for all 95 samples. Across samples, indel polymorphism varied from 0 to 22, with a total of 89 indel polymorphisms in the dataset.

There were 486 single nucleotide polymorphisms (SNPs) (one SNP per 5.3 bp) in the complete dataset, of which 403 were parsimony informative sites and 83 were singletons (Table 3). When considered separately, ECA harbored 560 SNPs (one SNP per 4.6 bp), including 433 parsimony informative sites and 117 singletons, whereas NIA harbored 649 SNPs (one SNP per 3.9 bp), including 447 parsimony informative sites and 202 singletons (Table 3).

Table 3. Summary statistics of nucleotide variability for existing Caribbean breadfruit accessions (ECA) and newly introduced breadfruit accessions (NIA).

Microsatellite Locus	Group	SNP (Parsimony Informative)	Nucleotide Diversity (π_T)	Nucleotide Polymorphism (θ_W)	Tajima's D Test	Fu's F Test	Harpending Index
	Total	40 (9)	0.334	0.16	−1.709	−2.338	0.004
MAA40	ECA	50 (7)	0.323	0.204	−1.269	−1.266	0.002
	NIA	61 (22)	0.349	0.182	−1.181	−2.285	0.004
	Total	50 (7)	0.328	0.154	−0.536	−2.468	0.046
MAA54A	ECA	51 (8)	0.338	0.171	−0.398	−2.518	0.032
	NIA	56 (30)	0.193	0.1	−2.069	−2.609	0.065
	Total	46 (8)	0.276	0.141	−0.643	−2.815	0.010
MAA71	ECA	70 (13)	0.34	0.239	0.911	−1.432	0.002
	NIA	37 (19)	0.185	0.108	−1.29	−2.221	0.021
	Total	55 (1)	0.303	0.253	1.48	−0.451	0.001
MAA85	ECA	60 (3)	0.286	0.251	−0.426	−0.329	0.002
	NIA	72 (9)	0.325	0.297	1.479	0.303	0.004
	Total	66 (13)	0.305	0.171	0.021	−1.969	0.002
MAA96	ECA	71 (17)	0.297	0.163	−1.567	−1.491	0.003
	NIA	90 (27)	0.324	0.209	0.172	−1.693	0.005
	Total	25 (10)	0.301	0.087	−2.236	−3.685	0.019
MAA140	ECA	33 (23)	0.302	0.096	−2.287	−4.752	0.030
	NIA	54 (27)	0.304	0.154	−1.853	−2.598	0.014
	Total	57 (12)	0.33	0.181	−1.498	−1.545	0.003
MAA178A	ECA	62 (12)	0.337	0.202	−1.392	−1.502	0.003
	NIA	63 (17)	0.28	0.192	−1.182	−1.396	0.007
	Total	83 (18)	0.259	0.176	−1.081	−1.874	0.001
MAA178B	ECA	86 (26)	0.23	0.167	−0.953	−1.798	0.002
	NIA	106 (36)	0.277	0.206	−0.978	−1.715	0.004
	Total	0 (0)	0	0	0	0	ND
MAA182	ECA	10 (0)	0.383	0.247	−1.084	0.167	0.082
	NIA	39 (1)	0.333	0.348	0.164	0.075	0.006
	Total	64 (50	0.355	0.269	−0.805	−1.356	0.001
MAA251	ECA	67 (8)	0.284	0.233	−0.625	−0.974	0.001
	NAC	71 (14)	0.397	0.3	−0.927	−1.222	0.004
	Total	48.6 (8.3)	0.159	0.279	−0.701	−1.850	0.010
Average	ECA	56 (11.7)	0.197	0.312	−0.909	−1.590	0.016
	NIA	64.9 (20.2)	0.210	0.297	−0.767	−1.536	0.013

Nucleotide diversity (π_T) for the dataset varied among the 10 loci from 0 to 0.269 (mean = 0.159), and nucleotide polymorphism (θ_W) ranged from 0 to 0.355 (mean = 0.279) (Table 3). The average nucleotide diversity (π_T) of the 10 loci was slightly lower among ECA compared to NIA (π_T = 0.197 vs. 0.209). However, the overall mean nucleotide polymorphism (θ_W) was slightly higher for ECA compared to NIA (θ_W = 0.312 vs. 0.297) (Table 3). Furthermore, comparing the ECA and NIA subsamples revealed no significant difference in π_T ($p > 0.690$) and θ_W ($p > 0.545$).

In terms of allele frequency distribution among the full dataset, Tajima's D value was significantly negative only at locus MAA140 (Table 3). When ECA and NIA were considered separately, the ECA accounted for a higher Tajima's D value in loci MAA54A, MAA71, MAA85, MAA178B, and MAA251, whereas the NIA showed higher values for loci MAA40, MAA96, MAA140, MAA178A, and MAA182 (Table 3). On average, negative values of Tajima's D and Fu's F tests were returned for the ECA, NIA, and all samples. Positive Tajima's D values were observed for 10% of loci in the ECA subset and 30% of loci in the NIA subset. A similar output for Fu's F test was returned, except that the NIA subset had 20% positive loci. The Harpending index was low in all loci in the two groups and subsequently for the overall ECA, NIA, and all samples.

3.2. Identity Analysis

A total of 48 cultivar names and eight unidentified accessions were recorded for the 94 breadfruit samples. Total haplotype number and haplotype diversity (h) varied among loci and between the ECA and NIA germplasm groups. Among the total dataset of 94 breadfruit accessions, the haplotype number ranged from 29 to 93 (Table 4). Loci MAA140 and MAA178B provided the lowest and highest number of haplotypes, respectively (Table 4). Among the 62 ECA, the mean number of haplotypes detected was 50.30 (SE ± 15.27), which ranged from 28 for loci MAA182 and MAA54A to 62 for loci MAA71 and MAA178B (Table 4). The 32 samples from the newly introduced accession group had a mean number of haplotypes of 27.9 (SE ± 6.57), which ranged from 13 for locus MAA54A to 32 for loci MAA85, MAA178A, MAA178B, and MAA251 (Table 5). Mean haplotype diversity (h) was similar for both accession groups but slightly higher among the NIA (Table 5).

Table 4. Summary of the observed number of unique haplotypes and haplotype diversity (h) within the existing Caribbean and newly introduced breadfruit accessions, as well as estimates of the minimum number of recombination events (R_M),) estimate of population recombination (ρ), and Wall's B statistic.

Loci	Accession Group	No. of Unique Haplotypes	Haplotype Diversity (h)	R_M	ρ	Walls' B Statistic
MAA40	ECA	58	0.996 ± 0.004	10	0.225	0.000
	NIA	31	0.998 ± 0.008	7	0.404	0.000
MAA54A	ECA	28	0.836 ± 0.039	7	0.623	0.000
	NIA	13	0.843 ± 0.043	5	1.196	0.000
MAA71	ECA	62	1.000 ± 0.003	9	0.164	0.125
	NIA	20	0.950 ± 0.024	8	0.559	0.071
MAA85	ECA	61	0.999 ± 0.003	17	0.129	0.000
	NIA	32	1.000 ± 0.008	23	0.138	0.024
MAA96	ECA	56	0.992 ± 0.007	10	0.458	0.129
	NIA	31	0.998 ± 0.008	11	0.286	0.075
MAA140	ECA	29	0.914 ± 0.024	1	1.173	0.375
	NIA	25	0.956 ± 0.029	5	0.534	0.207
MAA178A	ECA	59	0.998 ± 0.003	7	0.343	0.174
	NIA	32	1.000 ± 0.008	13	0.172	0.091
MAA178B	ECA	62	1.000 ± 0.003	17	0.241	0.125
	NIA	32	1.000 ± 0.008	16	0.154	0.127
MAA182	ECA	28	0.955 ± 0.011	1	0.171	0.000
	NIA	31	0.998 ± 0.008	15	0.087	0.044
MAA251	ECA	60	0.999 ± 0.003	15	0.332	0.000
	NIA	32	1.000 ± 0.008	8	0.156	0.044
Average	ECA	50.3 ± 5.268	0.969 ± 0.055	9.400 ± 5.778	0.386 ± 0.315	0.093 ± 0.121
	NIA	27.900 ± 6.574	0.974 ± 0.050	11.100 ± 5.724	0.369 ± 0.336	0.068 ± 0.063

Linkage disequilibrium patterns were evaluated in terms of frequency distribution and rates of decay. The results of patterns of LD analyses are presented in Table 4 and Figure 3. Among 95 accessions, there were 117 polymorphic sites with 6876 pairwise comparisons, of which 475 were significant ($p < 0.01$) by Fisher's exact test after Bonferroni corrections. When separated into their respective accession groups, 157 polymorphic sites were analyzed for ECA, which gave 12,246 pairwise comparisons, of which 876 were significant ($p < 0.01$) by Fisher's exact test after Bonferroni corrections. On the other hand, 211 polymorphic sites were analyzed for NIA, which gave 22,155 pairwise comparisons, of which 927 were significant ($p < 0.01$) based on Fisher's exact test after Bonferroni corrections. In general, pairwise sites of low association value (r^2)~0.1 and separated by a short distance were in high percentages and common to both accession groups (Figure 1).

Table 5. Estimate of F_{ST} and test of genetic differentiation in existing Caribbean and newly introduced breadfruit accessions.

Locus	All Accessions (n = 94)	
	F_{ST} [a]	S_{nn} [b]
MAA40	−0.0002	0.6084
MAA54A	0.0105	0.5533
MAA71	0.0147	0.6008
MAA85	0.0178	0.7018 ***
MAA96	0.0332	0.6939 ***
MAA140	0.0513	0.6024 **
MAA178A	0.0035	0.5829
MAA178B	0.0358	0.7128 **
MAA182	-	-
MAA251	0.1095	0.6720 **
Average (n = 9)	0.0306	0.6365

Significant at ** $p < 0.01$, *** $p < 0.001$; [a] Wright's fixation index [39]. [b] Statistical test of genetic differentiation [33].

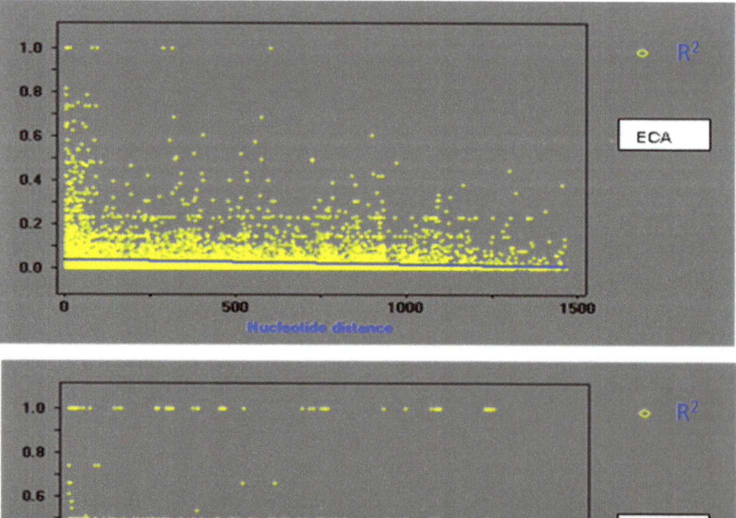

Figure 3. Plots of Linkage disequilibrium (LD) against the physical separation distance of existing Caribbean breadfruit accessions (ECA) and newly introduced breadfruit accessions (NIA).

3.3. Population Recombination

Estimates of the population recombination parameter (ρ) based on Hudson [40] ranged from 0.129 to 1.173 (0.386 ± 0.315) in ECA and 0.087 to 1.196 (0.369 ± 0.336) in NIA. Similarly, the minimum number of recombination events ranged from 1 to 17 (9.400 ± 5.778) in ECA and 5 to 23 in NIA (11.100 ± 5.724), and Wall's B ranged from 0 to 0.375 (0.093 ± 0.121) in ECA and 0 to 0.207 (0.068 ± 0.063) in NIA. There were no significant differences ($p > 0.05$) between the two groups for population recombination rates based on these three statistical tests.

3.4. Population Structure and Demographic Analysis

The extent of geographic structure was estimated for ECA and NIA. F_{ST} values were obtained for nine of the ten loci used. Among the nine loci, F_{ST} varied between −0.0002 (MAA40) and 0.01095 (MAA251) with an overall mean of 0.0306 and revealed only slight differentiation among the two accession groups (Table 5). Significant ($p < 0.05$) values of Snn were detected in five loci (Table 5). Thus, at least 50% of loci demonstrated significant geographic structure. However, inspection of Bayesian clustering implemented by BAPS (Figure 2, did not show any distinction based on the assigned accession groups. Nevertheless, in both figures, two distinct clusters were formed among the accessions. They were classified as Cluster I, which contained 67 breadfruit accessions (66% of all accessions), and Cluster II, which contained 27 breadfruit accessions (27.8% of all accessions). Cluster I comprised 79% ECA and 21% NIA, while cluster II comprised 66.7% NIA and 33.3% ECA. Cluster I contained only triploid accessions that bore smooth or sandpapery-skinned fruit. Cluster II comprised all diploid (fertile and sterile) and all hybrid (*A. altilis* × *A. mariannensis*) accessions. However, cluster II also comprised rough-skinned triploid accessions.

Not all individuals with the same cultivar names (Table 1) shared the same cluster or sub-clusters (Figure 4). Most of the individuals in Cluster I did not form any sub-clusters. However, there were a few exceptions. A sub-cluster of eight individuals was formed with the newly introduced cultivar 'Roiha'a' and one unknown cultivar from the group of existing Caribbean accessions. Other individuals in this sub-cluster included two samples of the cultivar 'Timor', two samples of the cultivar 'Yellow' (one from Jamaica and one from Trinidad), two samples of the cultivar 'White' (one from Jamaica and one from Trinidad), and one 'Captain Bligh' from St. Vincent.

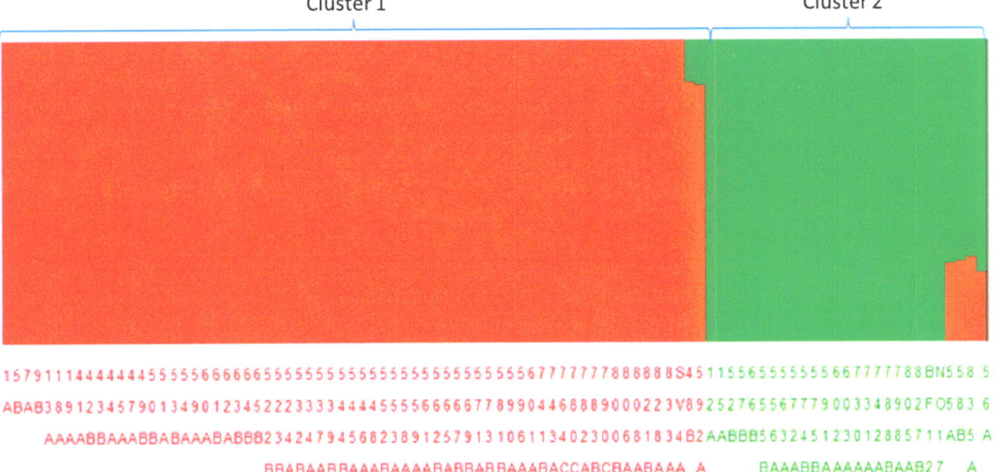

Figure 4. Clustering of 91 *A. altilis* and three *A. altilis* × *A. mariannensis* hybrid samples. Sample names run vertically, and their details are presented in Table 1. An optimal fit of K = 2 was found with Bayesian analysis in the software BAPS [36,37].

Most ECA with rough skin were grouped along with two newly introduced accessions, 'Pu'upu'u' and one unidentified sample. 'Pu'upu'u' is a NIA, which is a triploid, rough-skinned cultivar. The ECA in this sub-cluster were 'Soursop' and 'Waterloo/Cotton' from St. Vincent, 'Macca', 'Couscous', 'Manbread', 'Brambram', and 'Monkey Breadfruit' from Jamaica. Both 'Choufchouf' and 'Kashee' were both rough-skinned cultivars collected in Tobago and St. Vincent, respectively, and were unexpectedly not included in this cluster. The UWI breadfruit germplasm collection accession 'Macca' was also not included in this cluster but instead was grouped with smoothed-skinned triploid cultivars.

Four of five 'Ma'afala' samples were grouped together. The fifth sample was grouped with a sample called 'Masunwa.' Sample 58B, which is an unidentified diploid Pacific cultivar, formed a sub-cluster with the lone *A. camansi* sample. The three hybrids 'Huehue','Meinpadahk' and 'Piipiia' were clustered together. The newly introduced diploid cultivars 'Uluea' and 'Puou' were grouped together (Figure 4).

4. Discussion

This study aimed to assess the level of genetic diversity of breadfruit at the molecular level and compare the diversity of existing Caribbean accessions (ECA) with newly introduced accessions (NIA) using SSR sequence. Based on sequence data, there was no significant difference between ECA and NIA, although the NIA had a slightly higher overall mean nucleotide diversity (π_T) and the group of ECA had a slightly higher mean nucleotide polymorphism (θ_W) (Table 3). The overall means for both nucleotide diversity and nucleotide polymorphism indicated a moderate but substantial genetic diversity for breadfruit in this study, but although there was considerable locus-to-locus variation among the 10 loci evaluated, there was no significant difference in the level of genetic variation between the two accession groups. This lack of significant differences in genetic diversity among the two accession groups was surprising given the fact that the group consisting of NIA included triploids, fertile diploids, and hybrid accessions and showed greater morphological variations compared to the ECA, which consisted of only triploids [11,24]. However, it is important to note that both accession groups were originally collected in the same sub-region of the Pacific, although they were collected at widely different times, and this could be a factor in the similarity observed in genetic diversity. In the first successful introduction of breadfruit to the Caribbean in 1793, all except one cultivar were collected in Tahiti, in Eastern Polynesia [41]. Many of the newly introduced accessions used in this study were also collected in Eastern Polynesia and maintained at the National Tropical Botanical Gardens (NTBG) in Hawaii, from where they were collected and brought to the Caribbean [29,42]. Furthermore, previous studies using SSR markers reported that *A. altilis* triploids, regardless of current distribution, and hybrids of *A. altilis* × *A. marianensis* were shown to belong to one genetic lineage and consist of a single genotype [22,24]. Also, although there have been several attempts, studies on breadfruit characterization have been unable to separate Caribbean triploids from triploids in Eastern Polynesia [22,24,26]. The diversity of breadfruit accessions in the Caribbean may reflect both its historical genetic base and the selection of types based on distinct sociocultural phenomena in this region [11].

Breadfruit showed higher genetic diversity than estimates for the outcrossing tropical tree species, avocado (*Persea americana* Mill) (θ_W =0.007, π_T = 0.0066) [43], and the highly outcrossed tree species *Populus tremula* (π_T =0.0111) [44]. This higher level of genetic diversity for breadfruit compared with outcrossing species *Persea americana* and *Populus tremula* was unexpected, especially for existing Caribbean breadfruit accessions, which are propagated using vegetative means due to triploidy [9,24]. However, it is not unusual to find high levels of genetic diversity in triploid genotypes, which are vegetatively propagated. For example, within the genus Musa, triploid commercial cultivars, which are always propagated by vegetative means, showed higher levels of expected heterozygosity than wild diploid and improved hybrid diploid cultivars [45].

This is the first study that examined the genetic diversity of breadfruit based on the pattern of nucleotide diversity and nucleotide polymorphism using SSR amplicon sequence data. Nucleotide diversity is comparable to heterozygosity [46,47]. In a study on the genetic diversity of breadfruit in different regions of the world, Caribbean accessions showed the highest level of expected heterozygosity (H_e) (0.729) compared with those from East Polynesia (0.582), West Polynesia (0.659), Micronesia (0.684), Melanesia (0.686), and non-Oceania (0.677) [24]. The unexpectedly high H_e in Caribbean breadfruit germplasm was attributed to the small sample size (n = 5) in the Caribbean [24]. The high level of expected heterozygosity obtained in that study was also higher than the nucleotide diversity and nucleotide polymorphism reported in the present study, which used a much larger sample size of 65 existing Caribbean accessions. However, both studies suggest a higher-than-expected level of genetic diversity in existing Caribbean breadfruit accessions. Zerega, et al. [22] posit that a possible explanation for the high level of genetic diversity in triploid breadfruit is that the original triploid cultivars may have resulted from the capture of a wide diversity that became fixed and passed from generation to generation without genetic recombination. To further support this view, it has been shown that although most triploid breadfruit cultivars from Polynesia were derived solely from *A. camansi*, a significant number of cultivars also contained *A. mariannensis*-specific markers, which would likely have contributed to an increase in the overall genetic diversity of that sub-region [1].

Linkage disequilibrium is characterized as the nonrandom association of alleles at different loci and can be affected by most of the processes observed in population genetics, including mating pattern, frequency of recombination, and population history [48,49] Triploid and sterile breadfruit cultivars are propagated only by asexual methods such as cuttings and layering, while fertile diploids can reproduce sexually. The latter is expected to have an impact on the effective rate of recombination in populations where fertile diploid cultivars occur naturally [50]. Furthermore, breadfruit is an outcrossing species and hybridizes with *A. mariannensis*, which would support higher rates of recombination in populations or groups with fertile diploids compared to populations or groups having only triploids or sterile diploids [51,52]. For this reason, linkage disequilibrium, which is rarely used as a measure of genetic diversity in studies of asexually propagated species, was assessed in this study, in which the group of newly introduced accessions had four diploids and two hybrids (*A. altilis* × *A. mariannensis*) samples. The results showed no significant differences in the recombination events using the Hudson [53] recombination parameter and Wall's B statistics between the two accession groups evaluated. Furthermore, the r^2 was low, and the rate of decline in linkage disequilibrium was negligible in both accession groups. This result could also be a consequence of the *A. mariannensis*-specific marker in existing Caribbean breadfruit accessions. In other outcrossing species such as sunflower (*Helianthus annus*), Liu and Burke [54] reported extremely rapid decay of linkage disequilibrium to negligible levels at short distances. In contrast, Zhu, et al. [55] reported that there was little decay in linkage disequilibrium in the autogamous soybean (*Glycine max*).

The estimate of overall population differentiation between ECA and NIA using F_{ST} (0.0306) also indicated a strong similarity between both groups of accessions. Because most breadfruit cultivars are asexually reproduced, there is little or no opportunity for gene flow from one population to another; therefore, the similarity between both groups is most likely because of their common origin. This slight differentiation between both groups could also suggest that the duration of breadfruit cultivation in the Caribbean was not long enough to permit the development of Caribbean breadfruit germplasm with distinct genetic backgrounds. Similar results (F_{ST} = 0.04) have been reported for the outcrossing tropical tree *Guaiacum sanctum*, which suggests a high frequency of interpolation migration [56]. However, in a study with six breadfruit populations in the Western Ghats of India using five AFLP markers, Sreekumar, et al. [20] obtained a F_{ST} value of 0.574 and concluded that there was relatively high genetic differentiation between the populations studied. The use of AFLP markers in that study and the composition of the different breadfruit populations used are possible reasons for the high F_{ST} value reported.

The lack of separation of accessions based on the assigned groupings in the current study was also demonstrated in the clustering of accessions. The two main clusters that formed contained accessions from both accession groups. However, there was some differentiation of accessions, and some commonality could be observed among accessions within each cluster and sub-cluster. Based on previously reported ploidy and hybridization levels for accessions used in this study [22], it was observed that diploid and hybrid accessions formed tighter clusters and were more distinguishable from triploid accessions. Similar findings were also reported by Zerega, et al. [22]. In the present study, there were 12 diploid samples, which represented four cultivars and one *A. camansi* sample, and all the *A. altilis* samples were included in cluster II (Figure 4). Although cluster II also included some triploid cultivars, the diploid cultivars were grouped closer. The clustering of some triploid cultivars, along with diploid and hybrid cultivars, could possibly be linked to the existence of *A. mariannensis* specific markers in some triploid breadfruit cultivars because of their genetic background.

There was an inability to distinguish among many of the triploid accessions, including some that appeared morphologically distinct. For example, cultivars 'Meitehid', 'Timor', and 'Cassava', which showed clear differences in leaf shape, lobing, and leaf apex shape [11], were all grouped together in the same cluster. Zerega, et al. [22] also reported the inability of SSR markers to distinguish among many individual breadfruit accessions, which displayed clear differences in fruit and leaf morphological characteristics under the same growing conditions. Therefore, morphological diversity in breadfruit is not consistent with genetic diversity. Furthermore, previous studies, showed that many cultivars were misclassified based on discriminant analysis and other methods used to analyze the morphological data of breadfruit cultivars [11,12]. The UWI breadfruit germplasm collection accession 'Macca' was not grouped with other named 'Macca' cultivars collected throughout the Caribbean. It was felt that this accession was misnamed in the collection. Although not conclusive, the results of this study support the view that the accession in the UWI breadfruit germplasm collection called 'Macca' is different from other 'Macca' cultivars in the Caribbean. Ten SSR markers were used in this study, which is a relatively small number, and there is the possibility that they are not able to detect differences among many of these accessions at the genetic level. However, other researchers using a larger number of markers, also reported the inability to distinguish among some triploid breadfruit cultivars, which indicates that other factors must be considered for future assessments [22].

Cultivar names are essential in communicating plant species diversity. However, numerous names and synonyms can cause confusion and obscure the true diversity of a species. They can also contribute to wasteful duplication in gene banks and in conducting basic studies. In the current study, 43 cultivar names were used to represent the 94 breadfruit samples. Many of the cultivars, especially those recently introduced, were represented by a single sample, while others were deliberately represented by several samples. The inability to separate or distinguish among some cultivars could be based on the fact that, in some cases, there were different names representing the same cultivars [14]. The current study has helped to provide further insight into the genetic identities of breadfruit cultivars, especially those in the Caribbean. Yet, the inability to distinguish among some cultivars that are known to be morphologically distinct suggests that the markers were not targeting the genic regions of these morphological traits or were non-polymorphic in these regions. The negative estimates for Fu's F test and Tajima's D test in both ECA and NIA indicated an excess of alleles and of those rare alleles, respectively, following a population expansion after a bottleneck. This was corroborated by the low Harpending index. These findings are concomitant with a population size expansion after domestication by *A. camansi*. Furthermore, the inability to separate morphologically distinct cultivars may be linked to a few changes in the genome that govern these traits. Whole genome sequencing of triploid and diploid cultivars with variable morphological features should be undertaken to better understand the evolution of this tropical tree crop and identify SNP markers that would be useful for germplasm management and unambiguous cultivar identification.

5. Conclusions

The results of this study clearly showed that there was moderate but substantial genetic diversity in the breadfruit germplasm of both existing Caribbean accessions and newly introduced accessions. Some existing Caribbean accessions with clear morphological differences, as reported in previous studies [11,57], showed genetic similarity and remained unseparated in this study. Nevertheless, the results showed that microsatellite markers can be a useful tool in helping to map breadfruit genetic diversity and for linking genetic diversity to morphological and other phenotypic expressions. Understanding the genetic diversity along with morphological and other characteristics will assist with the selection of new cultivars to meet specific purposes. For example, it was determined that in terms of genetic diversity, newly introduced cultivars such as 'Roiha'a' were more genetically similar to the preferred local Caribbean cultivar 'Yellow' as compared to cultivars such as 'Ma'afala.' However, genetic similarity should not be the only basis for selecting new cultivars to add to the existing germplasm base for increased production and utilization of breadfruit. There are many factors that influence consumer choice and preference, and these are important for identifying new cultivars for commercial production and utilization.

Author Contributions: Conceptualization, O.O.D., A.T.A. and L.B.R.-N.; Data Curation, O.O.D.; Formal analysis, O.O.D.; Methodology, O.O.D., A.T.A. and L.A.M.; Supervision; L.B.R.-N. and A.T.A.; Writing (Original), O.O.D.; Writing (Review and Editing) O.O.D., A.T.A., L.B.R.-N. and L.A.M. All authors have read and agreed to the published version of the manuscript.

Funding: This research was funded by the Office of Graduate Studies and Department of Food Production, The University of the West Indies, St. Augustine Campus, Trinidad and Tobago (CRP.5.NOV11.6) and the Jamaica Agricultural Development Foundation (JADF) (Breadfruit FUND/ORG:11037

Data Availability Statement: Data supporting reported results can be requested by contacting the corresponding author. The data are not publicly available due to compliance with data protection regulations.

Acknowledgments: The authors would like to thank the Jamaica Agricultural Development Foundation (JADF) and the University of the West Indies (UWI) for their financial support in establishing and maintaining the breadfruit germplasm collection in Trinidad and Tobago.

Conflicts of Interest: The authors declare no conflicts of interest.

References

1. Zerega, N.J.C.; Ragone, D.C.; Motley, T.J. Complex Origins of Breadfruit (*Artocarpus altilis*, Moraceae): Implications for Human Migrations in Oceania. *Am. J. Bot.* **2004**, *91*, 760–766. [CrossRef] [PubMed]
2. Zerega, N.J.C.; Ragone, D.C.; Motley, T.J. Breadfruit Origins, Diversity, and Human-Facilitated Distribution. In *Darwin's Harvest: New Approaches to Origins, Evolution, and Conservation of Crop Plants*; Motley, T.J., Zerega, N.J.C., Ragone, D.C., Cross, H., Eds.; Columbia University Press: New York, NY, USA, 2006; pp. 213–238.
3. Roberts-Nkrumah, L.B. An Overview of Breadfruit (*Artocarpus altilis*) in the Caribbean. *Acta. Hort.* **2007**, *757*, 51–60. [CrossRef]
4. Powell, D. The Voyage of the Plant Nursery, H.M.S. Providence, 1791–1793. *Econ. Bot.* **1977**, *31*, 387–431. [CrossRef]
5. Jones, A.M.P.; Ragone, D.; Tavana, N.G.; Bernotas, D.W.; Murch, S.J. Beyond the Bounty: Breadfruit (*Artocarpus altilis*) for Food Security and Novel Foods in the 21st Century. *Ethnobot. Res. Appl.* **2011**, *9*, 129–149. [CrossRef]
6. Roberts-Nkrumah, L.B. Enhancing Breadfruit Contribution to Food Security in the Caribbean Through Improved Supply. In Proceedings of the Caribbean Food Crop Society, Frigate Bay, Federation of St. Kitts and Nevis, 12–17 July 2009.
7. Ragone, D. Breadfruit: Diversity, Conservation and Potential. *Acta. Hort.* **2007**, *757*, 19–30. [CrossRef]
8. Zerega, N.J.C.; Ragone, D.C.; Motley, T.J. Systematics and Species Limits of Breadfruit (*Artocarpus*, Moraceae). *Syst. Bot.* **2005**, *30*, 603–615. [CrossRef]
9. Ragone, D. Chromosome Numbers and Pollen Stainability of Three Species of Pacific Island Breadfruit (*Artocarpus*, Moraceae). *Am. J. Bot.* **2001**, *88*, 693–696. [CrossRef]
10. Brantjes, N.B.M. Nectar and the Pollination of Breadfruit, *Artocarpus altilis* (Moraceae). *Acta Bot. Neerl.* **1981**, *30*, 345–352. [CrossRef]

11. Daley, O.O.; Roberts-Nkrumah, L.B.; Alleyne, A.T. Morphological diversity of breadfruit [*Artocarpus altilis* (Parkinson) Fosberg] in the Caribbean. *Sci. Hort.* **2020**, *266*, 109278. [CrossRef]

12. Jones, A.M.P.; Murch, S.J.; Wiseman, J.; Ragone, D.C. Morphological Diversity in Breadfruit (*Artocarpus*, Moraceae): Insights into Domestication, Conservation, and Cultivar Identification. *Genet. Resour. Crop Evol.* **2013**, *60*, 175–192. [CrossRef]

13. Ragone, D.C. Description of Pacific Island Breadfruit Cultivars. *Acta. Hort.* **1995**, *413*, 93–98. [CrossRef]

14. Daley, O.O.; Roberts-Nkrumah, L.B.; Alleyne, A.T.; Gloster, M.C. Folk nomenclature and traditional knowledge of breadfruit [*Artocarpus altilis* (Parkinson) Fosberg] diversity in four Anglophone Caribbean countries. *J. Ethnobiol. Ethnomed.* **2022**, *18*, 65. [CrossRef]

15. Ragone, D.; Tavana, G.; Stevens, J.M.; Stewart, P.A.; Stone, R.; Cox, P.M.; Cox, P.A. Nomenclature of Breadfruit Cultivars in Samoa: Saliency, Ambiguity and Monomiality. *J. Ethnobiol.* **2004**, *24*, 33–49.

16. Korir, N.K.; Han, J.; Shangguan, L.; Wang, C.; Kayesh, E.; Zhang, Y.; Fang, J. Plant Variety and Cultivar Identification: Advances and Prospects. *Crit. Rev. Biotechnol.* **2013**, *33*, 111–125. [CrossRef] [PubMed]

17. Biswas, M.K.; Xu, Q.; Mayer, C.; Deng, X. Genome Wide Characterization of Short Tandem Repeat Markers in Sweet Orange (*Citrus sinensis*). *PLoS ONE* **2014**, *9*, e104182. [CrossRef] [PubMed]

18. Hoshino, A.A.; Bravo, J.P.; Morelli, K.A.; Nobile, P.M. Microsatellites as Tools for Genetic Diversity Analysis. In *Genetic Diversity in Microorganisms*; Caliskan, M., Ed.; INTECH Open Access Publisher: Rijeka, Croatia, 2012; pp. 149–170.

19. Kanzaki, S.; Yonemori, K.; Sugiura, A.; Subhadrabandhu, S. Phylogenetic Relationships between the Jackfruit, the Breadfruit and Nine Other *Artocarpus* spp. from RFLP Analysis of an Amplified Region of cpDNA. *Sci. Hort.* **1997**, *70*, 57–66. [CrossRef]

20. Sreekumar, V.B.; Binoy, A.M.; George, S.T. Genetic and Morphological Variation in Breadfruit (*Artocarpus altilis* (Park.) Fosberg) in the Western Ghats of India Using AFLP Markers. *Genet. Resour. Crop Evol.* **2007**, *54*, 1659–1665. [CrossRef]

21. Witherup, C.; Ragone, D.; Wiesner-Hanks, T.; Irish, B.; Scheffler, B.; Simpson, S.; Zee, F.; Zuberi, M.I.; Zerega, N.J.C. Development of Microsatellite Loci in *Artocarpus altilis* (Moraceae) and Cross-Amplification in Congeneric Species. *Appl. Plant Sci.* **2013**, *1*, 1200423. [CrossRef] [PubMed]

22. Zerega, N.J.C.; Wiesner-Hanks, T.; Ragone, D.; Irish, B.; Scheffler, B.; Simpson, S.; Zee, F. Diversity in the Breadfruit Complex (*Artocarpus*, Moraceae): Genetic Characterization of Critical Germplasm. *Tree Genet. Genomes* **2015**, *11*, 4. [CrossRef]

23. Gardner, E.M.; Laricchia, K.M.; Murphy, M.; Ragone, D.; Scheffler, B.E.; Simpson, S.; Williams, E.W.; Zerega, N.J. Chloroplast Microsatellite Markers for *Artocarpus* (Moraceae) Developed from Transcriptome Sequences. *Appl. Plant Sci.* **2015**, *3*, 1500049. [CrossRef] [PubMed]

24. Zerega, N.J.C.; Ragone, D. Toward a Global View of Breadfruit Genetic Diversity. *Trop. Agr.* **2016**, *93*, 77–91.

25. De Bellis, F.; Malapa, R.; Kagy, V.; Lebegin, S.; Billot, C.; Labouisse, J.-P. New Development and Validation of 50 SSR Markers in Breadfruit (*Artocarpus altilis*, Moraceae) by Next-Generation Sequencing. *Appl. Plant Sci.* **2016**, *4*, 1600021. [CrossRef]

26. Audi, L.; Shallow, G.; Robertson, E.; Bobo, D.; Ragone, D.; Gardner, E.M.; Jhurree-Dussoruth, B.; Wajer, J.; Zerega, N.J.C. Linking Breadfruit Cultivar Names across the Globe Connects Histories after 230 Years of Separation. *Curr. Biol.* **2023**, *33*, 287–297. [CrossRef]

27. Barthe, S.; Gugerli, F.; Barkley, N.A.; Maggia, L.; Cardi, C.; Scotti, I. Always Look on Both Sides: Phylogenetic Information Conveyed by Simple Sequence Repeat Allele Sequences. *PLoS ONE* **2012**, *7*, e40699. [CrossRef]

28. Estoup, A.; Jarne, P.; Cornuet, J.-M. Homoplasy and Mutation Model at Microsatellite Loci and Their Consequences for Population Genetics Analysis. *Mol. Ecol.* **2002**, *11*, 1591–1604. [CrossRef] [PubMed]

29. Roberts-Nkrumah, L.B. *The Breadfruit Germplasm Collection at the University of the West Indies, St Augustine Campus*; University of the West Indies Press: Kingston, Jamaica, 2018.

30. Roberts-Nkrumah, L.B. A Review of the Potential of Breadfruit Cultivar 'Ma'afala' for Commercial Production in Trinidad and Tobago. *Trop. Agr.* **2014**, *91*, 284–299.

31. Kearse, M.; Moir, R.; Wilson, A.; Stones-Havas, S.; Cheung, M.; Sturrock, S.; Buxton, S.; Cooper, A.; Markowitz, S.; Duran, C.; et al. Geneious Basic: An Integrated and Extendable Desktop Software Platform for the Organization and Analysis of Sequence Data. *Bioinformatics* **2012**, *28*, 1647–1649. [CrossRef] [PubMed]

32. Librado, P.; Rozas, J. DnaSP v5: A Software for Comprehensive Analysis of DNA Polymorphism Data. *Bioinformatics* **2009**, *25*, 1451–1452. [CrossRef] [PubMed]

33. Hudson, R.R. A New Statistic for Detecting Genetic Differentiation. *Genetics* **2000**, *155*, 2011–2014. [CrossRef] [PubMed]

34. Remington, D.L.; Thornsberry, J.M.; Matsuoka, Y.; Wilson, L.M.; Whitt, S.R.; Doebley, J.; Kresovich, S.; Goodman, M.M.; Buckler, E.S.T. Structure of Linkage Disequilibrium and Phenotypic Associations in the Maize Genome. *Proc. Natl. Acad. Sci. USA* **2001**, *98*, 11479–11484. [CrossRef]

35. Vaidya, G.; Lohman, D.J.; Meier, R. SequenceMatrix: Concatenation Software for the Fast Assembly of Multi-Gene Datasets with Character set and Codon Information. *Cladistics* **2011**, *27*, 171–180. [CrossRef] [PubMed]

36. Corander, J.; Marttinen, P.; Sirén, J.; Tang, J. Enhanced Bayesian Modelling in BAPS Software for Learning Genetic Structures of Populations. *BMC Bioinform.* **2008**, *9*, 1–14. [CrossRef] [PubMed]

37. Corander, J.; Marttinen, P.; Sirén, J.; Tang, J. *Baps: Bayesian Analysis of Population Structure*; Department of Mathematics and Statistics, University of Helsinki: Helsinki, Finland, 2009; pp. 1–27.

38. Swofford, D.L. *PAUP* Version 4.0 b10. Phylogenetic Analysis Using Parsimony (* and Other Methods)*; Sinauer: Sunderland, MA, USA, 2002.

39. Hudson, R.R.; Slatkin, M.; Maddison, W.P. Estimation of Levels of Gene Flow from DNA Sequence Data. *Genetics* **1992**, *132*, 583. [CrossRef] [PubMed]
40. Hudson, R.R. Estimating the Recombination Parameter of a Finite Population Model without Selection. *Genet. Res.* **1987**, *50*, 245–250. [CrossRef] [PubMed]
41. Bligh, W. *The Log of H. M. S. Providence 1791–1793*; Genesis Publications Limited: Guildford, UK, 1976.
42. Ragone, D.C. Collection, Establishment, and Evaluation of a Germplasm Collection of Pacific Island Breadfruit. Ph.D. Thesis, University of Hawai'i at Manoa, Honolulu, HI, USA, 1991.
43. Chen, H.; Morrell, P.L.; Cruz, M.d.l.; Clegg, M.T. Nucleotide Diversity and Linkage Disequilibrium in Wild Avocado (*Persea americana* Mill.). *J. Hered.* **2008**, *99*, 382–389. [CrossRef] [PubMed]
44. Ingvarsson, P.K. Nucleotide Polymorphism and Linkage Disequilibrium within and among Natural Populations of European Aspen (*Populus tremula* L., Salicaceae). *Genetics* **2005**, *169*, 945–953. [CrossRef] [PubMed]
45. Creste, S.; Tulmann Neto, A.; Vencovsky, R.; de Oliveira Silva, S.; Figueira, A. Genetic Diversity of *Musa* Diploid and Triploid Accessions from the Brazilian Banana Breeding Program Estimated by Microsatellite Markers. *Genet. Resour. Crop Evol.* **2004**, *51*, 723–733. [CrossRef]
46. Hedrick, P.W. *Genetics of Populations*, 4th ed.; Jones and Bartlett Publishers: Sadbury, MA, USA, 2011.
47. Hartl, D.L.; Clark, A.G. *Principles of Population Genetics*, 4th ed.; Sinaer Associates, Inc. Publishers: Sunderland, MA, USA, 2007.
48. Garris, A.J.; McCouch, S.R.; Kresovich, S. Population Structure and its Effect on Haplotype Diversity and Linkage Disequilibrium Surrounding the xa5 Locus of Rice (*Oryza sativa* L.). *Genetics* **2003**, *165*, 759. [CrossRef]
49. Flint-Garcia, S.A.; Thornsberry, J.M.; Buckler, E.S., IV. Structure of Linkage Disequilibrium in Plants. *Annu. Rev. Plant Biol.* **2003**, *54*, 357–374. [CrossRef]
50. Bamberg, J.B.; del Rio, A.H. Genetic Heterogeneity Estimated by RAPD Polymorphism of Four Tuber-Bearing Potato Species Differing by Breeding System. *Am. J. Pot Res.* **2004**, *81*, 377–383. [CrossRef]
51. Simko, I.; Haynes, K.G.; Jones, R.W. Assessment of Linkage Disequilibrium in Potato Genome with Single Nucleotide Polymorphism Markers. *Genetics* **2006**, *173*, 2237–2245. [CrossRef]
52. Nordborg, M. Linkage Disequilibrium, Gene Trees and Selfing: An Ancestral Recombination Graph with Partial Self-Fertilization. *Genetics* **2000**, *154*, 923–929. [CrossRef]
53. Hudson, R.R. Linkage Disequilibrium and Recombination. In *Handbook of Statistical Genetics*; Balding, D.J., Bishop, M., Cannings, C., Eds.; John Wiley and Sons, Ltd.: Chichester, UK, 2001.
54. Liu, A.; Burke, J.M. Patterns of Nucleotide Diversity in Wild and Cultivated Sunflower. *Genetics* **2006**, *173*, 321–330. [CrossRef] [PubMed]
55. Zhu, Y.L.; Song, Q.J.; Hyten, D.L.; Van Tassell, C.P.; Matukumalli, L.K.; Grimm, D.R.; Hyatt, S.M.; Fickus, E.W.; Young, N.D.; Cregan, P.B. Single-Nucleotide Polymorphisms in Soybean. *Genetics* **2003**, *163*, 1123–1134. [CrossRef] [PubMed]
56. Fuchs, E.J.; Hamrick, J.L. Genetic Diversity in the Endangered Tropical Tree, *Guaiacum sanctum* (Zygophyllaceae). *J. Hered.* **2010**, *101*, 284–291. [CrossRef] [PubMed]
57. Roberts-Nkrumah, L.B. Towards a Description of the Breadfruit Germplasm in St Vincent. *Fruits* **1997**, *52*, 27–35.

horticulturae

Article

Exploring the PpEXPs Family in Peach: Insights into Their Role in Fruit Texture Development through Identification and Transcriptional Analysis

Yakun Guo [1,†], Conghao Song [1,†], Fan Gao [1], Yixin Zhi [1], Xianbo Zheng [1,2,3], Xiaobei Wang [1,2,3], Haipeng Zhang [1,2,3], Nan Hou [1,2,3], Jun Cheng [1,2,3], Wei Wang [1,2,3], Langlang Zhang [1,2,3], Xia Ye [1,2,3], Jidong Li [2,3,4], Bin Tan [1,2,3], Xiaodong Lian [1,2,3,*] and Jiancan Feng [1,2,3,*]

[1] College of Horticulture, Henan Agricultural University, Zhengzhou 450046, China; songconghao773x@163.com (C.S.); gaofan19980301@outlook.com (F.G.); zyx20240201@163.com (Y.Z.); haipengzhang@henau.edu.cn (H.Z.); xye@henau.edu.cn (X.Y.)
[2] International Joint Laboratory of Henan Horticultural Crop Biology, Zhengzhou 450046, China; lijidong@henau.edu.cn
[3] Henan Engineering and Technology Center for Peach Germplasm Innovation and Utilization, Zhengzhou 450046, China
[4] College of Forestry, Henan Agricultural University, Zhengzhou 450046, China
* Correspondence: lianxd@henau.edu.cn (X.L.); jcfeng@henau.edu.cn (J.F.)
[†] These authors contributed equally to this work.

Citation: Guo, Y.; Song, C.; Gao, F.; Zhi, Y.; Zheng, X.; Wang, X.; Zhang, H.; Hou, N.; Cheng, J.; Wang, W.; et al. Exploring the PpEXPs Family in Peach: Insights into Their Role in Fruit Texture Development through Identification and Transcriptional Analysis. *Horticulturae* **2024**, *10*, 332. https://doi.org/10.3390/ horticulturae10040332

Academic Editor: Adriana F. Sestras

Received: 1 February 2024
Revised: 22 March 2024
Accepted: 24 March 2024
Published: 28 March 2024

Abstract: Expansins (EXPs) loosen plant cell walls and are involved in diverse developmental processes through modifying cell-walls; however, little is known about the role of PpEXPs in peach fruit. In this study, 26 PpEXP genes were identified in the peach genome and grouped into four subfamilies, with 20 PpEXPAs, three PpEXPBs, one PpEXPLA and two PpEXPLBs. The 26 PpEXPs were mapped on eight chromosomes. The primary mode of gene duplication of the PpEXPs was dispersed gene duplication (DSD, 50%). Notably, *cis*-elements involved in light responsiveness and MeJA-responsiveness were detected in the promoter regions of all PpEXPs, while ethylene responsive elements were observed in 12 PpEXPs. Transcript profiling of PpEXPs in the peach fruit varieties of MF (melting), NMF (non-melting) and SH (stony hard) at different stages showed that PpEXPs displayed distinct expression patterns. Among the 26 PpEXPs, 15 PpEXPs were expressed in the fruit. Combining the expressing patterns of PpEXPs in fruits with different flesh textures, PpEXPA7, PpEXPA13 and PpEXPA15 were selected as candidate genes, as they were highly consistent with the patterns of previous reported key genes (PpPGM, PpPGF and PpYUC11) in regard to peach fruit texture. The genes with different expression patterns between MF and NMF were divided into 16 modules, of which one module, with pink and midnightblue, negatively correlated with the phenotype of fruit firmness and was identified as PpEXPA1 and PpEXPA7, while the other module was identified as PpERF in the pink module, which might potentially effect fruit texture development by regulating PpEXPs. These results provide a foundation for the functional characterization of PpEXPs in peach.

Keywords: pach (*Prunus persica*); expansins; genome-wide; fruit texture

1. Introduction

The peach (*Prunus persica*) is an important economic tree species that originated in China. Fruit texture is a vital characteristic of quality in ripe fruit. In peaches, the fruit texture can be classified into three types: melting flesh (MF), non-melting flesh (NMF) and stony hard flesh (SH) [1,2]. Two genes encoding endopolygalacturonase (PG) were responsible for the MF/NMF flesh type [3], and PpYUC11 controls the stony hard phenotype in peaches [4]. As a climacteric fruit, peaches generally soften rapidly during maturation and especially after harvest, which causes difficulties in postharvest handling and quality maintenance during storage and transportation. Improving the fruit texture quality of peaches is one of the most important goals of cultivar development.

Biochemical and molecular studies have revealed that fruit softening is a consequence of cell wall disassembly, including decline in cell wall strength and cell-to-cell adhesion [5,6]. Numerous genes have been verified as being critical to cell wall disassembly in tomatoes (*Solanum lycopersicum*) [7], apples (*Malus domestica*) [8], strawberries (*Fragaria vesca*) and others [9–11]. These genes were mainly related to primary cell wall (PCW) degrading and modifying, such as PG, pectin methylesterases (PME), pectate lyases (PL) and so on. In strawberries, the ripened fruits of *FaPG1* antisense transgenic plants were on average 163% firmer than that of the control [9]. Similarly, downregulation of *MdPG1* increased the firmness and intercellular adhesion in apples [8]. Numerous studies regarding fruit softening have focused on tomatoes as a model system [12]. There is increasing evidence that softening is a complex process that involves multiple genes. PG, PME, PL, β-galactosidase arabinofuranosidase and expansin have been proven to be involved in the softening of tomatoes [13–18].

Expansins (EXPs), regarding classic cell wall remodeling agents, could restore cell wall extensions through nonenzymatic reversible disruption of non-covalent bonds in cell wall polymers [19,20]. Expansins were reported first in cucumbers, with extension activity to inactivated walls, the activity of which were associated with cell growth [19]. Plant EXPs harbor two functional domains: a six-stranded double-psi beta-barrel (DPBB) domain and a Pollen_allerg_1 domain. The EXP super-family was grouped into four subfamilies, including α-expansin (EXPA), β-expansin (EXPB), expansin-like A (EXLA), and expansin-like B (EXLB) [20]. The EXP family has been identified in various plant species, such as *Arabidopsis thaliana*, rice (*Oryza sativa*), maize (*Zea mays*), grape (*Vitis vinifera*), apple, poplar (*Populus trichocarpa*), cotton (*Gossypium hirsutum*), etc. [21–26].

In recent years, a few studies have indicated that EXPs play a critical role in fruit firmness and ripening. *SlExp1* loss-of-function mutants showed firmer and later ripening fruits [27,28]. Overexpressing *MdEXLB1* lines in tomatoes showed lower fruit firmness and an accelerated fruit ripening process [29]. Four EXPs in strawberry were selected as candidate genes involved in fruit softening [30]. However, the functions of EXPs in regulating of peach fruit development remains unknown.

In this study, PpEXPs were identified in the peach genome, and their molecular and physiological characters, phylogenetic relationships and expression profiles were further examined to better understand the PpEXPs family. To probe the potential role of *PpEXPs* in flesh phenotypes, the expression of *PpEXPs* among different flesh phenotypes (MF, NMF and SH) was examined. The results provide a functional basis for further investigation of PpEXPs in fruit textures.

2. Materials and Methods

2.1. Plant Materials

The peach cultivars 'HuangShuimi' (HSM, MF) and 'Zhongyoutao14' (CN14, NMF) were planted in the experimental station of the College of Horticulture, Henan Agricultural University (Zhengzhou, China, 34.86° N, 113.60° E). Samples of each fruits' flesh was collected for transcriptome and qRT-PCR analysis at four critical growth stages, namely S1 (fruit setting stage), S2 (fruit expansion period), S3 (fruit color change period) and S4 (ripening and softening stage). All samples were quickly frozen in liquid nitrogen and stored at −80 °C.

2.2. Measurement of Peach Fruit Firmness

The peach fruit firmness was assessed using the TA.XT Plus Texture Analyzer (Stable Micro Systems Ltd., Godalming, UK) by inserting a P/2 probe into the fruit at a depth of 5 mm, with each fruit being pierced in three places. The data on firmness were collected by selecting and assessed 10 fruits in each period.

2.3. Identification of PpEXPs in Peach Genome

The hidden markov model (HMM) of DPBB domain (PF03330) and Pollen_allergens domain (PF01357), two conserved domains in expansin proteins, were downloaded from the Pfam website (http://pfam.xfam.org, accessed on 6 June 2023). The hmmsearch was performed against the peach genome protein sequences (v2.1) with a threshold of $e < 1 \times 10^{-5}$. Candidate proteins were further identified by Pfam analysis, with each protein containing both conserved domains.

2.4. Chromosome Location, Gene Structure and cis-Elements Analysis of the PpEXP Genes

The chromosomal physical position of each expansin gene was mapped by TBtools. The gene duplication of all the PpEXPs was analyzed using MCScanX and drawn by TBtools [31]. The exon−intron structure of the *PpEXP* genes were graphically displayed by TBtools. The conserved motif of each PpEXPs protein was pretreated by MEME online software (https://meme-suite.org/meme/ accessed on 22 February 2024). The up-stream 2-Kb genomic DNA sequences of PpEXPs genes were used to predict the *cis*-element in the PlantCARE database (http://bioinformatics.psb.ugent.be/webtools/plantcare/html/ accessed on 22 February 2024).

2.5. Phylogeny Analysis of PpEXPs

Amino acid sequences of 35 AtEXPs (*Arabidopsis thaliana*), 58 OsEXPs (*Oryza sativa*) and 26 PpEXPs (*Prunus persica*) were gathered using ClustalW with system default settings. The phylogenetic trees were formulated by the maximum likelihood method (ML) with 1000-bootstrap replicates in MEGA 7.0 (v7.0, Sudhir kumar, Hachioji, Tokyo, Japan).

2.6. Transcriptomic Analysis

Total RNA was extracted using the Total RNA Rapid Extraction Kit (Sangon, Shanghai, China). For each sample, 2 μg of high-quality RNA was used for library construction and sequencing. Eight libraries, each generated from a sample of CN14 and HSM fruit at S1–4 stages, were sequenced on the BGI-SEQ-500 platform in paired-end 100 bp mode.

The low-quality reads and adapters of the RNA-Seq raw data were removed for all samples using Trimmomatic-0.38. Clean reads were mapped to the CN14 reference genome using hisat2 (v2.1), and transcript abundance was estimated using HTSeq (v0.12.3) based on the alignments.

2.7. Expression Pattern Analysis of PpEXP Genes in Peach Fruit and qRT-PCR Analysis

The expression levels of the PpEXP genes were shown via a heatmap generated by TBtools [31]. To validate the results of RNA sequencing, the expression level of six PpEXP genes was detected using qRT-PCR. The RNA was used to transcribe cDNA with a Prime-Script RT kit (TaKaRa, Kusatsu, Japan), then the cDNA was subjected to qRT-PCR using a Step One Plus Real-Time PCR System (Applied Biosystems, Foster, CA, USA). PpGAPDH (Prupe.1G234000) was used as the endogenous control gene. The primer sequences were listed in Table S7, and we also produced another heatmap through the melting flesh of CN13 and stony hard flesh of CN16 at four development stages (S3, S4I, S4II, S4III), as obtained from our previous study [32].

2.8. Weighted Gene Co-Expression Network Analysis (WGCNA)

A total of 17,395 genes with FPKM values >1 was selected to perform WGCNA on ImageGP (https://www.bic.ac.cn/ImageGP/, accessed on 22 February 2024). Modules were identified using default settings, with the soft power being 30, the minimal module size being 25, and only select top genes with maximum mean absolute deviation (MAD) reaching 2000 [33]. In addition, the expression of the genes of each module were correlated with the change in fruit firmness at four stages of CN14 and HSM.

3. Results

3.1. Identification and Phylogenetic Analysis of the PpEXPs Family in Peaches

A total of 26 *PpEXP* genes containing two conserved domains (DPBB_1 and Pollen_allerg) were identified in the peach genome. The PpEXP protein showed divergent lengths (241–288 amino acids, aa), molecular weights (26.12–31.56 kDa) and isoelectric points (4.69 to 9.83) (Table 1).

Table 1. The characteristics of EXP genes in peach.

Gene ID	Gene Name	No. of Amino Acid (aa)	Molecular Weight(KDa)	Predicted Isoelectric Point (PI)	Subcellular Localization
Prupe.1G276700	PpEXPA1	252	26.74	9.33	Cell wall
Prupe.1G360700	PpEXPA2	257	27.78	9.14	Cell wall
Prupe.1G516600	PpEXPA3	253	28.13	9.10	Cell wall
Prupe.1G516700	PpEXPA4	262	29.35	9.18	Cell wall
Prupe.2G121800	PpEXPA5	260	28.74	8.59	Cell wall
Prupe.2G136500	PpEXPB1	272	28.52	4.72	Cell wall
Prupe.2G237000	PpEXPA6	259	27.83	9.41	Cell wall
Prupe.2G263600	PpEXPA7	252	26.74	6.92	Cell wall
Prupe.2G274400	PpEXPB2	277	29.55	6.58	Cell wall
Prupe.3G014800	PpEXPA8	265	28.67	8.60	Cell wall
Prupe.3G155600	PpEXPB3	266	28.58	8.83	Cell wall
Prupe.3G256600	PpEXPA9	259	28.51	8.89	Cell wall
Prupe.3G265800	PpEXPA10	241	26.12	9.32	Cell wall
Prupe.4G183400	PpEXPA11	265	29.11	9.83	Cell wall
Prupe.5G047300	PpEXPLB1	255	27.70	4.69	Cell wall
Prupe.5G057900	PpEXPLB2	252	27.98	7.46	Cell wall
Prupe.5G087100	PpEXPA12	263	28.99	9.39	Cell wall
Prupe.5G195200	PpEXPA13	249	26.22	9.08	Cell wall
Prupe.6G042000	PpEXPA14	254	27.27	8.43	Cell wall
Prupe.6G075100	PpEXPA15	260	28.02	9.58	Cell wall
Prupe.6G256500	PpEXPA16	258	27.96	8.59	Cell wall
Prupe.6G292300	PpEXPA17	257	27.51	8.92	Cell wall
Prupe.7G124900	PpEXPA18	266	29.50	8.84	Cell wall
Prupe.7G125000	PpEXPA19	288	31.56	8.57	Cell wall
Prupe.8G174500	PpEXPLA1	261	28.22	8.84	Cell wall
Prupe.8G182000	PpEXPA20	259	28.57	8.28	Cell wall

Phylogenetic analysis showed that 26 PpEXPs were grouped into four subfamilies based on the full length protein sequences, including expansins from Arabidopsis (35) and rice (58). EXPA, as the largest subfamily, contained 79 genes, including 20 PpEXPAs, 25 AtEXPAs and 34 OsEXPAs (Figure 1). The second largest subfamily, EXPB, had 28 genes, with three PpEXPBs, six AtEXPBs and 19 OsEXPBs. There were four, three and one EXPLAs in rice, Arabidopsis and peach, respectively. In peach, two EXPLBs were identified, while only one EXPLB was found in rice and Arabidopsis, respectively.

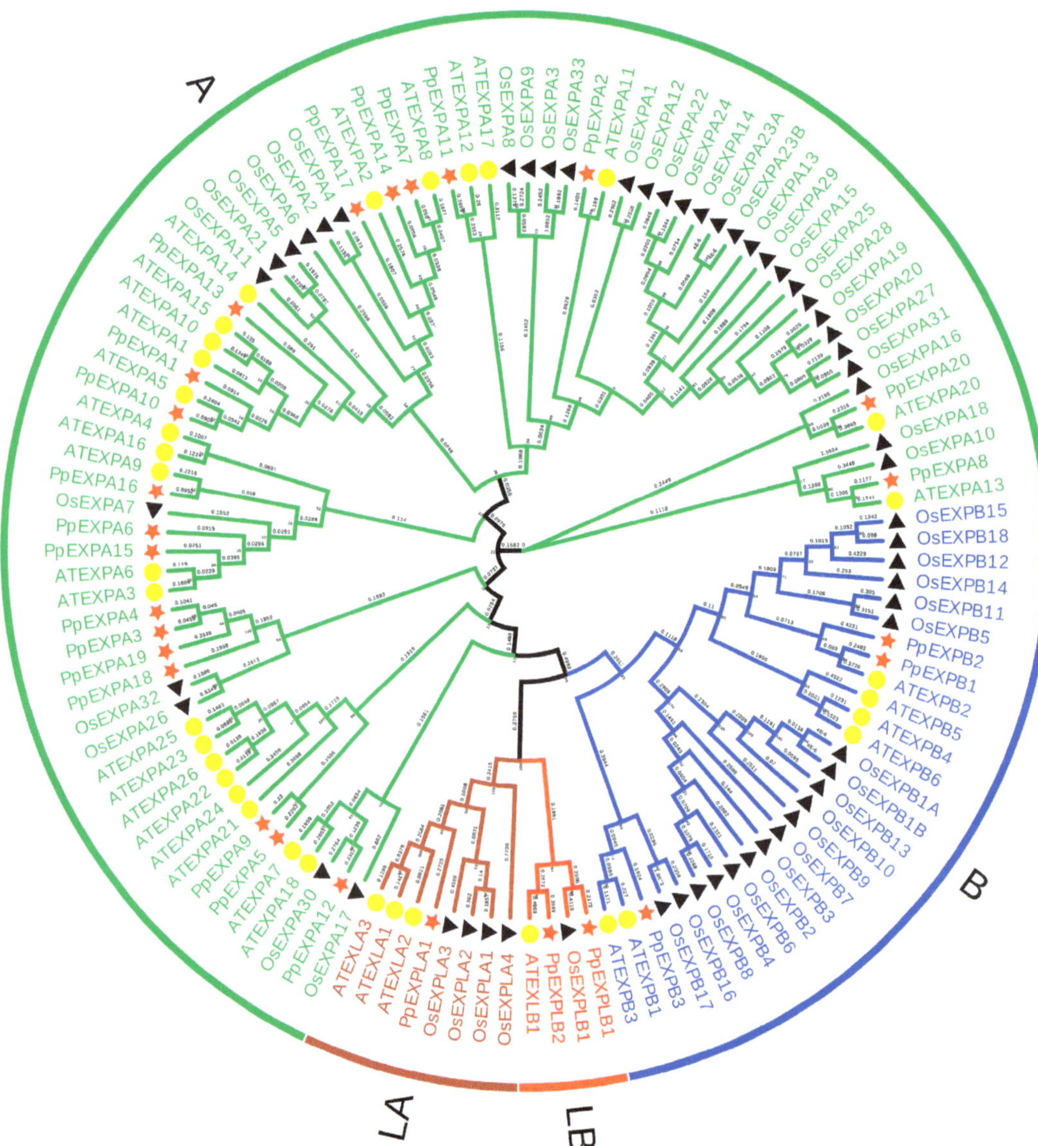

Figure 1. Phylogenetic analysis showed that EXPs in *Arabidopsis thaliana* (yellow round), rice (*Oryza sativa*, green triangle), Peach (*Prunus persica*, red star). A: EPXA; B: EXPB; LA:EXPLA; LB: EXPLB.

3.2. Gene Structure and Motif Analysis of PpEXPs

The PpEXPs that clustered in the same subfamily showed similar gene structures and motif composition (Figure 2). All the members of the PpEXPB subfamily had four exons (Figure 2A,B). Most PpEXPA genes (80%) contained three exons, while others had two exons (20%). The PpEXPLB1 had five exons and PpEXPLB2 had four exons (Figure 2B). This indicated that PpEXPA and PpEXPB were more conserved in an exon−intron structure than PpEXPLB and reflected the gene structure diversity of different subfamilies.

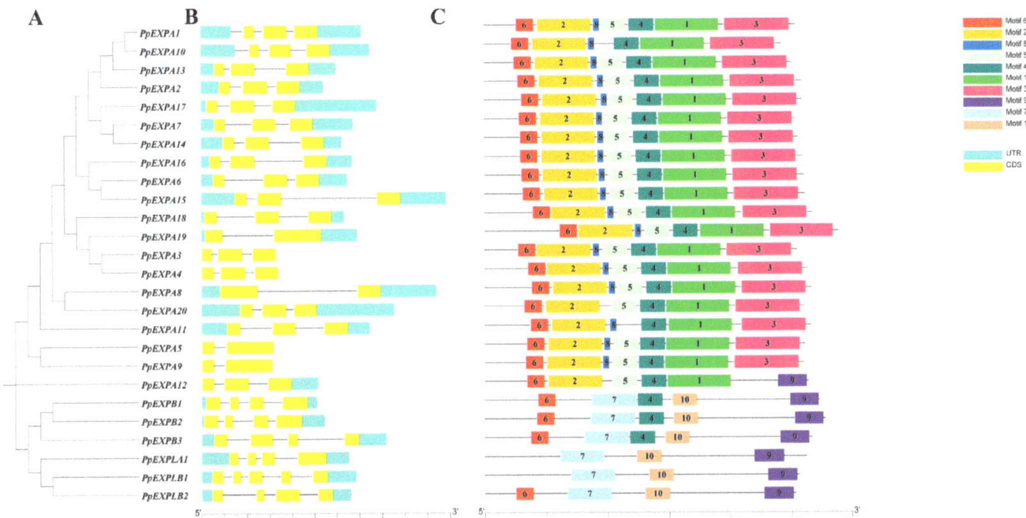

Figure 2. Gene structure and motif analysis of PpEXPs. (**A**): Phylogenetic relationships of PpEXPs; (**B**): Gene structure of PpEXPs. Yellow box, turquoise box and black line represent exon, UTR and intron, respectively; (**C**): Conserved motif of PpEXPs. Different color boxes represent different motifs, noted in upper-right.

The subfamily of PpEXPA apparently differed from the other subfamilies in motif composition. All the PpEXPAs contained Motifs 1, 2, 4 and 6, and most of the PpEXPAs also contained Motifs 3, 5 and 8, except PpEXPA10, PpEXPA11, PpEXPA12 and PpEXPA20 (Figure 2C, Table S1). All the PpEXPBs contained the same motifs, including Motif 1, Motif 4, Motif 7, Motif 9 and Motif 10. The motif analysis was consistent with the phylogenetic analyses, showing similar motif composition in members in the same subfamily.

3.3. Chromosome Distribution and Gene Duplication

All the PpEXPs were unevenly distributed on eight chromosomes (Chr). Chr 1, Chr3, Chr5 and Chr6 each contained four PpEXPs, while, 5, 1, 2 and 2 were mapped on Chr2, Chr4, Chr7 and Chr8, respectively (Figure 3). To investigate the gene duplication model, the duplication events were detected (Figure 3, Table S2). The results showed that 13 dispersed duplication genes (DSD, 50%) were detected, PpEXPA2, PpEXPA5, PpEXPA8-9, PpEXPA11-12, PpEXPA20, PpEXPB1-3, PpEXPLA1 and PpEXPLB1-2. Four tandem duplication (TD, 15%) genes were detected, with a PpEXPA3−PpEXPA4 pair on Chr1 and a PpEXPA18−PpEXPA19 pair on Chr7. Nine genes evolved from whole genome duplication (WGD, 35%). Proximal duplication (PD) and singleton were both unidentified in PpEXPs.

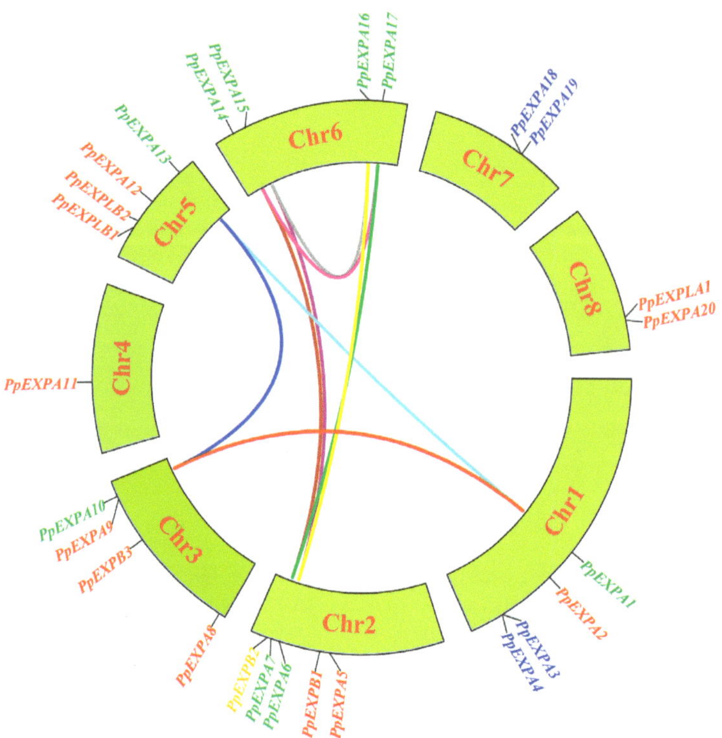

Figure 3. Chromosome distribution and gene duplication of PpEXPs. The genes with the same color represented a type of gene duplication: red, DSD; blue, TD; green, WGD. Syntenic pairs were linked with lines, with colors representing different pairs.

3.4. Cis-Acting Elements Detection in the Promoter Regions of PpEXPs

Twenty-one types of *cis*-elements were identified in the 2-kb upstream regions of PpEXPs (Figure 4A, Table S3). The elements involved in light responsiveness and MeJA-responsiveness presented in the promoters of all PpEXPs, especially light responsive elements appearing at least two times in each promoter (Figure 4B). The copies of elements in the promoter region might enhance the regulating effects of their corresponding factors. In addition, elements related to the MYB binding site, abscisic acid responsiveness and anaerobic induction elements were apparently abundant, being detected in most PpEXPs promoters. These elements were related to many processes, such as plant hormone (MeJA respone, auxin response, abscisic acid response, ethylene response), environmental factors (light response, low-temperature response) and stress (wound, cold and defense responses). The different *cis*-elements in the promoters of PpEXPs suggested that PpEXPs may participate in various processes.

Ethylene responsive elements were identified in the promoter regions of 12 PpEXPs, such as PpEXPA1, PpEXPA3, PpEXPA7, PpEXPA15, etc. Since ethylene plays an important role in fruit ripening [34], the expression of these genes might be involved in regulating peach fruit ripening.

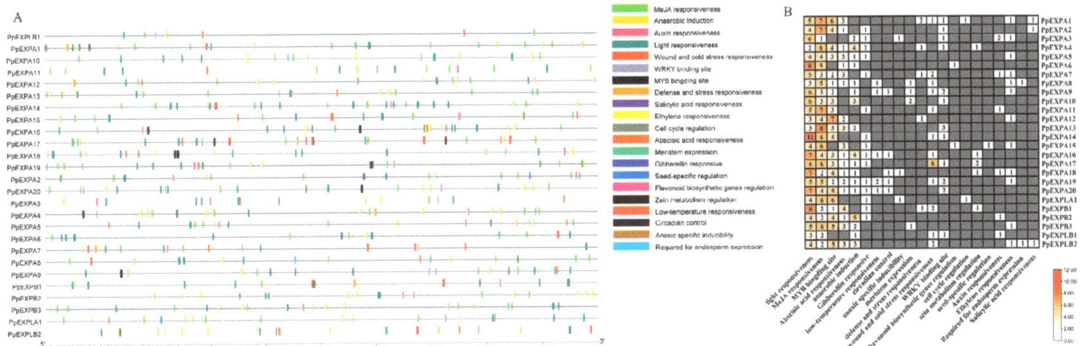

Figure 4. The *cis*-elements in the promoters of PpEXPs. (**A**): The different types of *cis*-elements, with different colors, were shown in the promoter region of each PpEXP. (**B**): The number of each *cis*-element in the promoter regions.

3.5. Expression Analysis of PpEXPs during Fruit Ripening in MF, NMF and SH Peach

To comprehensively reveal the expression patterns of PpEXPs in peach fruit, the transcriptome of the fruit was performed in HSM (MF) and CN14 (NMF) cultivars at four different stages (Figure 5A, Table S4). Among 26 PpEXP genes, 14 PpEXPs were expressed in the fruit of either MF or NMF during fruit ripening, including PpEXPA1, PpEXPA6-8, PpEXPA10, PpEXPA13-16, PpEXPA20, PpEXPB2-3, PpEXPLA1 and PpEXPLB2. Six PpEXPs were used to confirm the expression levels by quantitative real-time PCR (qRT-PCR) in CN14 and HSM at four stages. The results of the qRT-PCRs were consistent with those of the RNA-seq (Figure S1).

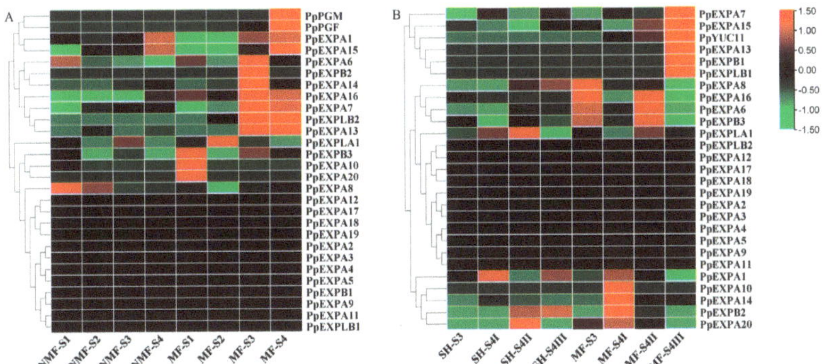

Figure 5. The expression pattern of PpEXPs in melting, non-melting and stony hard peaches. (**A**): The expression pattern of PpEXPs in melting and non-melting. (**B**): The expression pattern of PpEXPs in melting and non-melting NMF−S1−4: S1-S4 stages in non-melting peaches CN14; MF−S1−4: S1−S4 stages in melting peaches HSM; SH−S3−4III: S3−S4III in stony hard peaches CN16; MF−S3−4III: S3−S4III in melting peaches CN13.

PpEXPA1, PpEXPA6-8, PpEXPA15-16 and PpEXPLA1 showed high expression levels during the development and ripening stages in all samples. This suggested that these genes might be related to fruit development. The PpPGM and PpPGF were responsible for the flesh texture of MF and NMF fruit, promoting fruit softening [3]. According to the expression patterns, PpEXPA1, PpEXPA6-7, PpEXPA13-16, PpEXPB2 and PpEXPLB2 showed similar expression patterns with PpPGM and PpPGF. Moreover, PpEXPA1, Pp-EXPA7, PpEXPA13 and PpEXPA15 showed higher expression levels than the other genes, indicating that these four genes might regulate flesh texture.

In addition, the expression pattern of PpEXPs in CN13 (MF) and CN16 (SH) were also analyzed across four stages of ripening. The results showed that 15 PpEXPs were expressed in CN13 and CN16, including PpEXPA1, PpEXPA6-8, PpEXPA10, PpEXPA13-16, PpEXPA20, PpEXPB1-3, PpEXPLA1 and PpEXPLB1. PpYUC11 was a candidate gene for the control of the SH phenotype, of which the expression pattern was associated with flesh texture of MF, not expressed in SH [4]. PpEXPA7, PpEXPA13, PpEXPA15, PpEXPB1 and PpEXPLB1 showed a clustered expression pattern with PpYUC11. Combining the above results of the expression patterns of PpEXPs in the two sets of transcriptome data, PpEXPA7, PpEXPA13 and PpEXPA15 were the common candidate genes, suggesting that they might participate in fruit texture development.

3.6. Weighted Gene Co-Expression Network Analysis (WGCNA) of Fruit Development

The development of fruit texture is a complex and dynamic process consisting of a diverse collection of genes with a variety of functional relationships. In order to detect relationships underlying modules and fruit texture development, we performed a weighted gene co-expression network analysis (WGCNA). As a result, 2000 genes (FPKM > 1) in eight samples of four stages were clustered into 16 modules (Figure 6A, Table S5). The module-trait association showed that pink and midnightblue modules were negatively correlated with fruit firmness (Figure 6B, Table S6). There were 89 genes in the pink module and 36 genes were in the midnightblue module. These genes were annotated in the three GO (Gene ontology) categories (biological process, cellular component and molecular function) (Table S5). In the biological process category, two genes, annotated as cell wall organization (GO:0009664), encoded the PpEXPA1 and PpEXPA7. Meanwhile, PpEXPA7 was identified as the hub gene. Notably, a transcription factor, *PpERF* (Ethylene response factor, Prupe.1G130300), was grouped into the pink module. Based on the results, *PpEXPS*, *PpEXPA1* and *PpEXPA7*, play a critical role in fruit texture, which might be regulated by the *PpERF* transcription factor.

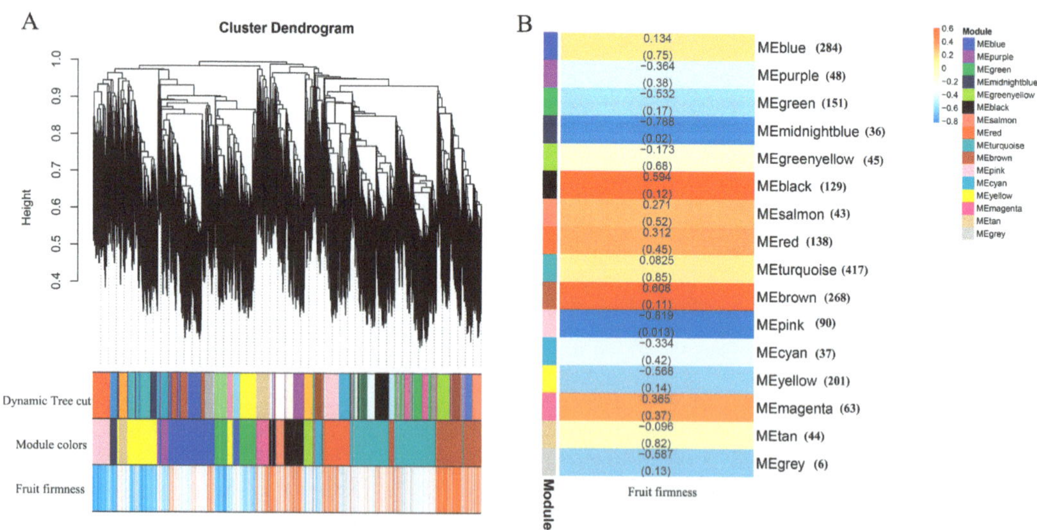

Figure 6. WGCNA of expressed genes at four stages of HSM and CN fruit. (**A**): Hierarchical cluster tree of WGCNA analysis ($R^2 = 0.85$). (**B**): Module–trait correlation analysis. Sixteen modules were labeled with different colors. The numbers in each row represent the correlation of every module, the numbers in parentheses indicate the *p*-values of correlations, and the gene numbers were displayed in the right of each module.

4. Discussion

EXPs, as plant cell-wall loosening proteins, participate in cell growth and other development processes by modifying cell-walls [35]. EXPs in peach were identified as having 26 members, of which all contained two conserved domains (DPBB and Pollen_allerg). Canonical expansin proteins in the plant were displayed as torpedo-shaped and consisted of the two domains (DPBB and Pollen_allerg) [20,36]. There were proteins, such as G2As, p12 and plant natriuretic peptide, which contained a domain homologous only to expansin and one domain in the plant [37,38], while they lacked the wall-extension activity of expansins and could not be considered part of the expansin superfamily [39,40]. In this study, we identified 27 genes featuring a DPBB domain or Pollen_allerg domain (Prupe.7G203100 only with Pollen_allergens domain). Hence, 26 genes were confirmed as expansins, containing both of these two domains.

The number of PpEXPs was less than that in other species. For example, 56 and 88 members were detected in maize and rice, respectively [21,25]. In addition, 75, 53, 41, 36, 36, 29 EXP genes were detected in soybeans (*Glycine max*) [41], Chinese cabbage (*Brassica rapa* ssp. *pekinensis*) [42], apples [24], *Arabidopsis thaliana* [25], poplar (*Populus trichocarpa*) [22] and grapes (*Vitis vinifera*) [23], respectively. It was speculated that the different number of EXP genes is associated with genome evolution and EXP functions. Similar to other plants, the PpEXPs were divided into four subfamilies based on the phylogenic analysis, including PpEXPA, PpEXPB, PpEXPLA and PpEXPLB. The genes of the same subfamily, but not the same species, clustered together, indicating that their ancestors differentiated before the divergence of different plant species. In this study, PpEXPA were the largest subfamily, which showed a significant expansion; in contrast, the number of the other three subfamilies was less than ten. This phenomenon was also observed in other eudicots, for example, apples (34 EXPAs, 1 EXPB, 2 EXPLAs, 4 EXPLBs), grapes (20 EXPAs, 4 EXPBs, 1 EXPLA, 4 EX-PLBs) and *Arabidopsis thaliana* (26 EXPAs, 6 EXPBs, 3 EXPLAs, 1 EXPLB) [23–25]. However, EXPBs proved to be more numerous in monocots, e.g., the number of EXPB were 18 and 48 in rice and maize, respectively [25]. The divergence of EXPB between monocots and eudicots might be associated with the distinctive cell wall composition [43].

The PpEXPs within the subfamily contained a similar gene structure and motif composition, indicating that PpEXPs in the same subfamily were highly conserved. Motif 1 and Motif 2 were present only in PpEXPAs, while all PpEXPBs and PpEXPLs contained Motif 7 and Motif 10. Gene duplication analysis showed that DSD was the major pattern, which was consistent with the other gene families in peaches, such as E3 ubiquitin ligase genes [32], HSF [44] and GRAS [45]. The model of gene duplication might furnish evidence for the evolution of species and was considered a driving force in evolution [46]. The peach has not undergone a recent whole-genome duplication [47], which might lead to the duplication pattern of gene families, including PpEXPs.

The expression of gene response to developmental or environmental signals were mainly regulated through *cis*-acting elements. The divergence of the promoter regions was associated with the expression pattern of genes [48]. In this study, the potential *cis*-acting elements in the 2 Kb region upstream of the start codons for each PpEXP were detected, with 21 types of *cis*-elements. It was found that light responsive elements and hormone responsive elements were apparently abundant for the PpEXPs, similar to other studies of EXPs [26,49,50]. In soybeans, four GmEXPBs were induced by IAA and eight GmEXPs responded to 6-BA correspondingly, with the promoters containing hormone-related *cis*-elements [51]. The wheat expansin gene *TaEXPB23* only showed improved water-stress tolerance with the stress-inducible promoter *RD29A* while also showing negative effects on plant growth and development under the *35S* promoter, such as earlier flowering and shorter plant height [48]. The results showed a diversity of *cis*-acting regulatory elements in the promoter region of PpEXPs, which might participate in various processes.

It has been widely reported that EXPs participate in root hair formation [52,53], biotic and abiotic stresses [54,55] and organ development [56]. Cell wall loosening catalyzed by EXP was central during fleshy fruit ripening [7]. To identify EXPs related to peach fruit development, we identified the expression pattern of *PpEXPs* in fruits of different flesh types, with MF, NMF and SH featuring at four stages. Only half of 26 *PpEXPs* were expressed in peach fruit (Table S4). The flesh of melting, non-melting and stony hard peaches showed significantly differences in regard to firmness and texture, and *PpPGM*, *PGF* and *PpYUC11* showed positive correlations with fruit firmness [3,4]. The expression of nine *PpEXPs* clustered with *PpPGM* and *PpPGF* in melting and non-melting peaches, and six *PpEXPs* showed similar expression patterns to *PpYUC11* in melting and stony hard peaches. Combining the two pairs of expression patterns, three *PpEXPs*, *PpEXPA7*, *PpEXPA13* and *PpEXPA15*, were highly expressed during ripening. The expression of the four candidate EXPs dramatically increased when firmness decreased in strawberries [30]. In tomatoes, *SlEXP1* promoted fruit softening cooperating with SlCel2 [18]. It is worth noting that ethylene responsive elements and abscisic acid responsive elements were detected in the promoter region of PpEXPA7, PpEXPA13 and PpEXPA15; therefore, we propose that PpEXPs might be involved in regulating fruit textures in peaches.

In this study, we identified the gene modules co-expressed with fruit firmness. Of 16 modules, 2 modules, pink and midnightblue, showed significantly negatively correlations in terms of fruit firmness, including 125 genes. PpEXPA1 and PpEXPA7 were confirmed to be highly related to fruit texture based on the WGCNA. In addition, a transcription factor, PpERF, was identified in the correlation module, indicating a potential role for them in regulating fruit texture. WGCNA has been used to identify the coexpression modules related to volatile organic compounds (VOCs) in *Osmanthus fragrans* [57]. It has been reported that three genes (Prupe.7G234800, Prupe.8G079200 and Prupe.8G082100) were obtained as candidate genes for single fruit weight traits based on the WGCNA in peaches [58]. ERF has been demonstrated to activate the expression of genes involved in fruit softening, such as PG [59], PME [60], and EXP [61]. In peaches, the PpEXP gene (ppa017982m) might control fruit weight [62]. After a homologous comparison, ppa017982m was the PpEXPLB1 homologous gene. PpExp3, which is homologous with PpEXPA15, showed an increase in the expression of ethylene-treated 'Manami' (SH) [63]. These findings implied that PpEXPs play an important role in fruit texture and might be regulated by the PpERF transcription factor.

5. Conclusions

In this study, we identified 26 *PpEXP* genes in the peach genome which can be divided into four subfamilies (PpEXPA, PpEXPB, PpEXPLA and PpEXPLB) based on the phylogenic analysis. The gene structure and motif of members proved to be highly diverse among different subfamilies. The model of gene duplication of PpEXPs were DSD (50%), WGD (35%) and TD (15%). Moreover, 15 PpEXP genes were expressed in peach fruit at four stages. Among them, PpEXPA7, PpEXPA13 and PpEXPA15 were coexpressions with PpPGM, PpPGF and PpYUC11, key genes for peach fruit texture. Meanwhile, the *cis*-element of ethylene responsiveness and abscisic acid responsiveness were detected in the promoter of these three genes. Based on a WGCNA, PpEXPA1, PpEXPA7 and PpERF were identified in the correlation module; thus, the results lay a functional foundation for the PpEXP gene family and for further exploring the roles of candidate *PpEXPs* in peach fruit texture and their regulatory work.

Supplementary Materials: The following supporting information can be downloaded at: https://www.mdpi.com/article/10.3390/horticulturae10040332/s1, Figure S1: The relative expression level of PpEXPs in the fruit of MF (HSM) and NMF (CN14); Table S1: Motif sequences identified by MEME tools in PpEXPs; Table S2: The gene duplication of EXP genes in peach; Table S3: The responsive elements of EXP genes in peach; Table S4: The expression of PpEXPs in fruit of melting, non-melting and stony hard peaches at differernt stages; Table S5: The number of genes in each module was identified by WGCNA method and gene annotation of GO; Table S6: Fruit firmness across fruit development of MF and NMF; Table S7: List of primers used in this study.

Author Contributions: J.F., B.T. and X.L. conceived and designed the research. Y.G., C.S., F.G. and Y.Z. carried out most experiments. X.Z., X.W., J.C. and W.W. performed the bioinformatics analyses. X.Y., L.Z. and J.L. analyzed the data. B.T. and X.L. wrote the paper. All authors contributed, read and approved the final manuscript. All authors have read and agreed to the published version of the manuscript.

Funding: This work was supported by the Modern Agricultural Industry Technology of Henan Province (HARS-22-09-G1), the Henan Scientific and Technological Research Project (232102110211), the Young Talents Project of Henan Agricultural University (30501343).

Data Availability Statement: Data are contained within the Supplementary Materials.

Conflicts of Interest: The authors declare no conflicts of interest.

References

1. Yoshida, M. Genetical studies on the fruit quality of peach varieties. III. Texture and keeping quality. *Bull. Fruit Tree Res. Sta.* **1976**, *3*, 1–16.
2. Haji, T.; Yaegaki, H.; Yamaguchi, M. Inheritance and expression of fruit texture melting, non-melting and stony hard in peach. *Sci. Hortic.* **2005**, *105*, 241–248. [CrossRef]
3. Gu, C.; Wang, L.; Wang, W.; Zhou, H.; Ma, B.; Zheng, H.; Fang, T.; Ogutu, C.; Vimolmangkang, S.; Han, Y. Copy number variation of a gene cluster encoding endopolygalacturonase mediates flesh texture and stone adhesion in peach. *J. Exp. Bot.* **2016**, *67*, 1993–2005. [CrossRef] [PubMed]
4. Pan, L.; Zeng, W.; Niu, L.; Lu, Z.; Liu, H.; Cui, G.; Zhu, Y.; Chu, J.; Li, W.; Fang, W.; et al. PpYUC11, a strong candidate gene for the stony hard phenotype in peach (*Prunus persica* L. Batsch), participates in IAA biosynthesis during fruit ripening. *J. Exp. Bot.* **2015**, *66*, 7031–7044. [CrossRef] [PubMed]
5. Li, X.; Xu, C.; Korban, S.S.; Chen, K. Regulatory Mechanisms of Textural Changes in Ripening Fruits. *Crit. Rev. Plant Sci.* **2010**, *29*, 222–243. [CrossRef]
6. Wang, D.; Yeats, T.H.; Uluisik, S.; Rose, J.K.C.; Seymour, G.B. Fruit Softening: Revisiting the Role of Pectin. *Trends Plant Sci* **2018**, *23*, 302–310. [CrossRef]
7. Wang, D.; Seymour, G.B. Molecular and biochemical basis of softening in tomato. *Mol. Hortic.* **2022**, *2*, 5. [CrossRef] [PubMed]
8. Atkinson, R.G.; Sutherland, P.W.; Johnston, S.L.; Gunaseelan, K.; Hallett, I.C.; Mitra, D.; Brummell, D.A.; Schroder, R.; Johnston, J.W.; Schaffer, R.J. Down-regulation of POLYGALACTURONASE1 alters firmness, tensile strength and water loss in apple (*Malus × domestica*) fruit. *BMC Plant Biol.* **2012**, *12*, 129. [CrossRef]
9. Quesada, M.A.; Blanco-Portales, R.; Posé, S.; García-Gago, J.A.; Jiménez-Bermúdez, S.; Muñoz-Serrano, A.S.; Caballero, J.L.; Pliego-Alfaro, F.; Mercado, J.A.; Muñoz-Blanco, J. Antisense Down-Regulation of the FaPG1 Gene Reveals an Unexpected Central Role for Polygalacturonase in Strawberry Fruit Softening. *Plant Physiol.* **2009**, *150*, 1022–1032. [CrossRef] [PubMed]
10. Paniagua, C.; Blanco-Portales, R.; Barcelo-Munoz, M.; Garcia-Gago, J.A.; Waldron, K.W.; Quesada, M.A.; Munoz-Blanco, J.; Mercado, J.A. Antisense down-regulation of the strawberry beta-galactosidase gene FabetaGal4 increases cell wall galactose levels and reduces fruit softening. *J. Exp. Bot.* **2016**, *67*, 619–631. [CrossRef]
11. Zhang, W.W.; Zhao, S.Q.; Gu, S.; Cao, X.Y.; Zhang, Y.; Niu, J.F.; Liu, L.; Li, A.R.; Jia, W.S.; Qi, B.X.; et al. FvWRKY48 binds to the pectate lyase FvPLA promoter to control fruit softening in *Fragaria vesca*. *Plant Physiol.* **2022**, *189*, 1037–1049. [CrossRef] [PubMed]
12. Seymour, G.B.; Østergaard, L.; Chapman, N.H.; Knapp, S.; Martin, C. Fruit Development and Ripening. *Annu. Rev. Plant Biol.* **2013**, *64*, 219–241. [CrossRef] [PubMed]
13. Smith, C.J.S.; Watson, C.F.; Ray, J.; Bird, C.R.; Morris, P.C.; Schuch, W.; Grierson, D. Antisense RNA inhibition of polygalacturonase gene expression in transgenic tomatoes. *Nature* **1988**, *334*, 724–726. [CrossRef]
14. Tieman, D.M.; Harriman, R.W.; Ramamohan, G.; Handa, A.K. An Antisense Pectin Methylesterase Gene Alters Pectin Chemistry and Soluble Solids in Tomato Fruit. *Plant Cell* **1992**, *4*, 667–679. [CrossRef] [PubMed]
15. Uluisik, S.; Chapman, N.H.; Smith, R.; Poole, M.; Adams, G.; Gillis, R.B.; Besong, T.M.D.; Sheldon, J.; Stiegelmeyer, S.; Perez, L.; et al. Genetic improvement of tomato by targeted control of fruit softening. *Nat. Biotechnol.* **2016**, *34*, 950–952. [CrossRef] [PubMed]
16. Yang, L.; Huang, W.; Xiong, F.; Xian, Z.; Su, D.; Ren, M.; Li, Z. Silencing of SlPL, which encodes a pectate lyase in tomato, confers enhanced fruit firmness, prolonged shelf-life and reduced susceptibility to grey mould. *Plant Biotechnol. J.* **2017**, *15*, 1544–1555. [CrossRef]
17. Sozzi, G.O.; Fraschina, A.A.; Navarro, A.A.; Cascone, O.; Greve, L.C.; Labavitch, J.M. α-L-arabinofuranosidase activity during development and ripening of normal and ACC synthase antisense tomato fruit. *Hortscience* **2002**, *37*, 564–566. [CrossRef]
18. Su, G.; Lin, Y.; Wang, C.; Lu, J.; Liu, Z.; He, Z.; Shu, X.; Chen, W.; Wu, R.; Li, B.; et al. Expansin SlExp1 and endoglucanase SlCel2 synergistically promote fruit softening and cell wall disassembly in tomato. *Plant Cell* **2023**, *36*, 709–726. [CrossRef] [PubMed]
19. McQueen-Mason, S.; Durachko, D.M.; Cosgrove, D.J. Two endogenous proteins that induce cell wall extension in plants. *Plant Cell* **1992**, *4*, 1425–1433.
20. Cosgrove, D.J. Plant expansins: Diversity and interactions with plant cell walls. *Curr Opin Plant Biol* **2015**, *25*, 162–172. [CrossRef]

21. Zhang, W.; Yan, H.; Chen, W.; Liu, J.; Jiang, C.; Jiang, H.; Zhu, S.; Cheng, B. Genome-wide identification and characterization of maize expansin genes expressed in endosperm. *Mol. Genet. Genom.* **2014**, *289*, 1061–1074. [CrossRef] [PubMed]
22. Sampedro, J.; Carey, R.E.; Cosgrove, D.J. Genome histories clarify evolution of the expansin superfamily: New insights from the poplar genome and pine ESTs. *J. Plant Res.* **2006**, *119*, 11–21. [CrossRef] [PubMed]
23. Dal Santo, S.; Vannozzi, A.; Tornielli, G.B.; Fasoli, M.; Venturini, L.; Pezzotti, M.; Zenoni, S. Genome-Wide Analysis of the Expansin Gene Superfamily Reveals Grapevine-Specific Structural and Functional Characteristics. *PLoS ONE* **2013**, *8*, e62206. [CrossRef] [PubMed]
24. Zhang, S.; Xu, R.; Gao, Z.; Chen, C.; Jiang, Z.; Shu, H. A genome-wide analysis of the expansin genes in Malus × Domestica. *Mol. Genet. Genom.* **2014**, *289*, 225–236. [CrossRef]
25. Sampedro, J.; Lee, Y.; Carey, R.E.; DePamphilis, C.; Cosgrove, D.J. Use of genomic history to improve phylogeny and understanding of births and deaths in a gene family. *Plant J.* **2005**, *44*, 409–419. [CrossRef] [PubMed]
26. Lv, L.-M.; Zuo, D.-Y.; Wang, X.-F.; Cheng, H.-L.; Zhang, Y.-P.; Wang, Q.-L.; Song, G.-L.; Ma, Z.-Y. Genome-wide identification of the expansin gene family reveals that expansin genes are involved in fibre cell growth in cotton. *BMC Plant Biol.* **2020**, *20*, 223. [CrossRef] [PubMed]
27. Minoia, S.; Boualem, A.; Marcel, F.; Troadec, C.; Quemener, B.; Cellini, F.; Petrozza, A.; Vigouroux, J.; Lahaye, M.; Carriero, F.; et al. Induced mutations in tomato SlExp1 alter cell wall metabolism and delay fruit softening. *Plant Sci.* **2016**, *242*, 195–202. [CrossRef] [PubMed]
28. Jiang, F.; Lopez, A.; Jeon, S.; de Freitas, S.T.; Yu, Q.; Wu, Z.; Labavitch, J.M.; Tian, S.; Powell, A.L.T.; Mitcham, E. Disassembly of the fruit cell wall by the ripening-associated polygalacturonase and expansin influences tomato cracking. *Hortic. Res.* **2019**, *6*, 17. [CrossRef] [PubMed]
29. Chen, Y.-H.; Xie, B.; An, X.-H.; Ma, R.-P.; Zhao, D.-Y.; Cheng, C.-G.; Li, E.-M.; Zhou, J.-T.; Kang, G.-D.; Zhang, Y.-Z. Overexpression of the apple expansin-like gene MdEXLB1 accelerates the softening of fruit texture in tomato. *J. Integr. Agric.* **2022**, *21*, 3578–3588. [CrossRef]
30. Mu, Q.; Li, X.; Luo, J.; Pan, Q.; Li, Y.; Gu, T. Characterization of expansin genes and their transcriptional regulation by histone modifications in strawberry. *Planta* **2021**, *254*, 21. [CrossRef]
31. Chen, C.; Wu, Y.; Li, J.; Wang, X.; Zeng, Z.; Xu, J.; Liu, Y.; Feng, J.; Chen, H.; He, Y.; et al. TBtools-II: A "one for all, all for one"; bioinformatics platform for biological big-data mining. *Mol. Plant* **2023**, *16*, 1733–1742. [CrossRef] [PubMed]
32. Tan, B.; Lian, X.; Cheng, J.; Zeng, W.; Zheng, X.; Wang, W.; Ye, X.; Li, J.; Li, Z.; Zhang, L.; et al. Genome-wide identification and transcriptome profiling reveal that E3 ubiquitin ligase genes relevant to ethylene, auxin and abscisic acid are differentially expressed in the fruits of melting flesh and stony hard peach varieties. *BMC Genom.* **2019**, *20*, 892. [CrossRef] [PubMed]
33. Chen, T.; Liu, Y.-X.; Huang, L. ImageGP: An easy-to-use data visualization web server for scientific researchers. *Imeta* **2022**, *1*, e5. [CrossRef]
34. Fenn, M.A.; Giovannoni, J.J. Phytohormones in fruit development and maturation. *Plant J.* **2021**, *105*, 446–458. [CrossRef] [PubMed]
35. Cosgrove, D.J. Loosening of plant cell walls by expansins. *Nature* **2000**, *407*, 321–326. [CrossRef] [PubMed]
36. Georgelis, N.; Yennawar, N.H.; Cosgrove, D.J. Structural basis for entropy-driven cellulose binding by a type-A cellulose-binding module (CBM) and bacterial expansin. *Proc. Natl. Acad. Sci. USA* **2012**, *109*, 14830–14835. [CrossRef] [PubMed]
37. Yennawar, N.H.; Li, L.-C.; Dudzinski, D.M.; Tabuchi, A.; Cosgrove, D.J. Crystal structure and activities of EXPB1 (Zea m 1), a β-expansin and group-1 pollen allergen from maize. *Proc. Natl. Acad. Sci. USA* **2006**, *103*, 14664–14671. [CrossRef] [PubMed]
38. Ceccardi, T.L.; Barthe, G.A.; Derrick, K.S. A novel protein associated with citrus blight has sequence similarities to expansin. *Plant Mol. Biol.* **1998**, *38*, 775–783. [CrossRef] [PubMed]
39. Sampedro, J.; Cosgrove, D.J. The expansin superfamily. *Genome Biol.* **2005**, *6*, 242. [CrossRef]
40. Rafudeen, S.; Gxaba, G.; Makgoke, G.; Bradley, G.; Pironcheva, G.; Raitt, L.; Irving, H.; Gehring, C. A role for plant natriuretic peptide immuno-analogues in NaCl- and drought-stress responses. *Physiol. Plant* **2003**, *119*, 554–562. [CrossRef]
41. Zhu, Y.; Wu, N.; Song, W.; Yin, G.; Qin, Y.; Yan, Y.; Hu, Y. Soybean (*Glycine max*) expansin gene superfamily origins: Segmental and tandem duplication events followed by divergent selection among subfamilies. *BMC Plant Biol.* **2014**, *14*, 93. [CrossRef]
42. Krishnamurthy, P.; Hong, J.K.; Kim, J.A.; Jeong, M.-J.; Lee, Y.-H.; Lee, S.I. Genome-wide analysis of the expansin gene superfamily reveals *Brassica rapa*-specific evolutionary dynamics upon whole genome triplication. *Mol. Genet. Genom.* **2015**, *290*, 521–530. [CrossRef] [PubMed]
43. Sampedro, J.; Guttman, M.; Li, L.C.; Cosgrove, D.J. Evolutionary divergence of beta-expansin structure and function in grasses parallels emergence of distinctive primary cell wall traits. *Plant J.* **2015**, *81*, 108–120. [CrossRef]
44. Tan, B.; Yan, L.; Li, H.; Lian, X.; Cheng, J.; Wang, W.; Zheng, X.; Wang, X.; Li, J.; Ye, X.; et al. Genome-wide identification of HSF family in peach and functional analysis of PpHSF5 involvement in root and aerial organ development. *PeerJ* **2021**, *9*, e10961. [CrossRef] [PubMed]
45. Jiang, C.; Gao, F.; Li, T.; Chen, T.; Zheng, X.; Lian, X.; Wang, X.; Zhang, H.; Cheng, J.; Wang, W.; et al. Genome-wide analysis of the GRAS transcription factor gene family in peach (*Prunus persica*) and ectopic expression of PpeDELLA1 and PpeDELLA2 in Arabidopsis result in dwarf phenotypes. *Sci. Hortic.* **2022**, *298*, 111003. [CrossRef]
46. Zhang, J. Evolution by gene duplication: An update. *Trends Ecol. Evol.* **2003**, *18*, 292–298. [CrossRef]

47. International Peach Genome, I.; Verde, I.; Abbott, A.G.; Scalabrin, S.; Jung, S.; Shu, S.; Marroni, F.; Zhebentyayeva, T.; Dettori, M.T.; Grimwood, J.; et al. The high-quality draft genome of peach (*Prunus persica*) identifies unique patterns of genetic diversity, domestication and genome evolution. *Nat. Genet.* **2013**, *45*, 487–494.

48. Li, F.; Han, Y.; Feng, Y.; Xing, S.; Zhao, M.; Chen, Y.; Wang, W. Expression of wheat expansin driven by the RD29 promoter in tobacco confers water-stress tolerance without impacting growth and development. *J. Biotechnol.* **2013**, *163*, 281–291. [CrossRef] [PubMed]

49. Ding, A.; Marowa, P.; Kong, Y. Genome-wide identification of the expansin gene family in tobacco (*Nicotiana tabacum*). *Mol Genet Genom.* **2016**, *291*, 1891–1907. [CrossRef]

50. Chen, S.; Luo, Y.; Wang, G.; Feng, C.; Li, H. Genome-wide identification of expansin genes in *Brachypodium distachyon* and functional characterization of *BdEXPA27*. *Plant Sci.* **2020**, *296*, 110490. [CrossRef]

51. Li, X.; Zhao, J.; Walk, T.C.; Liao, H. Characterization of soybean beta-expansin genes and their expression responses to symbiosis, nutrient deficiency, and hormone treatment. *Appl. Microbiol. Biotechnol.* **2014**, *98*, 2805–2817. [CrossRef] [PubMed]

52. Cho, H.-T.; Cosgrove, D.J. Regulation of Root Hair Initiation and Expansin Gene Expression in Arabidopsis[W]. *Plant Cell* **2002**, *14*, 3237–3253. [CrossRef]

53. Lin, C.; Choi, H.-S.; Cho, H.-T. Root Hair-Specific EXPANSIN A7 Is Required for Root Hair Elongation in Arabidopsis. *Mol. Cells* **2011**, *31*, 393–398. [CrossRef]

54. Wu, Y.; Thorne, E.T.; Sharp, R.E.; Cosgrove, D.J. Modification of Expansin Transcript Levels in the Maize Primary Root at Low Water Potentials. *Plant Physiol.* **2001**, *126*, 1471–1479. [CrossRef] [PubMed]

55. Xu, Q.; Xu, X.; Shi, Y.; Xu, J.; Huang, B. Transgenic Tobacco Plants Overexpressing a Grass PpEXP1 Gene Exhibit Enhanced Tolerance to Heat Stress. *PLoS ONE* **2014**, *9*, e100792. [CrossRef] [PubMed]

56. Goh, H.-H.; Sloan, J.; Dorca-Fornell, C.; Fleming, A. Inducible Repression of Multiple Expansin Genes Leads to Growth Suppression during Leaf Development. *Plant Physiol.* **2012**, *159*, 1759–1770. [CrossRef]

57. Yue, Y.; Shi, T.; Liu, J.; Tian, Q.; Yang, X.; Wang, L. Genomic, metabonomic and transcriptomic analyses of sweet osmanthus varieties provide insights into floral aroma formation. *Sci. Hortic.* **2022**, *306*, 111442. [CrossRef]

58. Bie, H.; Wang, H.; Wang, L.; Li, Y.; Fang, W.; Chen, C.; Wang, X.; Wu, J.; Cao, K. Mining Genes Related to Single Fruit Weight of Peach (*Prunus persica*) Based on WGCNA and GSEA. *Horticulturae* **2023**, *9*, 1335. [CrossRef]

59. Liu, M.; Diretto, G.; Pirrello, J.; Roustan, J.-P.; Li, Z.; Giuliano, G.; Regad, F.; Bouzayen, M. The chimeric repressor version of an Ethylene Response Factor (ERF) family member, Sl-ERF.B3, shows contrasting effects on tomato fruit ripening. *New Phytol.* **2014**, *203*, 206–218. [CrossRef]

60. Han, Y.C.; Kuang, J.F.; Chen, J.Y.; Liu, X.C.; Xiao, Y.Y.; Fu, C.C.; Wang, J.N.; Wu, K.Q.; Lu, W.J. Banana Transcription Factor MaERF11 Recruits Histone Deacetylase MaHDA1 and Represses the Expression of MaACO1 and Expansins during Fruit Ripening. *Plant Physiol.* **2016**, *171*, 1070–1084. [CrossRef]

61. Fu, C.C.; Han, Y.C.; Qi, X.Y.; Shan, W.; Chen, J.Y.; Lu, W.J.; Kuang, J.F. Papaya CpERF9 acts as a transcriptional repressor of cell-wall-modifying genes CpPME1/2 and CpPG5 involved in fruit ripening. *Plant Cell Rep.* **2016**, *35*, 2341–2352. [CrossRef] [PubMed]

62. Cao, K.; Zhao, P.; Zhu, G.; Fang, W.; Chen, C.; Wang, X.; Wang, L. Expansin genes are candidate markers for the control of fruit weight in peach. *Euphytica* **2016**, *210*, 441–449. [CrossRef]

63. Hayama, H.; Ito, A.; Moriguchi, T.; Kashimura, Y. Identification of a new expansin gene closely associated with peach fruit softening. *Postharvest Biol. Technol.* **2003**, *29*, 1–10. [CrossRef]

Article

Effects of Gibberellic Acid on Soluble Sugar Content, Organic Acid Composition, Endogenous Hormone Levels, and Carbon Sink Strength in Shine Muscat Grapes during Berry Development Stage

Xiujie Li [1], Zhonghui Cai [2], Xueli Liu [3], Yusen Wu [1], Zhen Han [1], Guowei Yang [1], Shaoxuan Li [4], Zhaosen Xie [2], Li Liu [1,*] and Bo Li [1,*]

[1] Shandong Academy of Grape, Shandong Academy of Agricultural Sciences, Jinan 250100, China; lixiujie-2007@163.com (X.L.); senwy886@163.com (Y.W.); yanhui65@163.com (Z.H.); guoweiyang66@163.com (G.Y.)
[2] College of Horticulture and Garden, Yangzhou University, Yangzhou 225009, China; caizhonghui2023@163.com (Z.C.); xiezhaosen@yzu.edu.cn (Z.X.)
[3] Taishan Institute of Science and Technology, Taian 271000, China; lily0538@163.com
[4] Qingdao Academy of Agricultural Sciences, Qingdao 266000, China; lsxsdau@163.com
* Correspondence: gssliuli@shandong.cn (L.L.); sdtalibo@163.com (B.L.)

Abstract: The phytohormone gibberellic acid (GA_3) is widely used in the table grape industry. However, there is a paucity of information concerning the effects of GA_3 on fruit quality and sink strength. This study investigated the effects of exogenous GA_3 treatments (elongating cluster + seedless + expanding, T1; seedless + expanding, T2; expanding, T3; and water, CK) on the content of sugars, organic acids, and endogenous hormones and sink strength. Results showed that T2 treatment displayed the highest fructose and glucose levels at 100 days after treatment (DAT), whereas its effect on tartaric acid, malic acid, and citric acid concentrations at 80 and 100 DAT was relatively weak. Under GA_3 treatments, GA_3, IAA, and CTK contents increased, whereas ABA content decreased at 1, 2, 4, 8, and 48 h. Analysis of sugar phloem unloading revealed that T2 treatment exhibited the highest values during softening and ripening stages. Our findings indicate that appropriate GA_3 application can positively influence sink strength by regulating sink size and activity, including berry size enlargement, sugar phloem unloading, and sugar accumulation in grape sink cells.

Keywords: gibberellin; grape; soluble sugar; organic acid; sugar unloading; sink strength

Citation: Li, X.; Cai, Z.; Liu, X.; Wu, Y.; Han, Z.; Yang, G.; Li, S.; Xie, Z.; Liu, L.; Li, B. Effects of Gibberellic Acid on Soluble Sugar Content, Organic Acid Composition, Endogenous Hormone Levels, and Carbon Sink Strength in Shine Muscat Grapes during Berry Development Stage. *Horticulturae* **2024**, *10*, 346. https://doi.org/10.3390/horticulturae10040346

Academic Editor: Stefano Poni

Received: 17 February 2024
Revised: 20 March 2024
Accepted: 20 March 2024
Published: 30 March 2024

1. Introduction

Gibberellin, a pivotal phytohormone that interacts synergistically with other plant hormones, functions as a central regulator of diverse developmental processes within plants. Gibberellic acid 3 (GA_3), a specific form of gibberellin, plays a critical role in promoting parthenocarpy and has garnered extensive attention for application in the cultivation of seedless fruits, augmentation of berry dimensions, and elongation of rachis across a wide spectrum of grape cultivars [1–4]. GA_3, in conjunction with auxin, plays a pivotal role in stimulating cell division and expansion. This interaction regulates fruit development and subsequent enlargement following fertilization [5]. Recently, the consumption of fruits has increased, owing to their high internal and external appearance quality. This trend is associated with economic growth and changing consumer preferences. To meet market demands, producing grapes with standardized cluster length and uniform and large berry size has become crucial [6]. Consequently, gibberellin finds widespread application in production.

Presently, a number of studies have demonstrated the efficacy of gibberellin as an effective agent in enhancing the elongation of grape inflorescence. Sun [7] reported that the application of gibberellin to Cabernet Franc grapes led to a proportional increase in peduncle length with an increase in gibberellin concentration. Similarly, Wang et al. [8] observed that grape varieties such as Midnight Beauty, Zaoheibao, and Summer Black exhibited elongated fruit pedicels following gibberellin treatments at a concentration of 10 mg/kg. Yang et al. [6] revealed that the application of exogenous GA_3 not only inhibited the synthesis of endogenous gibberellin, but also orchestrated the modulation of gibberellin signal transduction, thereby promoting inflorescence elongation.

Prior research has highlighted the utilization of GA_3 as a viable strategy in grape cultivation, particularly for seed management. Notably, the application of GA_3 solution at a concentration of 100 mg L^{-1} prior to full bloom has demonstrated its potential to produce seedless cultivars and prompt seed abortion in seeded cultivars [1]. Han and Lee [9] observed that GA_3 exhibited efficacy in not only fostering fruit enlargement, but also augmenting cluster length, cluster weight, and berry weight. Korkutal et al. [4] further substantiated the multifaceted impacts of gibberellins, revealing their role in promoting stem elongation, triggering flowering induction, stimulating pollen tube growth, yielding seedless fruits, and increasing the size of seedless berries. Zhao [10] expounded upon the benefits of employing GA_3 and streptomycin before the full-bloom stage, coupled with GA_3 application alone post this stage. This approach was found to notably increase berry size of "Shenxiangwuhe" and "12–17", concurrently facilitating seedlessness in the latter. For the grape variety "Zhuosexiang", the combined administration of GA_3 and forchlorfenuron (CPPU) before full bloom, followed by GA_3 application post full bloom, exhibited comparable benefits, yielding a remarkable seedless rate of 71%. Notably, GA_3 application emerged as the most effective approach in enhancing single berry weight and seedless rate, as underscored by Zhao et al. [10].

Sink strength, which has been elucidated as the competitive capacity of an organ to efficiently intake photoassimilates, can be considered as a product of two critical components: sink size and sink activity [11]. Sink size pertains to the physical limitation represented by the total biomass of the sink organ. Notably, the quantification of cells, both in terms of their number and size in the sink organ, can serve as a suitable metric for evaluating sink size. Sink activity is regarded as the physiological limitation that governs the import of assimilates into a sink organ. It is determined by three pivotal physiological processes. Firstly, it involves the unloading of assimilates from the phloem, followed by the subsequent transport of sugars beyond the phloem, leading to their absorption by the sink organ. Secondly, the sink organ's own respiratory consumption contributes to sink activity. Thirdly, the accumulation of carbohydrates within sink organs also influences sink activity [12]. Prior research has indicated that phytohormones can modulate sink strength. Notably, hormones such as GA_3, cytokinin (CTK), abscisic acid (ABA), and auxin have been recognized as participants in this regulatory process [13,14]. However, the underlying mechanism through which GA specifically influences sink strength in perennial fruit crops, including grapes, remains elusive.

Shine Muscat is a prominent table grape variety extensively cultivated in China. This variety is widely known for its excellent characteristics, such as large size, sweet taste, good storage capacity, and aromatic qualities; however, certain issues, such as the occurrence of seeded fruit, bitter seeds and pericarp, and tightly packed bunches, remain persistent. GA is widely employed to cultivate larger, seedless berries and enhance overall fruit quality. Previous investigations have demonstrated that GA can induce inflorescence elongation, foster the development of seedless fruits, and enhance the size of seedless berries in different grape varieties. Our previous research has also indicated that GA influences the aromatic components in Shine Muscat berries [15]. However, a research gap remains, particularly in understanding the interrelationship among the contents of soluble sugars, organic acids, and endogenous hormones and sink strength in Shine Muscat grapes subjected to GA_3 treatments. Therefore, the present study aimed to investigate how GA_3 influences the contents of soluble sugars and organic acids, preliminarily explore the interrelationship among these components, and clarify the mechanism underlying the fruit sink strength modulation by GA$_3$.

2. Materials and Methods

2.1. Field Conditions and Materials

The experiment was conducted at the Jinniushan vineyard in Tai'an, Shandong, China (36.127° N, 117.004° E). For the study, Shine Muscat (*Vitis labruscana*) grapes, which were self-rooted and planted in 2012, were used. The planting arrangement involved a plant density of 3.0 m^2 (with a spacing of 1.0 m between plants and 3.0 m between rows) within a rain shelter cultivation system. For the experimentation, a total of 30 healthy and well-developed vines were meticulously selected, with each treatment group comprising three replicates. Randomly selected branches with moderate vigor and a consistent number of leaves were chosen on the vines where one inflorescence on each branch was left and selected as the test object. Each inflorescence was trimmed one week before flowering to comprise only 5 cm of the apex. Trimmed inflorescences were assigned to one of three treatments or one control. The vines were subjected to GA$_3$ (ProGibb 40, Valent BioScience, Walnut Creek, CA, USA) treatment at varying concentrations and during distinct periods. The experimentation comprised four distinct treatment groups, including three GA$_3$ treatment groups and one control (CK) group, as detailed in Table 1.

Table 1. Experimental design of different gibberellin treatments.

No.	Treatment	Note
CK	Water	1. Rachis elongation treatment: twenty days prior to anthesis, inflorescences were immersed in a solution containing 5 mg/L GA$_3$ for 5 s. 2. Seedlessness treatment: a day after full bloom, inflorescences were immersed in a solution containing 25 mg/L GA$_3$ for 5 s.
T1	Rachis elongation treatment + Seedlessness treatment + Expansion treatment	3. Expansion treatment: within two weeks after bloom withering, inflorescences were immersed in a solution containing 30 mg/L GA$_3$ for 5 s.
T2	Seedlessness treatment + Expansion treatment	4. CK treatment: inflorescences were treated with water (treated at the same time as T1).
T3	Expansion treatment	

2.2. Tissue Collection

For each treatment, 30 berries were randomly chosen from 30 vines at 7:00 am every 14 days [14, 28, 42, 56, 70, 84, and 98 days after treatment (DAT)] for analysis of longitudinal diameter (LD) and transverse diameter (TD). For each treatment, 120 berries were randomly chosen from 30 vines on a regular basis at 7:00 am every 20 days [20, 40, 60, 80, and 100 days after treatment (DAT)] post expansion treatment. The samples were segregated into two groups. The samples from the first group, comprising 60 berries from each treatment, were collected and transported in an ice box to the laboratory. The berries were then analyzed for berry weight, soluble solids (SS), and titratable acidity (TA). The samples from the second group, comprising 60 berries from each treatment, were frozen in liquid nitrogen and immediately stored in a refrigerator at −80 °C. These samples were intended for the assessment of soluble sugars (glucose, fructose, and sucrose) and organic acids (tartaric acid, citric acid, and malic acid) in the grape berries. Moreover, samples comprising 30 berries with consistent fruit diameter, obtained from 30 berries of each treatment, were collected at specific intervals: 0, 1, 2, 4, 8, 24, and 48 h post expansion treatment. Following liquid nitrogen treatment, these samples were enveloped in tinfoil, refrigerated, and subsequently brought back to the laboratory. The samples were then stored in a low-temperature refrigerator (−80 °C) for the determination of indole-3-acetic acid (IAA), CTK, GA, and ABA contents in grape berries within one month.

2.3. Measurements of Berry Growth

The weight of individual fruits was determined using an electronic balance, with an accuracy of 0.01 g. For each analysis, a total of 30 berries were randomly selected from 30 clusters and each of the three replications. LD and TD of the fruit were measured using a vernier caliper. SS was assessed by utilizing an aliquot of grapevine juice with the aid of a digital refractometer (Automatic Refractometer SMART-1, Atago, Tokyo, Japan), which is denoted as Brix. Total acidity was quantified by an ATAGO (PAL-1) hand-held digital refractometer.

2.4. High-Performance Liquid Chromatography Analysis of Glucose, Fructose, and Sucrose

Sugars were extracted following the procedure: for each treatment, 1 g of homogenized grape berry material was accurately weighed and then diluted to 10.0 mL using ultrapure water (Millipore, Bedford, MA, USA). The solution was subsequently incubated for 20 min in a water bath set at 35 °C. Afterward, the supernatant was subjected to centrifugation at $21,000 \times g$ for 10 min at room temperature (BHG-Hermle Z 365, Wehingen, Germany). The extraction process was repeated three times, and the resulting supernatants were pooled. The liquid supernatant underwent filtration through a 0.22 μm, 13 mm diameter syringe filter (Shanghai Xingya Purification Material Factory, Shanghai, China). Subsequently, the filtered solution was employed for the analysis of glucose, sucrose, and fructose contents. High-performance liquid chromatography (HPLC; Waters, Milford, MA, USA) was employed for the analysis. The separation conditions used for the soluble sugar analysis were as follows: detector, differential refractive index detector (RID); column, YMC-Pack Polyamine II (4.6 mm × 250 mm); phase, acetonitrile/water = 75:25 (*v*/*v*); flow rate, 0.8 mL/min; injection amount, 10 μL; column temperature, 40 °C; analysis time, 20 min. The eluted peaks were detected utilizing an RID-10A differential refractive index detector (Shimadzu Co., Ltd., Kyoto, Japan). The quantification of glucose, fructose, and sucrose was conducted using standard curves of authentic compounds. Each treatment was replicated three times.

2.5. High-Performance Liquid Chromatography Analysis of Tartaric Acid, Citric Acid, and Malic Acid

Organic acid analyses were conducted utilizing a Waters series 515 chromatography unit, which was equipped with two 515 pumps and a 2487 dual UV detector (Waters Alliance 2695 HPLC) operating at a wavelength of 210 nm. The separation conditions used for organic acid analysis were as follows: column, Thermo Hypersil COLD aQ (4.6 mm × 250 mm, 5 μm); phase, 10 mmol/L $NH_4H_2PO_4$ (pH = 2.3)/methanol = 98/2 (*v*/*v*); flow rate, 0.8 mL/min; injection amount, 10 μL; column temperature, 25 °C; analysis time, 20 min. Quantification of tartaric acid, citric acid, and malic acid was performed using standard curves of authentic compounds. The analysis encompassed extracts from three replicate tissue samples.

2.6. Extraction and Determination of GA_3, IAA, CTK, and ABA

HPLC (Nexera LC-30A, Shimadzu, Japan) was employed for analysis of GA_3, IAA, CTK, and ABA. For each treatment, 5 g of grape berry was ground to powder under liquid nitrogen. After extraction, 50 mL of 80% (*v*/*v*) MeOH (methanol) and 50 μL of 30 mg/mL sodium diethyldithiocarbamate were added. After full oscillation, samples were kept at 0 °C overnight. After filtration, the residue was washed twice with 40 mL and 20 mL 80% methanol, respectively. The filtrate was dried at 40 °C on a rotary evaporator. The distillation bottle was flushed with 10 mL of petroleum ether and 10 mL of phosphoric acid buffer twice, and the flushed solution was passed through a 0.45 μm filter membrane. The water phase was decolorized 3 times with petroleum ether (equal volume). The pH value was adjusted to pH = 8, and the samples were extracted with ethyl acetate (equal volume) 3 times. Then, when pH was adjusted to pH = 3, the water phase was extracted with ethyl acetate (equal volume) 3 times, and the extract was dried at 40 °C. The extract was dissolved with 1 mL of 50% MeOH, filtered with 0.45 μm filter membrane, and 20 μL of the sample was taken for HPLC analysis. The standard hormone sample that was utilized was a product of Sigma Company. The chromatographic conditions were as follows: mobile phase, methanol: 0.8% glacial acetic acid solution = 55:45 (v/v); flow rate, 0.8 mL/min; injection amount, 10 μL; column temperature, 30 °C; detection wavelength, 254 nm. The analysis comprised extracts from three replicate tissue samples.

2.7. Analysis of Sugar Phloem Unloading

The sugar phloem unloading was assessed using the fruit cup method [16]. During the softening stage (August 19) and ripening stage (September 28), one sunward-facing grain in grape berries was randomly selected for each treatment. A cross of about 1 cm was delicately marked at the grape's umbilical region using a scalpel. This process ensured that the flesh remained unharmed and the grape flesh tissue was not cut. Subsequently, the peel was gently separated from the pedicel using tweezers and carefully cut with anatomical scissors. The prepared Mes buffer, comprising 5 mmol/L Mes (2-(n-morph) ethanesulfonic acid), 2 mol/L $CaCl_2$, 100 mmol/L mannitol (D-mannitol), and 0.2% (w/v) polyvinylpyrrolidone (PVPP), was then used to rinse the peel. The peeled grape fruit was meticulously placed into a 20 mm infusion tube filled with Mes buffer. The opening was sealed with a film, and the fruit cup was secured on the cluster. The outer layer of the fruit cup was covered with tin foil to prevent temperature fluctuations from affecting the test results and extending the "berry cup's" longevity. Sampling times were designated as 9:30–10:00, 11:30–12:00, 13:30–14:00, 15:30–16:00, and 17:30–18:00, during which the replaced buffer was discharged from the cup. Each time, the buffer liquid from the fruit cup was collected, and it was filled using a syringe. The collected samples were immediately frozen using liquid nitrogen and stored at −80 °C for subsequent analysis of glucose, fructose, and sucrose.

2.8. Statistical Analysis

Data analysis was performed using Microsoft Excel 2010 and SPSS 26.0. Graphical representations were created using Prism 9. Pearson correlation analysis was conducted using R statistical software (Version 3.0.3).

3. Results

3.1. Effects of GA_3 on Physicochemical Characteristics of Grape Berries

To investigate the effects of GA_3 on Shine Muscat grapes, various key attributes of grape berries during development were evaluated in both GA_3-treated and CK groups. These parameters included berry weight, berry size, LD, TD, vertical/horizontal ratio (a measure of fruit shape index), TA (g/L), and SS (Brix) of the berries. As illustrated in Figure 1A, the trend of changes in berry weight was analogous between GA_3-treated groups and the CK group. Notably, at each developmental stage, significantly higher berry weights were observed under T1, T2, and T3 treatments than those under the CK treatment. Furthermore, berry weights were substantially higher under T1 and T2 treatments than those under the T3 treatment. At 20–100 DAT, the berry size exhibited a comparable trend to the berry weight for both GA_3-treated groups and the CK group. The berry weight and volume were more rapidly increased under the T2 treatment than those under any other treatment (as shown in Figure 1A,B). Specifically, the berry weights were 11.3 g and 12.05 g in 2022 and 2021, respectively, marking a remarkable 52.2% (2022) and 42.8% (2021) increase, respectively, over those in the CK group at 100 DAT (Figure 1A). Similarly, the berry volumes were 10.59 cm^3 and 10.63 cm^3 in 2022 and 2021, respectively, representing a substantial 21.8% (2021) and 59.2% (2022) increase, respectively, compared to those in the CK group at 100 DAT (Figure 1B). Moreover, bunch compactness decreased in response to GA_3 treatments (T2 and T3 treatments). The berry size was significantly higher in T1, T2, and T3 treatment groups than in the CK group (Figure 1C), which suggests an expansion effect of GA_3.

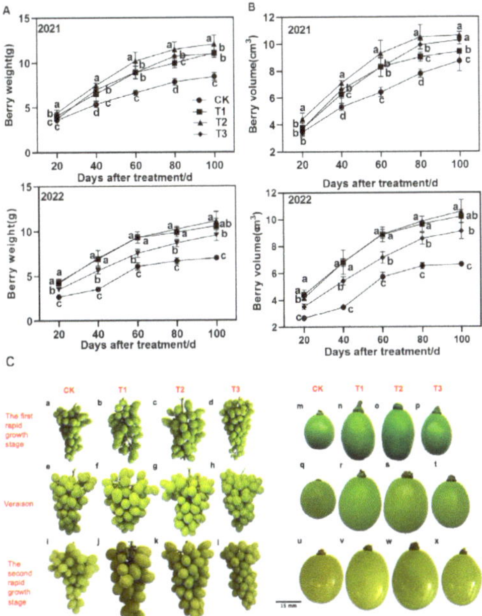

Figure 1. Berry weight (**A**) and volume (**B**) in 2021 and 2022, and berry growth at three stages: hard-core stage, softening stage, and harvest stage (**C**) of "Shine Muscat" under various GA$_3$ treatments. a–x: berry corresponding to respective treatments and periods. Different letters denote statistically significant differences among treatments at the same period, as determined by Duncan's test ($p < 0.05$).

Moreover, no significant increase in TD was observed between GA$_3$-treated groups and the CK group (Figure 2A). In contrast, LD was substantially higher in the three GA$_3$-treated groups than in the CK group at 20–100 DAT (Figure 2B). However, similar LD values were observed for T1 and T2 treatment groups. The fruit shape index was evidently higher in T1 and T2 treatment groups than in the other groups, with that in the CK group being the lowest. At 21–70 DAT, a slight increase in fruit shape index in the Tl group was evident compared to that in the T2 group (Figure 2C).

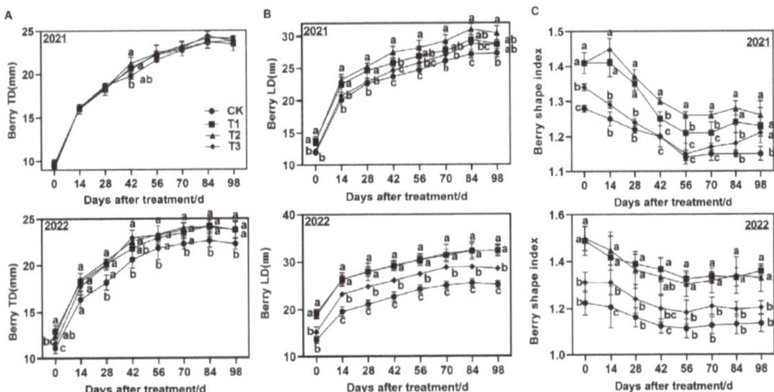

Figure 2. Transverse diameter (**A**), longitudinal diameter (**B**), and berry shape index (**C**) in 2021 and 2022 of "Shine Muscat" under various GA$_3$ treatments. Different letters denote statistically significant differences among treatments at the same period, as determined by Duncan's test ($p < 0.05$).

The SS content in grape berries is illustrated in Figure 3A. It displayed an increasing trend as the fruits ripened. In 2021, at 20 DAT, remarkably higher SS contents were observed in GA-treated groups than those in the CK group. At 40 DAT, the SS content demonstrated the following trend: T2 > T3 > T1 > CK. At 60, 80, and 100 DAT, significantly higher SS contents were observed under T1 and T2 treatments than those under T3 and CK treatments. Moreover, in 2022, SS contents increased across all treatments as the fruits matured. At 60 DAT, a significant difference in SS content was observed between the T3 and CK groups. However, at 20, 40, 80, and 100 DAT, no significant difference was observed between GA-treated groups and the CK group.

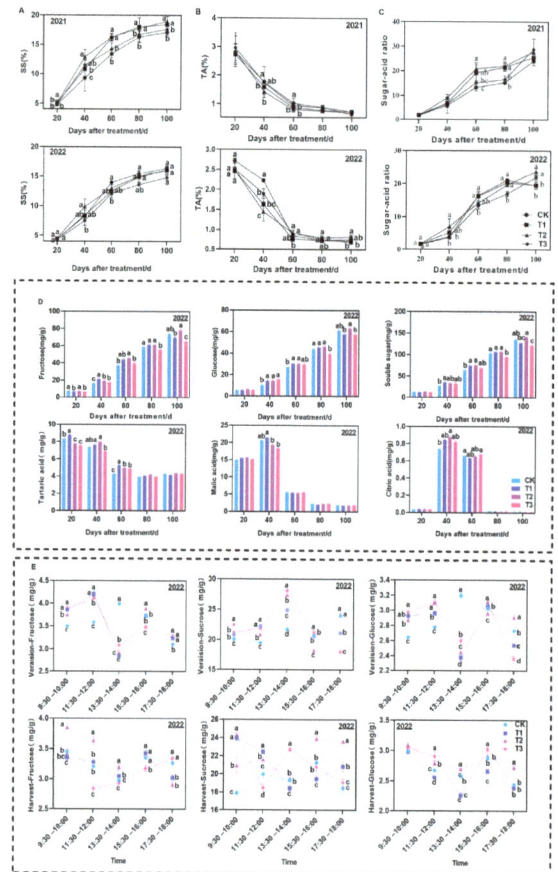

Figure 3. Soluble solid (SS) content (**A**), titratable acidity (TA) (**B**), and SS/TA ratio (**C**) in 2021 and 2022. Contents of soluble sugars and organic acids (**D**), and sugar phloem unloading (**E**) in "Shine Muscat" grapes during veraison and harvest stages under various gibberellin treatments in "Shine Muscat" grapes. Different letters denote statistically significant differences among treatments at the same period, as determined by Duncan's test ($p < 0.05$).

An evident declining trend in TA during fruit development was observed, particularly at 20–60 DAT (Figure 3B). In 2021, no significant differences were observed among the treatments at 20, 80, and 100 DAT. However, at 40 DAT, CK and T3 groups exhibited higher TA compared to T1 and T2 groups. At 60 DAT, the CK group displayed the highest TA among the treatments. In 2022, no significant differences in TA were recorded among the treatments at 20 DAT. At 40 DAT, the overall trend observed was as follows:

CK > T3 > T1 > T2, with TA in the CK group being significantly higher than that in GA-treated groups. At 100 DAT, a significant difference in TA was observed under the T3 and T1 (or T2) treatments.

The SS/TA ratio demonstrated a consistent increasing trend during the course of fruit development. In 2021, varying GA treatments yielded minimal influence on the SS/TA ratio across different time periods. However, at 100 DAT, the SS/TA ratio in CK, T1, and T2 groups was significantly higher than that in the T3 group; however, no significant differences were observed among CK, T1, and T2 groups (Figure 3C).

3.2. Effect of GA_3 on Soluble Sugar Content in Grape Berries

As illustrated in Figure 3D, the soluble sugar composition of grape berries primarily consisted of glucose and fructose. Notably, the sucrose content was negligible. The glucose and fructose contents within all four groups exhibited a consistent increasing trend; however, throughout the hanging life, the fructose content consistently surpassed that of glucose. The increase in total sugar content was significantly higher at 40–60 DAT than during the maturation stage (80–100 DAT). Prior to reaching 60 DAT, the total sugar content in T1 and T2 treatment groups was significantly higher than that in the CK group. Moreover, the total sugar content in the T3 group was the lowest, at 80 DAT, among all four treatment groups. During the ripening stage, the T2 group exhibited the highest total sugar content among all four treatment groups.

The fructose and glucose contents within each treatment group gradually increased throughout the hanging life of the fruit. Notably, the changes in glucose and fructose contents in fruits at 80 DAT closely paralleled those observed in the total sugar content. Upon attaining complete maturation, the contents of fructose and glucose in the T2 treatment group exceeded those observed in other treatment groups (glucose: 78.66 $\pm$ 1.30 mg/g; fructose: 63.49 $\pm$ 1.03 mg/g). Notably, among all groups, the lowest glucose (65.36 $\pm$ 1.40 mg/g) and fructose (57.58 $\pm$ 1.01 mg/g) contents were observed in the T3 treatment group.

3.3. Effect of GA_3 on Organic Acid Content in Grape Berries

As illustrated in Figure 3D, the primary organic acids present in grape berries included tartaric acid, malic acid, and citric acid, with citric acid content being relatively minimal. During the period of hanging life, we noted a distinctive V-shaped trend in the tartaric acid content, which peaked at 20 DAT and plummeted at 80 DAT across all treatments. Prior to 60 DAT, significant differences were observed in the tartaric content among all four treatments; however, these differences were not evident at 100 DAT. At 100 DAT, we observed the lowest and highest tartaric acid contents under T1 (4.31 $\pm$ 0.08 mg/g) and T2 (4.38 $\pm$ 0.02 mg/g) treatments, respectively.

The malic acid content in grape berries initially increased and then decreased over the course of hanging life. Notably, a significant difference among the four treatments was evident at 40 DAT alone. During the maturation phase, the lowest malic acid content was observed under the T1 treatment (1.83 mg/g), which was 8.74% lower than that observed under the CK treatment.

The change trends in citric acid content paralleled those of malic acid content. At 40 DAT, the citric acid content in all groups reached its peak during the hanging life. Notably, significant differences in citric acid content were observed across all groups at 40 and 60 DAT. At 100 DAT, the lowest citric acid content was noted under the T2 treatment (0.012 mg/g), which was 8.33% lower than those observed under the T1 and T3 treatments.

3.4. Analysis of Sugar Unloading in Phloem

Our results demonstrated that the primary sugars in grape berries were fructose and glucose, with relatively minimal sucrose content at the veraison stage. Notably, the fructose content exceeded that of glucose. As depicted in Figure 3E, the changes in fructose and glucose unloading in the phloem exhibited a double-peak curve at five distinct time points following GA_3 treatment. In the morning, sugar unloading increased rapidly, reaching its peak at 12:00, closely aligned with the increase in photosynthetic rate. Subsequently, a sharp decline was observed at 14:00. Furthermore, a resurgence in sugar unloading was observed with an increase in photosynthetic rate, reaching a minor peak at 16:00, followed by a gradual decline at sunset. Notably, the morning unloading volume was approximately 1.1 times that of the afternoon unloading. However, the changes in fructose and glucose unloading in the phloem exhibited a unimodal curve at the five time points under the CK treatment. In the morning, sugar unloading increased rapidly, reaching its peak at 14:00, followed by a rapid decrement.

The maximum fructose unloading (3.89 mg/g) was observed between 9:30 and 10:00 in the T3 treatment fruits, significantly surpassing that observed in the CK treatment fruits (3.51 mg/g). For the T1 treatment, the peak fructose unloading (4.21 mg/g) occurred between 11:30 and 12:00. Similarly, the peak fructose unloading for the CK treatment reached 4.00 mg/g at 13:30–14:00 and that for the T2 treatment reached 3.89 mg/g and 3.30 mg/g at 15:30–16:00 and 17:30–18:00, respectively.

Furthermore, the peak glucose unloading for the T3 treatment occurred at 9:30–10:00 and 11:30–12:00, measuring 2.98 mg/g and 3.11 mg/g, respectively. For the CK treatment, the highest glucose unloading occurred at 13:30–14:00, following a pattern similar to that of fructose. Notably, the T3 treatment exhibited the highest glucose unloading (3.11 mg/g) at 15:30–16:00. In addition, at 17:30–18:00, glucose unloading was significantly higher in the T2 group than that observed in other groups. Overall, these findings indicate that the highest sugar (fructose+glucose) unloading (32.25 mg/g) was observed in the T2 treatment group during the veraison stage.

Similarly, in vivo analysis of sugar unloading at the maturation stage resembled that at the veraison stage. The primary sugars in grape berries were fructose and glucose, with relatively minimal amounts of sucrose owing to incomplete transformation. As illustrated in Figure 3E, the changes in fructose and glucose unloading in the phloem exhibited a unimodal curve at each time point during the maturation stage. Sugar unloading reached its peak at 10:00 in the morning, gradually declining to a minimum at 14:00. Subsequently, sugar unloading increased again with an increase in photosynthetic rate, achieving a minor peak at 16:00 and, finally, decreasing at sunset. The unloading value observed in the morning was one fold higher than that observed in the afternoon. Among the treatments, the T2 treatment demonstrated the highest fructose unloading at the first four time points, while the highest unloading under the T3 treatment was observed at 17:30–18:00. Moreover, the maximum glucose unloading under the T2 treatment was evident at 9:30–10:00 (3.09 mg/g), 13:30–14:00 (2.92 mg/g), and 17:30–18:00 (2.72 mg/g). The T3 treatment group exhibited the highest glucose unloading at 11:30–12:00 (2.91 mg/g) and 15:30–16:00 (3.03 mg/g).

Based on the sugar unloading data at various time points for each treatment, the T2 treatment displayed the highest sugar unloading (fructose + glucose) at the maturation stage, with a value of 31.26 mg/g.

3.5. Effect of GA₃ on Endogenous Hormone Contents in Grape Berries

Four types of endogenous hormones—GA, ABA, IAA, and CTK—were analyzed. The GA content demonstrated a fluctuating trend at 0–48 h after GA$_3$ expansion treatment. It reached a peak at 0–4 h, followed by a sharp decline and a rapid increase at 24 h. In contrast, the GA content under the CK treatment exhibited a gradual increase at 48 h. Notably, the overall trend in GA content at 1–24 h was as follows: T3 > T2 > T1 > CK (Figure 4A). The ABA contents were significantly higher in CK treatment fruits than those in fruits subjected to the GA treatment. The lowest and highest ABA contents in berries were observed at 4 h and 24 h after GA treatment, respectively. At 48 h, the overall trend in ABA content was as follows: CK > T1 > T3 > T2 (Figure 4B). The IAA content in fruits displayed a trend similar to that observed for GA. The highest IAA content in berries under the T1 treatment was observed at 8 h, and at 48 h, the following overall trend was noted: T3 > T2 > T1 > CK (Figure 4C). The CTK content in berries increased rapidly at 0–2 h under CK, T2, and T3 treatments, followed by a slower upward trend that gradually decreased at 4 h. Moreover, the CTK content showed an upward trend under the T1 treatment at 0–8 h. At 48 h, the following trend was observed: T2 > T3 > T1 > CK (Figure 4D).

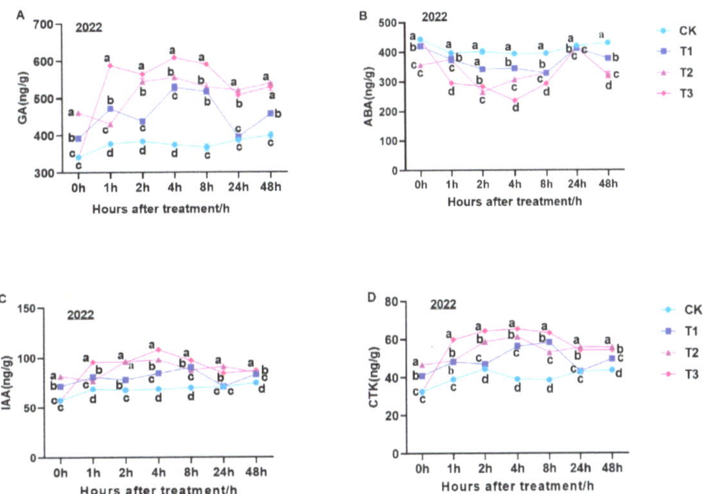

Figure 4. Contents of endogenous hormones—gibberellic acid (GA$_3$) (**A**), abscisic acid (ABA) (**B**), indole-3-acetic acid (IAA) (**C**), and cytokinin (CTK) (**D**)—in "Shine Muscat" grapes under various GA$_3$ treatments. Different letters denote statistically significant differences among treatments at the same period, as determined by Duncan's test ($p < 0.05$).

3.6. Correlation Analysis

The correlation analysis depicted in Figure 5 examines the relationships among the contents of endogenous GA, soluble sugars, organic acids, and key parameters, namely, berry weight, berry shape index, TD, LD, volume, SS, and TA. Through correlation analysis, we observed a significant positive correlation between the GA content and the key parameters, including berry weight, berry shape index, TD, LD, and volume, particularly with TD. Furthermore, a significant negative correlation was noted between GA and SS contents at 60, 80, and 100 DAT, as well as between the GA content and TA at 20 and 40 DAT. In contrast, a significant positive correlation was observed between the GA content and TA at 60 and 80 DAT (Figure 5A).

The contents of reducing sugars, glucose, and fructose in fruits displayed a significant positive correlation with GA contents at 40 and 60 DAT (Figure 5B); however, a significant negative correlation between them was observed at 80 and 100 DAT. Notably, the GA content was significantly negatively correlated with tartaric acid content at 20 DAT; malic acid content at 40, 60, 80, and 100 DAT; and citric acid content at 80 and 100 DAT (Figure 5B). In contrast, the GA content was significantly positively correlated with tartaric acid content at 60, 80, and 100 DAT; malic acid content at 20 DAT; and citric acid content at 20, 40, and 60 DAT (Figure 5B).

The heatmap presented in Figure 5C indicates that IAA and CTK contents were significantly positively correlated with GA$_3$ contents at 1, 2, 4, 8, 24, and 48 h. Furthermore, the correlation analysis highlighted that ABA contents were significantly negatively correlated with GA$_3$ contents and IAA and CTK contents at 1, 2, 4, 8, 24, and 48 h. Notably, a significant positive correlation was observed between IAA and CTK contents.

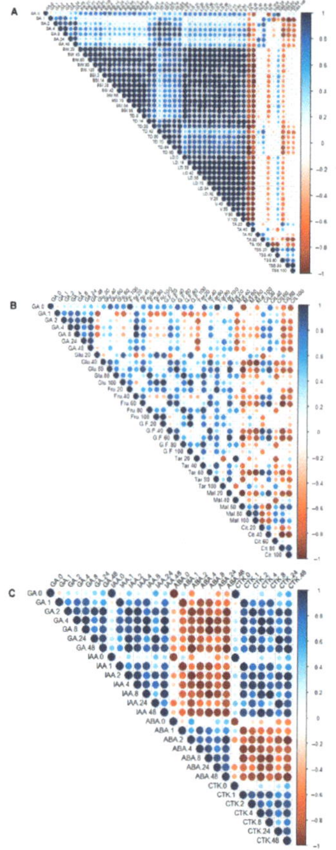

Figure 5. Correlation analysis of GA_3 content with key parameters (**A**), soluble sugar and organic acid contents (**B**), and endogenous hormone contents (**C**) from grapes under various GA_3 treatments.

4. Discussion

4.1. GA_3 Application by Cluster Dipping at Different Periods Results in Berry Size Enlargement in Grapes

Our study conclusively showed that GA_3 effectively enhanced the size of Shine Muscat berries. Intriguingly, we found that the effect of GA_3 on LD was more significant than that on TD. Subsequently, we conducted additional analyses to measure berry weight and volume in order to investigate the enlargement effect. As anticipated, GA_3 significantly enhanced both berry weight and volume, particularly under the seedlessness + expansion treatment (T2 treatment). This observation potentially indicates that the concentrations of compounds, such as sugars and acids, might have been regulated through expansion dilution.

4.2. GA_3 Application Alters the Concentrations of Sugars and Acids

To verify the hypothesis mentioned earlier, we conducted analyses of soluble sugar and organic acid contents. The application of GA led to a significant increase in SS content in berries during 2021, which is consistent with the findings of Tyagi et al. [17] in Sangiovese grapes at 79 d. However, this effect was not observed during 2022, consistent with previous findings for Sangiovese grapes at 49 d [17]. We further observed lower TA in GA-treated berries than that in the untreated control at 40 and 60 DAT in 2021 and 40 DAT in 2022. This finding aligns with that of Gao et al. [2] in Cabernet Sauvignon grapes. However, no significant difference in TA was observed at 80 and 100 DAT in

2021 and 80 DAT in 2022, aligning with the findings of previous research on Sangiovese grapes [17]. In commercial settings, the SS/TA ratio is considered the most reliable indicator of fruit flavor [18]. This study demonstrates that GA-treated berries [rachis elongation (28/4/2022) + seedlessness (24/5/2022) + expansion treatment (7/6/2022), or seedlessness (24/5/2022) + expansion treatment (7/6/2022)] exhibited a higher SS/TA ratio at 80 DAT in both 2021 and 2022, suggesting improved berry flavor. Subsequently, we investigated the effect of GA application on soluble sugar and organic acid contents in fruits to thoroughly explore the potential existence of the expansion dilution effect.

It is widely accepted that soluble sugars in grape berries primarily consist of glucose and fructose. Although some studies have reported the presence of sucrose in certain table grape varieties [19], sucrose has not been detected in Cabernet Sauvignon grapes [6]. In the present study, traces of sucrose were detectable, albeit in limited quantities. Monosaccharides serve as effective osmotica in plants. Herein, GA treatment [seedlessness (24/5/2022) + expansion treatment (7/6/2022)] increased the accumulation of monosaccharides (fructose and glucose) in grape berries at 100 DAT. However, berries treated with expansion alone exhibited the lowest values, suggesting an effect of expansion dilution. This phenomenon may be influenced by other phytohormones mediated by GA or from the competition for photosynthates due to seed development. Our findings indicate that appropriate GA$_3$ application (T2 treatment) can enhance sugar accumulation, overcoming the potential impact of expansion dilution.

4.3. Interactions between Signaling Pathways of GA$_3$ and Other Phytohormones Contribute to the Coordinated Development of Berries

GA, in conjunction with IAA and CTK, plays a role in promoting cell division and expansion, leading to increased cell growth and protein synthesis [20]. Therefore, these hormones coordinate the development and enlargement of fruits following fertilization. Notably, these three plant hormones, along with ABA, have significant regulatory functions in the establishment and maintenance of fruit sink strength [13]. The application of exogenous hormones could potentially trigger changes in endogenous hormone levels in plants [21]. Therefore, in this study, we assessed alterations in the levels of four endogenous hormones (GA$_3$, CTK, IAA, and ABA). Our findings indicate that GA$_3$ application led to increased levels of IAA and CTK compared with those in the CK group, while also exerting an inhibitory effect on ABA production. The correlation analysis unveiled a significant positive correlation between GA$_3$ and IAA/CTK contents and a significant negative correlation between GA$_3$ and ABA contents. These results, in conjunction with prior research, suggest a potential crosstalk between GA and other plant hormones, such as auxin and CTK [22]. Interestingly, the growth of grape berries induced by auxin and CK treatment is also partially dependent on GA biosynthesis [23], underscoring the significance of the synergy between these plant hormones. GA$_3$ has been substantiated to enhance fruit size by promoting cell division and expansion, contributing to the mechanisms of cell wall formation and relaxation [24]. Previous studies have revealed that auxin modulates grape berry ripening, which could be associated to cell expansion [25,26]. Numerous studies have also provided support for the notion that CTK could govern critical rate-limiting steps in nutrient distribution and utilization [27,28]. These findings collectively suggest the significant contributions of GA, IAA, and CTK in modulating fruit sink strength.

4.4. GA$_3$ Application Increases the Sink Strength in Grapes

The application of gibberellin can lead to an increase in auxin content. Gibberellin, along with auxin and CTK, plays a role in nutrient transport and attraction, ensuring that the fruit becomes a robust "reservoir" and is positioned favorably in the nutritional competition. Sink organs in plants serve as net importers of assimilates. Throughout various stages of plant development, all plant organs can function as sinks, thereby receiving assimilates. In terms of the transport of assimilates, the capacity of a sink organ to import assimilates is referred to as its sink strength. However, a substantial proportion of the imported assimilates can be utilized for respiratory processes within the sink organs. Consequently, measuring the sink strength of a sink organ solely based on parameters such as the absolute growth rate or net accumulation rate of dry matter falls short of accurately gauging the actual capacity of a sink organ to receive assimilates. This conventional approach instead represents an apparent sink strength. Measuring the import rate of assimilates, which takes into account the sum of the net carbon gain and the carbon loss through respiration within a sink organ, would provide a more accurate estimate of the true sink strength. In addition, CTK has the potential to enhance sink capacity by promoting cell proliferation or sustaining sink activity through the regulation of sucrolytic enzymes, thereby allowing the acquisition of more photoassimilates.

Sink strength is generally defined as the competitive ability of an organ to attract assimilates, represented by the product of sink size and sink activity [29]. In various fruit crops, including grapes, GA has been shown to increase fruit size by stimulating cell division and elongation [30]. Our study demonstrated that GA_3 application significantly increased berry size, berry weight, LD, and TD. This suggests that GA_3 can bolster sink strength by actively contributing to the enhancement of sink size, encompassing both cell number and size. This augmentation, in turn, ensures the accrual of a greater amount of photoassimilates and dry matter within the sink structure. Similar findings were observed in pear (*Pyrus pyrifolia*), where GA treatment led to a significant increase in sink demand and fruit size [31]. The transport of photosynthates occurs through a permeation-driven loading or unloading process within the phloem, which is governed by the pressure gradient that exists between the source and the phloem reservoir. Our findings demonstrate that an appropriate application of GA_3 can expedite phloem unloading, resulting in enhanced sugar unloading in berry fruits during both the veraison and maturation stages. This enhancement implies an increase in sink activity. Apart from sugar translocation, the accumulation of sugars stands as a pivotal determinant of sink strength. Our study shows that GA_3 application leads to the accumulation of substantial concentrations of glucose and fructose in berry fruits. This aligns with the findings of previous studies [32,33], which collectively support the notion that GA_3 can affect sink strength by expediting the accumulation of hexose sugars. These findings imply that the phytohormone GA_3 actively participates in the regulation of sink size and activity, encompassing an increase in berry size, sugar phloem unloading, and sugar accumulation in the sink cells. Altogether, these mechanisms significantly contribute to the modulation of sink strength in grapes (as depicted in Figure 6).

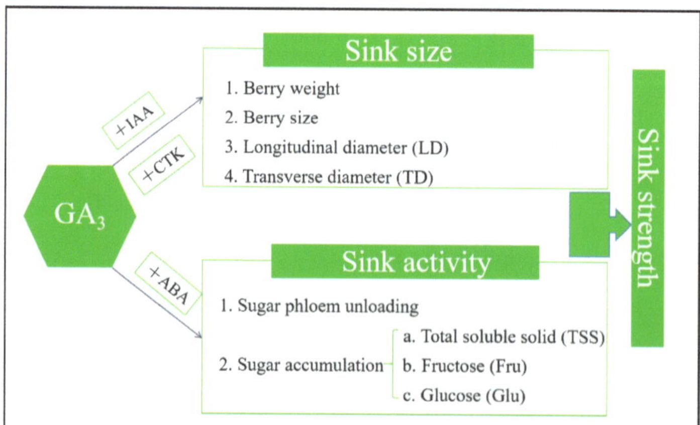

Figure 6. Mechanism through which GA_3 regulates sink strength via modulation of sink size and activity, encompassing an increase in berry size, sugar phloem unloading, and sugar accumulation in the sink cells in grapes.

5. Conclusions

Collectively, this study introduced two significant and novel findings that contribute to our understanding of grape physiology and growth regulation: (i) GA$_3$ exerted significant effects on the contents of soluble sugars, organic acids, and endogenous hormones (IAA, CTK, and ABA). The application of GA$_3$ led to enhanced sugar unloading during the softening and ripening stages. (ii) The appropriate application of GA$_3$ plays a crucial role in orchestrating the modulation of sink size and activity, including the enhancement of berry size, the facilitation of sugar phloem unloading, and the accumulation of sugars within sink cells. These insights collectively exert a robust effect on the overall sink strength in grape development. Moreover, these novel findings significantly enhance our comprehension of the complex interplay between GA$_3$ and soluble sugar, organic acid, and endogenous hormone contents. This study presents compelling empirical evidence that contributes to the broader body of knowledge supporting effective strategies for table grape cultivation and the proficient utilization of GA$_3$, ultimately benefiting the grape industry in China.

Author Contributions: Conceptualization, X.L. (Xiujie Li), Z.X., L.L. and B.L.; methodology, X.L. (Xiujie Li), Z.C. and Z.X.; sample preparation and analysis, Z.C.; investigation, X.L. (Xueli Liu) and G.Y.; data curation, Z.C. and S.L.; writing—original draft preparation, L.L.; writing—review and editing, B.L.; Supervision, X.L. (Xueli Liu), Y.W. and Z.H.; project administration, B.L.; funding acquisition, Z.H., L.L. and B.L. All authors have read and agreed to the published version of the manuscript.

Funding: This research was funded by the Natural Science Foundation of Shandong Province (Grant No. ZR2023MC101, ZR2021QC165, and ZR2022QC185), Agricultural Science and Technology Innovation Project of Shandong Academy of Agricultural Sciences (Grant No. CXGC2023A47), Agricultural Variety Project of Shandong Province—New Varieties Grape Cultivation of High Quality with Characteristic and Industrialization Technology Innovation and Promotion (Grant No. 2022LZGCQY019), Collaborative Extension Project of Major Agricultural Technology of Shandong Province (Grant No. SDNYXTTG-2023-18), Key R&D Program of Shandong Province (Grant No. 2022TZXD0011), and National Natural Science Foundation of China (32202462).

Data Availability Statement: The original contributions presented in the study are included in the article, further inquiries can be directed to the corresponding authors.

Conflicts of Interest: The authors declare no conflicts of interest.

References

1. Cheng, C.; Jiao, C.; Singer, S.D.; Gao, M.; Xu, X.; Zhou, Y.; Li, Z.; Fei, Z.; Wang, Y.; Wang, X. Gibberellin-induced changes in the transcriptome of grapevine (*Vitis labrusca* × *V. vinifera*) cv. Kyoho flowers. *BMC Genom.* **2015**, *16*, 128. [CrossRef] [PubMed]
2. Gao, X.T.; Wu, M.H.; Sun, D.; Li, H.Q.; Chen, W.K.; Yang, H.Y.; Liu, F.Q.; Wang, Q.C.; Wang, Y.Y.; Wang, J.; et al. Effects of gibberellic acid (GA$_3$) application before anthesis on rachis elongation and berry quality and aroma and flavour compounds in *Vitis vinifera* L. 'Cabernet Franc' and 'Cabernet Sauvignon' grapes. *J. Sci. Food Agric.* **2020**, *100*, 3729–3740. [CrossRef]
3. Grimplet, J.; Tello, J.; Laguna, N.; Ibáñez, J. Differences in flower transcriptome between grapevine clones are related to their cluster compactness, fruitfulness, and berry size. *Front. Plant Sci.* **2017**, *8*, 632. [CrossRef] [PubMed]
4. Korkutal, I.; Bahar, E.; Gokhan, O. The characteristics of substances regulating growth and development of plants and the utilization of gibberellic acid (GA$_3$) in viticulture. *World J. Agric. Sci.* **2008**, *4*, 321–325.
5. He, H.; Yamamuro, C. Interplays between auxin and GA signaling coordinate early fruit development. *Hortic. Res.* **2022**, *9*, uhab078. [CrossRef]
6. Yang, L.L.; Niu, Z.Z.; Wei, J.G.; Zhao, Y.Z.; Chen, Z.; Xuan, L.F.; Chu, F.J. Effects of exogenous GA$_3$ on inflorescence elongation and gibberellin anabolism related gene expression in grapevine. *Mol. Plant Breed.* **2020**, *18*, 5600–5606.
7. Sun, D. The Effects of Peduncle and Berry Quality with Gibberellin Treatments in the Cabernet Franc Grape (*Vitis vinifera* L. cv.). Master's Thesis, Northeast Forestry University, Harbin, China, 2017.
8. Wang, M.; Huang, L.P.; Liu, X.T.; Zhao, Q.F.; Ma, X.H. Effects of gibberellin on grape inflorescence elongation and berry growth treatment before flowering. *Anhui Nongxue Tongbao* **2018**, *24*, 35–36.
9. Han, D.H.; Lee, C.H. The effects of GA$_3$, CPPU and ABA applications on the quality of kyoho (*Vitis vinifera* L. × *V. Labrusca* L.) grape. *Acta Hortic.* **2004**, *653*, 193–197. [CrossRef]
10. Zhao, J. The Influence of GA3, CPPU and Streptomycin on the Denuclearization and Quality of Grape. Master's Thesis, Shenyang Agricultural University, Shenyang, China, 2018.

11. Ho, L.C. Metabolism and compartmentation of imported sugars in sink organs in relation to sink strength. *Annu. Rev.* **1988**, *39*, 355–378. [CrossRef]
12. Roitsch, T. Source-sink regulation by sugar and stress. *Curr. Opin. Plant Biol.* **1999**, *2*, 198–206. [CrossRef]
13. Li, Y.M.; Forney, C.; Bondada, B.; Leng, F.; Xie, Z.S. The molecular regulation of carbon sink strength in grapevine (*Vitis vinifera* L.). *Front. Plant Sci.* **2021**, *11*, 606918. [CrossRef] [PubMed]
14. Ma, Q.J.; Sun, M.H.; Lu, J.; Liu, Y.J.; Hu, D.G.; Hao, Y.J. Transcription factor AREB2 is involved in soluble sugar accumulation by activating sugar transporter and amylase genes. *Plant Physiol.* **2017**, *174*, 2348–2362. [CrossRef] [PubMed]
15. Wu, Y.; Li, X.; Zhang, W.; Wang, L.; Li, B.; Wang, S. Aroma profiling of Shine Muscat grape provides detailed insights into the regulatory effect of gibberellic acid and N-(2-chloro-4-pyridinyl)-N-phenylurea applications on aroma quality. *Food Res. Int.* **2023**, *170*, 112950. [CrossRef]
16. Wang, Z.P.; Carbonneau, A.; Deloir, A. A new method for consecutively measuring the accumulation of sugar in grape fruit. *J. Fruit Sci.* **2006**, *23*, 770–773. [CrossRef]
17. Tyagi, K.; Maoz, I.; Lapidot, O.; Kochanek, B.; Butnaro, Y.; Shlisel, M.; Lerno, L.; Ebeler, S.; Lichter, A. Effects of gibberellin and cytokinin on phenolic and volatile composition of Sangiovese grapes. *Sci. Hortic.* **2022**, *295*, 110860. [CrossRef]
18. Parson, S.; Paull, R.E. Pineapple organic acid metabolism and accumulation during fruit development. *Sci. Hortic.* **2007**, *112*, 297–303. [CrossRef]
19. Shiraishi, M.; Shinomiya, R.; Chijiwa, H. Preliminary genetic analysis of sucrose accumulation in berries of table grapes. *Sci. Hortic.* **2012**, *137*, 107–113. [CrossRef]
20. Spartz, A.K.; Lee, S.H.; Wenger, J.P.; Gonzalez, N.; Itoh, H.; Inzé, D.; Peer, W.A.; Murphy, A.S.; Overvoorde, P.J.; Gray, W.M. The SAUR19 subfamily of SMALL AUXIN UP RNA genes promote cell expansion. *Plant J.* **2012**, *70*, 978–990. [CrossRef]
21. El-Mergawi, R.A.; Abd El-Wahed, M.S.A. Effect of exogenous salicylic acid or indole acetic acid on their endogenous levels, germination, and growth in maize. *Bull. Natl. Res. Cent.* **2020**, *44*, 167. [CrossRef]
22. Wang, W.; Bai, Y.; Koilkonda, P.; Guan, L.; Zhuge, Y.; Wang, X.; Liu, Z.; Jia, H.; Wang, C.; Fang, J. Genome-wide identification and characterization of gibberellin metabolic and signal transduction (GA MST) pathway mediating seed and berry development (SBD) in grape (*Vitis vinifera* L.). *BMC Plant Biol.* **2020**, *20*, 384. [CrossRef]
23. Lu, L.; Liang, J.; Zhu, X.K.; Li, T.Z.; Hu, J.F. Auxin- and cytokinin-induced berries set in grapevine partly rely on enhanced gibberellin biosynthesis. *Tree Genet.* **2016**, *12*, 41. [CrossRef]
24. Upadhyay, A.; Maske, S.; Jogaiah, S.; Kadoo, N.Y.; Gupta, V.S. GA3 application in grapes (*Vitis vinifera* L.) modulates different sets of genes at cluster emergence, full bloom, and berry stage as revealed by RNA sequence-based transcriptome analysis. *Funct. Integr. Genom.* **2018**, *18*, 439–455. [CrossRef] [PubMed]
25. Wong, D.C.; Lopez Gutierrez, R.; Dimopoulos, N.; Gambetta, G.A.; Castellarin, S.D. Combined physiological, transcriptome, and cis-regulatory element analyses indicate that key aspects of ripening, metabolism, and transcriptional program in grapes (*Vitis vinifera* L.) are differentially modulated accordingly to fruit size. *BMC Genom.* **2016**, *17*, 416. [CrossRef] [PubMed]
26. Dal Santo, S.; Tucker, M.R.; Tan, H.T.; Burbidge, C.A.; Fasoli, M.; Böttcher, C.; Boss, P.K.; Pezzotti, M.; Davies, C. Auxin treatment of grapevine (*Vitis vinifera* L.) berries delays ripening onset by inhibiting cell expansion. *Plant Mol. Biol.* **2020**, *103*, 91–111. [CrossRef] [PubMed]
27. Body, M.; Kaiser, W.; Dubreuil, G.; Casas, J.; Giron, D. Leaf-miners co-opt microorganisms to enhance their nutritional environment. *J. Chem. Ecol.* **2013**, *39*, 969–977. [CrossRef] [PubMed]
28. Werner, T.; Holst, K.; Pörs, Y.; Guivarc'h, A.; Mustroph, A.; Chriqui, D.; Grimm, B.; Schmülling, T. Cytokinin deficiency causes distinct changes of sink and source parameters in tobacco shoots and roots. *J. Exp. Bot.* **2008**, *59*, 2659–2672. [CrossRef] [PubMed]
29. Farrar, J. Sink strength: What is it and how do we measure it? Introduction. *Plant Cell Environ.* **1993**, *16*, 1015–1016. [CrossRef]
30. Zhang, C.; Whiting, M. Pre-harvest foliar application of Prohexadione-Ca and gibberellins modify canopy source-sink relations and improve quality and shelf-life of 'Bing' sweet cherry. *Plant Growth Regul.* **2011**, *65*, 145–156. [CrossRef]
31. Li, J.; Yu, X.; Lou, Y.; Wang, L.; Slovin, J.P.; Xu, W.; Wang, S.; Zhang, C. Proteomic analysis of the effects of gibberellin on increased fruit sink strength in Asian pear (*Pyrus pyrifolia*). *Sci. Hortic.* **2015**, *195*, 25–36. [CrossRef]
32. Lecourieux, F.; Kappel, C.; Lecourieux, D.; Serrano, A.; Torres, E.; Arce-Johnson, P.; Delrot, S. An update on sugar transport and signalling in grapevine. *J. Exp. Bot.* **2014**, *65*, 821–832. [CrossRef]
33. Zhang, Y.; Zhen, L.; Tan, X.; Li, L.; Wang, X. The involvement of hexokinase in the coordinated regulation of glucose and gibberellin on cell wall invertase and sucrose synthesis in grape berry. *Mol. Biol. Rep.* **2014**, *41*, 7899–7910. [CrossRef] [PubMed]

horticulturae

Article

Identification of Laccase Genes in Grapevine and Their Roles in Response to *Botrytis cinerea*

Ran Wan [1,2,†], Zhenfeng Yang [1,2,†], Jun Liu [1,2], Mengxi Zhang [1,2], Jian Jiao [1,2], Miaomiao Wang [1,2], Kunxi Zhang [1,2], Pengbo Hao [1,2], Yu Liu [1,2], Tuanhui Bai [1,2], Chunhui Song [1,2], Shangwei Song [1,2], Jiangli Shi [1,2,*] and Xianbo Zheng [1,2,*]

1   College of Horticulture, Henan Agricultural University, Zhengzhou 450002, China;
    wanxayl@henau.edu.cn (R.W.); yzf15038898023@163.com (Z.Y.); lj13550345839@163.com (J.L.);
    zmx18337011095@126.com (M.Z.); jiaojian@henau.edu.cn (J.J.); wmm2018@henau.edu.cn (M.W.);
    kunxi66@163.com (K.Z.); hao_pb@henau.edu.cn (P.H.); yuliu@henau.edu.cn (Y.L.);
    tuanhuibai@henau.edu.cn (T.B.); songchunhui060305@henau.edu.cn (C.S.); songsw@henau.edu.cn (S.S.)
2   Henan Key Laboratory of Fruit and Cucurbit Biology, College of Horticulture, Henan Agricultural University,
    Zhengzhou 450002, China
*   Correspondence: sjli30@henau.edu.cn (J.S.); xbzheng@henau.edu.cn (X.Z.)
†   These authors contributed equally to this work.

Abstract: Laccases are the key enzymes responsible for plant lignin biosynthesis and responses to environment stress. However, the roles of LAC genes in plant disease resistance are still largely unknown, especially in grapevine, one of the most important horticultural crops in the world. Its quality and yield are very vulnerable to gray mold disease caused by *Botrytis cinerea*. In total, 30 *VvLAC* genes were identified and found to be unevenly distributed on seven chromosomes; they were classified into seven groups based on phylogenetic analysis according to the criteria applied in *Arabidopsis thaliana*. Collinearity and synteny analyses identified some orthologous gene pairs in Vitis vinifera and a few paralogous gene pairs among grape and peach. The *VvLAC* gene family has diverse gene structures and a highly conserved motif composition. The prominent presence of the MYB cis-elements in each *VvLAC* promoter highlighted MYB transcriptional factors as the main regulators of *VvLAC* genes. Furthermore, twenty-five *VvLAC* genes with functional redundancy are probably implicated in grape lignin biosynthesis. The expression patterns of the *LAC* genes in grape leaves of Chinese wild *V. amurensis* 'Shuangyou' (SY), a germplasm highly resistant to *B. cinerea*, were investigated through transcriptomic data and qRT-PCR verification. Combined with the phylogenetic analysis, with *AtLACs* participating in lignin metabolism, and the cis-element analysis, *VaLAC14*, *VaLAC19*, *VaLAC24* and *VaLAC30* were identified as key candidate genes for lignin biosynthesis in the grape response to *B. cinerea*. This study supplies a comprehensive understanding of the classification, evolution, structure and responses of the grape LAC genes against *B. cinerea*. It also provides valuable genetic resources for functional characterization towards enhancing grapevine disease resistance.

Keywords: grape laccase and lignin; the *VvLAC* gene family; gene expression; response to *Botrytis cinerea*

Citation: Wan, R.; Yang, Z.; Liu, J.; Zhang, M.; Jiao, J.; Wang, M.; Zhang, K.; Hao, P.; Liu, Y.; Bai, T.; et al. Identification of Laccase Genes in Grapevine and Their Roles in Response to *Botrytis cinerea*. *Horticulturae* 2024, 10, 376. https://doi.org/10.3390/horticulturae10040376

Academic Editor: Francesco Faretra

Received: 24 February 2024
Revised: 24 March 2024
Accepted: 3 April 2024
Published: 9 April 2024

1. Introduction

Grape is an important fruit crop worldwide with huge economic value. With the increase in grape planting years, the base number of pathogenic seedlings and insect population is gradually increasing [1]. The occurrence of *Botrytis cinerea* on grape, the ubiquitous necrotrophic fungal pathogen, is becoming more and more serious, resulting in a prominent problem of yield and quality decline [1,2]. *B. cinerea* causes serious damage to young leaves, young tendril, flower buds and ripened fruit of grape; it is generally called gray mold and is the second most important plant disease in the world [1–4].

B. cinerea has a very powerful 'arsenal' for causing disease in more than 1000 plant species which is especially characterized by massive secretion of cell wall degrading enzymes, such as BcPGs (endo-polygalacturonase), BcPMEs (pectin methylesterase), BcCBHs (cellobiohydrolase) and so on, as one of the primary infection strategies [5–7]. The cell wall matrix is one of the first and largest plant structures that pathogens encounter when interacting with potential hosts [8]. Lignin is one of the main components of the plant cell wall, and as a complex phenolic polymer produced by the phenylalanine/tyrosine metabolic pathway, it plays important roles in plant defense by enhancing plant cell wall rigidity, protecting cell wall integrity, acting as phytoalexins and acting as a physical barrier [9–11]. Previous studies have revealed that the cell wall strengthening through lignin accumulation mediated by the ethylene- and abscisic acid-associated pathways is significantly involved in the plant response to *B. cinerea* [12,13]. Lignin biosynthesis was also observed in grapevine defense against trunk diseases, downy mildew, Pierce's disease and postharvest infections [14–16]. Particularly, grapevine resistance to *B. cinerea* is also suggested to be associated with lignin metabolism. Higher lignin content was detected in Greek grape berries of the highly resistant 'Limnio' variety than in those of the susceptible 'Roditis' variety [17]. The genes associated with grapevine lignin resistant against *B. cinerea* remain unexplored.

In recent years, laccases (LACs), as one of the classes of key enzymes, have been found to have the important function of polymerizing lignin monomers into lignin in the cell wall [9,18]. A total of 17 laccase genes have been identified in *Arabidopsis thaliana*, and among them, *AtLAC4*, *AtLAC17*, *AtLAC2* and *AtLAC15* have been shown to participate in lignin polymerization in leaves, seed coat and roots [19–23]. The expression of *AtLAC4* and *AtLAC17* contributes to constitutive lignification in stems and to lignin monomer deposition in fibers; the *lac4 lac17* mutant increases their monolignol glucosides and display dwarfism, while lignin deposition in the roots is almost completely abolished in the *lac11 lac4 lac17* triple mutant [19–23]. For other species, *Miscanthus* MsLAC1 and *Pyrus bretschneideri* PbrLAC1, PbrLAC2 and PbrLAC18, as well as *Populus tomentosa* PtoLAC14 and *Cleome hassleriana* ChLAC8, were all reported to be significantly involved in lignin polymerization in various tissue types [24–27].

Laccases associated with lignin biosynthesis have been a prominent research focus in recent years for their significant disease resistance function. Among the 84 laccase genes in cotton (*Gossypium hirsutum*), GhLAC1, GhLAC4 and GhLAC15 are all implicated in disease resistance to verticillium wilt caused by *Verticillium dahliae* by enhancing G-lignin biosynthesis and lignification in the cell wall [28–31]. Laccase-induced lignification is suggested to be one of the major processes associated with the defense response in apple plant diseases, in which both *MhLAC7* and *MdLAC7* have an important function [32–34]. Tobacco overexpressing *EuLAC1* from *Eucommia ulmoides* Oliver. displayed significantly higher laccase activity and resistance to gray mold compared with wild-type tobacco [35]. However, the implication of laccase genes in grapevine lignin associated with disease resistance is little understood. Here, we characterized the laccase gene family in grapevine through bio-information analysis and analyzed the potential roles of these genes in the defense response based on the expression profile of grape leaves under infection by *B. cinerea*. The results highlight four important candidate genes, *VaLAC14*, *VaLAC19*, *VaLAC24* and *VaLAC30*, associated with lignin biosynthesis in the response against *B. cinerea*. Our work would strengthen the understanding of grape laccase genes and provide an important basis for further studying their implication in disease resistance mechanisms.

2. Materials and Methods

2.1. Plant and Fungus Materials

The leaves of Chinese wild grape (*V. amurensis* Rupr.) 'Shuangyou' (SY), a germplasm highly resistant to *B. cinerea* [36], were used as experimental plant materials. They were planted in a plant incubator at 23 °C and 75% HR (relative humidity) under a cycle of 16 h light/8 h darkness. The *B. cinerea* fungus was isolated and purified from infected grape berries. Conidia suspension, inoculation and detached leaf assay were all performed according to the description by Wan et al. [36]. The experiment was repeated three times, and 8–10 leaves were collected repeatedly in each experiment. All treated and control (inoculated with sterile water) leaves were collected at 4, 8, 18, 36 and 72 h post-inoculation (hpi). All collected samples were immediately frozen in liquid nitrogen and stored at $-80\,^{\circ}$C for further use.

2.2. Identification of the VvLAC Gene Family

The sequence distributions of *V. vinifera* L. and *Arabidopsis thaliana* were downloaded from Ensembl Plants (http://plants.ensembl.org/index.html, accessed on 18 November 2023) and the TAIR database (http://www.arabidopsis.org/, accessed on 18 November 2023). Then, according to the conserved domain protein sequences retrieved from the Pfam database (http://pfam.sanger.ac.uk/, accessed on 18 November 2023), HMMER and BLASTP tools were used to screen the grape candidate LAC genes. The NCBI CDD database (Conserved Domain Database; http://www.ncbi.nlm.nih.gov//Structure/bwrpsb/bwrpsb.cgi, accessed on 18 November 2023) was used to analyze and identify the conserved domains of candidate *VvLAC* genes. The chromosome location information of the identified *VvLAC* genes was mapped by genome annotation files, and all the identified *VvLAC* genes were named according to the chromosome (Chr) locations. For further exploring the protein characteristics, the ExPASy protein online tool (https://web.expasy.org/compute_pi/, accessed on 22 November 2023) was used to predict the isoelectric point (pI) and molecular weight (MW) of the VvLAC proteins. WoLF PSORT (https://www.genscript.com/wolf-psort.html, accessed on 22 November 2023) was used to predict the subcellular localizations of VvLACs.

2.3. Chromosomal Location, Phylogenetic Analysis and Collinearity Analysis

MEGA software (version 10.0; Mega Limited, Auckland, New Zealand) was used to construct a phylogenetic tree with the maximum likelihood (ML) method, and the bootstrap test was repeated 1000 times [37]. Finally, the tree was annotated and decorated by ITOL online tools (http://itol.embl.de/, accessed on 10 December 2023) [38]. The phylogenetic tree was constructed based on the full-length protein sequences of LACs in *Arabidopsis thaliana* [39], *Populus trichocarpa* [40], *Citrus reticulata* Blanco [41] and *V. vinifera*. TBtools software (SCAU; version 1.0687) was used to visualize the position of grape chromosomes. The Multiple Collinearity Scan tool kit with default parameters in TBtools (version 1.0687; SCAU, Guangzhou, Guangdong, China) was used to analyze the collinearity between grape and the following herbaceous plants: *Glycine max, Oryza sativa* Japonica, *Gossypium raimondii* and *Arabidopsis thaliana*, and ligneous plants, *Populus trichocarpa, Malus domestica* Golden and *Prunus persica*.

2.4. Gene Structures, Conserved Motifs and Cis-Elements

The exon–intron structures of the *VvLAC* genes were visualized by using TBtools software (version 1.0687; SCAU, Guangzhou, Guangdong, China) by referring to the downloaded genomic information. The Multiple Em for Motif Elicitation (MEME; https://meme-suite.org/meme/tools/meme, accessed on 13 December 2023) online tool was used to analyze the conserved motifs in the *VvLAC* genes. The NCBI CDD database (Conserved Domain Database; https://www.ncbi.nlm.nih.gov/Structure/bwrpsb/bwrpsb.cgi, accessed on 13 December 2023) was used to analyze and identify the conserved domains of VvLACs. To analyze the cis-acting elements of *VvLACs*, we intercepted a sequence of 2000 bp upstream of the start codon of each *VvLAC* and then used the PlantCARE (https://bioinformatics.psb.ugent.be/webtools/plantcare/html/, accessed on 20 December 2023) online tool [42] to predict and analyze it; finally, we visualized it with TBtools (version 1.0687; SCAU, Guangzhou, Guangdong, China).

2.5. Protein Interaction Network Analysis

The homologous proteins of the AtLAC genes in *Arabidopsis thaliana* were obtained by the STRING online tool (http://string-db.org, accessed on 22 January 2024), and protein interaction network (PPI) prediction analysis was carried out. The results were visualized by Cytoscape software (version 3.10.1; USA).

2.6. Expression Profile of Grape LAC Genes in SY Leaves against B. cinerea

RNA-seq data about the response of *VaLACs* in SY leaves against *B. cinerea* were derived from our previously published study [36].

2.7. RNA Extraction and qRT-PCR Analysis

The quantitative primers of the candidate genes were designed by Primer Premier 5.0 software (PREMIER Biosoft, San Francisco, CA, USA). The specificity of the primers was determined through NCBI (Table S1), and they were synthesized by Bioengineering Co., Ltd. (Shanghai, China). The RNA extraction kit E.Z.N.A.® Plant RNA Kit # R6827-0 (Omega Biotek, Dallas, TX, USA) was used to extract total RNA from SY leaves in the indicated infection stages. The NanoDrop 2000 instrument (Thermo Fisher Scientific, Shanghai, China) was used to check the purity and concentration of RNA, and 1% agarose gel electrophoresis was used to determine its quality. Then, according to the reverse transcription kit HiScript III qRT Super Mix for q PCR (+gDNA wiper) (Vazyme Biotech Co., Ltd., Nanjing, Jiangsu, China), the cDNAs of the corresponding samples were obtained by reverse transcription. Finally, the fluorescence quantitative PCR reaction system was configured according to the fluorescence quantitative kit ChamQ Universal SYBR qPCR Master Mix (Vazyme Biotech Co., Ltd., Nanjing, China), and qRT-PCR was performed on the CFX96 real-time system (Bio-Rad, Hercules, CA, USA). The reaction procedure was performed at 95 °C for 5 min, followed by 40 cycles at 95 °C for 15 s, 60 °C for 1 min and 72 °C for 5 min. The relative expression level of each gene was calculated by using the $2^{-\Delta\Delta t}$ method.

2.8. Determination of Laccase Activity and Lignin Content

Each leaf sample collected at the indicated time points was dried at 80 °C to constant weight, ground to powder, and sieved with a 0.425 mm aperture sieve. Then, 2 mg samples were evaluated in terms of lignin content (expressed as mg g^{-1}) by using the MZS-1-G Kit (Suzhou Comin Biotechnology Co. Ltd., Suzhou, Jiangsu, China) [43]. Then, homogenates made with 0.1 g samples with 1 mL of extracting solution in an ice bath were centrifuged at 1200 g at 4 °C for 30 min to obtain supernatant liquid. Afterwards, 45 μL of supernatant liquid was used to determine the laccase activity of each sample (expressed as nmol min^{-1} g^{-1}) by using the QM-1-G Kit (Suzhou Comin Biotechnology Co., Ltd., Suzhou, Jiangsu, China). The determination was repeated three times, and the results were expressed as means ± SEs.

2.9. Data Analysis

The data were analyzed by using SPSS version 20 (IBM SPSS Inc., Chicago, IL, USA). Statistical significance was analyzed by Duncan's test. Different lowercase letters indicated that the difference among treatments reached a significant level ($p \leq 0.05$).

3. Results

3.1. Laccase Activity and Lignin Content in SY Leaves in Response to B. cinerea

The phenotypic characteristics of grape leaves inoculated with *B. cinerea* conidia 4–72 hpi verified that SY was resistant against the pathogen (Figure 1a). In control samples, laccase activity and lignin biosynthesis almost showed no significant change. After infection, laccase activity showed a significant increase 18 hpi and was significant higher in the infected SY leaves than in the control samples from 8 hpi onwards. Lignin content in the *B. cinerea*-infected SY leaves was increased 4 hpi and was also significantly higher than in the control samples from then on, except at 36 hpi (Figure 1b,c).

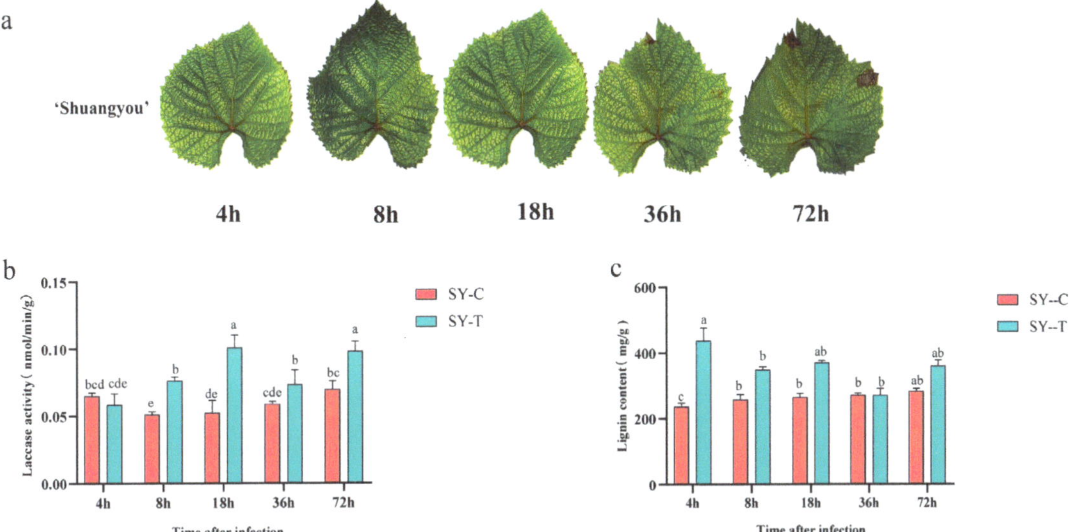

Figure 1. Changes in resistant 'Shuangyou' grape leaves. (**a**) Phenotypic observation. (**b**) Laccase activity and (**c**) lignin content. C: control (inoculation with sterile water); T: treatment (inoculation with *Botrytis cinerea*). SY: Chinese wild *V. amurensis* 'Shuangyou', a germplasm highly resistant to *B. cinerea*. The results are the means (SDs) of three biological replicates. There were significant differences between the results marked with different lowercase letters ($p \leq 0.05$).

3.2. Identification of VvLAC Genes in Grapevine

A total of 30 members of the *VvLAC* gene family were retrieved from the 'Pinot Noir' *Vitis* genome (PN40024.v4) and were renamed from *VvLAC1* to *VvLAC30* according to their positions on the Chrs (Table 1). The length of their encoding proteins ranged from 386 to 627 amino acids, and the molecular weights of the proteins ranged from 42.46 to 69.76 KDa (Table 1). The theoretical isoelectric points of these proteins ranged from 4.79 to 9.79, with VvLAC28 showing the lowest value and VvLAC11 the highest one; they comprised 24 basic proteins and 6 acidic proteins distributed on Chr 18, except for VvLAC14, which was found on Chr 8 (Table 1). Moreover, the subcellular localization displayed that 17 VvLAC proteins were located on chloroplasts, followed by the vacuolar membrane and extracellular matrix, each with 4 VvLAC proteins; the cytoskeleton, with 3 proteins; and the plasma membrane and peroxisome, with the lowest number of protein, with 1 VvLAC protein each.

Table 1. Information of the *VvLAC* gene family.

Name	Gene ID	Chromosome	Gene Length (bp)	pI	MW (Kda)	Amino Acid (aa)	Subcellular Localization
VvLAC1	Vitvi04g01317_t001	Chr4	1692	8.49	62.23	563	vacu
VvLAC2	Vitvi06g00378_t001	Chr6	1758	9.02	64.39	585	chlo
VvLAC3	Vitvi06g00405_t001	Chr6	1758	8.71	64.30	585	extr
VvLAC4	Vitvi06g00591_t001	Chr6	1665	9.26	60.70	554	chlo
VvLAC5	Vitvi06g00728_t001	Chr6	1725	8.59	63.59	574	vacu
VvLAC6	Vitvi08g01031_t001	Chr8	1689	8.37	62.49	562	extr
VvLAC7	Vitvi08g01223_t002	Chr8	1641	9.32	59.90	546	chlo
VvLAC8	Vitvi08g01223_t001	Chr8	1740	9.24	63.69	579	chlo
VvLAC9	Vitvi08g04233_t001	Chr8	1197	9.74	43.53	398	chlo
VvLAC10	Vitvi08g02201_t001	Chr8	1734	9.77	64.04	577	chlo
VvLAC11	Vitvi08g01228_t001	Chr8	1734	9.79	64.04	577	chlo

Table 1. *Cont.*

Name	Gene ID	Chromosome	Gene Length (bp)	pI	MW (Kda)	Amino Acid (aa)	Subcellular Localization
VvLAC12	Vitvi08g04234_t001	Chr8	1734	9.78	64.08	577	chlo
VvLAC13	Vitvi08g01229_t001	Chr8	1761	9.22	64.96	586	chlo
VvLAC14	Vitvi08g01299_t001	Chr8	1698	6.79	62.00	565	vacu
VvLAC15	Vitvi08g01335_t001	Chr8	1671	9.12	60.48	556	chlo
VvLAC16	Vitvi08g01735_t001	Chr8	1674	8.04	61.69	557	extr
VvLAC17	Vitvi13g00117_t001	Chr13	1734	7.63	63.18	577	cyto
VvLAC18	Vitvi13g00321_t001	Chr13	1752	9.44	64.30	583	chlo
VvLAC19	Vitvi13g00322_t001	Chr13	1752	9.53	64.39	583	chlo
VvLAC20	Vitvi13g00323_t001	Chr13	1755	9.35	64.61	584	vacu
VvLAC21	Vitvi13g00341_t001	Chr13	1755	9.4	64.50	584	chlo
VvLAC22	Vitvi13g00342_t001	Chr13	1755	9.22	64.45	584	chlo
VvLAC23	Vitvi13g00509_t001	Chr13	1662	8.67	60.40	553	chlo
VvLAC24	Vitvi15g00941_t001	Chr15	1884	8.19	69.76	627	chlo
VvLAC25	Vitvi17g00227_t001	Chr17	1752	8.5	65.26	583	chlo
VvLAC26	Vitvi18g01488_t001	Chr18	1776	5.3	66.34	591	extr
VvLAC27	Vitvi18g02924_t001	Chr18	1791	4.89	66.31	596	cyto
VvLAC28	Vitvi18g02922_t001	Chr18	1794	4.79	66.25	597	pero
VvLAC29	Vitvi18g02927_t001	Chr18	1779	5.52	66.58	592	plas
VvLAC30	Vitvi18g02928_t001	Chr18	1800	4.92	67.64	599	cyto

Note: vacu: vacuolar membrane; chlo: chloroplast; extr: eextracellular matrix; cyto: cytoskeleton; pero: peroxisome; plas: plasma membrance.

3.3. Chromosomal Locations and Phylogenetic Analysis

In the present study, 30 *VvLAC* genes were distributed on seven different chromosomes (Figure 2a). It was found that 11 of them were mainly located on Chr 8 (*VvLAC6–16*), 7 on Chr 13 (*VvLAC17–23*), 5 on Chr 18 (*VvLAC26–30*) and 4 on Chr 6 (*VvLAC2–5*), together containing 27 *VvLAC* genes, while Chr 4, Chr 15 and Chr 17 only contained 1 *VvLAC* gene each. The results show no positive correlation between chromosome length and number of *VvLAC* genes. Moreover, some members of the *VvLAC* family on Chr 8, Chr 13 and Chr 18 exist in the form of gene clusters.

To explore the evolutionary relationships of *VvLACs*, a phylogenetic analysis was performed based on the full-length amino acid sequences of 30 VvLACs and the LAC proteins from other plant species, which included 17 LACs from *Arabidopsis thaliana*, 53 from *Populus trichocarpa* and 27 from *Citrus reticulata*. According to the classification standard of *Arabidopsis thaliana* laccases, 30 VvLACs and 97 LACs were divided into seven groups, and their distribution in each group was rather uneven (Figure 2b). In detail, only 1 VvLAC was clustered with 2, 6 and 12 LACs in Group I, Group II and Group III, respectively. Also, Group IV contained 2 VvLACs, VvLAC5 and VvLAC17, together with 15 LACs (Figure 2b). Moreover, 22 and 29 LACs were classified in Group V and Group VI, each with 5 VvLAC, whereas Group VII contained the maximum number of LAC proteins, up to 37 LACs, in which there were 15 VvLACs (Figure 2b). In addition, it was found that all the 5 VvLACs clustered in Group V were located on Chr 18, while 8 out of the 11 VvLACs on Chr 13 all belonged to Group VII (Figure 2). These results indicate that specific evolutionary events occurred among *VvLAC* genes after the divergence of grapevine and the other three plants.

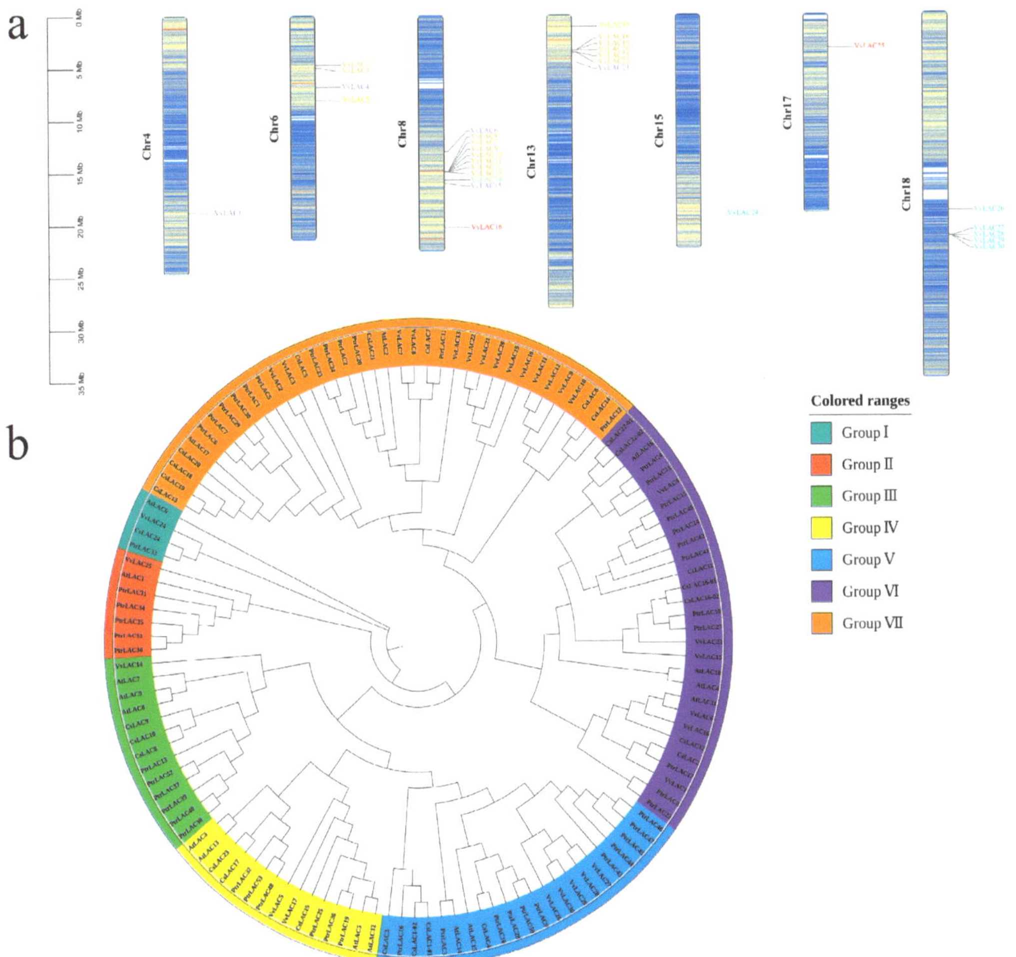

Figure 2. Chromosomal localization and phylogenetic analysis. (**a**) The chromosomal localization of the *VvLAC* gene family. Genes in the same subgroup are represented by the same color. The left ruler indicates the length of the grape chromosomes. (**b**) The phylogenetic tree was constructed based on the protein sequences from *Arabidopsis thaliana*, *Populus trichocarpa* and *Citrus reticulata* Blanco. The adjacency tree was constructed by using MEGA10.0. The seven groups are distinguished by different colors.

3.4. Gene Duplication and Syntenic Analysis

Generally, gene duplication, including tandem repeats, and segmental and tandem duplication, is one of the most important driving forces leading gene expansion and evolution during genome evolution [44,45]. It plays important roles in improving plant adaptability to various environmental stresses [45]. To understand the gene duplication pattern of *VvLACs*, a collinearity analysis was performed in this study. As a result, 11 *VvLAC* duplicated geneswere distributed on seven chromosomes (including Chr 6, Chr 13 and Chr 8), among which *VvlAC18–20* were found to be tandem repeat genes, and so were *VvlAC9*, *VvLAC11* and *VvLAC13* (Figure 3a).

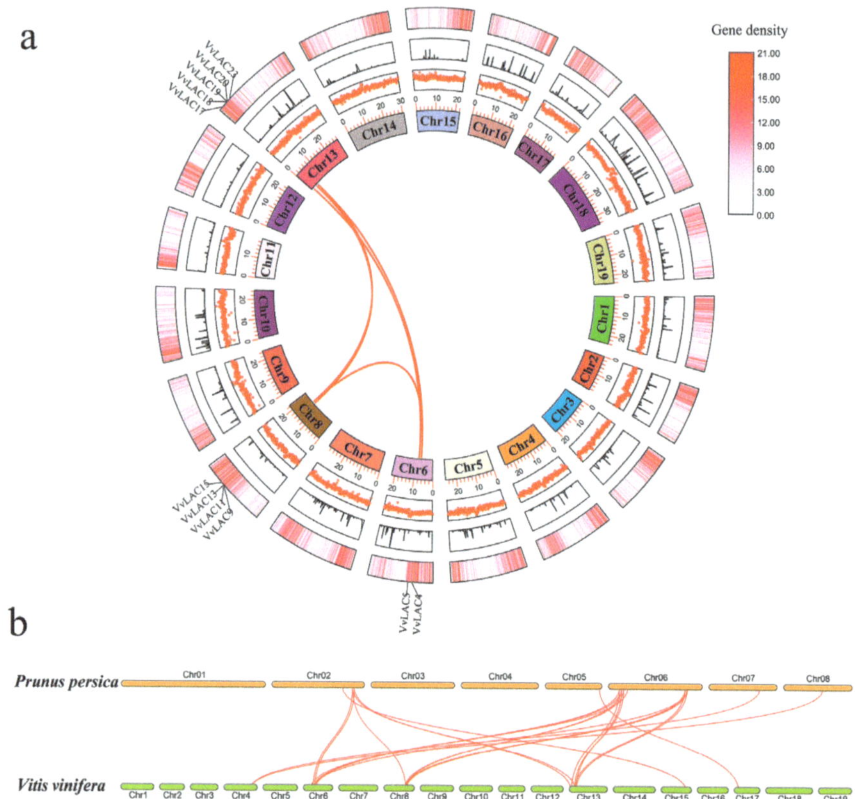

Figure 3. Interspecific and intraspecific collinearity analyses of the *VvLAC* genes. (**a**) Intraspecific collinearity analysis. The innermost circle shows the 19 chromosomes of *Vitis vinifera*. Different chromosomes are represented by different colors, followed by N ratio and G/C ratio. The outermost circle shows the gene densities on the corresponding chromosomers, and a redder color represents a higher gene densities. The identified *VvLAC* gene pairs are linked by red lines. (**b**) Syntenic analysis of *Prunus persica* and *Vitis vinifera LAC* genes. The different chromosomes of the two are represented by different colors. The gray lines represent the collinear lines in the genome of the two plants, while the red lines highlight the same line of the LAC gene pair.

Furthermore, a syntenic analysis was performed for predicting the function of *VvLACs* through the homology analysis of *LAC* genes from four herbaceous plants and three ligneous plants. In total, 22 syntenic gene pairs between grape and *Prunus persica* were observed; however, no syntenic gene pairs were present (Figure 3b), which may relate to the weaker phylogenetic relationships between grape and the other six plant species.

3.5. Motif Composition and Gene Structure

A total of 10 types of conserved motifs of the VvLAC family genes were explored by using the MEME program (Figure 4). Many classes of VvLAC proteins were found to have quite consistent motif compositions, except for VvLAC9, which was composed of only six motifs (Figure 4a). This suggests that functional redundancy possibly exist among these genes. Also, motif variations in numbers, amino acids and lengths across VvLACs were observed (Figure 4b), and as a result, their functional divergence was hypothesized. As expected, all the identified VvLAC proteins contained the conserved functional laccase domain (Figure 4c).

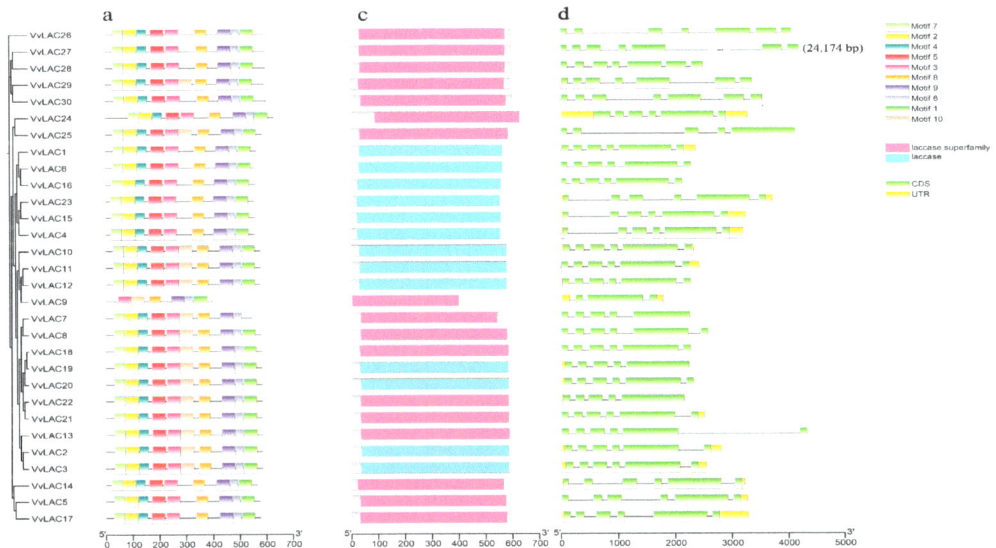

Figure 4. Motif composition and gene structure analyses of *VvLACs*. (**a**,**b**) Motif information. (**c**) Conserved domain. (**d**) Intron–exon patterns. UTR and CDS are represented by yellow boxes and green boxes, respectively.

To further investigate gene evolution, the exon–intron distribution of *VvLACs* was analyzed by aligning coding sequences against their corresponding genomic sequences. The results show that the gene structures of *VvLACs* exhibit diverse intron–exon patterns (Figure 4d). For example, except for *VvLAC9*, which has only four exons, the exon number varies from five to seven among the *VvLAC* genes. Notably, there are five genes, all in Group V, containing the highest number of exons, seven, with *VvLAC27* being the longest gene, with 24,174 bp in total. Generally, *VvLACs* displaying high homology have highly similar gene structures (Figure 4).

Subsequently, the amino acid sequences of all the AtLACs and VvLACs were aligned together. The results show that the VvLAC proteins have higher similarity with the AtLAC proteins and have three canonical plant laccase Cu oxidase domains, namely, Cu RO_1_LCC_plant, Cu RO_2_LCC_plant and Cu RO_3_LCC_plant (Figure 5).

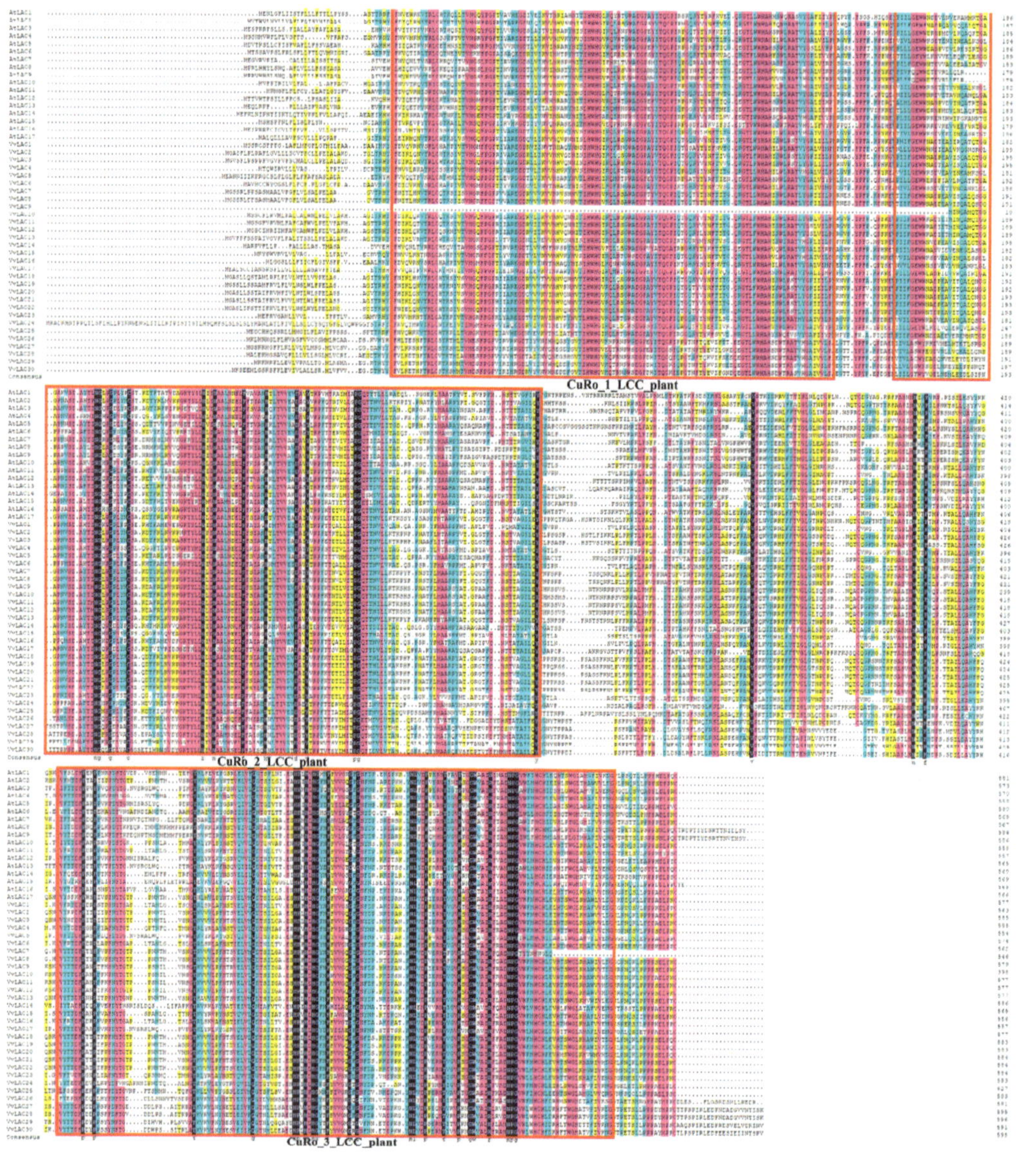

Figure 5. VvLACs and AtLAC protein sequence alignment analysis.

3.6. Cis-Elements in Promoters of VvLAC Genes

To explain their transcriptional differences, we analyzed the cis-elements in the promoter regions of the laccase gene family. Sixty–eight *cis*-elements were observed in *VvLAC* promoters and classified into the seven categories shown in Figure 6. The light-responsive elements, phytohormone elements, biotic/abiotic stress elements and MYB-TF (transcriptional factor) elements were the most frequently identified in the promoter regions of the grape laccase genes. Among all the elements, the MYB elements were the most prominent, and the only ones present in each *VvLAC* promoter, reaching the maximum of 11 in *VvLAC6*, *VvLAC16* and *VvLAC19* (Figure 6). Meanwhile, the MYC element was specially noticed, being present the most among phytohormone elements, except in the promoters of *VvLAC29* and *VvLAC30*. Also, STRE (stress response promoter element), W-box (WRKY-binding site [36]) and WRE3 were relatively abundant among biotic/abiotic stress elements (Figure 6). Moreover, there were 22 types of light-responsive elements, and Box 4, GT1-motif, G-box and GATA-motif were abundantly detected (Figure 6).

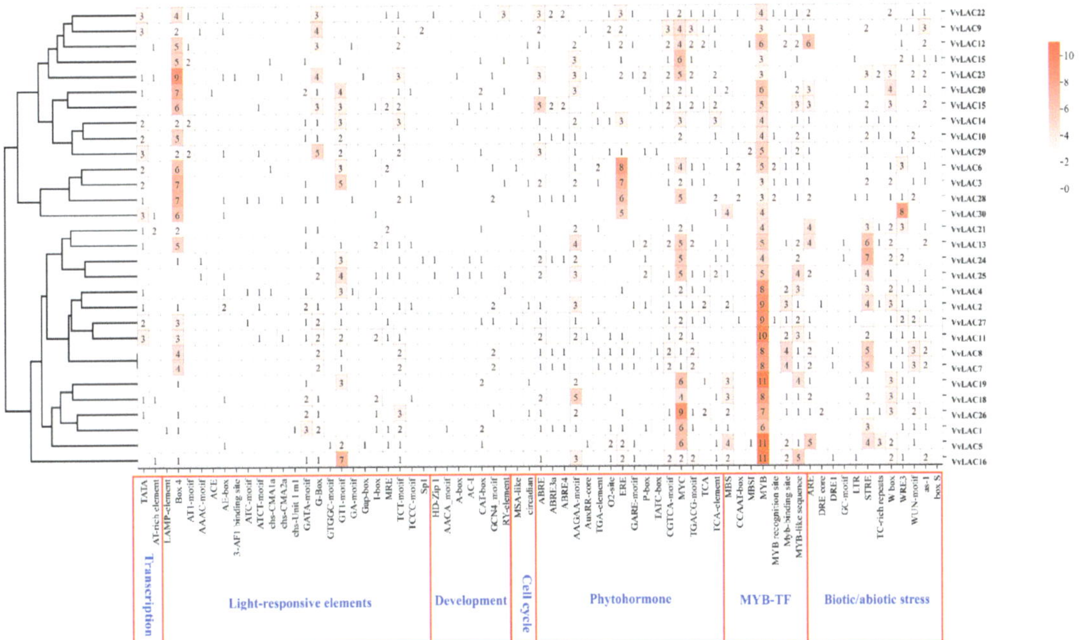

Figure 6. Identification of *cis*-elements in the upstream promoters of the indicated *VvLAC* genes. The heat map was generated based on the numbers and predicted functions of the *cis*-elements. Different colors and numbers in the grid represent the number of the corresponding *cis*-elements in each *VvLAC* gene promoter.

3.7. Protein–Protein Interaction Network

To investigate whether the VvLAC proteins might function by forming homo- or hetero-protein complexes, we constructed a protein interaction network for VvLACs based on their orthology with AtLAC proteins (Figure 7, Table S2). A total of 62 interacting protein pairs composed of 30 VvLACs and 11 VtLACs were predicted and divided into 11 subfamilies. There were no direct interactions between the VvLACs themselves, and the interacting proteins mainly included six types, PAL1-4 and CCR1/2, except for the 11th subfamily, which had distinct types, including PAL1/4, CESA4/7/8 and MYB63, T25K17.30, CESA4/7/8, PAL1, PAL1/2/4 and CCR1/2 were all predicted to interact with the VvLAC2, VvLAC7/8 and VvLAC1/6/16 proteins in the first, second and eighth subfamilies, respectively (Figure 7, Table S2). The fifth, sixth and seventh subfamilies, including VvLAC24, VvLAC14 and VvLAC15, respectively, revealed the same interaction proteins, PAL1-4 and CCR1. Additionally, both the third (VvLAC17) and tenth (VvLAC4 and VvLAC23) subfamilies contained T25K17.30 and PAL1-4. Additionally, PAL1-3 and CCR1/2 were predicted as the interaction proteins of VvLAC29-30 in the ninth subfamily, while VvLAC17 in the fourth subfamily was linked with the T25K17.30, PAL1/2 and CCR1/2 interaction proteins (Figure 7, Table S2).

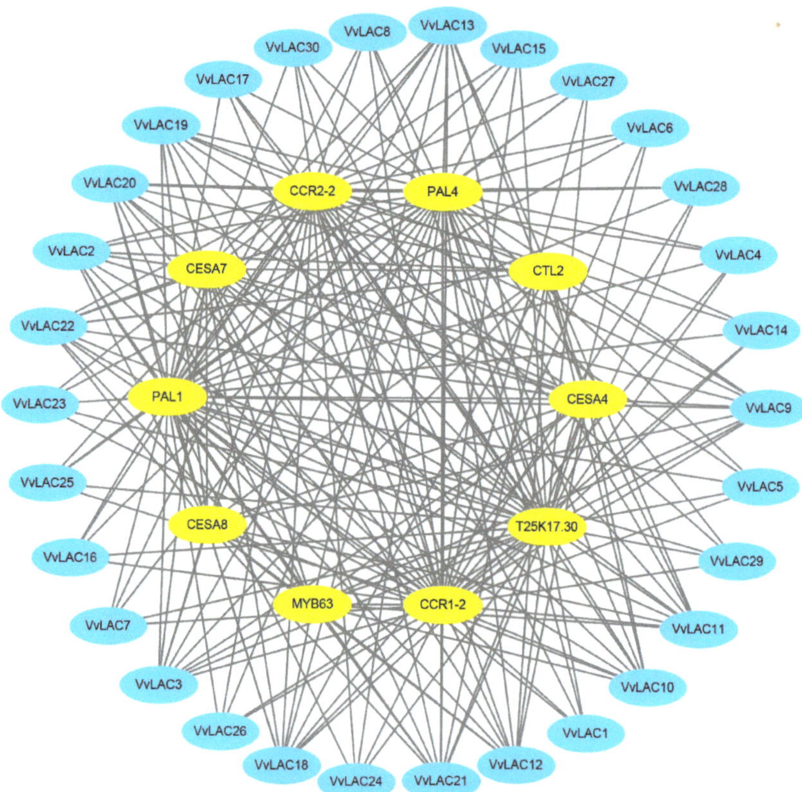

Figure 7. Prediction of protein–protein interaction network of VvLACs based on AtLACs in *Arabidopsis thaliana*. VvLAC proteins are shown in blue circles and the predicted interaction proteins are shown in yellow circles.

3.8. Expression Profiles of LAC Genes in SY Leaves in Response against Botrytis cinerea

The expression patterns of all the 30 *LAC* genes were analyzed based on the RNA-seq data of the SY leaves in the early interaction stages, 4–36 after inoculation with *B. cinerea* and sterile water [36]. There were four genes previously annotated as *LACs* according to the RNA-seq data; however, in the present study, each was actually identified to be containing two *LACs* according to the latest released grape genomic annotation version (PN40024. v4). These grape genes are *LAC7/8*, *LAC11/12*, *LAC19/20* and *LAC21/22*. Even so, we obviously found several genes with different expression in the infected SY leaves compared with the control leaves, for example, *VaLAC14*, *VaLAC24*, *VaLAC30*, *VaLAC17* and so on (Figure 8a). Relative quantification was used to confirm the expression levels of the selected genes as shown in Figure 8b. The results show that in contrast with other *VaLAC* genes, the similar trends of expression of *VaLAC14* and *VaLAC24* were significantly high in the SY leaves during infection by *B. cinerea* (Figure 8b). After *B. cinerea* infection, the expression levels of *VaLAC13*, *VaLAC30* and *VaLAC25* significantly increased, especially from 4 hpi to 8 hpi and following 36 hpi. Also, *VaLAC19* expression was significantly increased from 4 hpi to 18 hpi (Figure 8b). Despite displaying significant changes, the relative expression of the remaining *VaLAC* genes following infection was very low (Figure 8b).

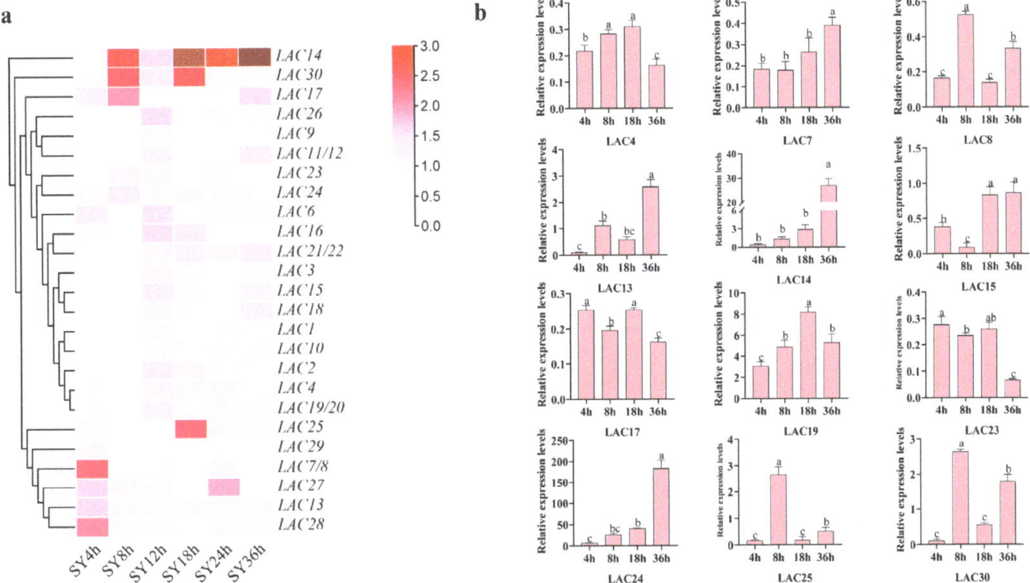

Figure 8. The expression analysis of the *VaLACs* in the SY response to *B. cinerea*. (**a**) The ratios of *VaLAC* expression after infection compared with the control, which were derived from previously published RNA-Seq data on the interaction between SY leaves and *B. cinerea* at the indicated time points after infection [36]. (**b**) The relative expression levels of the selected *VaLAC* genes determined through qRT-PCR. SY: Chinese wild *V. amurensis* 'Shuangyou', a germplasm highly resistant to *B. cinerea*. The expression value was reported as the mean ± standard deviation of three independent biological replicates, where each biological replicate contained three technical replicates. Different letters in the figure indicate significant differences, $p < 0.05$. C: control (inoculation with sterile water); T: treatment (inoculation with *B. cinerea*).

4. Discussion

Laccase is a multi-copper oxidase that catalyzes the oxidation of a wide range of phenolic compounds and has been widely and traditionally researched for potential applications in the food, pharmaceutical and environmental industries for its eco-friendly catalysis [46]. In recent decades, increasing numerous studies have demonstrated that plant LACs, a family of key enzymes polymerizing lignin monomers into lignins, play multiple roles in development and response to biotic and abiotic stresses [19,28,33,34,44]. Systematic analyses have been conducted to identify laccase gene families in many model plants, as well as crop, woody and horticultural plants. The laccase gene families in *Arabidopsis thaliana* (17) [47], *Oryza sativa* (30) [48], *Populus trichocarpa* (49) [40], *Brassica rapa* var. pekinensis (27) [49], *Brassica rapa* var. rapa (27) [49], *Bretschneideri* (45) [25], *Citrus sinensis* (41) [41] and *Punica granatum* (57) [43] have been revealed. Comparatively, *Triticum aestivum* (95) [50], soybean (93) and *Gossypium hirsutum* (84) [28] have more *LAC* members.

Grape is one of the most important horticultural crops in the world. We found that the *B. cinerea*-resistant grape germplasm SY significantly increased the lignin levels in accordance with the increase in laccase activity after infection. Considering the crucial roles of lignin in plant disease resistance [9,11], it might be an important characteristic for SY resistance to *B. cinerea*. Thus far, the laccase gene family members have been little studied in grapevine. Here, a total of 30 *VvLACs* were identified based on the grapevine genome in public databases, and a comprehensive analysis of the *VvLAC* gene family was performed. In grapevine, 30 *VvLACs* are unevenly distributed on seven chromosomes, including 6 tandem repeat genes and 5 segmentally duplicated genes on three chromosomes. Gene duplication events, including tandem repeat and segmental duplication, contributed to grape LAC gene family expansion and evolution. Moreover, syntenic pairs were only observed between grape and peach *LAC* genes, and they might have little evolution relationship with the other six species, including *Arabidopsis thaliana* and *Malus domestica*, in the present study. Motif and gene structure analyses showed that the closely related grape *LAC* members in the phylogenetic tree had similar exon–intron structures and common motif compositions, indicating a possible functional similarity. Diversities in gene structures and motif composition across different groups were also found, indicating functional differentiation in grape laccases.

Here, 30 VvLACs were classified into seven groups according to the classification criteria applied in *Arabidopsis thaliana* [39,47]. AtLAC15 in Group V, AtLAC4 and AtLAC11 in Group VI, and AtLAC2 and AtLAC17 in Group VII have been demonstrated to have a defined function in lignin biosynthesis [19–23]. PbrLAC1 and PbrLAC2 homologous with AtLAC4 and PbrLAC18 homologous with AtLAC17 have been proved to regulate lignin biosynthesis in pear stone cells [25]. This implies that as many as 25 VvLACs in the three groups are probably implicated in grape lignin biosynthesis. Combined for their quite consistent motif compositions and structures, these grape *LAC* genes possibly present functional redundancy. Recent research results also support the point that functional redundance exist in multiple *LAC* genes from one species. In *Arabidopsis thaliana* or pear, each single-laccase-gene-mutant or -silenced strain only slightly reduced the lignin content, whereas the simultaneous mutation or silencing of at least two laccase genes led to a significant decrease in lignin content [23,26].

V. amurensis, the most cold-tolerant *Vitis* species, is suggested to be highly resistant to *B. cinerea* and to share a common ancestor with *V. vinifera* dating approximately 2 million years ago [3,45,51]. *VaERF16*, *VaMYB306* and *VaERF20* from *V. amurensis* have been revealed to have a resistant function in *B. cinerea* infection [52,53]. Here, the expression profile of *V. amurensis* SY *LAC* genes highlighted the candidates *VaLAC14*, *VaLAC24*, *VaLAC30*, *VaLAC13*, *VaLAC19* and *VaLAC25* as being significantly involved in the grape response to *B. cinerea*. In Group III, VvLAC14 was clustered with AtLAC7, AtLAC8 and AtLAC9, whose expression is induced under environmental stress [39]. *VaLAC14* and *VaLAC24*, having similar expression trends, might both play potential roles in the response of grapevine to *B. cinerea*. In Group V, VvLAC30 was clustered with AtLAC15. In cotton, GhLAC15, which is phylogenetically related to AtLAC15, was reported to positively regulate disease-induced lignification to enhance resistance to broad-spectrum biotic stress response [30]. Here, *VaLAC30* was suggested to be involved in the SY defense against the fungus through lignin enhancement. Grape LAC13 and LAC19 were both clustered with AtLAC17 and PtrLAC23 in Group VII, implying their possible roles in grape lignin biosynthesis to resist to the fungus. PtrLAC23 was confirmed to have the function of lignin deposition in *Populus trichocarpa* stem transects through transgenic experiments [40]. VaLAC25 was in Group II with AtLAC1, PtrLAC31 and PtrLAC34, which are mainly expressed in plant roots and leaves [39,40], while their implication in lignin biosynthesis and plant disease, including *VaLAC25*, in the grape response to *B. cinerea* remains to be revealed in future.

To understand the potential regulation of grape *LAC* expression, we analyzed the *cis*-elements of *VvLACs*, and seven classes of cis-elements were obtained. The putative *cis*-elements suggest that *VvLACs* participate in many physiological processes, such as development, and light and stress responses. MYB and WRKY are significant regulators in the plant defense response to *B. cinerea*, and they can interact with genes associated with cell wall metabolism [36,54]. Notably, the MYB *cis*-element was present in all *VvLAC* promoters. AtMYB58 and AtMYB63, directly interacting with *AtLAC4*, are specific transcriptional activators of lignin biosynthesis [55]. Therefore, we predicted that MYB may have crucial roles in *B. cinerea* resistance by targeting *LACs* in grape leaves, but further research is needed. The *VvLAC19* promoter had three WRKY-targeting W-box *cis*-elements, and *MdLAC7* can be bound by MdWRKY75 to improve apple resistance to *A. alternata* by increasing the biosynthesis of laccase and lignin [33]. There might be other regulators besides MYB and

WRKY in this response of grapevine. The stress *cis*-element STRE was also observed in candidate *VvLAC13*, *VvLAC24* and *VvLAC25* promoters. The STRE *cis*-element is a stress response promoter element widely involved in responding to various stressors, e.g., salt, water and heat, and is regulated by the transcriptional factors of zinc finger proteins and heat shock factors [56,57].

5. Conclusions

Laccases have been revealed to play important roles in plant disease resistance associated with lignin biosynthesis in recent years. However, information on grape laccase activity in the defense against *B. cinerea* is largely unknown. In the present study, a total of 30 grape laccase genes from the *V. vinifera* genome were identified and divided into seven groups. The proteins containing three typical plant Cu-oxidase domains displayed high conservation among each other. The phylogenetic analysis indicated that up to 25 *VvLACs* might be related to grape lignin biosynthesis, also implying a possible redundant function. The *cis*-element analysis highlighted MYB's and WRKY's regulating roles in grape *LAC* expression. Further investigation of *LAC* gene expression in *B. cinerea*-resistant Chinese wild *V. amurensis* 'Shuangyou' revealed that *VaLAC14*, *VaLAC19*, *VaLAC24* and *VaLAC30* could potentially act as the key candidate genes in lignin biosynthesis in the grape response to *B. cinerea*. This study lays the foundation for understanding the grape *LAC* genes and their roles in the response to *B. cinerea*. How these candidate genes participate in *B. cinerea* resistance by regulating lignification needs to be further verified.

Supplementary Materials: The following supporting information can be downloaded at: https://www.mdpi.com/article/10.3390/horticulturae10040376/s1, Table S1: Primers of selected *VaLAC* genes used for qRT-PCR, Table S2: Information on protein–protein network prediction.

Author Contributions: Conceptualization, R.W., Z.Y., J.S. and X.Z.; formal analysis, Z.Y., J.L., M.Z. and J.J.; visualization, R.W., Z.Y., M.W. and C.S.; investigation, Z.Y., J.L., J.S., Y.L. and K.Z.; resources, R.W., X.Z., P.H. and T.B.; writing—original draft, R.W. and Z.Y.; writing—review and editing, R.W., S.S., J.S. and X.Z. All authors have read and agreed to the published version of the manuscript.

Funding: We appreciate the support from Natural Science Foundation of Henan province (222300420182) and National Natural Science Foundation of China (32302495).

Data Availability Statement: The original contributions presented in the study are included in the article and supplementary materials.

Conflicts of Interest: The authors declare no conflicts of interest.

References

1. Shen, F.; Wu, W.; Han, X.; Wang, J.; Li, Y.; Liu, D. Study on the occurrence law and green control of grape gray mold from the perspective of ecological balance. *Bioengineered* **2021**, *12*, 779–790. [CrossRef] [PubMed]
2. De Simone, N.; Pace, B.; Grieco, F.; Chimienti, M.; Tyibilika, V.; Santoro, V.; Capozzi, V.; Colelli, G.; Spano, G.; Russo, P. *Botrytis cinerea* and table grapes: A review of the main physical, chemical, and bio-based control treatments in post-harvest. *Foods* **2020**, *9*, 1138. [CrossRef] [PubMed]
3. Wan, R.; Hou, X.; Wang, X.; Qu, J.; Singer, S.D.; Wang, Y.; Wang, X. Resistance evaluation of Chinese wild *Vitis* genotypes against *Botrytis cinerea* and different responses of resistant and susceptible hosts to the infection. *Front. Plant Sci.* **2015**, *6*, 160149. [CrossRef] [PubMed]
4. Harper, L.A.; Paton, S.; Hall, B.; McKay, S.; Oliver, R.P.; Lopez-Ruiz, F.J. Fungicide resistance characterized across seven modes of action in *Botrytis cinerea* isolated from Australian vineyards. *Pest Manag. Sci.* **2022**, *78*, 1326–1340. [CrossRef] [PubMed]
5. Bi, K.; Liang, Y.; Mengiste, T.; Sharon, A. Killing softly: A roadmap of *Botrytis cinerea* pathogenicity. *Trends Plant Sci.* **2023**, *28*, 211–222. [CrossRef]
6. Blanco-Ulate, B.; Morales-Cruz, A.; Amrine, K.C.H.; Labavitch, J.M.; Powell, A.L.T.; Cantu, D. Genome-wide transcriptional profiling of *Botrytis cinerea* genes targeting plant cell walls during infections of different hosts. *Front. Plant Sci.* **2014**, *5*, 98301. [CrossRef] [PubMed]
7. Veloso, J.; van Kan, J.A.L. Many shades of grey in *Botrytis*-host plant interactions. *Trends Plant Sci.* **2018**, *23*, 613–622. [CrossRef] [PubMed]
8. Goetz, G.; Fkyerat, A.; Métais, N.; Kunz, M.; Tabacchi, R.; Pezet, R.; Pont, V. Resistance factors to grey mould in grape berries: Identification of some phenolics inhibitors of *Botrytis cinerea* stilbene oxidase. *Phytochemistry* **1999**, *52*, 759–767. [CrossRef]

9. Liu, Q.; Luo, L.; Zheng, L. Lignins: Biosynthesis and biological functions in plants. *Int. J. Mol. Sci.* **2018**, *19*, 335. [CrossRef] [PubMed]
10. Lazniewska, J.; Macioszek, V.K.; Kononowicz, A.K. Plant-fungus interface: The role of surface structures in plant resistance and susceptibility to pathogenic fungi. *Physiol. Mol. Plant Pathol.* **2012**, *78*, 24–30. [CrossRef]
11. Lee, M.-H.; Jeon, H.S.; Kim, S.H.; Chung, J.H.; Roppolo, D.; Lee, H.-J.; Cho, H.J.; Tobimatsu, Y.; Ralph, J.; Park, O.K. Lignin-based barrier restricts pathogens to the infection site and confers resistance in plants. *Embo J.* **2019**, *38*, e101948. [CrossRef] [PubMed]
12. Asselbergh, B.; Curvers, K.; Franca, S.C.; Audenaert, K.; Vuylsteke, M.; Van Breusegem, F.; Hoefte, M. Resistance to *Botrytis cinerea* in sitiens, an abscisic acid-deficient tomato mutant, involves timely production of hydrogen peroxide and cell wall modifications in the epidermis. *Plant Physiol.* **2007**, *144*, 1863–1877. [CrossRef] [PubMed]
13. Lloyd, A.J.; Allwood, J.W.; Winder, C.L.; Dunn, W.B.; Heald, J.K.; Cristescu, S.M.; Sivakumaran, A.; Harren, F.J.M.; Mulema, J.; Denby, K.; et al. Metabolomic approaches reveal that cell wall modifications play a major role in ethylene-mediated resistance against *Botrytis cinerea*. *Plant J.* **2011**, *67*, 852–868. [CrossRef] [PubMed]
14. Xue, M.; Yi, H. Induction of disease resistance providing new insight into sulfur dioxide preservation in *Vitis vinifera* L. *Sci. Hortic.* **2017**, *225*, 567–573. [CrossRef]
15. Hamiduzzaman, M.M.; Jakab, G.; Barnavon, L.; Neuhaus, J.M.; Mauch-Mani, B. β-Aminobutyric acid-induced resistance against downy mildew in grapevine acts through the potentiation of callose formation and jasmonic acid signaling. *Mol. Plant-Microbe Interact.* **2005**, *18*, 819–829. [CrossRef] [PubMed]
16. Wallis, C.M.; Chen, J. Grapevine phenolic compounds in xylem sap and tissues are significantly altered during infection by *Xylella Fastidiosa*. *Phytopathology* **2012**, *102*, 816–826. [CrossRef] [PubMed]
17. Tziros, G.T.; Ainalidou, A.; Samaras, A.; Kollaros, M.; Karamanoli, K.; Menkissoglu-Spiroudi, U.; Karaoglanidis, G.S. Differences in defence-related gene expression and metabolite accumulation reveal insights into the resistance of Greek grape wine cultivars to *Botrytis* bunch rot. *Oeno One* **2022**, *56*, 111–123. [CrossRef]
18. Perkins, M.L.; Schuetz, M.; Unda, F.; Chen, K.T.; Bally, M.B.; Kulkarni, J.A.; Yan, Y.; Pico, J.; Castellarin, S.D.; Mansfield, S.D.; et al. Monolignol export by diffusion down a polymerization-induced concentration gradient. *Plant Cell* **2022**, *34*, 2080–2095. [CrossRef] [PubMed]
19. Yu, Y. LACCASE2 negatively regulates lignin deposition of *Arabidopsis* Roots. *Plant Physiol.* **2020**, *182*, 1190–1191. [CrossRef] [PubMed]
20. Zhao, Q.; Nakashima, J.; Chen, F.; Yin, Y.; Fu, C.; Yun, J.; Shao, H.; Wang, X.; Wang, Z.-Y.; Dixon, R.A. LACCASE is necessary and nonredundant with PEROXIDASE for lignin polymerization during vascular development in *Arabidopsis*. *Plant Cell* **2013**, *25*, 3976–3987. [CrossRef] [PubMed]
21. Liang, M.; Davis, E.; Gardner, D.; Cai, X.; Wu, Y. Involvement of AtLAC15 in lignin synthesis in seeds and in root elongation of *Arabidopsis*. *Planta* **2006**, *224*, 1185–1196. [CrossRef] [PubMed]
22. Berthet, S.; Demont-Caulet, N.; Pollet, B.; Bidzinski, P.; Cezard, L.; Le Bris, P.; Borrega, N.; Herve, J.; Blondet, E.; Balzergue, S.; et al. Disruption of LACCASE4 and 17 results in tissue-specific alterations to lignification of *Arabidopsis thaliana* stems. *Plant Cell* **2011**, *23*, 1124–1137. [CrossRef] [PubMed]
23. Perkins, M.L.; Schuetz, M.; Unda, F.; Smith, R.A.; Sibout, R.; Hoffmann, N.J.; Wong, D.C.J.; Castellarin, S.D.; Mansfield, S.D.; Samuels, L. Dwarfism of high-monolignol *Arabidopsis* plants is rescued by ectopic *LACCASE* overexpression. *Plant Direct* **2020**, *4*, e00265. [CrossRef] [PubMed]
24. He, F.; Machemer-Noonan, K.; Golfier, P.; Unda, F.; Dechert, J.; Zhang, W.; Hoffmann, N.; Samuels, L.; Mansfield, S.D.; Rausch, T.; et al. The in vivo impact of MsLAC1, a *Miscanthus* laccase isoform, on lignification and lignin composition contrasts with its in vitro substrate preference. *BMC Plant Biol.* **2019**, *19*, 552. [CrossRef]
25. Xue, C.; Yao, J.-L.; Qin, M.-F.; Zhang, M.-Y.; Allan, A.C.; Wang, D.-F.; Wu, J. PbrmiR397a regulates lignification during stone cell development in pear fruit. *Plant Biotechnol. J.* **2019**, *17*, 103–117. [CrossRef] [PubMed]
26. Qin, S.; Fan, C.; Li, X.; Li, Y.; Hu, J.; Li, C.; Luo, K. LACCASE14 is required for the deposition of guaiacyl lignin and affects cell wall digestibility in poplar. *Biotechnol. Biofuels* **2020**, *13*, 197. [CrossRef] [PubMed]
27. Wang, X.; Zhuo, C.; Xiao, X.; Wang, X.; Docampo-Palacios, M.; Chen, F.; Dixon, R.A. Substrate specificity of LACCASE8 facilitates polymerization of caffeyl alcohol for C-lignin biosynthesis in the seed coat of *Cleome hassleriana*. *Plant Cell* **2020**, *32*, 3825–3845. [CrossRef] [PubMed]
28. Balasubramanian, V.K.; Rai, K.M.; Thu, S.W.; Hii, M.M.; Mendu, V. Genome-wide identification of multifunctional laccase gene family in cotton (*Gossypium* spp.); expression and biochemical analysis during fiber development. *Sci. Rep.* **2016**, *6*, 34309. [CrossRef] [PubMed]
29. Wei, T.; Tang, Y.; Jia, P.; Zeng, Y.; Wang, B.; Wu, P.; Quan, Y.; Chen, A.; Li, Y.; Wu, J. A Cotton lignin biosynthesis gene, *GhLAC4*, fine-tuned by Ghr-miR397 modulates plant resistance against *Verticillium dahliae*. *Front. Plant Sci.* **2021**, *12*, 743795. [CrossRef] [PubMed]
30. Zhang, Y.; Wu, L.; Wang, X.; Chen, B.; Zhao, J.; Cui, J.; Li, Z.; Yang, J.; Wu, L.; Wu, J.; et al. The cotton laccase gene *GhLAC15* enhances *Verticillium* wilt resistance via an increase in defence-induced lignification and lignin components in the cell walls of plants. *Mol. Plant Pathol.* **2019**, *20*, 309–322. [CrossRef] [PubMed]
31. Hu, Q.; Min, L.; Yang, X.; Jin, S.; Zhang, L.; Li, Y.; Ma, Y.; Qi, X.; Li, D.; Liu, H.; et al. Laccase GhLac1 modulates broad-spectrum biotic stress tolerance via manipulating phenylpropanoid pathway and jasmonic acid synthesis. *Plant Physiol.* **2018**, *176*, 1808–1823. [CrossRef] [PubMed]

32. Yu, X.; Gong, H.; Cao, L.; Hou, Y.; Qu, S. MicroRNA397b negatively regulates resistance of *Malus hupehensis* to *Botryosphaeria dothidea* by modulating *MhLAC7* involved in lignin biosynthesis. *Plant Sci.* **2020**, *292*, 110390. [CrossRef] [PubMed]

33. Hou, Y.; Yu, X.; Chen, W.; Zhuang, W.; Wang, S.; Sun, C.; Cao, L.; Zhou, T.; Qu, S. MdWRKY75e enhances resistance to *Alternaria alternata* in *Malus domestica*. *Hortic. Res.* **2021**, *8*, 225. [CrossRef] [PubMed]

34. Zhu, Y.; Li, G.; Singh, J.; Khan, A.; Fazio, G.; Saltzgiver, M.; Xia, R. Laccase directed lignification is one of the major processes associated with the defense response against *Pythium ultimum* infection in apple roots. *Front. Plant Sci.* **2021**, *12*, 629776. [CrossRef] [PubMed]

35. Zhao, Y.; Liu, Y.; Dong, X.; Liu, J.-J.; Zhao, D.-G. Identification of a novel laccase gene *EuLAC1* and its potential resistance against *Botrytis cinerea*. *Transgenic Res.* **2022**, *31*, 215–225. [CrossRef] [PubMed]

36. Wan, R.; Guo, C.; Hou, X.; Zhu, Y.; Gao, M.; Hu, X.; Zhang, S.; Jiao, C.; Guo, R.; Li, Z.; et al. Comparative transcriptomic analysis highlights contrasting levels of resistance of *Vitis vinifera* and *Vitis amurensis* to *Botrytis cinerea*. *Hortic. Res.* **2021**, *8*, 103. [CrossRef] [PubMed]

37. Kumar, S.; Stecher, G.; Li, M.; Knyaz, C.; Tamura, K. MEGA X: Molecular evolutionary genetics analysis across computing platforms. *Mol. Biol. Evol.* **2018**, *35*, 1547–1549. [CrossRef] [PubMed]

38. Letunic, I.; Bork, P. Interactive tree of life (iTOL) v3: An online tool for the display and annotation of phylogenetic and other trees. *Nucleic Acids Res.* **2016**, *44*, W242–W245. [CrossRef] [PubMed]

39. McCaig, B.C.; Meagher, R.B.; Dean, J.F.D. Gene structure and molecular analysis of the laccase-like multicopper oxidase (LMCO) gene family in *Arabidopsis thaliana*. *Planta* **2005**, *221*, 619–636. [CrossRef] [PubMed]

40. Liao, B.; Wang, C.; Li, X.; Man, Y.; Ruan, H.; Zhao, Y. Genome-wide analysis of the *Populus trichocarpa* laccase gene family and functional identification of PtrLAC23. *Front. Plant Sci.* **2023**, *13*, 1063813. [CrossRef]

41. Xu, X.; Zhou, Y.; Wang, B.; Ding, L.; Wang, Y.; Luo, L.; Zhang, Y.; Kong, W. Genome-wide identification and characterization of laccase gene family in *Citrus sinensis*. *Gene* **2019**, *689*, 114–123. [CrossRef] [PubMed]

42. Lescot, M.; Déhais, P.; Thijs, G.; Marchal, K.; Moreau, Y.; Van de Peer, Y.; Rouzé, P.; Rombauts, S. PlantCARE, a database of plant cis-acting regulatory elements and a portal to tools for in silico analysis of promoter sequences. *Nucleic Acids Res.* **2002**, *30*, 325–327. [CrossRef] [PubMed]

43. Shi, J.; Yao, J.; Tong, R.; Wang, S.; Li, M.; Song, C.; Wan, R.; Jiao, J.; Zheng, X. Genome-wide identification of laccase gene family from *Punica granatum* and functional analysis towards potential involvement in lignin biosynthesis. *Horticulturae* **2023**, *9*, 918. [CrossRef]

44. Zhu, J.; Zhang, H.; Huang, K.; Guo, R.; Zhao, J.; Xie, H.; Zhu, J.; Gu, H.; Chen, H.; Li, G.; et al. Comprehensive analysis of the laccase gene family in tea plant highlights its roles in development and stress responses. *BMC Plant Biol.* **2023**, *23*, 129. [CrossRef] [PubMed]

45. Wan, R.; Song, J.; Lv, Z.; Qi, X.; Han, X.; Guo, Q.; Wang, S.; Shi, J.; Jian, Z.; Hu, Q.; et al. Genome-wide identification and comprehensive analysis of the *AP2/ERF* gene family in pomegranate fruit development and postharvest preservation. *Genes* **2022**, *13*, 895. [CrossRef]

46. Mayolo-Deloisa, K.; Gonzalez-Gonzalez, M.; Rito-Palomares, M. Laccases in food industry: Bioprocessing, potential industrial and biotechnological applications. *Front. Bioeng. Biotechnol.* **2020**, *8*, 222. [CrossRef] [PubMed]

47. Cai, X.; Davis, E.J.; Ballif, J.; Liang, M.; Bushman, E.; Haroldsen, V.; Torabinejad, J.; Wu, Y. Mutant identification and characterization of the laccase gene family in *Arabidopsis*. *J. Exp. Bot.* **2006**, *57*, 2563–2569. [CrossRef] [PubMed]

48. Liu, Q.; Luo, L.; Wang, X.; Shen, Z.; Zheng, L. Comprehensive Analysis of Rice Laccase Gene (OsLAC) Family and ectopic expression of *OsLAC10* enhances tolerance to copper stress in *Arabidopsis*. *Int. J. Mol. Sci.* **2017**, *18*, 209. [CrossRef] [PubMed]

49. Wen, J.; Liu, Y.; Yang, S.; Yang, Y.; Wang, C. Genome-wide characterization of laccase gene family from turnip and chinese cabbage and the role in xylem lignification in hypocotyls. *Horticulturae* **2022**, *8*, 522. [CrossRef]

50. Sun, Z.; Zhou, Y.; Hu, Y.; Jiang, N.; Hu, S.; Li, L.; Li, T. Identification of wheat LACCASEs in response to *Fusarium graminearum* as potential deoxynivalenol trappers. *Front. Plant Sci.* **2022**, *13*, 832800. [CrossRef] [PubMed]

51. Ramirez, V.; Garcia-Andrade, J.; Vera, P. Enhanced disease resistance to Botrytis cinerea in myb46 *Arabidopsis* plants is associated to an early down-regulation of *CesA* genes. *Plant Signal. Behav.* **2011**, *6*, 911–913. [CrossRef] [PubMed]

52. Wang, Y.; Xin, H.; Fan, P.; Zhang, J.; Liu, Y.; Dong, Y.; Wang, Z.; Yang, Y.; Zhang, Q.; Ming, Y.; et al. The genome of Shanputao (*Vitis amurensis*) provides a new insight into cold tolerance of grapevine. *Plant Journal* **2021**, *105*, 1495–1506. [CrossRef] [PubMed]

53. Zhu, Y.; Zhang, X.; Zhang, Q.; Chai, S.; Yin, W.; Gao, M.; Li, Z.; Wang, X. The transcription factors VaERF16 and VaMYB306 interact to enhance resistance of grapevine to *Botrytis cinerea* infection. *Mol. Plant Pathol.* **2022**, *23*, 1415–1432. [CrossRef] [PubMed]

54. Wang, M.; Zhu, Y.; Han, R.; Yin, W.; Guo, C.; Li, Z.; Wang, X. Expression of *Vitis amurensis* VaERF20 in *Arabidopsis thaliana* improves resistance to *Botrytis cinerea* and *Pseudomonas syringae* pv. tomato DC3000. *Int. J. Mol. Sci.* **2018**, *19*, 696. [CrossRef] [PubMed]

55. Zhou, J.; Lee, C.; Zhong, R.; Ye, Z. MYB58 and MYB63 are transcriptional activators of the lignin biosynthetic pathway during secondary cell wall formation in *Arabidopsis*. *Plant Cell* **2009**, *21*, 248–266. [CrossRef] [PubMed]

56. Fu, J.; Huang, S.; Qian, J.; Qing, H.; Wan, Z.; Cheng, H.; Zhang, C. Genome-Wide identification of *Petunia HSF* genes and potential function of *PhHSF19* in benzenoid/phenylpropanoid biosynthesis. *Int. J. Mol. Sci.* **2022**, *23*, 2974. [CrossRef] [PubMed]

57. Schmitt, A.P.; McEntee, K. Msn2p, a zinc finger DNA-binding protein, is the transcriptional activator of the multistress response in *Saccharomyces cerevisiae*. *Proc. Natl. Acad. Sci. USA* **1996**, *93*, 5777–5782. [CrossRef] [PubMed]

Article

Insights into the *PYR/PYL/RCAR* Gene Family in Pomegranates (*Punica granatum* L.): A Genome-Wide Study on Identification, Evolution, and Expression Analysis

Ke Yin [1,2], Fan Cheng [1,2], Hongfang Ren [1,2], Jingyi Huang [1,2], Xueqing Zhao [1,2] and Zhaohe Yuan [1,2,*]

[1] Co-Innovation Center for Sustainable Forestry in Southern China, Nanjing Forestry University, Nanjing 210037, China; keyin@njfu.edu.cn (K.Y.); chengfan990427@njfu.edu.cn (F.C.); fangfang@njfu.edu.cn (H.R.); nanlinhjy@njfu.edu.cn (J.H.); zhaoxq402@njfu.edu.cn (X.Z.)
[2] College of Forestry, Nanjing Forestry University, Nanjing 210037, China
* Correspondence: zhyuan88@hotmail.com

Abstract: The response of plants to abiotic stress is intricately mediated by *PYR/PYL/RCARs*, key components within the ABA signal transduction pathway. Despite the widespread identification of *PYL* genes across diverse plant species, the evolutionary history and structural characteristics of these genes within the pomegranate (*Punica granatum* L.) remained unexplored. In this study, we uncovered, for the first time, 12 *PgPYLs* from the whole genome dataset of 'Tunisia', mapping them onto five chromosomes and categorizing them into three distinct subgroups (Group I, Group II, and Group III) through phylogenetic analysis. Detailed examination of the composition of these genes revealed similar conserved motifs and exon–intron structures among genes within the same subgroup. Fragment duplication emerged as the primary mechanism driving the amplification of the *PYL* gene family, as evidenced by intra-species collinearity analysis. Furthermore, inter-species collinearity analysis provided insights into potential evolutionary relationships among the identified *PgPYL* genes. Cis-acting element analysis revealed a rich repertoire of stress and hormone response elements within the promoter region of *PgPYLs*, emphasizing their putative roles in diverse signaling pathways. Upon treatment with 100 μmol/L ABA, we investigated the expression patterns of the *PgPYL* gene family, and the qRT-PCR data indicated a significant up-regulation in the majority of *PYL* genes. This suggested an active involvement of *PgPYL* genes in the plant's response to exogenous ABA. Among them, *PgPYL1* was chosen as a candidate gene to explore the function of the gene family, and the CDS sequence of *PgPYL1* was cloned from pomegranate leaves with a full length of 657 bp, encoding 218 amino acids. Tobacco transient expression analysis demonstrated a consistent trend in the expression levels of pBI121-*PgPYL1* and the related genes of the ABA signaling pathway, both of which increased initially before declining. This study not only contributes to the elucidation of the genomic and structural attributes of *PgPYL* genes but also provides a foundation for understanding their potential functions in stress responses. The identified conserved motifs, evolutionary relationships, and expression patterns under ABA treatment pave the way for further research into the *PgPYL* gene family's role in pomegranate biology, offering valuable insights for future studies on genetic improvement and stress resilience in pomegranate cultivation.

Keywords: pomegranate; *PYL* gene family; bioinformatics; gene expression

Citation: Yin, K.; Cheng, F.; Ren, H.; Huang, J.; Zhao, X.; Yuan, Z. Insights into the *PYR/PYL/RCAR* Gene Family in Pomegranates (*Punica granatum* L.): A Genome-Wide Study on Identification, Evolution, and Expression Analysis. *Horticulturae* **2024**, *10*, 502. https://doi.org/10.3390/horticulturae10050502

Academic Editor: Adriana F. Sestras

Received: 1 April 2024
Revised: 10 May 2024
Accepted: 11 May 2024
Published: 13 May 2024

1. Introduction

The pomegranate (*Punica granatum* L.), a member of the genus *Punica* within the family Lythraceae stands as a globally significant economic forest species [1]. Renowned for its nutrient-rich fruit, containing vitamins, proteins, minerals, tannins, and flavonoids, pomegranates enjoy increasing popularity among consumers [2]. Despite their cultivation in both northern and southern regions of China, pomegranates, like many fruit trees, grapple with various stresses ranging from high salinity, drought, and low temperatures to nutritional deficiencies and pathogenic bacterial infestations during cultivation. These challenges substantially curtail their yield and quality, resulting in notable economic losses. Addressing this issue necessitates the identification of stress-resistant and high-quality genes in pomegranates, pursued through molecular methods and functional studies. The contemporary focus on excavating anti-stress genes in molecular breeding, facilitated by genome sequencing advancements, amplifies the significance of such investigations. The recent completion of pomegranate genome sequencing, including varieties like 'Taishanhong' [1], 'Dabenzi' [3], and 'Tunisia' [4], has furnished high-quality genomic maps, forming a critical knowledge foundation for comprehensive studies on pomegranate gene function, paving the way for systematic advancements in pomegranate molecular biology and genetic improvement.

Abscisic acid (ABA), a growth regulator of plants, orchestrates essential physiological functions, including embryo development, stomatal regulation, seed germination, and dormancy [5,6]. Notably, ABA assumes a central role in plants' responses to stress and adversity. When plants encounter abiotic stressors such as drought, hypersalinity, or low temperatures, a rapid increase in intracellular ABA content ensues. This surge regulates water balance and osmotic pressure, concurrently adjusting gene expression linked to adversity in response to environmental pressures [7]. The ABA signal transduction mechanism finely regulates ABA content in plants. This intricate pathway involves protein phosphatases 2C (*PP2C*) as negative regulators and sucrose nonfermenting 1-related protein kinases2 (*SnRK2*) as positive regulators [8–10]. In the absence of ABA, the *PYR/PYL/RCAR* receptors fail to interact with *PP2C*, allowing *PP2C* to exhibit high phosphatase activity. This, in turn, inhibits *SnRK2* kinase activity, impeding downstream gene expression in the ABA signal transduction pathway [11,12]. Conversely, under adverse conditions like hypersalinity, drought, or low temperatures, a rapid increase in ABA content prompts interaction with ABA receptors, initiating a cascade that inhibits *PP2C*, unleashing *SnRK2* kinase activity. Activated *SnRK2* phosphorylates downstream transcription factors, including *ABI5*, *AREB1*, and *AREB2*, which then attach to cis-acting elements in the downstream ABA-responsive genes' promoters, initiating the expression of ABA signaling-related genes [13–17].

The *PYR/PYL/RCAR* gene family, characterized by a conserved START structural domain [18], comprises fourteen members in *Arabidopsis*, divided into three subgroups, among which *AtPYL7*, *AtPYL8*, *AtPYL9*, and *AtPYL10* were classified under Group I, while *AtPYL4*, *AtPYL5*, *AtPYL6*, *AtPYL11*, *AtPYL12*, and *PgPYL13* were categorized under Group II, and *AtPYR1*, *AtPYL1*, *AtPYL2*, and *AtPYL3* were assigned to Group III [19]. Notably, overexpressing specific members, such as *AtPYL4*, *AtPYL5*, and *AtPYL9*, has demonstrated enhanced resistance to drought while simultaneously increasing sensitivity to ABA in *Arabidopsis* [12,20,21]. In tobacco, this gene family has been identified in 29 *NtPYLs* genes which played an important role in seed development, germination, and response to ABA, particularly in drought tolerance [22]. This gene family has been discovered in various plants, such as rice (13) [23], maize (13) [24], strawberry (11) [25], tomato (14) [26], and grape (6) [27], with evidence suggesting that the overexpression of ABA receptors heightens transgenic plants' ABA sensitivity and tolerance to abiotic stresses. For instance, in maize, *ZmPYL8*, *ZmPYL9*, and *ZmPYL12* play pivotal roles in drought tolerance [28]. The overexpression of *PtPYRL5* and *PtPYRL1* in poplar enhances reactive oxygen species scavenging [29], while *OsPYL/RCAR5* positively regulates ABA signal transduction, enhancing rice's resistance to drought and salinity stress and increasing seeds' sensitivity to ABA during seed germination and seedling growth stages [30,31].

Despite extensive exploration of the *PYL* gene family in various plants, its investigation in pomegranates (*Punica granatum* L.) has been limited. In alignment with the genomic information of the 'Tunisia' variety, we identified members of the *PYL* gene family in pomegranate and conducted a comprehensive analysis encompassing the chromosomal location, physicochemical properties of proteins, phylogenetic relationships, exon–intron gene structures, cis-acting elements, and collinearity. Additionally, we scrutinized the expression patterns of each member under ABA treatment. Finally, according to the results of the evolutionary tree, the amino acid sequence homology of *PgPYL1* and *AtPYL4* proteins is high, suggesting that their gene function may be similar, whereas *AtPYL4* has previously been reported to exhibit increased drought resistance in *Arabidopsis*. Therefore, the *PgPYL1* gene was cloned and analyzed for transient expression in tobacco. This study lays a foundational groundwork for further research into the functional aspects of the *PgPYL* gene family, offering prospective candidate genes for genetic engineering and pomegranate breeding.

2. Materials and Methods

2.1. Experimental Plant Materials and Treatment

The focus of the study was the 'Tunisia' variety of pomegranates. Hard branches that were one year old were chosen and clipped to a height of 10 to 15 cm. Subsequently, the cuttings were placed in pots measuring 32 cm by 25 cm and covered with a 1:1 mixture of perlite and charcoal soil for potting. The cuttings were cultivated under a 16 h light and 8 h dark photoperiod in a controlled environment chamber at Nanjing Forestry University. The diurnal temperature was sustained at 25 °C/18 °C, with 70% relative humidity. After 65 days of incubation, pomegranate cuttings exhibiting uniform, robust, and disease-free growth were chosen as the experimental materials for ABA treatment. Pomegranate cuttings were treated with 100 µmol/L ABA. The solution was evenly sprayed on the pomegranate leaves, and the spraying was stopped when water droplets were about to drop from the leaf edge. Three biological replications were required for the treatment, and 15 pomegranate cuttings were used as one biological replication. Pomegranate leaves were randomly collected at 0 (control check), 2, 6, 12, 24, and 48 h after treatment; quick-frozen in liquid nitrogen; and then brought back to the laboratory and stored in an ultra-low-temperature refrigerator at −80 °C for subsequent experiments.

2.2. Genome-Wide Identification of PgPYL gene Family

We downloaded the pomegranate whole genome information ('Tunisia' ASM765513v2) and the *PYL* sequences of *Arabidopsis* from the online program NCBI (http://www.ncbi.nlm.nih.gov/, accessed on 5 September 2023) and TAIR (http://www.Arabidopsis.org/, accessed on 5 September 2023), respectively. To identify the *PYL* gene family members in pomegranates, we employed two different methods. Firstly, the *PYL* sequences of *Arabidopsis* were used to find *PgPYLs* in the pomegranate genome with BLAST [32] software. The E-value was set to default (E < e^{-10}). Second, the Pfam (http://pfam-legacy.xfam.org/, accessed on 6 September 2023) was used to acquire the Hidden Markov Model (HMM) that represented the distinctive domain Polyketide_cyc2 (PF10604) of the *PYL* gene family proteins. The sequences of pomegranate were subsequently screened using the HMMER [33] software. Furthermore, the results obtained by the above two methods were pooled to acquire the candidate gene ID, and the protein sequence of the candidate gene was extracted by Tbtools software. Finally, we used the website SMART (http://smart.embl.de/, accessed on 6 September 2023) for candidate gene screening and revalidation of conserved structural domains and manually removed redundant sequences without Polyketide_cyc2 domains. Twelve confirmed members were finally obtained in pomegranate sequences. Based on the gene registration numbers, they were named *PgPYL1~PgPYL12*. Subsequently, the chromosomal coordinate result of the acquired 12 genes was submitted to the MG2C [34] (http://mg2c.iask.in/mg2cv2.1/, accessed on 7 September 2023) for visualization.

2.3. Proteins Physicochemical Properties, Structural Analysis, and Subcellular Localization Prediction

The fundamental physicochemical properties of the *PgPYL* gene family were recognized with Expasy-ProtParam [35] (http://web.expasy.org/protparam/, accessed on 7 September 2023). The prediction of signal peptides was conducted utilizing SignalP4.1 [36] (https://services.healthtech.dtu.dk/services/SignalP-4.1/, accessed on 7 September 2023). SoPMA [37] (http://npsa-pbil.ibcp.fr/cgi-bin/npsa_automat.pl?page=npsa_sopma.html, accessed on 7 September 2023) was applied to analyze the secondary structure of *PgPYL* proteins, and Plant-mPloc [38] (http://www.csbio.sjtu.edu.cn/bioinf/plant-multi/, accessed on 7 September 2023) was accessed to predict subcellular localization.

2.4. Phylogenetic Analysis

The *PYL* amino acid sequences in pomegranate were compared with those in *Arabidopsis*, grape, tomato, and strawberry plants with the Clustal W alignment tool. MEGA 11.0 [39] was employed to frame an evolutionary tree with the maximum likelihood (ML) method (Bootstrap = 1000). Subsequently, the files containing the evolutionary tree data were uploaded to the iTOL online platform (https://itol.embl.de/, accessed on 8 September 2023) for visualization enhancement.

2.5. Conserved Domains, Conserved Motif, and Exon–Intron Structure Analysis

We used bioinformatics software DNAMAN 9.0 to compare the 12 acquired *PgPYL* sequences. The structural domains of the *PgPYL* proteins were predicted with Batch-CD-Search (https://www.ncbi.nlm.nih.gov/Structure/bwrpsb/bwrpsb.cgi/, accessed on 9 September 2023). We used the program MEME [40] (http://meme-suite.org/meme/tools/meme, accessed on 9 September 2023) to obtain a conserved motif analysis with motifs set to 12 and default parameters for other settings. In order to analyze the gene structure, the 12 *PgPYL* genes' annotation data were extracted from genome-wide gff files in pomegranates and submitted to the GSDS [41] (http://gsds.gao-lab.org/, accessed on 9 September 2023). Subsequently, the protein structural domains, conserved motifs, and gene structures of *PgPYLs* were visualized with TBtools [42].

2.6. Promoter Cis-Acting Elements Analysis

We used software TBtools to extract sequences located 2000 bp upstream of the pomegranate genome start codon. Subsequently, potential cis-acting elements were identified using PlantCARE [43] (http://bioinformatics.psb.ugent.be/webtools/plantcare/html/, accessed on 10 September 2023), and the outcomes were also visualized by the software TBtools.

2.7. Collinearity and Selection Pressure Analysis

The MCScanX [44] program was applied to obtain the results of intraspecific collinearity in pomegranates for gene duplication pattern analysis. Circo plot of intraspecific collinearity was performed with TBtools and gene family member locations and collinearity were marked. The whole genome sequences in pomegranates were analyzed for interspecific collinearity with *Arabidopsis* and tomato by the MCScanX program, and the results of interspecific collinearity were presented with TBtools. To analyze the selection pressure of pomegranate, the Ka (non-synonymous substitution rate) and Ks (synonymous substitution rate) between *PgPYL* genes were calculated by DnaSP 5.0 [45] program.

2.8. RNA Extraction and Gene Clone

Total RNA was extracted from pomegranate leaf samples with the SteadyPure Plant RNA Extraction Kit (Accurate, Changsha, China). Subsequently, RNA was transcribed into cDNA by using cDNA synthesis kit (HiScript III RT SuperMix for qPCR (+gDNA wiper), Vazyme, Nanjing, China). Primer design was conducted using Primer Premier 5.0 software (Table S1), and primers were synthesized by Nanjing Tsingke Biotech Company, Nanjing, China.

The PCR amplification system comprised 50 μL total volume, consisting of 25 μL of 2× Rapid Taq Master Mix (Vazyme, Nanjing), 20 μL of ddH$_2$O, 2 μL of DNA template, and 1.5 μL of forward and reverse primers. The PCR reaction program was as follows: 95 °C, 5 min; 95 °C, 15 s; 54 °C, 20 s; 72 °C, 30 s with a total of 33 cycles; and 72 °C, 5 min. After the PCR products were separated by 1% agarose gel electrophoresis, the target bands were cut, and the products were recovered by DNA gel extraction kit (Tsingke, Nanjing).

Moreover, the recovered product was ligated to the laboratory-preserved vector pBI121 (constitutive promoter CaMV35S) using Clon Express® II One Step Cloning Kit (Vazyme, Nanjing). The ligation product was put into a 2 mL centrifuge tube with *Escherichia coli* DH5α receptor cell (Tsingke, Nanjing, China) at a ratio of 1:100 and transformed by heat shock method. Then antibiotic-free LB liquid medium was added to the centrifuge tube and incubated at 37 °C, 220 rpm for 40 min. The cultured fluid was evenly spread on the LB solid medium containing Kana antibiotic (concentration 50 μg/mL) and incubated at 37 °C for 16 h. When a single colony appeared in the medium, it was picked out and identified by PCR, the same PCR reaction program as above, and if a positive band appeared, it was sent to Nanjing Tsingke Biotech Company for sequencing. Finally, the sequence results obtained were translated into amino acid sequences using the ExPaSy-Translate website (https://web.expasy.org/translate/, accessed on 20 September 2023) and compared with DNAMAN software.

2.9. Tobacco Transient Expression

The correctly sequenced pBI121-*PgPYL1* plasmid was transferred into *Agrobacterium tumefaciens* GV3101 (Tsingke, Nanjing, China) by freeze–thaw method. The operation was as follows: the plasmid and GV3101 receptor cell were put into the centrifuge tube at a ratio of 1:50, which was put into an ice bath for 10 min, followed by liquid nitrogen for 5 min, then a 37 °C water bath for 5 min and ice bath for 10 min. In addition, used an oscillation rate of 220 r/min, the activated Agrobacterium was injected at a ratio of 1:100 into 50 mL of LB liquid medium and allowed to incubate for 16 h at 28 °C. The cultured bacterial solution was centrifuged at 4000 r/min for 10 min to collect the bacteria, which were resuspended in the prepared resuspension solution (10 mmol/L MES + 10 mmol/L $MgCl_2.6H_2O$ + 100 mmol/L AS, pH = 5.7). After standing for 3 h, they were injected into tobacco leaves, while wild-type (WT) tobacco was used as a control. The tobacco leaves were picked on days 1, 3, 5, and 7 and stored at −80 °C for subsequent gene expression studies.

2.10. Analysis of Quantitative Real-Time PCR (qRT-PCR)

Expression patterns of PgPYLs in pomegranate and tobacco leaves were detected using qRT-PCR. The Primer Premier 5.0 was used to design the specific quantitative primers (Table S1). The internal reference gene selected PgActin and NtActin. In addition, the online website Primer-BLAST (https://www.ncbi.nlm.nih.gov/tools/primer-blast/index.cgi?LINK_LOC=BlastHome, accessed on 5 September 2023) was used to check the specificity of the primers. The melting curve of the primers was verified as unimodal before the experiment.

The quantitative real-time PCR reaction system comprised 20 μL total volume, consisting of 10 μL of 2X SYBR® Green Pro Taq HS Premix (Accurate, Nanjing, China), 8.2 μL of RNase-free water, 1 μL of cDNA template (the RNA viscosity was about 150 ng/μL), and 0.4 μL of forward and reverse primers. The experiment was completed with Applied Biosystems 7500. The qRT-PCR reaction protocol was as follows: 95 °C, 30 s; 40 cycles of 95 °C, 5 s; and 60 °C, 30 s; and we collected melting curves with the instrument's default program. For each treatment, three technical and biological replicates had to be conducted. The relative expressions of *PgPYL* genes were analyzed by $2^{-\Delta\Delta CT}$ method [46]. The results were analyzed and plotted with SPSS 26.0 and GraphPad Prism 9 software, respectively.

2.11. Data Analysis

The data are shown as the mean ± SD of the three biological replicates. Duncan's test ($p < 0.05$) in SPSS 26.0 was selected to analyze the data differences' significance, and GraphPad Prism 9.0 was used for data visualization.

3. Results

3.1. Genome-Wide Identification and Chromosomal Location of PgPYLs

Through homology matching and de-redundancy, 12 *PgPYL* genes were identified in pomegranates. They were named *PgPYL1~PgPYL12* based on their gene accession numbers. The chromosomal localization of the *PgPYLs* was examined (Figure 1). The outcomes showed that these genes were sporadically dispersed over five pomegranate chromosomes. Specifically, four *PgPYLs* (*PgPYL3*, *PgPYL6*, *PgPYL10*, and *PgPYL12*) were located on chromosome 1, three *PgPYLs* (*PgPYL8*, *PgPYL9*, and *PgPYL11*) were distributed on chromosome 4, two *PgPYLs* (*PgPYL4*, *PgPYL5* and *PgPYL1*, *PgPYL2*) were located on chromosome 2 and 8 separately, and only *PgPYL* (*PgPYL7*) was distributed on chromosome 3. Notably, chromosomes 5, 6, and 7 did not exhibit any distribution of *PgPYLs*.

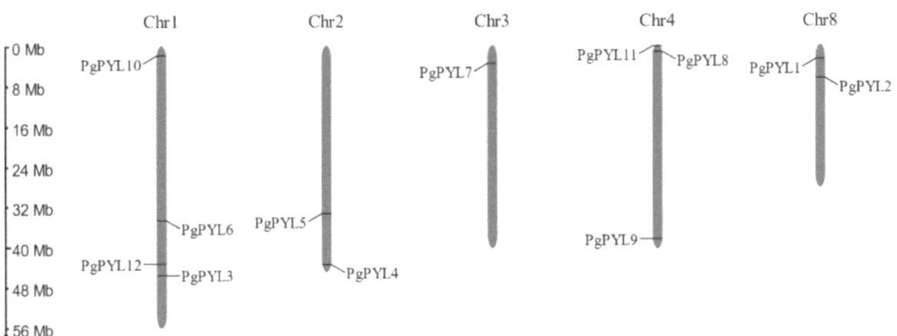

Figure 1. Chromosome distribution and localization of *PgPYL* genes.

3.2. Protein Physicochemical Properties, Structural Analysis, and Subcellular Localization Prediction of PgPYLs

The results of the basic physical and chemical properties, as displayed in Table S2, show that the *PgPYL* proteins contained 135 (*PgPYL4*) to 248 (*PgPYL5*) amino acids, an average of 203. *PgPYL5* was the largest protein; in contrast, *PgPYL4* was the shortest protein, which showed significant variation from the other family members. The molecular weight (MW) varied from 14.78 kDa (*PgPYL4*) to 26.92 kDa (*PgPYL5*), with an average of 20.52 kDa. The theoretical isoelectric point (PI) ranged from 5.06 (*PgPYL2*) to 6.54 (*PgPYL1*), with an average of 6.02. These values fall within the acidic range, indicating that the *PgPYL* gene family consisted of acidic proteins. The instability index of the *PgPYL* proteins ranged from 30.61 (*PgPYL7*) to 53.99 (*PgPYL10*), with an average of 41.25. The instability index of *PgPYL2*, *PgPYL3*, *PgPYL7*, and *PgPYL8* is less than 40, indicating that they are stable proteins, while the stability index of the other eight *PgPYL* proteins is more than 40, indicating that they are unstable proteins. The aliphatic index ranged from 79.12 (*PgPYL8*) to 95.23 (*PgPYL1*), with a mean value of 88.44. The grand average of hydrophilicity ranged from −0.409 (*PgPYL11*) to −0.032 (*PgPYL1*), with an average of −0.257, which are negative, suggesting that *PgPYLs* are hydrophilic proteins. Moreover, none of the twelve *PgPYL* proteins had signal peptides according to the results of the signal peptide prediction, indicating that they are non-secretory proteins.

The SoPMA and Plant-mPloc were used to predict protein secondary structure and subcellular localization. As shown in Table S2, all *PgPYL* proteins consisted of four distinct forms of secondary structures, with α-helices, extended strands, and random coils dominating the secondary structures of all amino acid sequences. Regarding the structural elements, *PgPYL7* exhibited the most α-helices, while *PgPYL9* displayed the most β-turns. Additionally, *PgPYL10* demonstrated the highest number of random coils. Furthermore, among the *PgPYL* proteins, *PgPYL4* and *PgPYL10* had the highest numbers of extended strands in comparison to the other *PgPYL* proteins.

Subcellular localization prediction analyses revealed that *PgPYL1*, *PgPYL2*, *PgPYL5*, *PgPYL6*, *PgPYL7*, *PgPYL8*, and *PgPYL10* were localized in the cytoplasm. *PgPYL4* and *PgPYL9* were found to be localized in the chloroplasts and nucleus, while *PgPYL11* and

PgPYL12 were specifically localized in the chloroplasts. In contrast, *PgPYL3* differed from the other members of the *PgPYL* gene family in that it was exclusively localized in the nucleus.

3.3. Phylogenetic Analysis of the PgPYL gene Family

To investigate the phylogenetic and developmental connections among the *PgPYL* gene family members, we conducted an alignment of the *PYL* sequences from pomegranate, *Arabidopsis*, grape, tomato, and strawberry plants (Table S3). Subsequently, an evolutionary tree was built with the maximum likelihood method. The phylogenetic development outcome (Figure 2) revealed that the *PYL* gene family members were categorized into three subgroups (Group I, Group II, and Group III). According to the phylogenetic tree grouping, *PgPYL4*, *PgPYL9*, *PgPYL11*, and *PgPYL12* were classified under Group I; while *PgPYL1*, *PgPYL5*, *PgPYL6*, and *PgPYL10* were categorized under Group II; and *PgPYL2*, *PgPYL3*, *PgPYL7*, and *PgPYL8* were assigned to Group III. The phylogenetic tree indicated that all *PgPYLs* were dispersed among various subgroups, suggesting that they possess unique functions. It can be seen from the evolutionary tree that in Group I, the three *FvPYL* genes (*FvPYL7*, *FvPYL8*, and *FvPYL16*) and the five *FvPYL* genes (*FvPYL1*, *FvPYL3*, *FvPYL4*, *FvPYL13*, and *FvPYL14*) of strawberry plants each formed a more independent clustered evolutionary branch. In Group II, the three *SlPYL* genes (*SlPYL15*, *SlPYL17*, and *SlPYL18*) of tomatoes, the three *AtPYL* genes (*AtPYL11*, *AtPYL12*, and *AtPYL13*) of *Arabidopsis*, and the two *PgPYL* genes (*PgPYL3* and *PgPYL8*) of pomegranates in Group III showed the same phenomenon. Therefore, we speculate that different members of the same species may have experienced gene duplication events during evolution. In addition, through phylogenetic analysis, we could infer similar functions of the *PgPYL* gene from genes with high homology to those of pomegranates. For example, we found that the amino acid sequences of *PgPYL1*, *SlPYL9*, and *AtPYL4* have a high degree of identity percentage, and the genes *SlPYL9* and *AtPYL4* have also been reported to enhance drought resistance [21,47], so we hypothesize that *PgPYL1* may have similar functions.

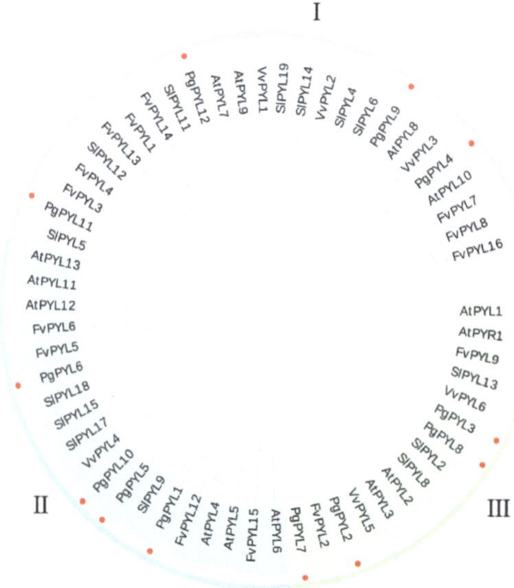

Figure 2. Phylogenetic analysis of *PYL* genes in pomegranate (*Punica granatum* L.), *Arabidopsis* (*Arabidopsis thaliana* L.), grape (*Vitis vinifera* L.), tomato (*Solanum lycopersicum* L.), and strawberry (*Fragaria vesca* L.). The different colors indicate different subgroups. The red dots represent *PgPYLs*.

3.4. Conserved Domains, Conserved Motif, and Intron–Exon Structure Analysis of PgPYLs

The *PgPYL* gene family sequences were analyzed using bioinformatic software DNA-MAN 9.0. The structural domain of the *PYL* proteins, which served as a direct receptor for ABA, was characterized by four highly conserved loops. Among them, the conserved structural domain CL2 was referred to as the gate or proline-cap. Similarly, the conserved structural domain CL4 was recognized as the latch or leucine-lucker. The GATE-LATCH structure of the *PYL* proteins changed their interaction with the ABA molecule, resulting in the formation of a surface that interacted with *PP2C* proteins while immobilizing the ABA molecule. Thus, the GATE-LATCH structure, which comprises two conserved structural domains, CL2 and CL4, is a distinctive characteristic of *PYL* proteins participating in the ABA signaling pathway, serving as receptors of ABA. The multiple amino acid sequence comparison of the *PgPYL* proteins (Figure 3) revealed that all 12 *PgPYL* proteins possess conserved structural domains, CL2 and CL4. Several amino acid sites within the structural domains were also observed to be conserved and invariant. Although some of the conserved structural domains of *PgPYL* genes exhibited base mutations, it was speculated that this phenomenon could be associated with their regulatory function. In the conserved CL2 structural domains of *PgPYL3* and *PgPYL6*, it was found that the serine(S) was substituted with threonine(T) in "SGLPA", which did not impact its function as a gatekeeper, as has been demonstrated in sorghum [48].

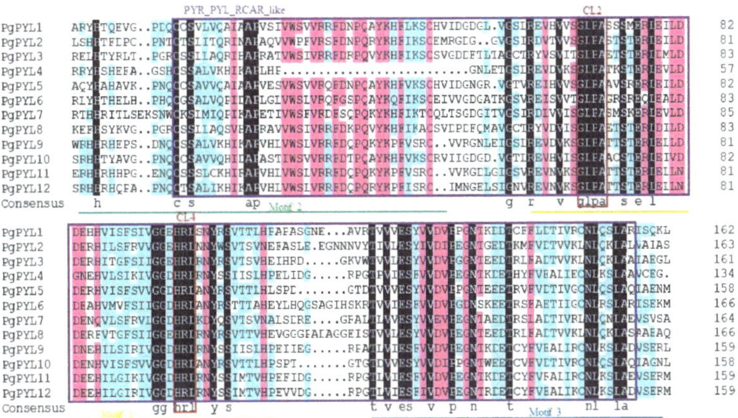

Figure 3. Multiple alignments of *PgPYL* proteins. The black, pink, and blue backgrounds indicate complete (100%), high (75%), and moderate (50%) conservatism, respectively. The purple square represents *PYR/PYL/RCAR* protein structural domains, and the red square represents conserved structural domains CL2 and CL4. The yellow, green, and blue lines represent motif1, motif2, and motif3, respectively.

The *PgPYL* proteins' conserved motifs were observed via the website MEME. Twelve conserved motifs are shown in Figure 4a, of which motif1, motif2, and motif3 have the most amino acids. The distribution of the conserved motifs in *PgPYLs* was established based on the motif analysis results (Figure 4b-B), and it was discovered that the number of motifs varied among the different *PgPYLs*, which confirmed the diversity of their functions. The *PgPYL* gene family encompasses 3–7 motifs, with *PgPYL7* exhibiting the fewest motifs, while *PgPYL9*, *PgPYL11*, and *PgPYL12* display the most motifs. All gene family members except *PgPYL4* contain motif1, motif2, and motif3. Therefore, it is presumed that motif1, motif2, and motif3 serve as the functional bases of *PgPYL* genes, and the positional regions of motif1, motif2, and motif3 also correspond to the positional regions of the conserved domains of PYR_PYL_RCAR_like protein. Except for *PgPYL7*, the majority of *PgPYLs* exhibit distinct motifs, indicating potential specificity in their biological functions. For example, *PgPYL3* and *PgPYL8* both contain motif4 and motif9, and *PgPYL6* features motif11.

Various arrangements and combinations of motifs constitute the specific structure of genes, and the allocation of these specific motifs serves as the theoretical foundation for gene categorization.

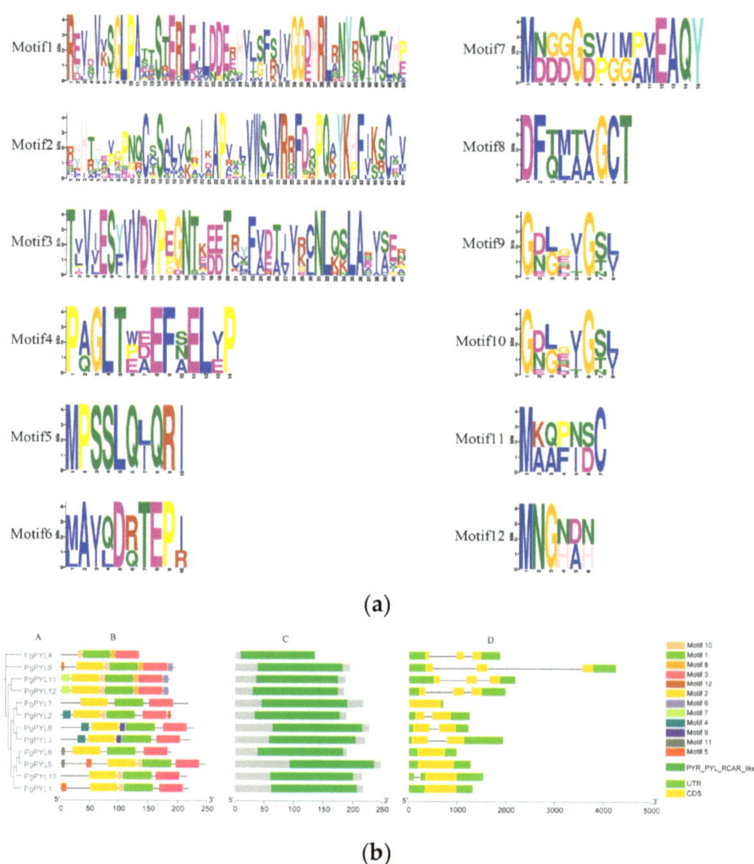

Figure 4. (**a**) Protein conserved motifs of the *PgPYL* genes. The horizontal coordinate indicates the position of the amino acid in the sequence, the vertical coordinate indicates the amino acid conservation, and the height of the amino acid letter indicates the frequency of occurrence; (**b**) conserved motifs and gene structure among *PgPYLs*. (A) The phylogenetic tree of *PgPYLs*; (B) conserved motifs of *PgPYLs*; (C) conserved domains of *PYR_PYL_RCAR* in *PgPYLs*; (D) exon–intron structure of *PgPYL* genes.

To explore the gene structure of the *PgPYL* gene family, the 12 *PgPYL* genes' structures were analyzed to attain deeper insights into the evolutionary process of the *PgPYL* gene family. The results are shown in Figure 4b-D; as far as gene structure is concerned, the *PgPYL* gene family does not fall into the category of exon-enriched gene families, which only contain 1–3 exons. It has been demonstrated that most genes exhibiting similar exon–intron structures within a specific gene family have akin evolutionary relationships. There are three distinct exon–intron structures in the *PgPYL* gene family. Specifically, four members of Group I (*PgPYL4*, *PgPYL9*, *PgPYL11*, and *PgPYL12*) possess three exons and two introns. Another member of Group II (*PgPYL10*) exhibits two exons and one intron, while the remaining three members of Group II (*PgPYL1*, *PgPYL5*, and *PgPYL6*) display one exon and no introns. Additionally, three members of Group III (*PgPYL2*, *PgPYL3*, and *PgPYL8*) have two exons and one intron, while the remaining member of Group III

(*PgPYL7*) has one exon and no introns. The observed outcome may be attributed to the insertion or removal of introns within the gene structure.

3.5. Cis-Acting Elements Analysis of PgPYLs

We analyzed the cis-acting elements of 12 *PgPYLs* and visualized the results with TBtools to better comprehend the potential functions (Table S4). The results revealed that the promoter region contained 33 cis-acting elements (Figure 5), which divided into three classifications: stress-response, growth and developmental, and hormone-response. Certain *PgPYLs'* promoter regions exhibited MYB binding sites, which are important in drought induction (MBS) [49] and light response (MRE), respectively. Apart from the MRE, the promoter regions of *PgPYLs* contained G-box, Box 4, GATA-motif, and GT1-motif, which related to light response [50]. This observation suggests that *PgPYLs* may be regulated by light. Furthermore, the twelve *PgPYL* genes possessed low-temperature-response elements (LTR), and seven *PgPYL* genes contained wound-responsive elements (WUN-motif). We also identified several elements that are directly linked to the growth and development, including meristem expression regulatory elements (CAT-box), cell cycle regulation element (MSA-like), an element controlling circadian rhythms (circadian), and an element involved in zein metabolism regulation (O2-site). Furthermore, the *PgPYL* promoter region was found to contain five kinds of hormone-related components, including the abscisic acid response element, growth hormone response elements, methyl jasmonate response elements, gibberellin response elements, and salicylic acid response element. This finding indicates that other hormones also regulate the response of *PgPYL* to ABA and may entail a more intricate signal transduction network.

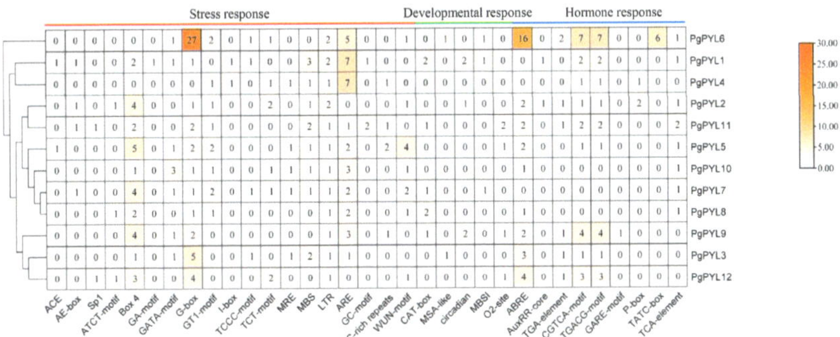

Figure 5. The cis-acting element analysis of *PgPYL* genes. The numbers represent the number of cis-acting elements. The red line represents stress response. The green line represents development response. The blue line represents hormone response.

3.6. Collinearity and Selection Pressure Analysis of PgPYLs

The creation of new genes and functional differentiation depended on gene duplication [51]. Analysis of whole-genome duplication could provide a more comprehensive exploration of the evolutionary relationships in the *PgPYL* genes. In the *PgPYLs'* duplication events analysis, six pairs of genome duplications were identified (Figure 6): *PgPYL1* and *PgPYL5*, *PgPYL3* and *PgPYL8*, *PgPYL4* and *PgPYL9*, *PgPYL4* and *PgPYL12*, *PgPYL9* and *PgPYL11*, and *PgPYL9* and *PgPYL12*. The outcome implies that gene duplication might have produced certain *PgPYL* genes and potentially facilitated the extended evolution of *PgPYL* gene family. In addition, we found that *PgPYL2*, *PgPYL6*, *PgPYL7* and *PgPYL10* appear to have no other genes with which they are collinear, suggesting that these genes may have evolved without replication events.

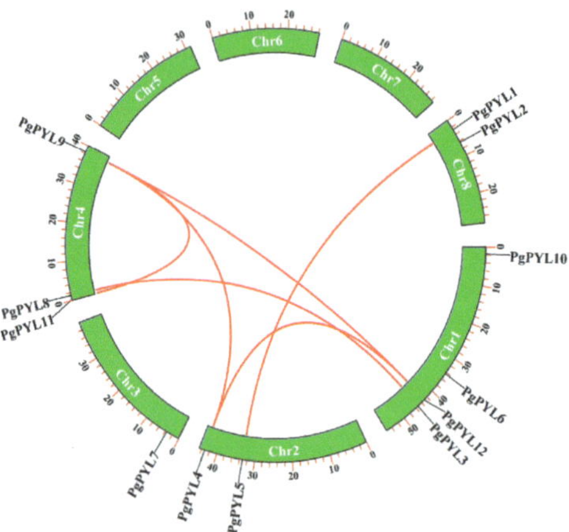

Figure 6. Collinearity analysis of the *PgPYL* genes. The red lines denote the *PgPYL*-duplicated gene pairs.

The collinearity among *Arabidopsis*, tomato, and pomegranate genes was analyzed (Figure 7) to ascertain the homology between the *PgPYL* genes and other plants. There were 21 pairs of collinearities observed in pomegranate and *Arabidopsis* and 28 pairs of collinearities in pomegranate and tomato. Several *PgPYL* genes were linked to at least two pairs of collinearities. In particular, *PgPYL1*, *PgPYL4*, *PgPYL5*, and *PgPYL9* were closely linked to the remainder of the genome and likely played significant components in the evolution of the *PYL* gene family. The collinearity analyses of the *PYLs* in pomegranate trees and other plants hold significant importance in establishing interspecific affinities and predicting gene functions.

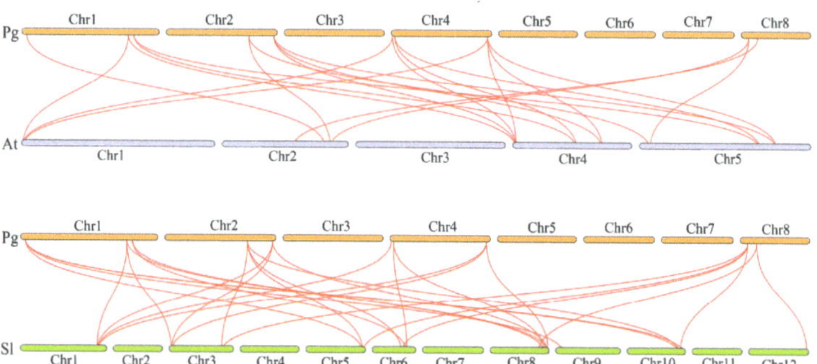

Figure 7. Syntenic relationships of the *PYL* gene family in pomegranate (*Punica granatum* L.) with Arabidopsis (*Arabidopsis thaliana* L.) and tomato (*Solanum lycopersicum* L.). Red lines denote the syntenic gene pairs of *PYLs*.

The Ka/Ks rate plays a crucial role in the evolutionary analysis of gene families and can be used to evaluate the existence of selective pressure on the gene encoding the protein. Ks denotes a synonymous substitution rate, which does not impact the type of amino acid, while Ka represents a nonsynonymous substitution rate, which modifies the expression of the amino acid. Nonsynonymous substitutions are predominantly detrimental, and purifying selection promotes their gradual decrease. A few nonsynonymous substitutions confer benefits to the organism, and base substitutions are retained by positive selection [52]. The Ka/Ks analysis between the coding regions of the *PgPYLs* (Table 1) revealed that the Ka/Ks is less than 1, suggesting that the *PgPYL* genes were subjected to purify selection.

Table 1. Selection pressure analysis of *PgPYL* gene family.

Gene Pairs	Nonsynonymous Substitution Rate, Ka	Synonymous Substitution Rate, Ks	Ka/Ks
PgPYL1 and *PgPYL5*	0.2414	0.8796	0.2744
PgPYL3 and *PgPYL8*	0.2350	0.8092	0.2904
PgPYL4 and *PgPYL9*	0.1065	1.3583	0.0784
PgPYL4 and *PgPYL12*	0.1860	1.9159	0.0971
PgPYL9 and *PgPYL11*	0.1634	1.2598	0.1297

3.7. Expression of PgPYLs under ABA Treatment

After 100 µmol/L ABA treatment, the relative expression levels of *PgPYLs* were detected using quantitative real-time PCR (qRT-PCR). The findings (Figure 8) indicate that the presence of the *PgPYL* genes exhibited diverse responses to ABA, accompanied by notable variations in relative expression levels. Eight genes (*PgPYL1*, *PgPYL2*, *PgPYL3*, *PgPYL4*, *PgPYL7*, *PgPYL8*, *PgPYL9*, and *PgPYL11*) exhibited up-regulation in expression. In comparison, four genes (*PgPYL5*, *PgPYL6*, *PgPYL10*, and *PgPYL12*) showed down-regulation. It was speculated that *PgPYL* genes had positive and negative responses to 100 µmol/L ABA treatment. Throughout the course of the treatment, the relative expression patterns of *PgPYL1*, *PgPYL2*, *PgPYL3*, *PgPYL7*, *PgPYL8*, and *PgPYL9* all displayed a similar tendency, with an initial rise, followed by a decline and then another increase. Notably, *PgPYL5* exhibited low expression or was hardly expressed from 2 h to 48 h. Besides *PgPYL5*, *PgPYL6*, *PgPYL10*, and *PgPYL12*, the relative expression of the remaining eight genes rose significantly after 2 h of ABA experiment. Subsequently, after 24 h of ABA treatment, the relative expression of the remaining nine genes demonstrated a decreasing trend, except for *PgPYL1*, *PgPYL3*, and *PgPYL6*.

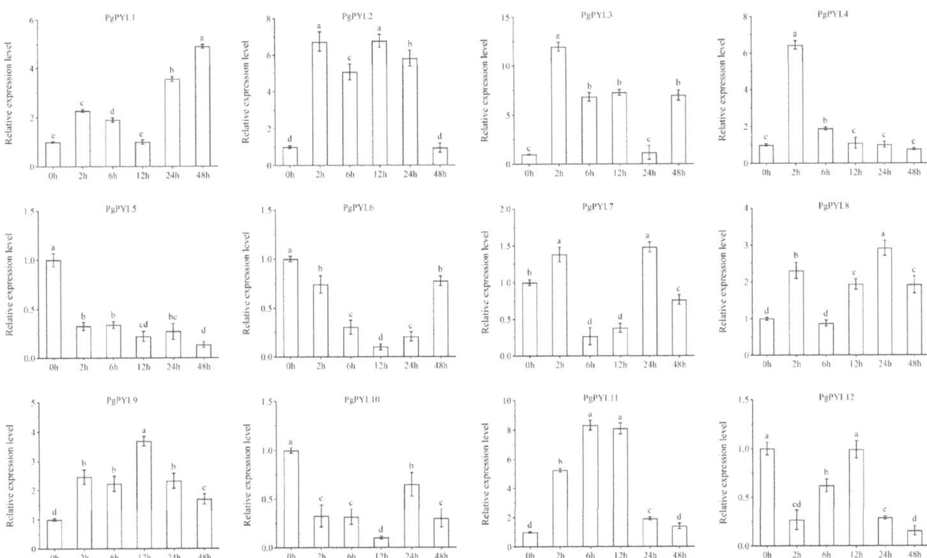

Figure 8. Expression patterns of the *PgPYL* gene family under 100 µmol/L ABA treatment. Pomegranate leaves were randomly collected at 0 (control check), 2, 6, 12, 24, and 48 h after treatment. Each gene's relative expression was calculated using the $2^{-\Delta\Delta Ct}$ method with *PgActin* serving as an internal control. The data show the mean ± SD of three biological replicates. The vertical bars show the standard error. Bars with different letters (a–e) indicate significant differences at $p < 0.05$ according to Duncan's test.

3.8. Gene Clone

We extracted the total RNA from 'Tunisia' pomegranate leaves and reverse-transcribed it into cDNA. The PCR-amplificated product was identified by 1% agarose gel electrophoresis. The results (Figure 9A) show that there was a clear and bright band between 500 and 750 bp for *PgPYL1*, which was consistent with the predicted size of the amplified fragment. Subsequently, the target fragment was recovered, ligated, transformed, and sequenced. The sequencing outcomes (Figure 9B) demonstrated that the cloned *PgPYL1* was identical to the expected results, and the coding region of *PgPYL1* was obtained, with a length of 657 bp, encoding 218 amino acids. The gene contained the complete *PYR_PYL_RCAR_like* conserved structural domain.

A
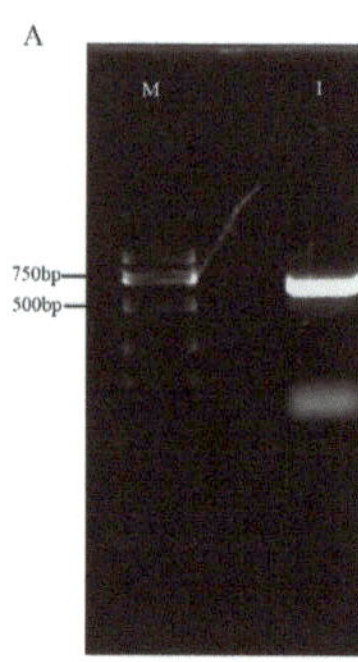

B
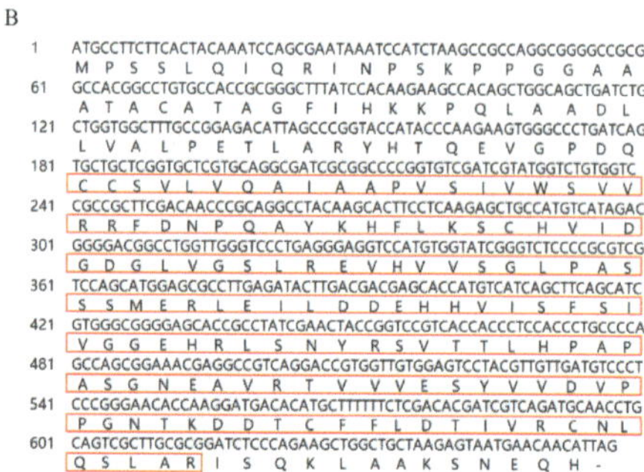

Figure 9. *PgPYL1* gene clone. (**A**) PCR amplification, M: DL ladder 2000 DNA Marker; 1: *PgPYL1*. (**B**) CDS sequence and coding amino acid sequence. The red square is the *PYR_PYL_RCAR_like* domain.

3.9. Tobacco Transient Expression Analysis

The constructed pBI121-*PgPYL1* vector was subjected to transient expression analysis in tobacco, as shown in Figure 10B; in tobacco, the *PgPYL1* gene was significantly more highly expressed than both the wild-type (WT) and the unloaded pBI121. The data showed a tendency of rising and then falling over time, with a notable increase in the expression level starting on day 3 and reaching the highest level on day 5. The expression level was 6.64 times more than that of wild-type (WT) and 2.09 times higher than that of unloaded pBI121, and following the fifth day, the expression level began to decline. Furthermore, we measured the expression patterns of genes associated with the ABA signaling system in tobacco leaves. The data demonstrated that the *NtPP2C* genes' relative expression levels were down-regulated (Figure 10C), and the gene expression was substantially lower than that of the wild-type (WT) and the unloaded pBI121. In contrast, the *NtSnRK2* and *NtAREB* genes' expression levels were up-regulated and exhibited a tendency of increasing and then decreasing with the change in time, which was basically consistent with the tendency of *PgPYL1* gene expression in tobacco leaves (Figure 10D,E), suggesting that the overexpression of *PgPYL1* in tobacco may lead to a considerable up-regulation of *NtSnRK2* and *NtAREB* expression and a down-regulation of *NtPP2C* expression in tobacco leaves.

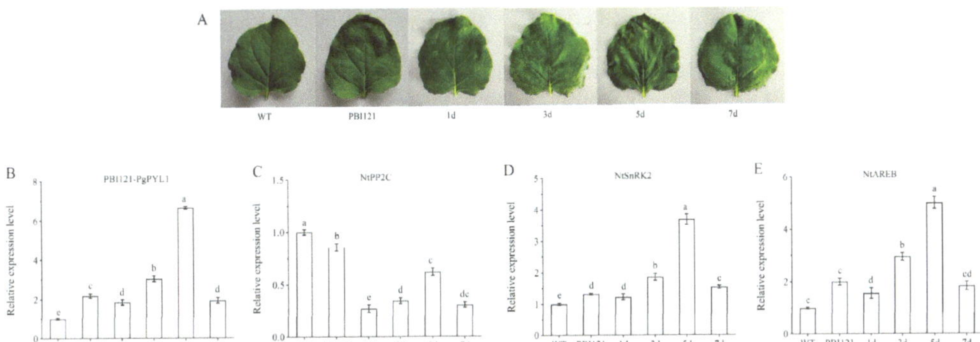

Figure 10. Tobacco transient expression. (**A**) Images of wild-type (WT), PBI121-overexpression, and *PgPYL1*-overexpression tobacco leaves. (**B**) *PgPYL1* expression, (**C**) *NtPP2C* expression, (**D**) *NtSnRK2* expression, and (**E**) *NtAREB* expression in WT, PBI121-overexpression, and *PgPYL1*-overexpression tobacco leaves. Bars with different letters (a–e) indicate significant differences at $p < 0.05$ according to Duncan's test.

4. Discussion

In orchestrating plant growth, development, and responses to both abiotic and biotic stresses, abscisic acid (ABA) assumes a pivotal role [53]. Positioned at the forefront of the ABA signaling regulation pathway, the *PYL* gene family stands out as an extensive group of phytohormone receptor genes primarily responsible for detecting ABA signals and initiating signaling [19]. Its significant contributions extend to influencing plant physiological and biochemical responses. Over recent years, the widespread presence of *PYL* genes has been unveiled in various species, such as tomato [26], grape [27], apple [54], and sweet cherry [55]. Environmental adversities, encompassing high salinity, drought, and low temperature, pose significant threats to normal plant growth, inhibiting photosynthesis, leading to premature plant failure, and ultimately affecting yield. The pomegranate industry, including fruit, seedlings, and processing, holds promising potential for development. However, their susceptibility to unfavorable environmental conditions jeopardizes pomegranates' yield and quality, resulting in considerable economic losses. Therefore, a crucial imperative emerges to excavate the function of the *PYL* gene family in pomegranates.

The discovery of 12 *PgPYL* genes through homology matching with known *PYL* protein sequences in *Arabidopsis* represents a noteworthy contribution. Notably, the variability in the number of *PgPYL* genes is comparatively lower than that observed in other species, such as *Arabidopsis* [19] (14), rice [23] (13), tomato [26] (14), and sweet cherry [55] (11). Chromosomal localization analysis unveiled a higher-density distribution of *PgPYL* genes on chromosomes 1 and 4 (Figure 1), hinting at potential tandem gene duplication. Physicochemical properties analysis characterized *PgPYL* proteins as acidic, with a negative total average hydrophilicity, rendering them hydrophilic. Secondary structure analysis indicated prevalent α-helices, extended strands, and random coils, which is consistent with studies on maize [56] and grape [57]. Subcellular localization analysis suggests a predominant expression of *PgPYL* genes in the cytoplasm, aligning with observations on *Arabidopsis* [19] and hazel [58]. Combining sequences from pomegranate, *Arabidopsis*, grape, tomato, and strawberry plants, an evolutionary tree built with the maximum likelihood method categorized the *PYL* gene family in pomegranate into three subgroups, aligning with classification outcomes of *Arabidopsis* [19] and maize [56].

Insights into the *PgPYL* protein sequences reveal binding sites for ABA, specifically within the conserved structural domains CL2 (gate) and CL4 (latch). This GATE-LATCH structure interacts with the ABA molecule, inducing conformational changes that form a surface interacting with *PP2C* proteins [59]. Mutations in certain structural domains suggest potential associations with executive function. Conserved motif analysis identified 12 motifs within the *PgPYL* family, with subgroups exhibiting similar motif numbers and types, indicating close evolutionary relationships and probable functional similarities [60,61]. Examination of the exon–intron gene structure highlighted that four genes (*PgPYL1*, *PgPYL5*, *PgPYL6*, and *PgPYL7*) lack introns, signifying a high level of conservation and potential representation of a more primitive evolutionary state. The remaining members with 1–2 introns may have evolved from other family members through mechanisms like exon shuffling.

Analyses of promoter cis-acting elements are helpful for predicting the potential functions of genes. The promoter region of the *PgPYL* genes contains various cis-elements, mainly involving light-responsive elements, hormone-responsive elements, stress-responsive elements, and growth and development-related elements. So, we speculate that the *PgPYL* genes not only regulate growth, development, and stress tolerance, but also that their expression is influenced by hormones and the environment during growth. In pomegranates, *PgPYL* gene family members contain 33 cis-type elements, 13 elements of which are related to light response, accounting for a large proportion, suggesting that *PgPYL* genes are affected by light regulation during growth and development, which is consistent with the results of strawberry [25] and wild emmer wheat [62] studies. Transcription factor MYB binding sites were found in the promoter region, among which MBS mainly participates in drought induction [49]. Eight *PgPYL* genes contain 1–3 MBS elements, with *PgPYL1* having the most MBS elements, suggesting that this gene may be involved in regulating the drought-response mechanism. In addition, there are also some genes, such as *PgPYL9*, containing meristem-expression regulatory elements (CAT-box) and an element involved in zein metabolism regulation (O2-site); *PgPYL6* has a cell cycle regulation element (MSA-like) and involvement in the regulation of flavonoid biosynthesis (MBSI); and *PgPYL2* has an element controlling circadian rhythms (circadian), indicating that these *PgPYL* genes may be involved in regulating pomegranate development and secondary metabolism. Hormonal signals play an important role in plant resistance to drought and other stress conditions. Among the twelve members of the *PYL* gene family, nine *PgPYL* genes have ABRE elements, while a few genes do not have ABRE elements, a phenomenon also present in other species such as sweet cherry [55], soybean [63], and tea tree [64]. In addition, the promoter region of the *PgPYL* genes also contains response elements for various hormones such as auxin, gibberellin, jasmonic acid, and salicylic acid, further indicating that the *PgPYL* genes play an important role in stress tolerance and may be influenced by various hormones, involving a more complex multi-hormone signaling transduction network, consistent with the results of soybean research [65].

Gene duplication research within the *PgPYL* gene family had advanced our knowledge of the expansion and functionality of the specific gene family. The examination of the duplication of the *PgPYLs* reveals six pairs of colinear genes, suggesting that these colinear genes may have arisen from gene duplication. The process of gene duplication can lead to the creation of multiple duplicated genes in the genomes of plants. Furthermore, the existence of duplicate genes may facilitate the emergence of novel gene functions and improve plants' capacity for environmental adaptation [66]. Moreover, to enhance further knowledge of the evolutionary relationship of the *PYL* gene family in pomegranates, 21 and 28 homologous chromosome pairs have been identified in pomegranate, *Arabidopsis*, and tomato plants. Purifying selection in *PgPYL* genes implies their crucial role in plant adaptation to adverse conditions. However, in the phylogenetic analysis, some of these genes seemed to have no evolutionary origin consistent with gene duplication, such as *PgPYL9* and *PgPYL11*. We speculate that gene rearrangement events within the gene family may have occurred, leading to changes in the gene sequence. This change may affect the outcome of gene replication, or the evolutionary origin of some genes seems inconsistent with the gene duplication results due to the use of different algorithms and models to construct phylogenetic trees.

In this study, pomegranate leaves were treated with 100 µmol/L ABA. The qRT-PCR data showed that most *PgPYL* genes were significantly up-regulated, and it was speculated that the *PgPYL* genes may be involved in endogenous ABA signaling and plant response to exogenous ABA. The outcome was consistent with previous research on tomatoes [67]. In other plants, for example, applying 100 μmol/L ABA to pears found that the expression levels of the remaining nine *PbrPYL* genes except *PbrPYL4* and *PbrPYL6* were up-regulated [61]. Applying 200 μmol/L ABA to sweet cherries found that the expression levels of *PaPYL6*, *PaPYL7*, *PaPYL8* and *PaPYL10* genes were up-regulated [55]. It seems that the expression patterns of *PYL* genes are diverse in different species. It is worth noting that we found that *PgPYL6* has 16 ABRE elements, but the expression level showed a downward trend. We speculated whether the gene is sensitive to high-concentration ABA treatment, thereby inhibiting its expression level. In addition, we found that the *PgPYL6* gene's expression level began to increase after 48 h of ABA treatment. It is speculated that this gene may also be involved in its adaptive regulation in the later stages of hormone stress. This requires further verification in our future experiments. This phenomenon is also seen in sweet cherry [55], where *PaPYL5* has five AREB elements, but after applying 200 μmol/L ABA, the expression level of the gene is down-regulated; *PaPYL6* and *PaPYL10* do not have ABRE elements, but at 12 h after treatment, the expression level is significantly higher than 0 h, so we guess there appears to be no correlation between gene expression levels and the ABA response elements found. Combined with the phylogenetic tree, it was found that *PgPYL4*, *PgPYL9*, and *PgPYL11* whose expression levels increased after ABA treatment were located in Group I; *PgPYL2*, *PgPYL3*, *PgPYL7*, and *PgPYL8* were located in Group II; while *PgPYL5*, *PgPYL6*, and *PgPYL10* whose expression levels decreased were located in Group III, so it is speculated that there may be a certain correlation between the evolutionary tree classification results and the expression levels of genes. This finding is consistent with the research on pears [61] and sweet cherries [55].

Studies have found that overexpression of *AtPYL4*, *AtPYL5* and *AtPYL9* genes in *Arabidopsis* can not only significantly increase the sensitivity of plants to ABA in stages such as seed germination, seedling growth, and stomatal opening and closing, but also improve the plant's drought resistance [20,21]. In addition, some scholars have found that the overexpression of ABA receptors can enhance the sensitivity of transgenic plants to ABA and improve the tolerance of these plants to abiotic stress. For example, in maize, *ZmPYL8*, *ZmPYL9*, and *ZmPYL12* play an important role in plant drought tolerance [28]. In rice, *OsPYL/RCAR5* has been reported to positively regulate ABA signaling, including increasing the sensitivity of seeds to ABA during germination and seedling growth stages and improving rice's resistance to drought and salt-alkali stress [30,31]. In wheat, over-expression of *TaPYL4* increased the sensitivity of wheat to ABA, reduced stomatal pore size and transpiration, and improved wheat yield and WUE (Water Usage Effectiveness) under drought conditions [68]. In this study, *PgPYL1* was transiently expressed in tobacco, and qPCR technology was used to detect genes related to the ABA signaling pathway in tobacco. In the ABA signaling pathway, *SnRK2* is a positive regulator, *PP2C* is a negative regulator, and *AREB* is the core transcription factor of the ABA signaling pathway and participates in signal transduction. The data showed that the expression of *NtPP2C* was generally down-regulated, while the expression of *NtSnRK2* and *NtAREB* was generally up-regulated. The expression of *PgPYL1* was consistent with the expression of the *NtSnRK2* and *NtAREB* genes, showing a trend of first increasing and then decreasing (Figure 10). It is speculated that *PgPYL1* may be involved in the ABA signal transduction pathway; however, the specific functional information of *PgPYL1* needs to be further explored and verified in future experiments.

5. Conclusions

In this investigation, our scrutiny of the pomegranate 'Tunisia' genome led to the identification of 12 *PgPYL* genes, each uniquely positioned on five distinct chromosomes. Employing phylogenetic research, we successfully classified the *PgPYL* gene family into three discernible subgroups. The comprehensive analysis of conserved motifs and intron–exon structures within this gene family unearthed noteworthy structural similarities among genes within the same subgroup, suggesting potential functional congruence in their encoded proteins. A pivotal observation emerged from our exploration of the amplification mechanisms within the *PgPYL* gene family, emphasizing fragment duplication as the primary mechanism. This mechanism not only shed light on the evolutionary dynamics of the gene family but also underscored the significance of this process in shaping the genomics of pomegranates. Delving into the regulatory elements governing *PgPYL* gene expression, our investigation of cis-acting elements within the promoter region unveiled the genes' responsiveness to a spectrum of stimuli, encompassing hormonal cues and abiotic stressors. This finding adds a layer of complexity to our understanding of how *PgPYL* genes integrate various signals, contributing to the plant's adaptive responses. The culmination of our efforts in this study resides in the qRT-PCR data, which provide compelling evidence implicating *PgPYL* genes in the plant's response to exogenous ABA. This insight into their involvement in ABA signaling pathways positions *PgPYL* genes as key players in the intricate molecular machinery orchestrating the plant's reactions to environmental stimuli. The results of tobacco transient expression demonstrated that the expression of pBI121-*PgPYL1* and the related genes of the ABA signaling pathway showed a consistent tendency of rising and then decreasing.

In conclusion, our investigation serves as a foundational exploration into the *PgPYL* gene family function, unraveling key aspects of their genomic organization, evolutionary history, and responsiveness to crucial signaling pathways. This study not only expands our knowledge of pomegranate genetics but also presents a reservoir of potential candidate genes for future endeavors in genetic engineering and pomegranate breeding. As we delve deeper into the intricate network of plant molecular responses, this work paves the way for additional inquiries, promising a more nuanced understanding of the *PgPYL* gene family's role in shaping the adaptive strategies of pomegranate in dynamic environments.

Supplementary Materials: The following supporting information can be downloaded at: https://www.mdpi.com/article/10.3390/horticulturae10050502/s1, Table S1: Primer sequences; Table S2: Physicochemical properties, secondary structure and subcellular localization of PgPYL proteins; Table S3: Information of constructing phylogenetic tree; Table S4: The cis-acting elements of PgPYLs.

Author Contributions: Conceptualization, K.Y. and Z.Y.; methodology, K.Y.; software, K.Y. and F.C.; validation, K.Y., F.C. and H.R.; formal analysis, K.Y. and J.H.; data curation, K.Y.; writing—original draft preparation, K.Y.; writing—review and editing, H.R., X.Z. and Z.Y.; visualization, K.Y.; supervision, Z.Y.; funding acquisition, Z.Y. All authors have read and agreed to the published version of the manuscript.

Funding: This research received no external funding.

Data Availability Statement: The raw data supporting the conclusions of this article will be made available by the authors on request.

Conflicts of Interest: The authors declare no conflicts of interest.

References

1. Yuan, Z.; Fang, Y.; Zhang, T.; Fei, Z.; Han, F.; Liu, C.; Liu, M.; Xiao, W.; Zhang, W.; Wu, S.; et al. The pomegranate (*Punica granatum* L.) genome provides insights into fruit quality and ovule developmental biology. *Plant Biotechnol. J.* **2018**, *16*, 1363–1374. [CrossRef]

2. Ben, L.; Kim, K.; Quah, C.; Kim, W.; Shahimi, M. Anti-inflammatory potential of ellagic acid, gallic acid and punicalagin A&B isolated from Punica granatum. *BMC Complement. Altern. Med.* **2017**, *17*, 47–57. [CrossRef] [PubMed]

3. Qin, G.; Xu, C.; Ming, R.; Tang, H.; Guyot, R.; Kramer, E.M.; Hu, Y.; Yi, X.; Qi, Y.; Xu, X.; et al. The pomegranate (*Punica granatum* L.) geneme and the genomics of punicalagin biosynthesis. *Plant J.* **2017**, *91*, 1108–1128. [CrossRef] [PubMed]

4. Chen, L.; Zhang, J.; Li, H.; Niu, J.; Xue, H.; Liu, B.; Wang, Q.; Luo, X.; Zhang, F.; Zhao, D.; et al. Transcriptomic Analysis Reveals Candidate Genes for Female Sterility in Pomegranate Flowers. *Front. Plant Sci.* **2017**, *8*, 1430. [CrossRef] [PubMed]

5. Finkelstein, R.R.; Gampala, S.S.; Rock, C.D. Abscisic acid signaling in seeds and seedlings. *Plant Cell* **2002**, *14*, S15–S45. [CrossRef] [PubMed]

6. Sreenivasulu, N.; Radchuk, V.; Strickert, M.; Miersch, O.; Weschke, W.; Wobus, U. Gene expression patterns reveal tissue-specific signaling networks controlling programmed cell death and ABA-regulated maturation in developing barley seeds. *Plant J.* **2006**, *47*, 310–327. [CrossRef] [PubMed]
7. Zhu, J.K. Salt and drought stress signal transduction in plants. *Annu. Rev. Plant Biol.* **2002**, *53*, 247–273. [CrossRef]
8. Melcher, K.; Ng, L.M.; Zhou, X.E.; Soon, F.F.; Xu, Y.; Park, S.Y.; Weiner, J.J.; Fujii, H.; Chinnusamy, V.; Kovach, A.; et al. A gate-latch-lock mechanism for hormone signalling by abscisic acid receptors. *Nature* **2009**, *462*, 602–608. [CrossRef] [PubMed]
9. Santiago, J.; Rodrigues, A.; Saez, A.; Rubio, S.; Antoni, R.; Dupeux, F.; Park, S.Y.; Márquez, J.A.; Cutler, S.R.; Rodriguez, P.L. Modulation of drought resistance by the abscisic acid receptor PYL5 through inhibition of clade A PP2Cs. *Plant J.* **2009**, *60*, 575–588. [CrossRef]
10. Umezawa, T.; Sugiyama, N.; Mizoguchi, M.; Hayashi, S.; Myouga, F.; Yamaguchi, S.K.; Ishihama, Y.; Hirayama, T.; Shinozaki, K. Type 2C protein phosphatases directly regulate abscisic acid-activated protein kinases in Arabidopsis. *Proc. Natl. Acad. Sci. USA* **2009**, *106*, 17588–17593.
11. Schweighofer, A.; Hirt, H.; Meskiene, I. Plant PP2c phosphatases: Emerging functions in stress signaling. *Trends Plant Sci.* **2004**, *9*, 236–243. [CrossRef] [PubMed]
12. Saavedra, X.; Modrego, A.; Rodríguez, D.; GonzálezG, M.P.; Sanz, L.; Nicolás, G.; Lorenzo, O. The nuclear interactor PYL8/RCAR3 of Fagus sylvatica FsPP2C1 is a positive regulator of abscisic acid signaling in seeds and stress. *Plant Physiol.* **2010**, *152*, 133–150. [CrossRef] [PubMed]
13. Yuan, X.; Yin, P.; Hao, Q.; Yan, C.; Wang, J.; Yan, N. Single amino acid alteration between valine and isoleucine determines the distinct pyrabactin selectivity by PYL1 and PYL2. *J. Biol. Chem.* **2010**, *285*, 28953–28958. [CrossRef] [PubMed]
14. Fujita, Y.; Nakashima, K.; Yoshida, T.; Katagiri, T.; Kidokoro, S.; Kanamori, N.; Umezawa, T.; Fujita, M.; Maruyama, K.; Ishiyama, K.; et al. Three SnRK2 protein kinases are the main positive regulators of abscisic acid signaling in response to water stress in Arabidopsis. *Plant Cell Physiol.* **2009**, *50*, 2123–2132. [CrossRef] [PubMed]
15. Sirichandra, C.; Davanture, M.; Turk, B.E.; Zivy, M.; Valot, B.; Leung, J.; Merlot, S. The arabidopsis ABA-activated kinase OST1 phosphorylates the bZIP transcription factor ABF3 and creates a 14-3-3 binding site involved in its turnover. *PLoS ONE* **2010**, *5*, e13935. [CrossRef] [PubMed]
16. Hubbard, K.E.; Nishimura, N.; Hitomi, K.; Getzoff, E.D.; Schroeder, J.I. Early abscisic acid signal transduction mechanisms: Newly discovered components and newly emerging questions. *Genes Dev.* **2010**, *24*, 1695–1708. [CrossRef] [PubMed]
17. Yoshida, T.; Fujita, Y.; Sayama, H.; Kidokoro, S.; Maruyama, K.; Mizoi, J.; Shinozaki, K.; YamaguchiS, K. AREB1, AREB2, and ABF3 are master transcription factors that cooperatively regulate ABRE-dependent ABA signaling involved in drought stress tolerance and require ABA for full activation. *Plant J.* **2010**, *61*, 672–685. [CrossRef] [PubMed]
18. Ma, Y.; Szostkiewicz, I.; Korte, A.; Moes, D.; Yang, Y.; Christmann, A.; Grill, E. Regulators of PP2C phosphatase activity function as abscisic acid sensors. *Science* **2009**, *324*, 1064–1068. [CrossRef]
19. Park, S.Y.; Fung, P.; Nishimura, N.; Jensen, D.R.; Fujii, H.; Zhao, Y.; Lumba, S.; Santiago, J.; Rodrigues, A.; Chow, T.F.; et al. Abscisic acid inhibits type 2C protein phosphatases via the PYR/PYL family of START proteins. *Science* **2009**, *324*, 1068–1071. [CrossRef]
20. Santiago, J.; Dupeux, F.; Round, A.; Antoni, R.; Park, S.Y.; Jamin, M.; Cutler, S.R.; Rodriguez, P.L.; Márquez, J.A. The abscisic acid receptor PYR1 in complex with abscisic acid. *Nature* **2009**, *462*, 665–683. [CrossRef]
21. Shi, H.; Ye, T.; Zhu, J.K.; Chan, Z. Constitutive production of nitric oxide leads to enhanced drought stress resistance and extensive transcriptional reprogramming in Arabidopsis. *J. Exp. Bot.* **2014**, *65*, 4119–4931. [CrossRef] [PubMed]
22. Bai, G.; Xie, H.; Yao, H.; Li, F.; Chen, X.; Zhang, Y.; Xiao, B.; Yang, J.; Li, Y.; Yang, D.H. Genome-wide identification and characterization of ABA receptor PYL/RCAR gene family reveals evolution and roles in drought stress in Nicotiana tabacum. *BMC Genom.* **2019**, *20*, 575. [CrossRef] [PubMed]
23. He, Y.; Hao, Q.; Li, W.; Yan, C.; Yan, N.; Yin, P. Identification and characterization of ABA receptors in Oryza sativa. *PLoS ONE* **2014**, *9*, e95246. [CrossRef]
24. Wang, Y.; Fu, F.; Yu, H.; Hu, T.; Zhang, Y.; Tao, Y.; Zhu, J.; Zhao, Y.; Li, W. Interaction network of core ABA signaling components in maize. *Plant Mol. Biol.* **2018**, *96*, 245–263. [CrossRef]
25. Chai, Y.M.; Jia, H.F.; Li, C.L.; Dong, Q.H.; Shen, Y.Y. FaPYR1 is involved in strawberry fruit ripening. *J. Exp. Bot.* **2011**, *62*, 5079–5089. [CrossRef] [PubMed]
26. Sun, L.; Wang, Y.P.; Chen, P.; Ren, J.; Ji, K.; Li, Q.; Li, P.; Dai, S.J.; Leng, P. Transcriptional regulation of SlPYL, SlPP2C, and SlSnRK2 gene families encoding ABA signal core components during tomato fruit development and drought stress. *J. Exp. Bot.* **2011**, *62*, 5659–5669. [CrossRef] [PubMed]
27. Boneh, U.; Biton, I.; Zheng, C.; Schwartz, A.; Ben-Ari, G. Characterization of potential ABA receptors in Vitis vinifera. *Plant Cell Rep.* **2012**, *31*, 311–321. [CrossRef]
28. He, Z.; Zhong, J.; Sun, X.; Wang, B.; Terzaghi, W.; Dai, M. The maize ABA receptors ZmPYL8, 9, and 12 facilitate plant drought resistance. *Front. Plant Sci.* **2018**, *9*, 422. [CrossRef]
29. Yu, J.; Ge, H.; Wang, X.; Tang, R.; Wang, Y.; Zhao, F.; Lan, W.; Luan, S.; Yang, L. Overexpression of pyrabactin resistance-like abscisic acid receptors enhances drought, osmotic, and cold tolerance in transgenic poplars. *Front. Plant Sci.* **2017**, *8*, 1752. [CrossRef]

30. Kim, H.; Hwang, H.; Hong, J.W.; Lee, Y.N.; Ahn, I.P.; Yoon, I.S.; Yoo, S.D.; Lee, S.; Lee, S.C.; Kim, B.G. A rice orthologue of the ABA receptor, OsPYL/RCAR5, is a positive regulator of the ABA signal transduction pathway in seed germination and early seedling growth. *J. Exp. Bot.* **2012**, *63*, 1013–1024. [CrossRef]
31. Kim, H.; Lee, K.; Hwang, H.; Bhatnagar, N.; Kim, D.Y.; Yoon, I.S.; Byun, M.O.; Kim, S.T.; Jung, K.H.; Kim, B.G. Overexpression of PYL5 in rice enhances drought tolerance, inhibits growth, and modulates gene expression. *J. Exp. Bot.* **2014**, *65*, 453–464. [CrossRef] [PubMed]
32. Altschul, S.F.; Gish, W.; Miller, W.; Myers, E.W.; Lipman, D.J. Basic local alignment search tool. *J. Mol. Biol.* **1990**, *215*, 403–410. [CrossRef] [PubMed]
33. Finn, R.D.; Clements, J.; Eddy, S.R. HMMER web server: Interactive sequence similarity searching. *Nucleic Acids Res.* **2011**, *39*, W29–W37. [CrossRef] [PubMed]
34. Chao, J.; Li, Z.; Sun, Y.; Aluko, O.O.; Wu, X.; Wang, Q.; Liu, G. MG2C: A user-friendly online tool for drawing genetic maps. *Mol. Hortic.* **2021**, *1*, 16. [CrossRef] [PubMed]
35. Artimo, P.; Jonnalagedda, M.; Arnold, K.; Baratin, D.; Csardi, G.; Duvaud, S.; Flegel, V.; Fortier, A.; Gasteiger, E.; Grosdidier, A. ExPASy: SIB bioinformatics resource portal. *Nucleic Acids Res.* **2012**, *40*, W597–W603. [CrossRef] [PubMed]
36. Petersen, T.N.; Brunak, S.; Heijne, G.; Nielsen, H. SignalP 4.0: Discriminating signal peptides from transmembrane regions. *Nat. Methods* **2011**, *8*, 785–786. [CrossRef] [PubMed]
37. Geourjon, C.; Deléage, G. SOPMA: Significant improvements in protein secondary structure prediction by consensus prediction from multiple alignments. *Comput. Appl. Biosci.* **1995**, *11*, 681–684. [CrossRef] [PubMed]
38. Chou, K.C.; Shen, H.B. Plant-mPLoc: A top-down strategy to augment the power for predicting plant protein subcellular localization. *PLoS ONE* **2010**, *5*, e11335. [CrossRef] [PubMed]
39. Tamura, K.; Stecher, G.; Kumar, S. MEGA11: Molecular Evolutionary Genetics Analysis Version 11. *Mol. Biol. Evol.* **2021**, *38*, 3022–3027. [CrossRef]
40. Bailey, T.L.; Boden, M.; Buske, F.A.; Frith, M.; Grant, C.E.; Clementi, L.; Ren, J.; Li, W.W.; Noble, W.S. MEME SUITE: Tools for motif discovery and searching. *Nucleic Acids Res.* **2009**, *37*, 202–208. [CrossRef]
41. Hu, B.; Jin, J.; Guo, A.Y.; Zhang, H.; Luo, J.; Gao, G. GSDS 2.0: An upgraded gene feature visualization server. *Bioinformatics* **2015**, *31*, 1296–1297. [CrossRef] [PubMed]
42. Chen, C.; Chen, H.; Zhang, Y.; Thomas, H.R.; Frank, M.H.; He, Y.; Xia, R. TBtools: An Integrative Toolkit Developed for Interactive Analyses of Big Biological Data. *Mol. Plant* **2020**, *13*, 1194–1202. [CrossRef] [PubMed]
43. Rombauts, S.; Déhais, P.; Montagu, M.V.; Rouzé, P. PlantCARE, a plant cis-acting regulatory element database. *Nucleic Acids Res.* **1999**, *27*, 295–296. [CrossRef] [PubMed]
44. Wang, Y.; Tang, H.; Debarry, J.D.; Tan, X.; Li, J.; Wang, X.; Lee, T.H.; Jin, H.; Marler, B.; Guo, H.; et al. MCScanX: A toolkit for detection and evolutionary analysis of gene synteny and collinearity. *Nucleic Acids Res.* **2012**, *40*, e49. [CrossRef] [PubMed]
45. Librado, P.; Rozas, J. DnaSP v5: A software for comprehensive analysis of DNA polymorphism data. *Bioinformatics* **2009**, *25*, 1451–1452. [CrossRef]
46. Livak, K.; Schmittgen, T. Analysis of relative gene expression data using real-time quantitative PCR and the $2^{-\Delta\Delta Ct}$ method. *Methods* **2001**, *25*, 402–408. [CrossRef] [PubMed]
47. Kai, W.; Wang, J.; Liang, B.; Fu, Y.; Zheng, Y.; Zhang, W.; Li, Q.; Leng, P. PYL9 is involved in the regulation of ABA signaling during tomato fruit ripening. *J. Exp. Bot.* **2019**, *70*, 6305–6319. [CrossRef] [PubMed]
48. Dalal, M.; Inupakutika, M. Transcriptional regulation of ABA core signaling component genes in sorghum (*Sorghum bicolor* L. Moench). *Mol. Breed.* **2014**, *34*, 1517–1525. [CrossRef]
49. Liu, J.H.; Peng, T.; Dai, W. Critical cis-acting elements and interacting transcription factors: Key players associated with abiotic stress responses in plants. *Plant Mol. Biol. Rep.* **2014**, *32*, 303–317. [CrossRef]
50. Cui, G.; Hou, J.; Tong, L.; Xu, Z. Light responsive elements and binding proteins of plant genes. *Plant Physiol. J.* **2010**, *46*, 991–1000.
51. Flagel, L.E.; Wendel, J.F. Gene duplication and evolutionary novelty in plants. *New Phytol.* **2009**, *183*, 557–564. [CrossRef] [PubMed]
52. Xia, Y.; Yan, Y. Advances in positive selection sites and their computational software. *J. Yangtze Univ. (Nat. Sci. Ed.)* **2016**, *13*, 51–53. [CrossRef]
53. Lee, S.C.; Luan, S. ABA signal transduction at the crossroad of biotic and abiotic stress responses. *Plant Cell Environ.* **2012**, *35*, 53–60. [CrossRef]
54. Hou, H.; Lv, L.; Huo, H.; Dai, H.; Zhang, Y. Genome-wide identification of the ABA receptors genes and their response to abiotic stress in apple. *Plants* **2020**, *9*, 1028. [CrossRef]
55. Zhou, J.; An, F.; Sun, Y.; Guo, R.; Pan, L.; Wan, T.; Hao, Y.; Cai, Y. Genome-wide identification of the ABA receptor *PYL* gene family and expression analysis in *Prunus avium* L. *Sci. Hortic.* **2023**, *313*, 111919. [CrossRef]
56. Li, H.; Wang, Y.; Zhang, X.; Fu, F.; Li, W. Bioinformatic analysis for abscisic acid perceptor gene family in maize. *J. Nucl. Agric. Sci.* **2015**, *29*, 1657–1667. [CrossRef]
57. Ma, Z.; Chen, B.; Li, W.; Mao, J. Identification and expression analysis of *PYL* gene families in grape. *J. Fruit Sci.* **2018**, *35*, 265–274. [CrossRef]
58. Zhang, X.; Zhao, Y.; Chen, Y.; Sun, B.; Liu, J. Genome-wide identification and expression analysis of *PYL* gene family in Corylus heterophylla during fruit development. *Acta Agric. Univ. Jiangxiensis* **2021**, *43*, 435–444. [CrossRef]

59. Yin, P.; Fan, H.; Hao, Q.; Yuan, X.; Wu, D.; Pang, Y.; Yan, C.; Li, W.; Wang, J.; Yan, N. Structural insights into the mechanism of abscisic acid signaling by PYL proteins. *Nat. Struct. Mol. Biol.* **2009**, *16*, 1230–1236. [CrossRef]
60. Zhang, Z.; Luo, S.; Liu, Z.; Wan, Z.; Gao, X.; Qiao, Y.; Yu, J.; Zhang, G. Genome-wide identification and expression analysis of the cucumber *PYL* gene family. *PeerJ* **2022**, *10*, 12786. [CrossRef]
61. Wang, G.; Qi, K.; Gao, X.; Guo, L.; Cao, P.; Li, Q.; Qiao, X.; Gu, C.; Zhang, S. Genome-wide identification and comparative analysis of the *PYL* gene family in eight Rosaceae species and expression analysis of seeds germination in pear. *BMC Genom.* **2022**, *23*, 233. [CrossRef]
62. Wang, Z.Y.; Yang, G.; Sai, N.; Pan, W.Q.; Song, W.N.; Nie, X.J. Identification and expression analysis of *PYL* gene family in wild emmer wheat (*Triticum dicoccoides* L.) under abiotic stress. *J. Triticeae Crops* **2022**, *42*, 1–10.
63. Zhang, Z.; Ali, S.; Zhang, T.; Wang, W.; Xie, L. Identification, evolutionary and expression analysis of PYL-PP2C-SnRK2s gene families in soybean. *Plants* **2020**, *9*, 1356. [CrossRef] [PubMed]
64. An, Y.; Mi, X.; Xia, X.; Qiao, D.; Yu, S.; Zheng, H.; Jing, T.; Zhang, F. Genome-wide identification of the *PYL* gene family of tea plants (*Camellia sinensis*) revealed its expression profiles under different stress and tissues. *BMC Genom.* **2023**, *24*, 362. [CrossRef] [PubMed]
65. Zhang, Z.H.; Zhang, T.X.; Wang, W.P.; Xie, L.N. Identification, phylogenetic evolution and expression analysis of abscisic acid receptors gene family in *Glycine max* L. Merr. *J. South. Agric.* **2020**, *51*, 1904–1916.
66. Panchy, N.; Lehti-Shiu, M.; Shiu, S. Evolution of gene duplication in plants. *Plant Physiol.* **2016**, *171*, 2294–2316. [CrossRef]
67. Wang, A.X.; Meng, L.J.; Chen, X.L.; Mo, F.L.; Lv, R.; Xue, X.P.; Meng, F.Y.; Qi, H.N.; Zhang, Z.Z. Genome-wide identification and expression analysis of abscisic acid receptor *PYL* gene family in tomato. *J. Northeast Agric. Univ.* **2023**, *54*, 21–32. [CrossRef]
68. Mega, R.; Abe, F.; Kim, J.S.; Tsuboi, Y.; Tanaka, K.; Kobayashi, H.; Sakata, Y.; Hanada, K.; Tsujimoto, H.; Kikuchi, J.; et al. Tuning water-use efficiency and drought tolerance in wheat using abscisic acid receptors. *Nat. Plants* **2019**, *5*, 153–159. [CrossRef]

Article

Identification of S-RNase Genotypes of 65 Almond [*Prunus dulcis* (Mill.) D.A. Webb] Germplasm Resources and Close Relatives

Panyun Xu [1,2,3], Lirong Wang [4], Xinwei Wang [4], Yeting Xu [1,2,3], Yarmuhammat Ablitip [1,2,3], Chunmiao Guo [1,2,3] and Mubarek Ayup [1,2,3,*]

[1] Horticultural Crop Research Institute, Xinjiang Academy of Agricultural Sciences, Urumqi 830091, China; xupanyun1215@163.com (P.X.); cherishhave@xaas.ac.cn (Y.X.); yarmuhammatablitip@163.com (Y.A.); chunmiaoguo@126.com (C.G.)

[2] Key Laboratory of Genome Research and Genetic Improvement of Xinjiang Characteristic Fruits and Vegetables, Urumqi 830091, China

[3] Xinjiang Fruit Tree Scientific Observation and Experiment Station, Ministry of Agriculture and Rural Affairs, Qaghiliq 844900, China

[4] Zhengzhou Fruit Research Institute, Chinese Academy of Agricultural Sciences, Zhengzhou 450009, China; wanglirong@caas.cn (L.W.); morningtoyou@126.com (X.W.)

* Correspondence: mubarekayup@163.com

Abstract: Self-incompatibility (SI) systems in plants prevent self-pollination and mating among relatives, enhancing genetic diversity in nature but posing challenges in almond production and breeding. S-allele composition alongside the flowering periods of these cultivars enables the anticipation of cross-compatibility and optimal cultivar combinations for the allocation of pollinating trees in production. In the current study, 65 materials containing 61 almond (*Prunus dulcis*) germplasm resources, of which two were hybrids and the remaining four were peach (*Prunus persica*) germplasms, were used for the S-RNase genotype. The results showed that 55 genomic samples were amplified by PCR to obtain double-banded types, which identified their complete S-RNase genotypes, while the rest of the samples amplified only a single band, identifying one *S-RNase* gene in the *S* gene. A total of 30 *S-RNase* genes were identified in *Prunus dulcis*, *Prunus webbii*, *Prunus persica*, *Prunus armeniaca*, *Prunus salicina*, and *Prunus cerasifera*. Sequence analysis revealed polymorphisms spanning from 313 to 2031 bp within the amplified fragment sequence. The S_{57}-*RNase* gene exhibited the highest frequency at 31.75% among the identified materials, with S_1S_{57}, $S_{10}S_{57}$, and S_7S_{57} being the predominant S genotypes. A new *S-RNase* gene, named S_{65}, was identified with a sequencing length of 1483 bp. Its deduced amino acid sequence shared 98.24% similarity with the amino acid sequence of the *S-RNase* gene on GenBank, with the highest homology. Furthermore, according to the findings, 65 materials belong to eight S genotype cross-incompatibility groups (CIG) and one semi-compatibility or compatibility group (0). Among them, most of the seven main almond germplasm resources and 35 cultivars can be cross-pollinated. The results of the study can lay the foundation for pollinator tree allocation and breeding hybrid parent selection in almond production.

Keywords: almond germplasm; S-RNase genotype; cross incompatibility groups; molecular identification

Citation: Xu, P.; Wang, L.; Wang, X.; Xu, Y.; Ablitip, Y.; Guo, C.; Ayup, M. Identification of S-RNase Genotypes of 65 Almond [*Prunus dulcis* (Mill.) D.A. Webb] Germplasm Resources and Close Relatives. *Horticulturae* **2024**, *10*, 545. https://doi.org/10.3390/horticulturae10060545

Academic Editor: Radu E. Sestras

Received: 20 April 2024
Revised: 18 May 2024
Accepted: 20 May 2024
Published: 23 May 2024

1. Introduction

The almond [*Prunus dulcis* (Mill.) D.A. Webb] is a cultivated species classified within the almond group of the subgenus *Amygdalus* L. within the genus *Prunus* L. of the Rosaceae family. The almond is renowned globally as a dry fruit and as a woody oil-bearing tree species. The seed kernels of almonds hold a distinguished status as one of the four primary dry fruits worldwide, alongside walnut, hazelnut, and cashew. The almond group includes plants such as *Prunus tangutica* Batal., *Prunus pedunculata* Maxim., *Prunus mongolica* Maxim., *Prunus ledebouriana* Schleche., *Prunus dulcis* (Mill.) D.A. Webb (cultivated almonds), and *Prunus triloba* Lindl. [1]. Almonds are believed to have originated in the Eastern Mediterranean and Central Asian regions and were then distributed to other parts of the world. The Xinjiang region is the only main production area of almonds suitable for cultivation in China, with a cultivation history of more than 1300 years. Yarkant County, Yengisar County, and Shufu County in the Kashgar region of Xinjiang are mostly cultivated. The main cultivars are 'Zhipi', 'Shuangguo', 'Shuangruan', 'Aifeng', 'Wanfeng', 'Yingzui', and 'Yeerqiang'. Most of these seven local cultivars are of the early-flowering, early- to mid-maturing type. As of the end of 2021, the planting area of almonds in Xinjiang reached 977,100 mu (=65,140 ha), yielding a total output of approximately 45,000 tons (seed kernels).

Almonds are cross-pollinated plants and are a typical gametophytic self-incompatibility (GSI) tree species. Cross-pollination is required during flowering to achieve a certain fruit-setting rate [2]. The failure of fruit setting after cross-pollination of two different varieties is a common phenomenon in almonds [3]. At present, the almonds used in cultivation and production in many central Asian regions are self-incompatible varieties. The unstable fruit-setting rate and low yield of almonds in production are caused by factors such as sandstorm weather during the flowering period, low temperatures in spring, and possible unreasonable allocation of pollinating trees, which hinder economic benefits from almond production. Among them, the scientific planning of pollination trees is closely related to the S genotype encoding self-incompatibility. The gametophytic self-incompatibility phenotype is controlled by multiple alleles (S-alleles) at a single locus (S-locus) containing at least two genes, the pistil *S*-gene (*S-RNase*) and the pollen *S*-gene (*SFB/SLF*) [4]. Mutations in the *S-RNase* gene of the style or *SFB/SLF* gene of the pollen can affect the mechanism of gamete incompatibility, resulting in the formation of self-compatible materials [5]. Under normal circumstances, when the S alleles of both parents are different, they pollinate each other, and the fruit-setting rate is higher, which is manifested as cross compatibility, and vice versa. When the parents have one of the same *S* genes, the fruit-setting rate is low, which is manifested as semi-cross-compatibility [6]. Therefore, the identification of the germplasm S genotype is very important for the allocation of pollinating trees in production and the design of effective cross combinations in breeding. In the past, electron microscopy was used to observe the growth of pollen tubes after pollination and to count the fruit-setting rate of ripe fruits for determining the difference in the S genotype, but this method was time-consuming, laborious, and inaccurate. At present, genomic DNA extraction, sequence amplification, and sequencing based on *S* gene fragments have become mainstream technologies for determining the S genotype of materials, and are more accurate, which can efficiently clarify the S genotype of research materials. He et al. [7] identified 84 S genotypes of native Chinese pear accessions by allele-specific PCR using one pair of consensus primers and 29 pairs of *S*-allele-specific primers, including 49 new pear varieties (*P. pyrifolia*, *P. sinkiangensis*, and inter-specific hybrids) and 35 wild pear varieties (*P. ussuriensis*, *P. pseudopashia*, *P. betulifolia*, *P. calleryana*, and landraces). Jiang et al. [8] used three pairs of Rosaceae universal primer combinations for PCR amplification of specific alleles of leaf genomic DNA of 27 apricot varieties to obtain four registered *S-RNase* genes of the genus Apricot, as well as 15 new *S-RNase* genes. Liu et al. [9] identified the S genotype of 88 *Malus* germplasm resources using seven pairs of *S-RNase* gene-specific primers, and the results showed that 70 materials obtained complete S-RNase genotypes.

In terms of almonds, more than 180 almond cultivars have been identified for *S* gene research in foreign countries [10,11]. Researchers grouped varieties or germplasms with the same S genotype into a hybrid incompatibility group (CIG), and currently 48 CIGs have been established [12,13]. Chinese researchers have also conducted related studies on the identification of the S genotype of almonds. Ma et al. [14] designed primers based on the conserved sequences of the *S* gene of Rosaceae species and amplified 10 *S* gene fragments of different sizes from the genome of eight Xinjiang almond varieties. Jiang et al. [15] identified the S genotype of 10 almond varieties using two pairs of universal primers of Rosaceae. Guo et al. [16] PCR-amplified almond varieties with multiple pairs of universal primers from Rosaceae and identified the S genotypes of 15 materials.

As one of the origin centers of almonds in the world, Xinjiang's topography and climate are very suitable for the growth of almonds. There is abundant phenotypic diversity of almond germplasm resources in the woodland. At present, research on the S genotype of almonds and close relatives in China is not comprehensive. Although the S genotypes of some germplasms have been determined, this has been mainly limited to more than ten local cultivars and no CIG has been established for domestic almonds. More materials for almonds and close relatives need to be identified and systematically analyzed for the S genotype. The phenomenon of the same variety name but completely different genetic backgrounds in almond production requires further clarification and identification. Therefore, in this study, PCR combined with DNA sequencing technology was used to identify the S-RNase genotypes of 65 almond germplasm resources and close relatives in Xinjiang, establish an almond CIG in China for the first time, and analyze the compatibility between seven main almond cultivars and 35 cultivars in Xinjiang. This will provide a reference for enriching the S genotype and genetic information of almonds, studying the self-incompatibility mechanism, selecting suitable pollinating trees in production, designing new hybrid combinations, and breeding new cultivars.

2. Materials and Methods

2.1. Materials

A total of 65 materials were used in the current study, which includes 61 almond (*Prunus dulcis*) germplasm resources such as *Prunus communis* Fritsch, *Prunus triloba* Lindl., *Prunus Mongolica* Maxim, and almond × peach natural hybrids available in Xinjiang, China. Additionally, 4 peach (*Prunus persica*) germplasms, including *Prunus ferganensis* Kost. et Riab and *Prunus davidiana* Franch. Most of the materials (62) were collected from the almond germplasm resource garden in Yarkant county, Xinjiang, China, while 3 Jinbian varieties ('Jinbian 1', 'Jinbian 2', and 'Jinbian 3') were obtained from Shanxi Agricultural University. Tender leaves of the mentioned materials were collected in spring 2022, frozen in liquid nitrogen, and stored in a −80 °C refrigerator for future use.

2.2. Extraction of Leaf Genomic DNA

Leaf genomic DNA was extracted using a plant genomic DNA rapid extraction kit (Aidlab Biotechnologies Co.,Ltd., Beijing, China). An appropriate amount of genomic DNA was taken to test its quality and concentration. After 1.5% agarose gel electrophoresis and nucleic acid protein analyzer detection, it was stored at −80 °C for standby.

2.3. Primer

S-RNase genotype identification was performed using Rosaceae consensus primer pairs 1 to 3 (with minor adjustments), as reported in the literature [17–20]. For materials that did not amplify or only amplified a single band, primers C1-F and C5-R were designed using Primer Premier 5.0 based on the highly conserved amino acid sequence of the *S-RNase* gene of *Prunus* L. in the Rosaceae family. In addition, primer pair 5 "PaConsI-F/C5-R" and primer pair 6 "C1-F/PruC4R" were combined for further identification (Table 1).

2.4. PCR Amplification of the S-RNase Gene

The leaf genomic DNA was used as a template to amplify the S-RNase allele of the test genotypes. The reaction volume was 25 μL and contained 12.5 μL of 2×Taq Master Mix (Vazyme, Nanjing, China), 1 μL of primers (10 μmol/L), 1 μL of DNA template (100 ng), and 9.5 μL of RNase-free ddH$_2$O. Amplification was performed in the Mastercycler nexus GSX1 (Eppendorf, Germany) gradient PCR instrument using the following procedure: predenaturation at 95 °C for 3 min; denaturation at 95 °C for 30 s, annealing at 53 °C for 1 min, extension at 72 °C for 1 min, 32 cycles; finally, extension at 72 °C for 5 min and storage at 4 °C. The amplified product was detected in 1.5% agarose gel in 1×TAE buffer, and the stained gel was photographed under UV light with a UVitec gel recorder.

Table 1. Primers used for S-RNase genotype identification and sequences.

Primer Pair	Primer	Sequence 5′-3′	Reference
1	AS1II	TATTTTCAATTTGTGCAACAATGG	[17]
	AmyC5R	CAAAATACCACTTCATGTAACAAC	
2	PaConsI-F	YCTTGTTCTTGSTTTYGCTTTMTTC	[18]
	EM-PC1consRD	GCCAYTGTTGCACAAAYTGAA	[19]
3	PruC2	CTATGGCCAAGTAATTATTCAAACC	[20]
	PruC4R	GGATGTGGTACGATTGAAGCG	
4	C1-F	HCARTTTGTGCAACARTGGC	
	C5-R	CCACTTCATGTAACARYTS	
5	PaConsI-F	YCTTGTTCTTGSTTTYGCTTTMTTC	[18]
	C5-R	CCACTTCATGTAACARYTS	
6	C1-F	HCARTTTGTGCAACARTGGC	
	PruC4R	GGATGTGGTACGATTGAAGCG	[20]

2.5. Cloning and Sequencing of the S-RNase Gene

After the PCR expansion product was imaged by 1.5% agarose electrophoresis and gel, the purified product was ligated to the pCE2 TA/Blunt-Zero vector, the recombinant vector was transferred to E. coli DH5α-competent cells, and the positive clones were sequenced at least three times to confirm the correct sequence.

2.6. Sequence Analysis

Sample sequencing was conducted by the Zhengzhou Branch of Shenggong Biotechnology (Shanghai) Co., Ltd. (Shanghai, China) using the ABI 3730XL gene sequencer. The process involved utilizing Bio XM 2.7 to eliminate vector sequences, Chromas for correction, SeqManII for integrating three duplicate sequencing results, and finally, employing the Blast program to conduct homology sequence retrieval with the S-RNase sequences already registered in the NCBI database (http://blast.ncbi.nlm.nih.gov/Blast.cgi, accessed on 19 April 2024).

Nucleotide and amino acid sequence analysis of the *S-RNase* gene was carried out using DNAMAN software (Version V6; Lynnon Biosoft, San Ramon, CA, USA). If the amino acid sequence consistency reaches 99%, it is considered to be the same gene. Conversely, if there is inconsistency, it is regarded as a new *S-RNase* gene. This process aids in determining the S-RNase allele information of the test material.

2.7. Construction of the Phylogenetic Tree of the S Gene

The amino acid sequence of *Prunus* L. S-RNase was downloaded from the GenBank database (https://www.ncbi.nlm.nih.gov/genbank/, accessed on 19 April 2024). With MEGA 11 software, the Clustal W method was used to conduct multiple sequence comparisons between the S-RNase obtained from the experiment and the downloaded amino acid sequence, and the phylogenetic tree was constructed using the neighbor-joining (NJ) method. The bootstrap values were kept 1000 times.

2.8. Construction of a Database of Hybridization Combinations

According to the principle of S genotype incompatibility, if the hybrid parent shares the same S genotype, the hybrid is incompatible. If the S genotype of the hybrid parent contains one identical *S* gene, the hybrid is semi-compatible. When the S genotype of the hybrid parent is completely different, the hybrid is compatible. A database of almond hybrid incompatibility can be constructed based on this principle. Additionally, an analysis of hybrid incompatibility/compatibility between seven main almond varieties planted in Xinjiang and 35 other varieties was conducted. This analysis provides a reference basis for selecting parents in production and breeding.

2.9. Field Validation of Cross-Compatibility of Some Almond Cultivars

From late March to early April 2022–2023, according to the hybrid combination in Table 2, the balloon-stage flower buds were collected, the anthers were peeled, dried in the shade, completely pollinated, and then loaded into dry small tubes, which were placed at $-20\ ^{\circ}\text{C}$ for later use. No less than 300 large bud-stage flowers were selected for pollination in each combination. To improve the accuracy of the results, the unopened and already blooming flowers on the hybrid flower branches were removed, and finally, the hybrid branches were bagged. The bagging was removed after 2 weeks of pollination, and the number of fruits were counted two months later.

Table 2. Cross-pollination and fruit setting of some almond cultivars in 2022–2023.

2022					2023				
Hybrid Combination (♀× ♂)	Genotype Combination	Hybrid Strain	Fruit Number	Fruit Rate	Hybrid Combination (♀× ♂)	Genotype Combination	Hybrid Strain	Fruit Number	Fruit Rate
Jinsha Fruit × Taohuaguazi Badan	$S_6S_{16} \times \underline{S_{10}}S_{24}$	542	161	29.70	XX2 × Paizhui	$S_7S_{57} \times S_8S_{16}$	608	76	12.50
Jinsha Fruit × Yeerqiang	$S_6S_{16} \times S_5S_{57}$	581	122	21.00	Shache 29 × Zhipi	$S_1S_{55} \times S_{n3}S_{57}$	532	115	21.62
Jinsha Fruit × XX3	$S_6S_{16} \times S_1S_{27}$	517	29	5.61	Shache 29 × XX2	$\overline{S_1}S_{55} \times \overline{S_7}S_{57}$	532	91	17.11
Yeerqiang holy Fruit × Yeerqiang	$S_7S_8 \times \overline{S_5}S_{57}$	384	55	14.32	Amannisha × Zhipi	$\overline{S_5}\underline{S_{10}} \times S_{n3}S_{57}$	717	85	11.85

Note: The underline indicates that the gene is not *Prunus dulcis S-RNase*.

According to Audergon et al. [21], a fruit-setting rate higher than 5% indicates cross-compatibility, whereas a rate lower than 5% suggests cross-incompatibility. This criterion is utilized to determine the cross-compatibility between different almond cultivars.

3. Results

3.1. Identification of S-RNase Alleles

PCR amplification of leaf DNA using consensus primers and self-designed degenerate primers showed that two different fragments were detected in 55 samples, and only one fragment was amplified in the remaining 10 samples (Figure 1). The size of the product fragments showed a certain degree of polymorphism, and the sequencing results showed that the sequence fragments ranged from 313 to 2031 bp. Akram et al. [12] also showed that the size of S-RNase allele fragments in 22 Iranian almond cultivars ranged from 600 to 2400 bp. The sequences and sizes of introns in different S genes were different and showed high differences. The coding region of each S-RNase was almost separated by introns at the same position. Therefore, it causes high variability in fragment sizes.

After multiple sequence alignment of all nucleotide sequences obtained through DNAMAN, it was found that the bands of different varieties located at the same position have more than 99% similarity in the nucleic acid sequence, which can be considered as the same gene fragment. A total of 30 nucleotide sequences of different sizes were isolated from 65 germplasm: 313 bp ('Jinbian 3'), 347 bp ('XX3'), 350 bp (*Prunus triloba*), 408 bp ('Yeerqiang holy Fruit'), 419 bp ('Jinbian 3', etc.), 591 bp ('Amannisha', etc.), 621 bp (*Prunus triloba*), 624 bp ('Paizhui'), 653 bp ('Aifeng', etc.), 722 bp (*Prunus mongolica*), 754 bp ('Badanwang', etc.), 772 bp ('Shuangruan', etc.), 797 bp ('Shache 29'), 797 bp ('Shuangruan', etc.), 811 bp ('Dabadan', etc.), 821 bp ('Wanhua Taobadan'), 822 bp ('Jinbian 2', etc.), 857 bp ('Jinsha Fruit', etc.), 982 bp ('Hanfeng'), 993 bp ('Jinbian 2'), 994 bp ('Baishuang', etc.), 1086 bp ('A1', etc.), 1130 bp ('Tianren Taobadan', etc.), 1254 bp ('Yuanruan', etc.), 1274 bp ('Baiboke', etc.), 1483 bp ('Shuangbo', etc.), 1592 bp ('Shache 29', etc.), 1634 bp ('Huangshuang', etc.), 1650 bp ('Shuangguo', etc.), and 2031 bp ('Shitou Almond', etc.).

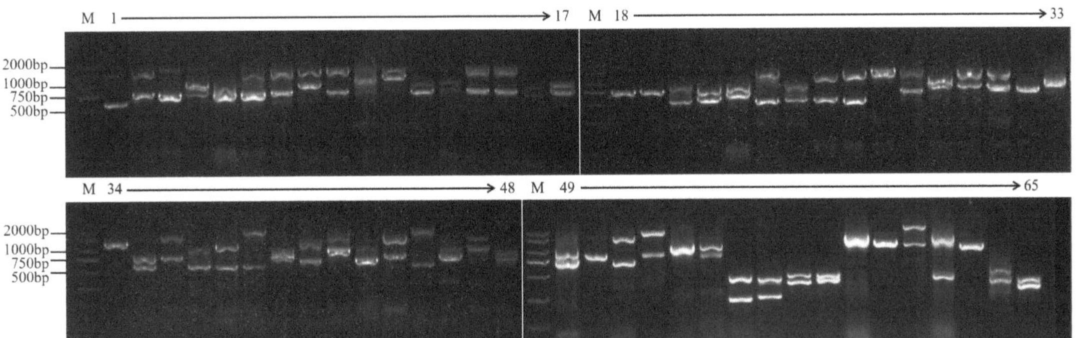

Figure 1. *S-RNase* gene amplification in 65 almond germplasm resources and close relatives. M: 2000 bp marker; The materials numbered 1–65 are shown in Table 3; Test materials 1–48 were amplified using primer pair 1, test materials 49–54, 60 using primer pair 4, test materials 55–59, 65 using primer pair 2, test materials 63–64 using primer pair 3, test material 61 using primer pair 5, and test material 62 using primer pair 6.

Table 3. S-RNase genotypes from 65 Xinjiang almond germplasms and close relatives.

No.	Sample	Fragment Length/bp	Genotype	No.	Sample	Fragment Length/bp	Genotype
1	Shuangbo	591/1483	S_5S_{65}	34	Kexi	1274	S_{54}
2	Huangshuang	1634/811	$S_{22}S_{57}$	35	Yeerqiang	591/811	S_5S_{57}
3	Shitou Almond	2031/811	S_7S_{57}	36	Shache 29	1592/797	$S_{\underline{1}}S_{55}$
4	Ku 1	977/747	$S_{\underline{10}}S_{63}$	37	A1	591/1086	S_5S_{16}
5	Ydl 1	977/811	$S_{\underline{10}}S_{57}$	38	Tianren Taobadan	591/1130	S_5S_{24}
6	Shache 28	754/1650	S_6S_{35}	39	Jianzuihuang	591/2031	S_5S_7
7	Jinsha Fruit × XX3	1592/1086	S_1S_{16}	40	Badanwang	591/754	$S_5S_{\underline{6}}$
8	Dabadan × XX3	1592/811	S_1S_{57}	41	Baiboke	653/1274	$S_{n3}S_{54}$
9	Taobadan	1592/811	S_1S_{57}	42	Aitelaisi Badan	977/1634	$S_{10}S_{22}$
10	Xiaokuren Taobadan	1592/1086	$S_{\underline{1}}S_{16}$	43	Paizhui	624/1086	S_8S_{16}
11	Threeling Taobadan	1592/1130	$S_{\underline{1}}S_{24}$	44	Wanhua Taobadan	1592/821	S_1S_2
12	Badanxing	994/747	$S_{23}S_{63}$	45	Wanfeng	591/2031	S_5S_7
13	Shache 79	977/811	$S_{\underline{10}}S_{57}$	46	Aifeng	653/994	$S_{n3}S_{23}$
14	Xinjiang Peach 1	1592/821	S_1S_2	47	XX1	1592/1086	$S_{\underline{1}}S_{16}$
15	Arele Taobadan	1592/811	S_1S_{57}	48	Houke Taobadan	977/811	$S_{\underline{10}}S_{57}$
10	Ydl 2	977/811	$S_{\underline{10}}S_{57}$	49	Zhipi	632/811	$S_{n3}S_{57}$
17	Jinsha Fruit	857/1086	S_6S_{16}	50	*Prunus mongolica*	722	S_3
18	Duoguo	747	S_{63}	51	Yingzui	632/1576	$S_{n3}S_1$
19	Dabadan	811	S_{57}	52	XX2	2021/811	S_7S_{57}
20	Amannisha	591/977	$S_5S_{\underline{10}}$	53	Shuangruan	797/772	$S_{n5}S_{63}$
21	Gongbadan	591/811	S_5S_{57}	54	Jinbian 2	822/993	$S_{49}S_f$
22	Xiaoshuangren	1086/811	$S_{16}S_{57}$	55	Ku Taobadan	455/190	$S_{\underline{2}}S_{57}$
23	Shuangguo	591/1650	S_5S_{35}	56	XX3	200/347	$S_{\underline{1}}S_{27}$
24	Kubadan	591/977	$S_5S_{\underline{10}}$	57	Jinbian 1	335/419	$S_{49}S_f'$
25	Baichang Badan	591/1274	$S_5S_{\underline{54}}$	58	Yeerqiang holy Fruit	355/408	S_7S_8
26	Shache 27	1592/591	$S_{\underline{1}}S_5$	59	Yuanruan	1254	S_{10}
27	Xinjiang Peach 2	1592	$S_{\underline{1}}$	60	Make	1212	S_{24}
28	Tianrenxiao Taobadan	1592/811	S_1S_{57}	61	Shache 48	1967/1008	S_7S_{57}
29	Taohuaguazi Badan	977/1130	$S_{10}S_{24}$	62	*Prunus davidiana*	1231/375	$S_{\underline{1}}S_{57}$
30	Xinjiang Peach 3	1592/977	S_1S_{10}	63	Hanfeng	982	S_{61}
31	Guazi Badan	811/1483	$S_{57}S_{65}$	64	*Prunus triloba*	350/621	S_7S_{11}
32	Changying	811	S_{57}	65	Jinbian 3	313/419	$S_{n1}S_f'$
33	Baishuang	994	S_{23}				

Note: The underline indicates that the gene is not *Prunus dulcis* S-RNase.

Homology sequence retrieval was performed on 30 nucleotide sequences of different fragment sizes in NCBI, and the results showed that the nucleotide sequences of the *S-RNase* genes highly homologous to the sequencing sequence were all *S-RNase* genes of the *Prunus* L. in the Rosaceae family, including 20 *Prunus dulcis*-style *S* gene sequences. *Prunus webbii*, *Prunus persica*, *Prunus armeniaca*, *Prunus salicina*, and *Prunus cerasifera* sequences were 4, 3, 1, 1, and 1, respectively (Table 4).

Table 4. Comparison results of 30 gene fragments in GenBank.

No.	Fragment Length/bp	*S-RNase* Genes with the Highest Homology on GenBank	Sequence ID	Similarity/%
1	313	*Prunus webbii* S_{n1}, exons 1-3	AM690352.1	99.36
2	347	*Prunus dulcis* S_{27}, partial cds	MH316093.1	99.71
3	350	*Prunus salicina* S_7, partial cds	AY781290.1	99.43
4	408	*Prunus dulcis* S_8, complete cds	AB481108.1	100.00
5	419	*Prunus dulcis* $S_f{}'$, complete cds	MH316088.1	99.52
6	591	*Prunus dulcis* S_5, complete cds	MH316078.1	98.98
7	621	*Prunus cerasifera* S_{11}, exons 1-2	AM992055.1	99.68
8	624	*Prunus dulcis* S_g, partial cds	DQ156219.1	99.84
9	653	*Prunus webbii* S_{n3}, exons 1-3	AM690354.1	100.00
10	722	*Prunus persica* S_3, complete cds	AB537563.1	99.86
11	754	*Prunus webbii* S_6, partial cds	EU294327.1	99.80
12	772	*Prunus dulcis* S_{63}, partial cds	AY613919.1	99.87
13	797	*Prunus dulcis* S_{55}, exons 1-3	FN599511.1	99.87
14	797	*Prunus dulcis* S_{n5}, exons 1-2	AM690355.1	99.87
15	811	*Prunus dulcis* S_{57}, exons 1-3	FN599513.1	100.00
16	821	*Prunus persica* S_2, complete cds	AB252417.1	100.00
17	822	*Prunus dulcis* S_{49}, partial cds	JX067634.2	99.51
18	857	*Prunus dulcis* S_6 precursor, exons 1-3	AM231657.1	100.00
19	982	*Prunus dulcis* S_{61}, partial cds	AY613917.1	98.07
20	993	*Prunus dulcis* S_f, partial cds	DQ156217.1	99.60
21	994	*Prunus dulcis* S_{23}, complete cds	AB488496.1	98.89
22	1086	*Prunus dulcis* S_{16} precursor, exons 1-3	AM231665.1	99.82
23	1130	*Prunus dulcis* S_{24}, exons 1-3	LN624639.1	100.00
24	1254	*Prunus webbii* S_{10}, partial cds	EU294324.1	99.60
25	1274	*Prunus dulcis* S_{54}, exons 1-3	FN599510.1	100.00
26	1483	*Prunus armeniaca* S_2, partial cds	AY587562.1	95.95
27	1592	*Prunus persica* S_1, complete cds	AB252415.1	99.87
28	1634	*Prunus dulcis* S_{22} precursor, exons 1-3	AM231671.1	96.22
29	1650	*Prunus dulcis* S_{35}, partial cds	KM225273.1	99.76
30	2031	*Prunus dulcis* S_7, complete cds	MH316075.1	100.00

Note: DQ156217.1 and MH316088.1 are *Prunus dulcis* S_f from two different sources. In order to distinguish the two, MH316088.1 uses *Prunus dulcis* $S_f{}'$, the same as above.

For nucleotide sequences with different fragment sizes, the amino acid sequences of nucleotide sequences with various fragment sizes were deduced by removing introns, utilizing the cDNA of the *S-RNase* gene with the highest homology found on GenBank. The amino acid sequences of the *S* genes with the highest homology rate in GenBank were compared for consistency by DNAMAN (V6) software. The results showed that the deduced amino acid sequences with amplified fragment sizes of 1483 bp and 1634 bp differed from the amino acid sequences of *Prunus armeniaca* S_2 (AY587562.1) and *Prunus dulcis* S_{22} (AM231671.1) by three and two amino acids, respectively, with consistencies of 98.24% and 98.57%, respectively. Among them, a sequence of 1483 bp was identified as a new *S-RNase* gene named S_{65}. The amino acid sequences of 722 bp, 772 bp, 1592 bp, and 1650 bp had consistencies of over 99% with the gene of the highest homology (the only difference in amino acid sequence). The remaining fragments were 100% consistent with the amino acid sequence of the gene with the highest homology. Based on the above results, the S-RNase genotypes of the test materials were determined (Table 3).

3.2. Structural Analysis of the Amino Acid Sequences of Different S-RNase Genes

Except for the sequence regions of $S_f{}'$, S_8, S_{27}, and S_{n1} alleles obtained from the signal peptide region to the C1 region, when comparing the deduced amino acid sequence structures, it was found that the amino acid sequences of 25 registered *S-RNases* and one new *S-RNase* all exhibit typical structural features of the S-RNase of *Prunus* L. Downstream of the N-terminal SP of the sequence are five conserved regions, namely C1, C2, C3, RC4, and C5. There is a high-variability region (RHV) between C2 and C3, as well as two histidine residues necessary for RNase activity and four conserved cysteine residues (Figure 2).

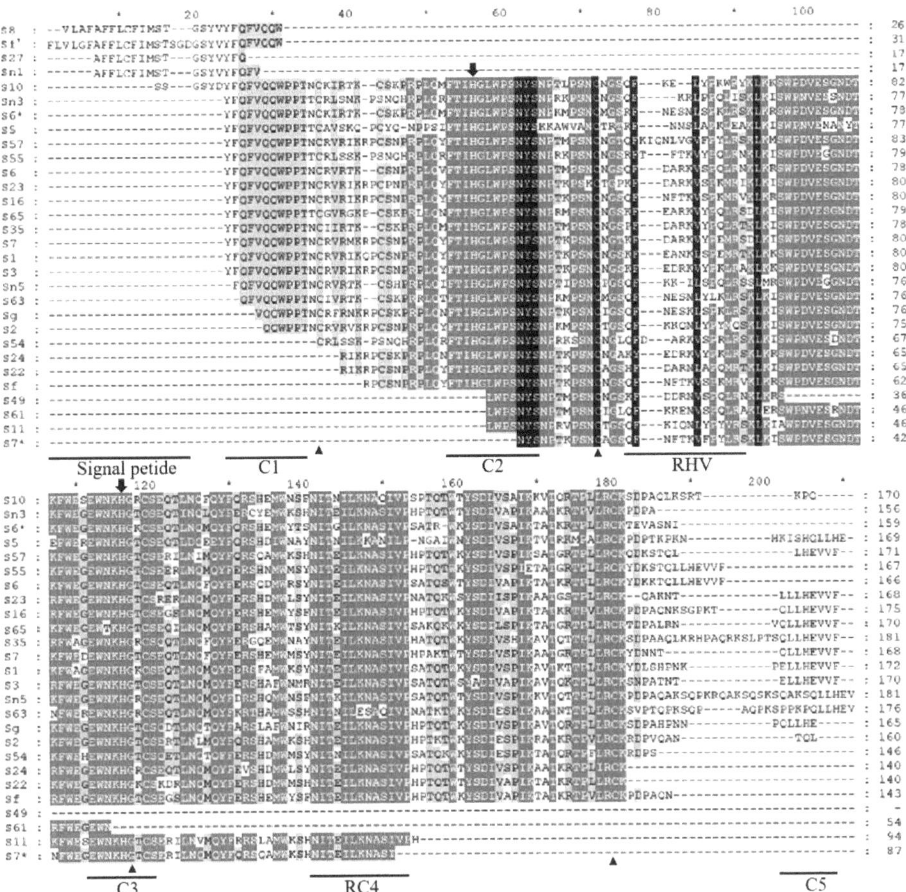

Figure 2. Amino acid sequence structures of 30 *S-RNase* genes. Conserved residues are located in the black area and dashes indicate gaps. The asterisk represents a difference of 10 residues from the previous number. The signal peptide region (SP), conserved regions (C1, C2, C3, RC4, and C5), and the hypervariable region (RHV) are underlined. The two histidine residues essential for RNase activity are labeled with the symbol ↓, and the four conserved cysteine residues are labeled with the symbol ▲. To distinguish the two S6s, S6* stands for *Prunus webbii* S_6.

3.3. Phylogenetic Analysis of the S Gene in Rosaceae Prunoideae

We removed four shorter gene amino acid sequences that cannot meet the requirements for constructing a phylogenetic tree, including S_f', S_8, S_{27}, and S_{n1}. The neighbor-joining (NJ) feature of MEGA 11 software was used to obtain 26 *S-RNase* gene amino acid sequences, combined with 19 registered *S-RNase* gene amino acid sequences of *Prunus webbii*, *Prunus salicina*, *Prunus armeniaca*, *Prunus avium*, *Prunus simonii*, and *Prunus mume* in the NCBI database to construct a phylogenetic tree of *Prunoideae* (Figure 3). Cluster analysis showed that 25 identified *S-RNase* genes and one new almond *S-RNase* gene clustered to varying degrees with 19 registered plum genes in six branches of the evolutionary tree, and their evolutionary relationships with *Prunus armeniaca*, *Prunus webbii*, and *Prunus avium* were relatively close in different branches. This result is consistent with the research results of Jiang et al. [8]. In addition, most of the *S-RNase* genes obtained from the experiment have close genetic relationships.

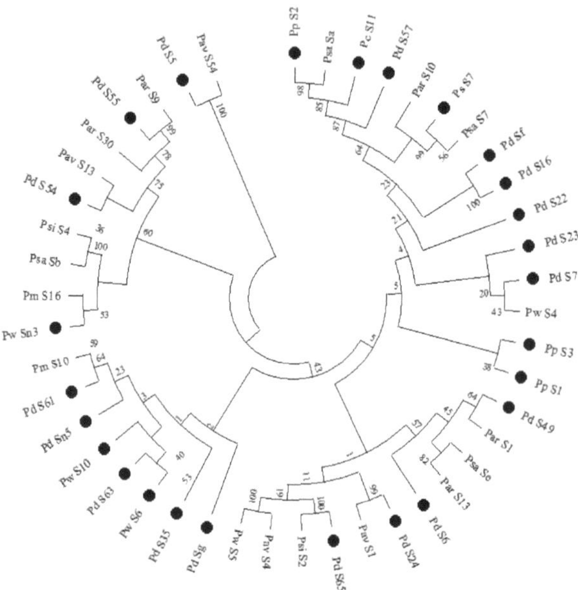

Figure 3. Phylogenetic tree of 50 S-RNase genes in the *Prunus* L. genus. Pd: Prunus dulcis; Pw: Prunus webbii; Pp: Prunus persica; Pc: Prunus cerasifera; Psa: Prunus salicina; Par: Prunus armeniaca; Pav: Prunus avium; Psi: Prunus simonii; Pm: Prunus mume. Those marked with circles are the S-RNase genes in the test.

3.4. Analysis of S-RNase Genes and Gene Frequencies

Among the 30 *S-RNase* gene fragments amplified from 65 germplasm resources, the highest gene frequency of S_{57}-*RNase* was 31.75%, which was amplified from 20 cultivars. The frequencies of S_1-, S_5-, S_7-, S_{10}-, and S_{16}-*RNase* are relatively high, ranging from 9.52% to 23.81%. The gene frequency of the remaining S-RNases ranged from 1.59% to 6.35%, appearing 1–4 times (Figure 4). The results also showed that high-frequency genes were commonly present in the seven main almond cultivars in Xinjiang, including 'Zhipi' $S_{n3}S_{57}$, 'Shuangguo' S_5S_{35}, 'Shuangruan' $S_{n5}S_{63}$, 'Aifeng' $S_{n3}S_{23}$, 'Wanfeng' S_5S_7, 'Yingzui' $S_{n3}S_1$, and 'Yeerqiang' S_5S_{57}. This resulted in different cultivars having the same *S-RNase* gene or even the same S genotype, which may be due to long-term natural selection or purposeful breeding by humans.

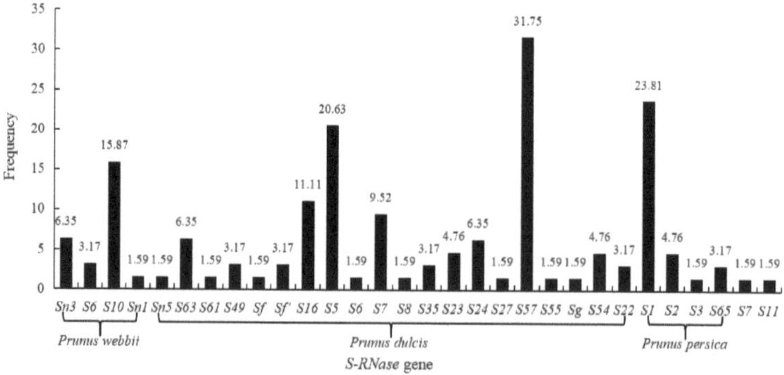

Figure 4. Gene frequencies of 30 *S-RNase* genes in the germplasm resources.

3.5. Establishment of Almond Hybrid Incompatibility Groups

In this study, the S genotype results of 55 out of 65 materials were successfully identified. Based on the theory that materials with the same S genotype are incompatible for cross-pollination, and there is semi-compatibility or compatibility for materials with one identical *S-RNase* gene [6], eight CIGs (I-VIII) and one semi-compatibility or compatibility group (0) were initially established (Table 5). Additionally, cross-incompatibility/compatibility between seven main cultivars of Xinjiang almond and 35 other cultivars was analyzed. The results showed that most varieties could pollinate each other, with only a few showing cross-incompatibility (Figure 5). In the later stages, further verification was needed through cross-pollination in the field.

Table 5. Identification results of almond cross-incompatibility groups and their S genotypes.

Cross Incompatibility Group/CIG	S-genotype	Sample
I	$\underline{S_{10}}S_{57}$	Houke Taobadan, Ydl 1, Ydl 2, Shache 79
II	$\underline{S_1}S_{57}$	Taobadan, Arele Taobadan, Tianrenxiao Taobadan
III	$\underline{S_5}S_{57}$	Gongbadan, Yeerqiang
IV	$S_5\underline{S_{10}}$	Amannisha, Kubadan
V	$S_{16}\underline{S_{57}}$	Xiaoshuangren
VI	S_7S_{57}	Shitou Almond, Shache 48, XX2
VII	S_5S_7	Jianzuihuang, Wanfeng
VIII	$\underline{S_1}S_{16}$	Xiaokuren Taobadan, XX1
0 [1]		Shuangbo (S_5S_{65}), Huangshuang ($S_{22}S_{57}$), Ku 1 ($\underline{S_{10}}S_{63}$), Shache 28 ($\underline{S_6}S_{35}$), Threeling Taobadan ($\underline{S_1}S_{24}$), Badanxing ($S_{23}S_{63}$), Jinsha Fruit (S_6S_{16}), Shuangguo (S_5S_{35}), Baichang Badan (S_5S_{54}), Shache 27 ($\overline{\underline{S_1}S_5}$), Taohuaguazi Badan ($\underline{S_{10}}S_{24}$), Guazi Badan ($S_{57}S_{65}$), Shache 29 ($\underline{S_1}S_{55}$), A1 ($S_5S_{16}$), Tianren Taobadan ($S_5S_{24}$), Badanwang ($S_5\underline{S_6}$), Baiboke ($\underline{S_{n3}}S_{54}$), Aitelaisi Badan ($\underline{S_{10}}S_{22}$), Paizhui ($S_8S_{16}$), Wanhua Taobadan ($\underline{S_1}S_2$), Aifeng ($\underline{S_{n3}}S_{23}$), Zhipi ($\underline{S_{n3}}S_{57}$), Yingzui ($\underline{S_{n3}}S_1$), Shuangruan ($\overline{S_{n5}}S_{63}$), Jinbian 1 ($S_{49}S_f'$), Jinbian 2 ($S_{49}S_f$), Jinbian 3 ($\underline{S_{n1}}S_f'$), Ku Taobadan ($\underline{S_2}S_{57}$), XX3 ($\underline{S_1}S_{27}$), Yeerqiang holy Fruit (S_7S_8), *Prunus triloba* ($\underline{S_7S_{11}}$)

[1] CIG 0 contains one identical or completely different S-RNase. The underline indicates that the gene is not *Prunus dulcis* S-RNase.

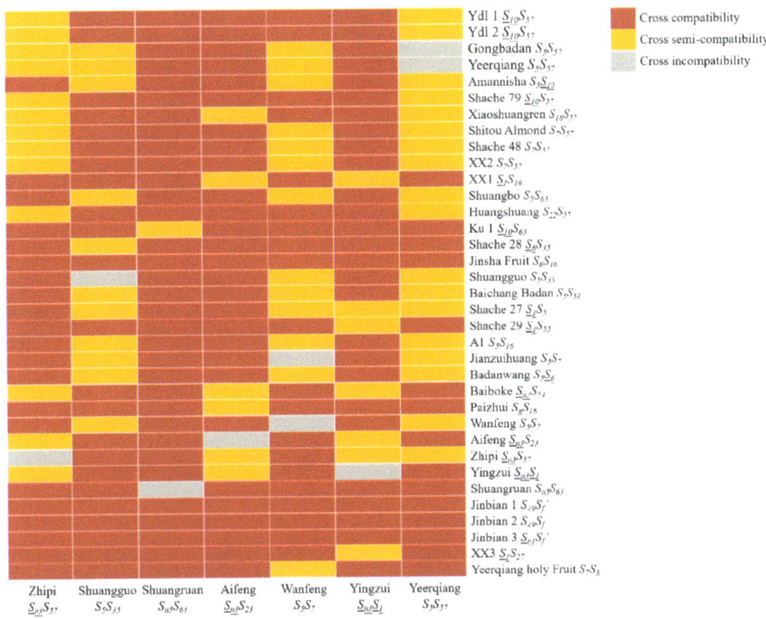

Figure 5. Heat map of cross-(non)compatibility between 7 local almond cultivars of the Xinjiang region and different almond cultivars.

3.6. Field Identification and Analysis of Some Almond Cultivars

Among the eight hybrid combinations from 2022 to 2023, all exhibited fruit-setting rates higher than 5%. Notably, the combination 'Jinsha Fruit' × 'Taohuaguazi Badan' yielded the highest fruit-setting rate at 29.70% (Table 2). Following the criteria outlined by Audergon et al. [21], all eight combinations demonstrated cross-compatibility. The identification results of parental genotypes in each cross-combination revealed that the S genotype of the parents differed across combinations, and these identifications were consistent with field observations.

4. Discussion

4.1. Analysis of the Results of S-RNase Allele Identification

PCR amplification combined with DNA sequencing has been widely used to identify S-RNase alleles in almonds and other *Prunus* L. plants [9,15,22]. It is only necessary to collect tissues such as the young leaves of plants, without long-term pollination observation. The S_1S_{57} and S_1S_{16} genotypes of hybrid offspring "Dabadan × XX3" and "Jinsha Fruit × XX3", as well as the genotypes of their parents 'Dabadan' S_{57}, 'Jinsha Fruit' (S_6S_{16}), and 'XX3' (S_1S_{27}), conform to Mendelian genetic inheritance rules, which to some extent indicates the reliability of this identification method. In addition, the field verification results of eight different hybrid combinations in this study also supported the identification of S genotypes, that is, the fruit-setting rates of all combinations were higher than 5%, and the S-RNase genotypes were different among the hybrid parents. Researchers conducted pollination experiments on some Xinjiang almond varieties, pointing out that when the flowering periods of 'Zhipi', 'Shuangguo', and 'Shuangruan' meet, they can become pollinating trees for each other [23]. In this study, the S-RNase genotypes of these three cultivars were different. In addition, a pollination experiment was conducted using 'Zhipi' and 'Yeerqiang' as male parents, and the results showed that 'Zhipi' had good fruit-setting rates with female parents 'Wanfeng', 'Shuangguo', 'Shuangbo', and 'Amannisha', while 'Yeerqiang' had better fruit-setting rates with female parents 'Zhipi', 'Shuangguo', 'Shuangbo', and 'Shuangruan' [24]. In this study, the S-RNase genotype of 'Zhipi' was completely different from the above maternal genotype, and the S-RNase genotype of 'Yeerqiang' had one identical *S-RNase* gene as the above maternal parent, which also showed that the identification method was accurate and reliable. In this study, the S-RNase genotypes of 85% of the materials were successfully identified using the consensus primers of Rosaceae and degeneracy primers designed on the basis of the conserved amino acid sequence of the S-allele. The electrophoretic map of 'Shuangruan' showed a single band, but the cloning and sequencing results revealed two different S-RNases with the genotype $S_{n5}S_{63}$ (797/772). For the 10 materials that only amplified single bands, it is speculated that the reason may be due to large introns in the target fragment causing amplification failure, or that the two S-RNase allele fragments of the cultivars were too close, and another *S-RNase* gene was not connected during cloning and sequencing. It is also possible that the variety carried a homozygous *S-RNase* gene, necessitating the use of specific primers for identification.

The nucleotide sequences of the 30 *S-RNase* genes obtained by amplification had 95.95~100% homology with the *S-RNase* genes registered in GenBank. The consistency of amino acid sequences was analyzed according to the GT/AG conserved features of the intron/exon splice site. The amino acid sequences of 29 *S-RNase* genes were more than 99% consistent with that of homologous genes, while the amino acid sequence of the 1483 bp amplified fragment was three amino acids different from that of homologous genes, and was identified as a new *S-RNase* gene named S_{65}. The discovery of the new *S-RNase* gene will help to expand the diversity of the local almond gene information database and enrich the hybrid parental material. At the same time, studies have shown that for SI populations, pollens carrying the new S gene are more likely to be successfully fertilized, thereby increasing their population size [25].

In the current research, it was found that Taobadan is a natural hybrid of almonds and peaches in Xinjiang, China [24], and was integrated into *Prunus persica* S_1 (AB252415.1) in the S genotype of most Taobadan resources. Although the names of the resources are similar, the S genotype is not completely consistent, indicating that the genetic background of the materials is different. In practical production applications, it is necessary to understand the S genotype of the materials. In addition, the identification results of the genotypes of 'Zhipi' and 'Yeerqiang' were $S_{n3}S_{57}$ and S_5S_{57}, respectively, while previous research results were $S_{50}S_{61}$ and $S_{15}S_{16}$ [16]. Sequence analysis showed that the nucleotide sequence consistency between *Prunus dulcis* cultivar *S50* RNase (AY613337) and S_{n3}, and *Prunus dulcis* S_{15} (AM231664.1) and S_5, was 98.92% and 99.33%, respectively. When the S genotype identification results of the same variety are different, it is often due to sampling error caused by different species having the same name, or the different ways researchers evaluate the sequencing results. Ortega et al. [19] proposed that the sequences of the *Prunus dulcis* alleles S_5 and S_{15} are identical, as are the sequences of S_4 and S_{20} and S_{13} and S_{19}. In the early studies, the researchers repeated the naming due to a lack of unified regulations and lack of communication. If the identification results are wrong, it will not only lead to the occurrence of different varieties with the same name or different names for the same variety, which will affect the evaluation and utilization of varieties and the research process but will also mislead production and utilization, resulting in serious economic losses. It can be seen that the identification of genotypes needs to be repeatedly compared and verified to be error-free before the accuracy of S-RNase allele identification and naming results can be guaranteed.

4.2. Analysis of the S Genotype

In the constructed phylogenetic tree, the *S-RNase* genes of *Prunus dulcis, Prunus salicina, Prunus armeniaca, Prunus avium, Prunus webbii, Prunus simonii, Prunus mume,* and *Prunus cerasifera* did not cluster into a single subclass, nor did they show intra-specific similarity higher than inter-specific similarity. The evolution of *S-RNase* genes may be out of sync with the development of plant systems within *Prunoideae*. In this study, the consistency between the 'Baiboke' 653 bp sequence, *Prunus webbii* S_{n3} (AM690354.1), the *Prunus mongolica* 722 bp sequence, and the *Prunus persica* S_3 (AB537563.1) amino acid sequence is higher than 99%. This suggests that the *S-RNase* gene has evolved across species. These findings align with previous studies indicating that the inter-specific homology of *S-RNase* genes in Rosaceae is often greater than intra-specific homology [26]. The evolutionary tree reveals relatively close phylogenetic relationships between tree species, particularly evident in the proximity of *Prunus dulcis, Prunus webbii,* and *Prunus avium*. Correspondingly, the genetic relationship among *S-RNase* genes obtained from experiments is also relatively close, narrowing the genetic range of almonds and potentially hindering the inheritance of S genes in crossbreeding parents. Therefore, it is imperative to identify the S genotype of almond germplasm resources and update the information base to broaden the genetic background of almonds, identify excellent genes, and provide precise guidance continuously and accurately for practical production.

The results of the pollination test performed by Heng et al. [27] showed that both self-pollination and cross-pollination among varieties of the same S genotype produced low fruiting rates (0%), except 'Hongding' which yielded 0.64%. Further, the results of Liang [28] showed that there was one identical *S-RNase* gene. The style S genotype of the two varieties showed semi-affinity, and the offspring only inherited 1/2 of the pollen of the parents. When the parent S genotype is completely different, four genotypes of offspring can be produced. Similarly, foreign researchers have placed varieties with the same S genotype in a cross-incompatibility group (CIG) based on this characteristic, and currently, 48 CIGs have been established [12,13]. Based on the identification results, this study established, for the first time, eight almond CIGs (I-VIII) and one semi-compatible or compatible group (0) in China. Materials within the same CIG group should be excluded from almond production and hybrid breeding. The cross-(in)compatibility between seven main almond cultivars and 35 other cultivars in Xinjiang was analyzed, and the heat map showed that most of the varieties could pollinate each other, with only a few showing hybridization incompatibilities. Further field cross-pollination is needed to verify this later. Understanding the affinity relationship between CIG and materials is very necessary for breeding programs. This understanding can provide a reference for enriching the genetic information of varieties, studying the mechanism of self-incompatibility, selecting suitable pollination trees, designing new hybrid combinations, and selecting and breeding new cultivars.

5. Conclusions

This study utilized 65 germplasm resources from 61 almond and four peach germplasms to investigate the genotype composition of S-RNase and its impact on the cross-compatibility of almond cultivars. Through PCR-based amplification and sequence analysis, 55 of the identified materials obtained a complete S-RNase genotype, while only one *S-RNase* gene was found in 10 of them. A total of 30 *S-RNase* genes were identified, including the discovery of a new S-RNase named S_{65}. Furthermore, this study classified the materials into eight S-genotype cross-incompatibility groups (CIG) and one semi-compatibility or compatibility group (0). The research results will contribute to pollinator tree allocation and breeding hybrid parent selection in almond production, providing valuable insights into the potential cross-pollination compatibility among almond cultivars and providing practical applications for improving almond production and breeding strategies.

Author Contributions: Conceptualization, P.X. and M.A.; methodology, P.X. and X.W.; software, Y.A.; investigation, Y.X.; resources, L.W. and M.A.; data curation, Y.A.; writing—original draft preparation, P.X.; writing—review and editing, P.X. and X.W.; supervision, L.W.; project administration, Y.X.; funding acquisition, P.X., Y.X., C.G. and M.A. All authors have read and agreed to the published version of the manuscript.

Funding: This research was funded by the Youth Science and Technology Backbone Project of Xinjiang Academy of Agricultural Sciences, grant number xjnkq-2023009; Natural Science Foundation of Xinjiang Uygur Autonomous Region Sponsored Programs, grant number 2022D01A90; National Region Fund Science Project of China, grant number 32260738; Central Government Guide Local Science and Technology development special fund project "Characteristic Fruit Tree Germplasm Innovation and Breeding Capacity Enhancement"; Basic Research Business of Public Welfare Research Institutes in Xinjiang Uygur Autonomous Region, grant number KY2022039.

Data Availability Statement: The datasets supporting the results presented in this manuscript are included within the article.

Conflicts of Interest: The authors declare no conflicts of interest.

References

1. Wang, L. *Chinese Peach Genetic Resources*, 1st ed.; China Agriculture Press: Beijing, China, 2012.
2. Martinez-Gomez, P.; Sanchez-Perez, R.; Dicenta, F.; Howard, W.; Arus, P.; Gradziel, T.M. Fruits and Nuts. In *Genome Mapping and Molecular Breeding in Plants*, 1st ed.; Kole, C., Ed.; Springer: Berlin, Germany, 2007; Volume 3, pp. 229–242.
3. Boskovic, R.; Tobutt, K.R.; Batlle, I.; Duval, H.; Martinez-Gomez, P.; Gradziel, T.M. Stylar ribonucleases in almond: Correlation with and prediction of incompatibility genotypes. *Plant Breed.* **2003**, *122*, 70–76. [CrossRef]
4. He, M.; Gu, C.; Wu, J.Y.; Zhang, S.L. Recent advances on self-incompatibility mechanism in fruit trees. *Acta Hortic. Sin.* **2021**, *48*, 759–777.
5. Tao, R.; Watari, A.; Hanada, T.; Habu, T.; Yaegaki, H.; Yamaguchi, M.; Yamane, H. Self-compatible peach (*Prunus persica*) has mutant versions of the S haplotypes found in self-incompatible *Prunus* species. *Plant Mol. Biol.* **2007**, *63*, 109–123. [CrossRef] [PubMed]
6. Liu, M. Study on Pollination Characteristic and S Genotype Identification of Self-Incompatibility of Kernel-Apricot. Ph.D. Thesis, Chinese Academy of Forestry, Beijing, China, 2015.
7. He, M.; Li, L.; Xu, Y.; Mu, J.; Xie, Z.; Gu, C.; Zhang, S. Identification of S-genotypes and a novel S-RNase in 84 native Chinese pear accessions. *Hortic. Plant J.* **2022**, *8*, 713–726. [CrossRef]
8. Jiang, X.; Cao, X.; Wang, D.; Feng, J.; Liu, Y.; Fan, X. Identification of self-incompatibility S-RNase genotypes for apricot cultivars in South of Xinjiang area. *J. Fruit Sci.* **2012**, *29*, 569–576.
9. Liu, Z.; Gao, Y.; Wang, K.; Feng, J.; Sun, S.; Lu, X.; Wang, L.; Tian, W.; Wang, G.; Li, Z.; et al. Identification of S-RNase genotype and analysis of its origin and evolutionary patterns in *Malus* plants. *J. Integr. Agric.* **2024**, *23*, 1205–1221. [CrossRef]
10. Lopez, M.; Vargas, F.J.; Batlle, I. Self-(in)compatibility almond genotypes: A review. *Euphytica* **2006**, *150*, 1–16. [CrossRef]

11. Kodad, O.; Alonso, J.M.; Sanchez, A.; Oliveira, M.; Company, R.S.I. Evaluation of genetic diversity of S-alleles in an almond germplasm collection. *J. Hortic. Sci. Biotech.* **2008**, *83*, 603–608. [CrossRef]

12. Akram, H.; Behrouz, S.; Bahram, M.; Ali, I.; Bojana, B. Identification of new S-RNase self-incompatibility alleles and characterization of natural mutations in Iranian almond cultivars. *Trees* **2013**, *27*, 497–510.

13. Gomez, E.M.; Dicenta, F.; Batlle, I.; Romero, A.; Ortega, E. Significance of S-genotype determination in the conservation of genetic resources and breeding of almond. *Ciheam* **2016**, *119*, 83–86.

14. Ma, Y.; Ma, R. Analysis on polymorphism of *S* gene in almond. *Acta Hortic. Sin.* **2006**, *33*, 137–139.

15. Jiang, J.; Feng, J.; Cao, X.; Li, W.; Wang, X.; Chen, L. Identification of self-Incompatibility S-RNase genotypes for almond cultivars in Xinjiang. *Xinjiang Agric. Sci.* **2011**, *48*, 1177–1182.

16. Guo, C.; Yang, B.; Tudi, M.; Tang, Y.; Li, N.; Wang, J.; Gong, P. Comparison of amplification effects on almond resources in China using different primer combinations. *Xinjiang Agric. Sci.* **2016**, *53*, 641–646.

17. Tamura, M.; Ushijima, K.; Sassa, H.; Hirano, H.; Tao, R.; Gradziel, T.M.; Dandekar, A.M. Identification of self-incompatibility genotypes of almond by allele-specific PCR analysis. *Theor. Appl. Genet.* **2000**, *101*, 344–349. [CrossRef]

18. Sonneveld, T.; Tobutt, K.R.; Robbins, T.P. Allele-specific PCR detection of sweet cherry self-incompatibility (S) alleles S1 to S16 using consensus and allele-specific primers. *Theor. Appl. Genet.* **2003**, *107*, 1059–1070. [CrossRef]

19. Ortega, E.; Sutherland, B.G.; Dicenta, F.; Boskovic, R.; Tobutt, K.R. Determination of incompatibility genotypes in almond using first and second intron consensus primers: Detection of new S alleles and correction of reported S genotypes. *Plant Breed.* **2005**, *124*, 188–196. [CrossRef]

20. Tao, R.; Yamane, H.; Sugiura, A. Molecular Typing of S-alleles through Identification, characterization and cDNA cloning for S-RNases in sweet cherry. *J. Am. Soc. Hortic. Sci.* **1999**, *24*, 224–233. [CrossRef]

21. Audergon, J.M. Conribution to the study of inheritance of the character self incompatibility in apricot. *Acta Hortic.* **1999**, *488*, 275–279. [CrossRef]

22. Sharafi, Y.; Hajilou, J.; Mohammadi, S.A.; Dadpour, M.R.; Eskandari, S. Pollen-pistil compatibility relationships in some Iranian almond (*Prunus dulcis*, Batch) genotypes as revealed by PCR analysis. *Afr. J. Biotechnol.* **2010**, *9*, 3338–3341.

23. Zeng, B.; Gao, Q.; Tian, J.; Li, J. Research Progress on Self-Incompatibility of Almond Plants. In Proceedings of the Dried Fruit Branch of Chinese Horticultural Society, Shijiazhuang, China, 29 November 2013.

24. Tian, J. *China Almond*, 1st ed.; China Agriculture Press: Beijing, China, 2008.

25. Charlesworth, D.; Guttman, D.S. Seeing selection in S allele sequence. *Curr. Biol.* **1997**, *7*, 34–37. [CrossRef]

26. Ortega, E.; Bonkovi, R.I.; Sargent, D.J.; Tobutt, K.R. Analysis of S-RNase alleles of almond (*Prunus dulcis*): Characterization of new sequences, resolution of synonyms and evidence of intragenic recombination. *Mol. Gen. Genomics.* **2006**, *276*, 413–426. [CrossRef] [PubMed]

27. Heng, W.; Wu, H.Q.; Chen, Q.X.; Wu, J.; Huang, S.; Zhang, S.L. Identification of S-genotypes and novel S-RNase alleles in *Prunus mume*. *J. Hortic. Sci. Biotech.* **2008**, *83*, 689–694. [CrossRef]

28. Liang, M. Identification and Evolution of Citrus Self-Incompatibility Related Genes. Master's Thesis, Huazhong Agricultural University, Wuhan, China, 2019.

 *horticulturae*

Article

Selective Retention of Cross-Fertilised Fruitlets during Premature Fruit Drop of Hass Avocado

Nimanie S. Hapuarachchi [1], Wiebke Kämper [2], Shahla Hosseini Bai [1], Steven M. Ogbourne [3,4], Joel Nichols [1], Helen M. Wallace [5] and Stephen J. Trueman [1,*]

[1] Centre for Planetary Health and Food Security, School of Environment and Science, Griffith University, Nathan, QLD 4111, Australia; nimanie.hapuarachchi@alumni.griffithuni.edu.au (N.S.H.); s.hosseini-bai@griffith.edu.au (S.H.B.); j.nichols2@griffith.edu.au (J.N.)
[2] Functional Agrobiodiversity and Agroecology, Department of Crop Sciences, University of Göttingen, 37077 Göttingen, Germany; wiebke.kaemper@uni-goettingen.de
[3] Centre for Bioinnovation, University of the Sunshine Coast, Maroochydore DC, QLD 4558, Australia; sogbourn@usc.edu.au
[4] School of Science, Technology & Engineering, University of the Sunshine Coast, Maroochydore DC, QLD 4558, Australia
[5] School of Biology and Environmental Science, Queensland University of Technology, GPO Box 2434, Brisbane, QLD 4001, Australia; helen.wallace@qut.edu.au
* Correspondence: s.trueman@griffith.edu.au

Citation: Hapuarachchi, N.S.; Kämper, W.; Hosseini Bai, S.; Ogbourne, S.M.; Nichols, J.; Wallace, H.M.; Trueman, S.J. Selective Retention of Cross-Fertilised Fruitlets during Premature Fruit Drop of Hass Avocado. *Horticulturae* **2024**, *10*, 591. https://doi.org/10.3390/horticulturae10060591

Academic Editors: Zhaohe Yuan, Bo Li and Yujie Zhao

Received: 16 May 2024
Revised: 30 May 2024
Accepted: 4 June 2024
Published: 5 June 2024

Abstract: The productivity of many tree crops is limited by low yield, partly due to high rates of fruitlet abscission during early fruit development. Early studies suggested that cross-pollinated fruitlets may be selectively retained during fruit development, although paternity testing of fruitlets to test this hypothesis was technically challenging. We used MassARRAY genotyping to determine the effects of pollen parentage on fruitlet retention and fruit quality of Hass avocado. We identified the paternity of abscised and retained fruitlets at 6 and 10 weeks, and mature fruit at 36 weeks, after peak anthesis. We measured the embryo mass, pericarp mass, total mass and nutrient concentrations of fruitlets, and the seed mass, flesh mass, total mass, diameter, length, nutrient concentrations and fatty-acid composition of mature fruit. The percentages of progeny on the tree that were cross-fertilised increased from 4.6% at 6 weeks after peak anthesis to 10.7% at fruit maturity. Only 1.0% of freshly abscised fruitlets on the ground at 10 weeks after peak anthesis were cross-pollinated even though 6.5% of retained fruitlets on the tree were cross-pollinated. At this stage, cross-pollinated fruitlets had similar nutrient concentrations to self-pollinated fruitlets, but they had higher total contents of P, K, Al, Ca, Fe, Mn and Zn due to having greater fruitlet mass. At maturity, cross-pollinated fruit were 6% heavier and had 2% greater diameter than self-pollinated fruit, without significant differences in flesh nutrient concentrations or fatty acid composition. The results demonstrate that Hass avocado trees selectively retain cross-pollinated fruitlets, which are larger than self-pollinated fruitlets and ultimately produce larger mature fruit. Avocado growers can increase fruit size and yield by improving the opportunities for cross-pollination, possibly by closely interplanting type A and type B cultivars and introducing more beehives into orchards.

Keywords: fatty acids; fruit drop; Lauraceae; nutrition; *Persea americana*; pollination; selective abortion; self-compatibility; self-incompatibility; xenia

1. Introduction

Tree crops contribute about 10% of global food production and they provide many fruits and nuts of high nutritional value [1,2]. High tree-crop yields can increase the total production of nutrient-rich food and strengthen food security [3]. However, tree yields can be limited by premature fruit drop, with many fruitlets abscising before they can reach maturity [4–9]. Premature fruit drop is often observed in mass-flowering tree crops even after successful pollination [8–10].

Selective abscission of inferior genotypes could be a cause of premature fruit abscission in tree crops [4,10]. Self-pollinated fruitlets can be shed selectively during the period of premature fruit drop due to either late-acting self-incompatibility or inbreeding depression [11–15]. Late-acting self-incompatibility is a genetic self-rejection mechanism that prevents the growth of self-pollen tubes from the lower style into the ovule or that arrests embryo development following self-fertilisation [13–16]. Self-fertilised embryos may also experience inbreeding depression due to an excess of non-lethal, deleterious, recessive alleles that results in slower development and poor capacity to compete with larger embryos [4,12,17]. Cross-fertilised fruitlets may outcompete self-pollinated fruitlets for maternal resources, allowing the cross-pollinated fruitlets to be retained selectively on trees until maturity [4,10,11]. Self-pollinated cashew, lychee and mango fruitlets appear to abscise at higher rates than cross-pollinated fruitlets [5,6,8,18,19]. However, many early studies could not definitively determine embryo paternity, especially during early fruitlet development.

The genetic contribution from the pollen parent can affect the "discernible characteristics, especially of size, color, shape, chemical composition, and developmental timing, of parts of seeds and other fruits, including the embryo, endosperm, and maternal tissues", a phenomenon termed "xenia" [20]. Thus, pollen-parent effects are not confined to the early stages of endosperm and embryo development but can also result in differences in the final size or quality of fruit [20,21]. Cross-pollinated mature fruit often have greater final mass, firmness, skin colour, sugar content, or oil content than self-pollinated mature fruit [8,22–26]. However, the influence of embryo paternity on early fruitlet growth and how this is translated into final fruit size has not been studied in most tropical crops.

Avocado is a mass-flowering tree crop that produces mature fruit from less than 1% of its flowers [27–32]. Many avocado fruitlets abscise during the first two months after peak flowering [28,29,32,33]. Avocado flowers are hermaphrodite but functionally unisexual, with a separation of the female and male phases within the one flower that functions to reduce the incidence of self-pollination [28,34,35]. The flowers of Type A cultivars such as Hass often open as females on the morning of one day, close for the afternoon and the following morning, and re-open as males in the afternoon [28,34,35]. The flowers of Type B cultivars such as Shepard often open as females in the afternoon, close overnight, and re-open as males for the morning of the next day [28,34,35]. This alternation of genders between Type A and Type B avocado cultivars can promote cross-pollination (i.e., pollination between cultivars) as there is a long overlap each day between the female-phase flowers of one cultivar and the male-phase flowers of alternate-type cultivars [28,34,35]. Avocado flowers are not autogamous (i.e., they are not capable of self-pollination within the same flower without a pollen vector). However, there can be a short overlap each day between female-phase and male-phase flowers within the one cultivar, allowing self-pollination (i.e., pollination within the same clonally propagated cultivar) by pollinators such as honeybees [28,34,35]. Avocado flowers are self-compatible; thus, they can set fruit through either self-pollination or cross-pollination [29,36–40]. Early studies on premature fruit drop in avocado using isozyme methods indicated that cross-pollinated fruitlets were selectively retained over self-pollinated fruitlets [40–42]. These early studies often could not definitely determine paternity, especially in small fruitlets, due to limitations of the existing technology. However, later research using DNA-microsatellite methods confirmed that self-fertilised fruitlets were selectively abscised, possibly because they were fertilised late in the flowering season and were, therefore, smaller than cross-fertilised fruitlets [43]. The percentage of cross- and self-pollinated avocado fruitlets during early fruit development and how paternity affects fruitlet retention during the early stages of fruit set require further study, especially for the most widely grown cultivar, Hass.

In addition, xenia effects have been observed on avocado fruit-quality characteristics including whole-fruit mass, flesh mass, calcium concentration and fatty acid composition [33,36,37,39,40]. Cross-fertilisation of Fuerte by Ettinger, Teague, or Topa-Topa increases whole-fruit mass and pericarp mass by 9–17% and 5–13%, respectively, when compared with self-fertilisation [36]. Cross-fertilisation of Hass by Shepard can increase whole-fruit mass by 12% [37] and flesh calcium concentration by 10% [39] while reducing the ratio of unsaturated to saturated fatty acids by 4% [39]. However, there is little understanding of paternity effects on the size and nutrient concentrations of avocado fruitlets during premature fruit drop, and whether these effects may be related to selective abscission of self-pollinated fruitlets. A better understanding of embryo paternity and fruitlet development may help to distinguish between late-acting self-incompatibility and inbreeding depression as the underlying causes of premature fruit drop from avocado trees.

Here, we aimed to determine how embryo paternity influences fruitlet retention and fruit quality in Hass avocado. We used novel markers, based on single nucleotide polymorphisms (SNPs) identified by high-throughput genotyping [39,44], to assign paternity to embryos. Specifically, we assessed: (a) whether the percentages of progeny in the tree that were cross-fertilised changed during fruit development; (b) whether there were differences in the percentages of progeny that were cross-fertilised among the abscised fruitlets vs. the fruitlets that were retained on the tree; and (c) how embryo paternity affects the size and nutrient composition of the fruitlets and mature fruit. We hypothesised that cross-fertilised fruitlets would be retained selectively during the period of premature fruit drop and that embryo paternity would affect the size and nutrient composition of the fruitlets and mature fruit.

2. Materials and Methods

2.1. Study Site

The experiment was conducted in Eastridge avocado orchard (25°13′25″ S 152°18′54″ E), near Childers, Queensland, Australia. The orchard contained cultivars Hass, Maluma Hass and Shepard in large single-cultivar blocks [37]. We selected 30 trees in the 82nd and 83rd rows of a 132-row-wide block of Hass trees, which were planted 5 years prior to the experiment, as described previously [29]. The nearest other cultivar was a block of Shepard trees at the northern end of the Hass rows, approximately 190 m from the experimental trees. The experimental trees were part of a boron-fertiliser experiment, with 30 experimental trees located in ten plots, with each of the three experimental trees within a plot being allocated randomly to one of three boron treatments (0, 15, or 30 g B per tree). The boron treatments had no significant effect on fruit paternity [29].

2.2. Experimental Design and Sample Processing

Peak anthesis in the experimental trees occurred in late August 2018. We collected ten retained fruitlets from one side of each tree, selecting five fruitlets from the outside of the canopy and five fruitlets from the inside of the canopy, at each of 6 and 10 weeks after peak anthesis; i.e., 10 fruitlets per tree × 30 trees × 2 time points = 600 retained fruitlets from the canopy. We also collected ten freshly-abscised fruitlets under each tree, selecting five fruitlets close to the tree trunk and five distant from the trunk, at each of 6 and 10 weeks after peak anthesis; i.e., 10 fruitlets per tree × 30 trees × 2 time points = 600 freshly-abscised fruitlets from the ground. All fruitlets were stored at $-18\,^\circ$C until further processing. We harvested 16 mature fruit per tree at 36 weeks after peak anthesis using a stratified sampling design, with each tree divided into eight sectors, four on each side of the canopy. Two fruit were sampled per sector, one from the outside and one from the inside of the canopy; i.e., 16 fruit per tree × 30 trees = 480 mature fruit. Mature fruit were stored in a cold room at $4\,^\circ$C until further processing.

We weighed all fruitlets that were collected at 6 and 10 weeks after peak anthesis. The fruitlets were then dissected longitudinally into two halves (Figure 1). We measured the mass of the following sub-samples from retained fruitlets collected from the canopy: (1) total embryo, possibly with endosperm, hereafter referred to as the embryo ($E_1 + E_2$); (2) a combined half-fruitlet comprising pericarp and embryo ($M_1 + E_1$); and (3) the remaining half of the pericarp without embryo (M_2).

Figure 1. A dissected Hass avocado fruitlet at 10 weeks after peak anthesis, with E_1 and E_2 (embryo, possibly including endosperm), and M_1 and M_2 (maternal tissue).

We randomly selected 8 of the 16 mature fruit per tree after 7–21 days at 4 °C, weighed each fruit individually, and recorded length and diameter. Each fruit was dissected, and the flesh (without skin) and the seed were weighed. The other eight mature fruit per tree were stored for a total of 22–28 days at 4 °C and then stored at 21 °C until ripe. Ripeness was measured using an FR-5120 fruit hardness tester with an 8 mm head (Lutron Electronic, Taipei, Taiwan). A fruit was considered ripe if the maximum force required to press the 8 mm head of the hardness tester approximately 1 mm deep was less than 20 N. We measured the length, diameter and mass of each ripe fruit, then dissected the fruit and weighed the flesh (without skin) and the seed.

2.3. Paternity Analysis

We used one half of the embryo (E_2) of each retained fruitlet and each abscised fruitlet collected at 6 and 10 weeks after peak anthesis, and approximately 70 mg fresh mass of each mature seed, for genotyping. DNA extraction from embryos and mature seeds followed the glass-fibre plate DNA extraction protocol for plants [45]. We added disposable 2.3 mm and 0.1 mm zirconia/silica beads and liquid nitrogen to the samples prior to shaking on a TissueLyser II (Qiagen, Hilden, Germany). Multiplex PCR reactions were conducted for each extracted DNA sample to amplify regions with unique homozygous SNPs (Table 1) that we identified from ten avocado cultivars [39,44]. High-throughput genotyping was performed using the Agena MassARRAY platform (Agena Bioscience, San Diego, CA, USA) to assign paternity using 28 recently designed assays for avocado [39,44].

Table 1. Examples of 8 of the 52 DNA sequences used for identification of pollen parents by MassARRAY analysis of unique homozygous SNPs from each avocado cultivar [44].

Cultivar	Common Avocado Allele	Private Cultivar Allele	DNA Sequence with SNP	Hass	Maluma Hass	Shepard
Hass	T	A	TGCAGAAAGCAGTTTCAGATGCCAGA GTCGGAGAACCATTTTTGCAGTCTTAT ATGAGAGGAGGTAGAAAAACACTAG GGAGGAGAAATTGTTTTGCCA[T/A]AC CCAACAGTAGAATATGTTACTATTGA AAAAGAAATCTTCA	AA	TT	TT
Hass	G	T	TGCAGAGCAGAAAAGCACATGACTTG GAACTTCAACAAGAGCCTGCATCTTG AGTCTGTC[G/T]GGAGGAACTGATCGC CTGATGGCAAATTAGGAATCTTGTTGG TGATGAGAGAGGTTTTGCCCTCCCAA ATTGCATCAGTC	TT	GG	GG

Table 1. *Cont.*

Cultivar	Common Avocado Allele	Private Cultivar Allele	DNA Sequence with SNP	Hass	Maluma Hass	Shepard
Maluma Hass	T	C	TGCAGTAGAAGAAAGAAGGATGTAAT GATCTTGGCTCCAAAGAGAAACTCTC C[T/C]TTTCTTTTCCCTCTTTTTCTCCAC TTGAGAAGGAAAACCATAGTCACATC AATGAAAAATATACCTCCTTTTTTATT ATTCCTGTCT	TT	CC	TT
Maluma Hass	A	G	TGCAGCTATTATTTATATCACATGATT TTTTCCATTCTATCAGGCGTTGGAGAA AACCCATCACCTGAAAGCAAGAAT[A/ G]CATTACATTAGTCTACATCCAGTTT AGCCTGAGTGGGCCCCGCTATTGAGT GATCCAACTCA	AA	GG	AA
Maluma Hass	A	G	TGCAGCCTGGAGCTGTTGCTGTTATAG TTGTGTTTTGAGAGTGCGGCGAGGGA AGGACACAGCAGACAAGAGTACAGA CTAGACGAAACTCAAAACCCTCGGG[A/G]CAAATGGCTGTGTGTTTTCCCCAT TGCATTGCATTG	AA	GG	AA
Shepard	G	C	TGCAGGCCAAGCCGAAACTGAGCTCA AGGGAAAGCGTGGAGGAGGTGAAAA GGAAAAGCGTGCTAAAAAAAAGCAG GTCT[G/C]CAACTGTATATTCTTTATGT TCTTCATAGAGTATCTTTTCCATGGAT CCTTGACTCCTCG	GG	GG	CC
Shepard	A	G	TGCAGTTGTGCTATCTATGTGGTCCCT GCT[A/G]GCTAACTGTGTTTTATCATGT GTAGACTCTTTGGATGGTTGAGATGAG TGTGATTCTTCTACACAATTGAATGGT CAGAATTCATGAATGGTACTGGACCG GCCTAAGAT	AA	AA	GG
Shepard	C	T	TGCAGCAAAGCATCACGGTGCCTTCA TTTGCCCGTGTCTATATTTGGATGCCA AATTTTTATAGCAGTTAGAAGCACTG ATAACAGCAACCAAA[C/T]AAATAAT CTGGTGCATACAGATAAAATACAACC CAGGATATCTAC	CC	CC	TT

2.4. Mineral Nutrient Analysis

We used the following subsamples to analyse mineral nutrient concentrations: (1) half-fruitlets collected from the canopy, comprising the pericarp and embryo ($M_1 + E_1$); (2) representative subsamples of at least 300 mg from the seed of each ripe fruit; and (3) representative subsamples of at least 300 mg from the flesh of each ripe fruit. Nitrogen (N) concentrations in the seed and the flesh of ripe fruit were determined by combustion analysis using a LECO CNS 928 analyser (TruSpec®, LECO Corporation, St. Joseph, MI, USA) [46,47]. Aluminium (Al), boron (B), calcium (Ca), iron (Fe), magnesium (Mg), manganese (Mn), phosphorus (P), potassium (K), sodium (Na), sulphur (S) and zinc (Zn) concentrations in fruitlets and the seed and flesh of ripe fruit were analysed by inductively coupled plasma–optical emission spectroscopy (Vista Pro®, Varian Incorporation, Palo Alto, CA, USA) after digestion with a 5:1 mixture of nitric and perchloric acids [48,49].

We calculated the total content of each mineral nutrient in 6- and 10-week-old fruitlets using Equation (1):

$$\text{Nutrient content in total fruitlet} = [M_1 + E_1] \times \text{total fruitlet mass} \tag{1}$$

where $[M_1 + E_1]$ was the nutrient concentration in subsample $M_1 + E_1$.

The contents of each nutrient in the seed ($Content_{seed}$), flesh ($Content_{flesh}$) and whole mature fruit without skin ($Content_{fruit}$) were calculated using Equations (2)–(4):

$$Content_{seed} = [Seed] \times Mass_{seed} \tag{2}$$

$$Content_{flesh} = [Flesh] \times Mass_{flesh} \tag{3}$$

$$Content_{fruit} = Content_{seed} + Content_{flesh} \tag{4}$$

where [Seed] and [Flesh] were the nutrient concentrations, and $Mass_{seed}$ and $Mass_{flesh}$ were the masses, of the seed and the flesh of mature fruit, respectively.

2.5. Fatty Acid Analysis

We finely mashed one half of the flesh from each of the eight ripe fruit per tree to extract oil. Fatty acid methyl esters were derivatised from the extracted oil, and fatty acid composition was determined by gas chromatography–mass spectrometry (PerkinElmer Clarus 580 GC coupled to a PerkinElmer Clarus SQ8S MS) using methods described previously [50,51]. Quantitation of each compound was via integration of the peak area on the total ion current chromatogram. We calculated the relative abundance of individual fatty acids in each sample by dividing the peak area of each individual fatty acid by the total peak area of all fatty acids in the sample and multiplying by 100%.

2.6. Statistical Analysis

We compared the percentages of retained progeny that were cross-pollinated at 6 weeks after peak anthesis vs. 10 weeks after peak anthesis vs. maturity (36 weeks after peak anthesis) using a generalised linear mixed model (GLMM) with a Gaussian distribution and identity link function. Time (6 vs. 10 vs. 36 weeks after peak anthesis), boron treatment and plot were included as fixed effects. We also compared the percentages of progeny at 6 and 10 weeks after peak anthesis that were cross-pollinated among the abscised fruitlets on the ground vs. the retained fruitlets in the canopy using generalised linear models (GLMs) with a Gaussian distribution and identity link function. Sample location (ground or canopy), boron treatment and plot were included as fixed effects. We analysed pollen-parent effects on (1) embryo mass, pericarp mass and total mass of fruitlets at 6 and 10 weeks after peak anthesis; (2) diameter, length, seed mass, flesh mass and total mass of mature fruit; (3) nutrient levels in fruitlets, seeds, flesh or whole fruit; and (4) relative abundances of fatty acids in flesh, using GLMs. Pollen parent, boron treatment and plot were included as fixed effects. Boron application increased the diameter and flesh mass of mature fruit [29] but partitioning of the data into the three boron treatments accounted for potential effects of boron application on fruitlet or fruit parameters. We compared differences between means using a pairwise comparison procedure, with sequential Šidák corrections when significant differences were detected. Differences between means were considered significant at $p < 0.05$. Means are reported with standard errors. All statistical analyses were performed using SPSS version 26.0 (IBM, Armonk, NY, USA).

3. Results

3.1. Percentages of Retained Fruitlets That Were Cross-Fertilised

We identified selective retention of cross-fertilised Hass avocado fruitlets because the percentage of retained progeny that were cross-fertilised more than doubled during fruit development, from 4.6% at 6 weeks after peak anthesis to 10.7% at fruit maturity (Table 2). Cross-fertilised fruitlets represented 4.6% of the retained fruitlets at 6 weeks after peak anthesis, but only 1.6% of the freshly-abscised fruitlets at this stage were cross-fertilised (Table 2). Similarly, 6.5% of the retained fruitlets at 10 weeks after peak anthesis were cross-fertilised whereas only 1.0% of the freshly-abscised fruitlets were cross-fertilised

(Table 2). All cross-fertilised fruitlets and cross-fertilised mature fruit were pollinated by the nearest other cultivar, viz. Shepard.

Table 2. Percentages of Hass avocado fruitlets or fruit, collected from the ground vs. from the canopy at 6, 10, or 36 weeks after peak anthesis that were cross-fertilised.

Time after Peak Anthesis	Sample Location	
	Ground	Canopy
6 weeks	1.6 ± 0.8 a	4.6 ± 1.2 b *
10 weeks	1.0 ± 0.6 a	6.5 ± 1.4 b
36 weeks (mature)	–	10.7 ± 1.6 *

Means ($\pm$SE) within a row with different letters are significantly different; means marked by an asterisk (*) within a column are also significantly different (GLM, $p < 0.05$, $n = 30$ trees).

3.2. Cross-Fertilisation Effects on Fruitlet or Fruit Size

Embryo, pericarp and whole-fruitlet masses at 6 weeks after peak anthesis did not differ significantly between cross-fertilised and self-fertilised fruitlets (Figure 2a). However, the pericarp and whole-fruitlet masses at 10 weeks after peak anthesis were 51% and 50% higher, respectively, in cross-fertilised fruitlets than self-fertilised fruitlets (Figure 2b). Cross-fertilised fruit at maturity were 6% heavier (Figure 2c) and had 2% greater diameter than self-fertilised mature fruit (Figure 2c, Table 3).

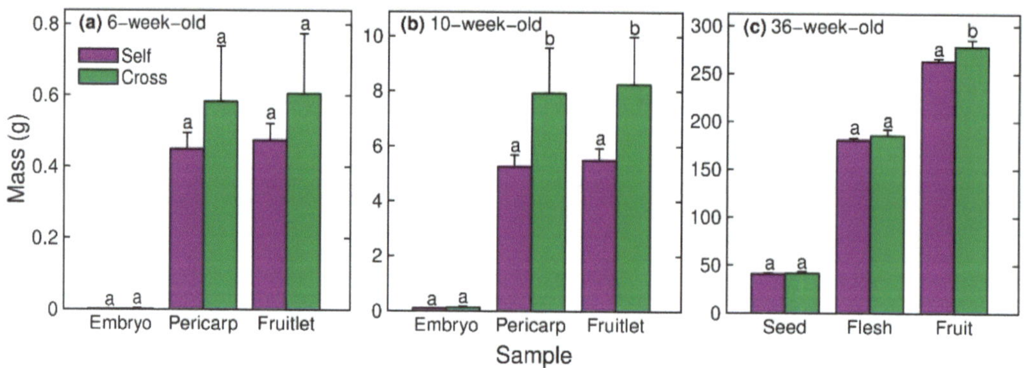

Figure 2. Mass of the (i) embryo or seed, (ii) pericarp or flesh, and (iii) total fruitlet or fruit of self-fertilised vs. cross-fertilised progeny of Hass avocado at (**a**) 6 weeks, (**b**) 10 weeks and (**c**) 36 weeks after peak anthesis. Means ($\pm$SE) with different letters within a sample type and time point are significantly different (GLM, $p < 0.05$, $n = 95$–325 selfs and $n = 13$–50 crosses).

Table 3. Length and diameter of self- and cross-fertilised mature fruit of Hass avocado.

Fruit Parameter	Pollen Parent	
	Hass (Self)	Shepard (Cross)
Length (cm)	9.93 ± 0.04 a	10.00 ± 0.11 a
Diameter (cm)	7.32 ± 0.02 a	7.49 ± 0.07 b

Means ($\pm$SE) within a row with different letters are significantly different (GLM, $p < 0.05$, $n = 322$ selfs and $n = 50$ crosses).

3.3. Cross-Fertilisation Effects on Mineral Nutrient Levels and Fatty Acid Composition

The concentrations of most mineral nutrients did not differ significantly between cross-fertilised and self-fertilised fruitlets at either 6 or 10 weeks after peak anthesis (Table 4). The total content of each mineral nutrient did not differ significantly between cross-fertilised and self-fertilised fruitlets at 6 weeks after peak anthesis (Table 5). However, the total contents of P, K, Al, Ca, Fe, Mn and Zn were higher in cross-fertilised fruitlets than self-fertilised fruitlets at 10 weeks after peak anthesis (Table 5).

Table 4. Mineral nutrient concentrations (mg/100 g) in self- and cross-fertilised fruitlets of Hass avocado at 6 and 10 weeks after peak anthesis.

Nutrient	Time after Peak Anthesis			
	6 Weeks		10 Weeks	
	Pollen Parent			
	Hass (Self)	Shepard (Cross)	Hass (Self)	Shepard (Cross)
P	95.7 ± 2.0 a	95.1 ± 6.0 a	70.7 ± 1.5 a	67.0 ± 3.9 a
K	364 ± 7 a	367 ± 17 a	317 ± 4 a	307 ± 13 a
Al	0.615 ± 0.048 a	0.478 ± 0.068 a	0.317 ± 0.019 a	0.276 ± 0.025 a
B	1.68 ± 0.20 a	1.84 ± 0.52 a	2.28 ± 0.21 a	2.10 ± 0.13 a
Ca	85.1 ± 3.0 a	68.4 ± 7.5 b	46.0 ± 2.4 a	40.3 ± 3.4 a
Fe	1.33 ± 0.06 a	1.32 ± 0.16 a	1.32 ± 0.08 a	1.27 ± 0.07 a
Mg	39.4 ± 1.0 a	37.3 ± 3.2 a	39.0 ± 6.0 a	31.2 ± 1.6 a
Mn	5.74 ± 0.30 a	4.51 ± 0.74 b	3.12 ± 0.16 a	3.01 ± 0.27 a
Na	1.63 ± 0.17 a	1.59 ± 0.34 a	1.78 ± 0.56 a	1.03 ± 0.19 a
S	50.5 ± 1.0 a	49.2 ± 3.3 a	41.1 ± 5.2 a	35.0 ± 2.1 a
Zn	1.51 ± 0.05 a	1.50 ± 0.16 a	0.94 ± 0.03 a	0.91 ± 0.06 a

Means (±SE) with different letters within the same row and time point are significantly different (GLM, $p < 0.05$, $n = 54$ selfs and $n = 12$ crosses at 6 weeks after peak anthesis, $n = 80$ selfs and $n = 19$ crosses at 10 weeks after peak anthesis).

Table 5. Mineral nutrient contents (mg) in self- and cross-fertilised fruitlets of Hass avocado at 6 and 10 weeks after peak anthesis.

Nutrient	Time after Peak Anthesis			
	6 Weeks		10 Weeks	
	Pollen Parent			
	Hass (Self)	Shepard (Cross)	Hass (Self)	Shepard (Cross)
P	0.51 ± 0.04 a	0.53 ± 0.12 a	3.03 ± 0.21 a	4.58 ± 0.85 b
K	1.98 ± 0.19 a	2.12 ± 0.52 a	14.30 ± 1.16 a	22.51 ± 4.43 b
Al	0.003 ± <0.001 a	0.002 ± 0.001 a	0.013 ± 0.001 a	0.020 ± 0.004 b
B	0.009 ± 0.002 a	0.008 ± 0.002 a	0.119 ± 0.026 a	0.168 ± 0.037 a
Ca	0.46 ± 0.04 a	0.36 ± 0.07 a	1.87 ± 0.13 a	2.81 ± 0.58 b
Fe	0.007 ± 0.001 a	0.006 ± 0.001 a	0.060 ± 0.006 a	0.102 ± 0.021 b
Mg	0.22 ± 0.02 a	0.27 ± 0.07 a	1.61 ± 0.19 a	2.22 ± 0.41 a
Mn	0.029 ± 0.002 a	0.022 ± 0.004 a	0.135 ± 0.012 a	0.207 ± 0.040 b
Na	0.008 ± 0.001 a	0.007 ± 0.001 a	0.069 ± 0.020 a	0.083 ± 0.023 a
S	0.27 ± 0.02 a	0.27 ± 0.06 a	1.77 ± 0.22 a	2.47 ± 0.48 a
Zn	0.008 ± 0.001 a	0.007 ± 0.002 a	0.041 ± 0.003 a	0.063 ± 0.012 b

Means (±SE) with different letters within the same row and time point are significantly different (GLM, $p < 0.05$, $n = 54$ selfs and $n = 12$ crosses at 6 weeks after peak anthesis, $n = 80$ selfs and $n = 19$ crosses at 10 weeks after peak anthesis).

The concentrations of most mineral nutrients in the flesh or seed did not differ significantly between cross-fertilised and self-fertilised mature Hass fruit (Table 6). Similarly, the total content of each mineral nutrient in the seed, flesh, or whole fruit did not differ significantly between cross-fertilised and self-fertilised Hass fruit (Table 7). Relative abundances of the six predominant fatty acids, the total unsaturated fatty acids (UFAs) and the total saturated fatty acids (SFAs) also did not differ significantly (Table 8).

Table 6. Mineral nutrient concentrations (mg/100 g) in the seed and flesh of self- and cross-fertilised mature fruit of Hass avocado.

Nutrient	Fruit Part			
	Seed		Flesh	
	Pollen Parent			
	Hass (Self)	Shepard (Cross)	Hass (Self)	Shepard (Cross)
N	522 ± 13 a	519 ± 24 a	472 ± 12 a	486 ± 21 a
P	67.3 ± 1.5 a	64.4 ± 2.8 a	58.0 ± 1.3 a	60.0 ± 2.7 a
K	570 ± 11 a	540 ± 18 a	572 ± 10 a	572 ± 23 a
Al	0.069 ± 0.006 a	0.072 ± 0.013 a	0.053 ± 0.004 a	0.049 ± 0.005 a
B	2.34 ± 0.14 a	2.65 ± 0.40 a	3.60 ± 0.13 a	3.39 ± 0.19 a
Ca	8.12 ± 0.49 a	7.77 ± 0.76 a	6.50 ± 0.19 a	6.37 ± 0.35 a
Fe	1.04 ± 0.03 a	1.11 ± 0.06 a	0.62 ± 0.02 a	0.60 ± 0.04 a
Mg	38.1 ± 1.2 a	37.3 ± 1.8 a	28.2 ± 0.4 a	28.4 ± 0.6 a
Mn	0.61 ± 0.02 a	0.53 ± 0.04 a	0.48 ± 0.02 a	0.50 ± 0.04 a
Na	1.07 ± 0.10 a	0.98 ± 0.25 a	7.88 ± 0.41 a	6.40 ± 0.48 a
S	36.1 ± 1.2 a	37.7 ± 2.5 a	29.3 ± 1.1 a	29.3 ± 2.0 a
Zn	0.631 ± 0.016 a	0.568 ± 0.029 b	0.759 ± 0.013 a	0.763 ± 0.027 a

Means (± SE) with different letters within the same row and tissue type are significantly different (GLM, $p < 0.05$, n = 103 selfs and n = 23 crosses).

Table 7. Mineral nutrient contents (mg) in the seed, flesh and whole fruit of self- and cross-fertilised Hass avocado.

Nutrient	Sample					
	Seed		Flesh		Fruit	
	Pollen Parent					
	Hass (Self)	Shepard (Cross)	Hass (Self)	Shepard (Cross)	Hass (Self)	Shepard (Cross)
N	224 ± 9 a	225 ± 18 a	878 ± 31 a	949 ± 68 a	1102 ± 37 a	1175 ± 82 a
P	28.5 ± 1.0 a	27.8 ± 2.1 a	107.7 ± 3.6 a	117.9 ± 9.6 a	136.2 ± 4.1 a	145.7 ± 11.2 a
K	241 ± 8 a	235 ± 17 a	1061 ± 32 a	1117 ± 80 a	1302 ± 37 a	1351 ± 90 a
Al	0.029 ± 0.002 a	0.033 ± 0.010 a	0.095 ± 0.006 a	0.095 ± 0.011 a	0.124 ± 0.008 a	0.128 ± 0.016 a
B	0.97 ± 0.07 a	1.08 ± 0.15 a	6.67 ± 0.29 a	6.54 ± 0.52 a	7.64 ± 0.34 a	7.62 ± 0.65 a
Ca	3.3 ± 0.2 a	3.2 ± 0.4 a	11.7 ± 0.3 a	11.9 ± 0.5 a	15.0 ± 0.4 a	15.1 ± 0.7 a
Fe	0.44 ± 0.02 a	0.48 ± 0.04 a	1.13 ± 0.05 a	1.15 ± 0.08 a	1.57 ± 0.05 a	1.63 ± 0.11 a
Mg	16.0 ± 0.6 a	15.9 ± 1.1 a	52.0 ± 1.4 a	55.0 ± 3.2 a	67.9 ± 1.6 a	70.9 ± 3.8 a
Mn	0.25 ± 0.01 a	0.22 ± 0.02 a	0.86 ± 0.03 a	0.94 ± 0.07 a	1.11 ± 0.03 a	1.16 ± 0.08 a
Na	0.44 ± 0.04 a	0.46 ± 0.13 a	16.9 ± 1.5 a	13.5 ± 1.6 a	17.3 ± 1.6 a	14.0 ± 1.7 a
S	15.3 ± 0.7 a	16.4 ± 1.5 a	53.5 ± 2.2 a	56.1 ± 4.6 a	68.8 ± 2.6 a	72.5 ± 5.9 a
Zn	0.27 ± 0.01 a	0.25 ± 0.02 a	1.40 ± 0.04 a	1.48 ± 0.10 a	1.67 ± 0.05 a	1.73 ± 0.11 a

Means (±SE) for self- and cross-fertilised fruit within the same row and tissue type do not differ significantly (GLM, $p > 0.05$, n = 103 selfs and n = 23 crosses).

Table 8. Relative abundances of individual fatty acids, total unsaturated fatty acids (UFAs) and total saturated fatty acids (UFAs) in self- and cross-fertilised Hass avocado fruit.

Fatty Acid (%)	Pollen Parent	
	Hass (Self)	**Shepard (Cross)**
Palmitic acid (C16:0)	26.1 ± 0.3 a	27.0 ± 0.4 a
Palmitoleic acid (C16:1 cis)	5.34 ± 0.14 a	5.65 ± 0.25 a
Stearic acid (C18:0)	0.186 ± 0.003 a	0.189 ± 0.007 a
Oleic acid (C18:1 cis)	48.6 ± 0.3 a	47.6 ± 0.5 a
Elaidic acid (C18:1 trans)	10.6 ± 0.2 a	10.7 ± 0.4 a
Linoleic acid (C18:2)	9.21 ± 0.16 a	8.82 ± 0.28 a
Total UFAs	73.8 ± 0.3 a	72.8 ± 0.4 a
Total SFAs	26.2 ± 0.3 a	27.2 ± 0.4 a

Means (±SE) within a row do not differ significantly (GLM, $p > 0.05$, $n = 103$ selfs and $n = 23$ crosses).

4. Discussion

The results of this study support the hypotheses that cross-fertilised avocado fruitlets would be retained selectively during premature fruit drop and that embryo paternity would affect the size of the fruitlets and fruit. Only 1.0% of the abscised fruitlets at 10 weeks after peak anthesis were cross-fertilised, while 99.0% of the abscised fruitlets were self-fertilised; thus, the percentage of progeny on the tree that were cross-fertilised more than doubled between 6 weeks after peak anthesis and fruit maturity. Cross-fertilised fruitlets were larger than self-fertilised fruit at 10 weeks after peak anthesis and cross-fertilised mature fruit were ultimately heavier, with greater diameter, than self-fertilised mature fruit. However, our results, for the most part, do not support the hypothesis that embryo paternity would affect the nutrient composition of the fruitlets and mature fruit. Differences in total nutrient accumulation between cross-fertilised fruitlets and self-fertilised fruitlets at 10 weeks after peak anthesis were not due to differences in nutrient concentrations but were due solely to differences in fruitlet mass. In addition, the flesh of cross-fertilised and self-fertilised mature fruit did not differ significantly in mineral nutrient concentrations, total contents of each mineral nutrient, or fatty acid composition.

Our results suggest that inbreeding depression and inter-fruit competition, rather than a strong post-zygotic self-incompatibility mechanism, may be involved in the selective abscission of self-fertilised avocado fruitlets during the period of premature fruit drop. The self-fertilised fruitlets were smaller than the cross-fertilised fruitlets; thus, they accrued a lower total mass of many mineral nutrients. Accumulation of non-lethal deleterious alleles can reduce the growth of selfed progeny, which then have a reduced capacity to compete with larger outcrossed progeny for maternal resources during the period of premature fruit drop [11,15,17]. Mineral nutrient transport to fruitlets occurs down a concentration gradient that follows, rather than regulates, the mass flow of water through the xylem and phloem [52–55]. This transport via mass flow explains why the self-fertilised fruitlets, which were smaller than the cross-fertilised fruitlets, did not have lower nutrient concentrations despite having lower total contents of many nutrients. The ultimately high levels of selfing among mature avocado fruit in this study and previous studies [36,37,39–41,43,56–58] also suggest that the selective abscission of self-fertilised fruitlets during premature fruit drop was not driven by a rigorous post-zygotic self-incompatibility mechanism. This type of genetic self-rejection system can cause embryo abortion at various stages during fruit development, although it usually operates soon after fertilisation [12,14–16,59].

The cross-fertilised mature fruit of Hass, which were all pollinated by the neighbouring cultivar Shepard, had 6% higher mass and 2% greater diameter than self-fertilised fruit. Similar xenia effects on avocado have been observed recently in Spain, with cross-fertilisation by Fuerte also increasing the mass of Hass fruit [33]. In addition, cross-fertilisation by Ettinger, Teague, or Topa-Topa has increased the mass of Fuerte fruit in Israel [36]. Larger fruit provide higher financial returns to avocado growers because premiums are paid for cartons that contain fruit of greater diameter [60–63]. The cross-fertilised Hass fruit in this study and a previous study [39] appeared to have the same nutritional quality as self-fertilised fruit, possessing similar flesh nutrient concentrations and fatty acid profiles. Both types of fruit had the characteristic fatty acid profile of avocado, with high levels of unsaturated oleic, elaidic, linoleic and palmitoleic acids, that are reported to make the fruit a valuable component of a healthy diet [64–68].

Increasing the opportunities for cross-pollination in avocado orchards can, therefore, provide larger fruit with higher commercial value for growers. We have shown recently that raising the levels of cross-fertilisation can also improve avocado yield by increasing the number of mature fruit produced by each tree [37]. Outcrossing rates often decline with increasing distance to trees of another avocado cultivar [37,39–41,43]. The Hass trees in the current study were located at least 190 m from the polliniser, Shepard, which explains why only 11% of mature Hass fruit were cross-fertilised. Much higher rates of outcrossing (30–52%) have been observed among Hass trees in the same orchard that were located only 20–40 m from the Shepard trees [37]. Cross-pollination and yield can be increased by closer interplanting of type A and type B cultivars [37,38,43] and by increasing the number and distribution of beehives in orchards [30,69–72]. Managing the surrounding habitat to enhance pollinator abundance and pollinator diversity may also increase the frequency of successful pollination of avocado flowers [72–75].

5. Conclusions

This study demonstrated that cross-fertilised Hass avocado fruitlets are retained selectively during the period of premature fruit drop. Only 1.0% of freshly-abscised fruitlets on the ground at 10 weeks after peak anthesis were cross-fertilised, while 99.0% of the freshly-abscised fruitlets at this stage were self-fertilised. Cross-fertilised fruitlets at this stage were larger than self-fertilised fruitlets, which indicates that inbreeding depression resulted in the self-fertilised fruitlets growing more slowly and having reduced capacity to compete for maternal resources during the period of premature fruit drop. Furthermore, cross-fertilised mature fruit were heavier and had greater diameter than self-fertilised fruit; thus, they can provide higher financial returns. Avocado growers might increase fruit size and yield by improving the opportunities for cross-pollination. This could be achieved by closely interplanting type A and type B cultivars, introducing more beehives into orchards and maintaining habitat that sustains an abundance and diversity of wild pollinators.

Author Contributions: Conceptualization, W.K., S.H.B., S.M.O., H.M.W. and S.J.T.; formal analysis, N.S.H.; funding acquisition, S.H.B., S.M.O., H.M.W. and S.J.T.; investigation, N.S.H., W.K. and J.N.; methodology, N.S.H., W.K., S.H.B., S.M.O., J.N., H.M.W. and S.J.T.; project administration, S.J.T.; supervision, W.K., S.H.B., S.M.O., H.M.W. and S.J.T.; writing—original draft, N.S.H.; writing—review and editing, W.K., S.H.B., S.M.O., J.N., H.M.W. and S.J.T. All authors have read and agreed to the published version of the manuscript.

Funding: This research was funded by project PH16001 of the Hort Frontiers strategic partnership initiative developed by Hort Innovation, with co-investment from Griffith University, Plant & Food Research Ltd., University of the Sunshine Coast, and contributions from the Australian Government. Nimanie Hapuarachchi was supported by a Griffith University International Postgraduate Research Scholarship and a Griffith University Postgraduate Research Scholarship.

Data Availability Statement: The data will be made available upon request with permission from Hort Innovation.

Acknowledgments: We thank the growers for assistance and access to the orchard. We thank Anushika De Silva, Marta Gallart, Sascha Kämper, Tsvakai Gama, Tracey McMahon, Rachele Wilson, Peter Brooks, Jack Royle and David Appleton for field and laboratory assistance.

Conflicts of Interest: The authors declare no conflicts of interest. The funders had no role in the design of the study; in the collection, analyses, or interpretation of data; or in the writing of the manuscript; however, they approved the publication of the results.

References

1. Food and Agriculture Organisation. *Fruit and Vegetables—Your Dietary Essentials*; FAO: Rome, Italy, 2020.
2. Production Volume of the Most Produced Food Commodities Worldwide in 2019, by Product. Available online: https://www.statista.com/statistics/1003455/most-produced-crops-and-livestock-products-worldwide/ (accessed on 25 March 2024).
3. Ickowitz, A.; McMullin, S.; Rosenstock, T.; Dawson, I.; Rowland, D.; Powell, B.; Mausch, K.; Djoudi, H.; Sunderland, T.; Nurhasan, M.; et al. Transforming food systems with trees and forests. *Lancet Planet. Health* **2022**, *6*, e632–e639. [CrossRef] [PubMed]
4. Pallardy, S.G. Reproductive growth. In *Physiology of Woody Plants*; Pallardy, S.G., Ed.; Academic Press: Cambridge, MA, USA, 2008; pp. 87–103.
5. Dag, A.; Eisenstein, D.; Gazit, S.; El-Batsri, R.; Degani, C. Effect of pollinizer distance and selective fruitlet abscission on outcrossing rate and yield in 'Tommy Atkins' mango. *J. Am. Soc. Hortic. Sci.* **1998**, *123*, 618–622. [CrossRef]
6. Holanda-Neto, J.P.; Freitas, B.M.; Bueno, D.M.; Araújo, Z.B. Low seed/nut productivity in cashew (*Anacardium occidentale*): Effects of self-incompatibility and honeybee (*Apis mellifera*) foraging behavior. *J. Hortic. Sci. Biotechnol.* **2002**, *77*, 226–231. [CrossRef]
7. Trueman, S.J.; Kämper, W.; Nichols, J.; Ogbourne, S.M.; Hawkes, D.; Peters, T.; Bai, S.H.; Wallace, H.M. Pollen limitation and xenia effects in a cultivated mass-flowering tree, *Macadamia integrifolia* (Proteaceae). *Ann. Bot.* **2022**, *129*, 135–146. [CrossRef] [PubMed]
8. Degani, C.; Stern, R.A.; El-Batsri, R.; Gazit, S. Pollen-parent effect on the selective abscission of 'Mauritius' and 'Floridian' Lychee Fruitlets. *J. Am. Soc. Hortic. Sci.* **1995**, *120*, 523–526. [CrossRef]
9. Pérez, V.; Herrero, M.; Hormaza, J.I. Self-fertility and preferential cross-fertilization in mango (*Mangifera indica*). *Sci. Hortic.* **2016**, *213*, 373–378. [CrossRef]
10. Stephenson, A.G. Flower and fruit abortion proximate causes and ultimate functions. *Annu. Rev. Ecol. Syst.* **1981**, *12*, 253–279. [CrossRef]
11. Xiong, H.; Zou, F.; Guo, S.; Yuan, D.; Niu, G. Self-sterility may be due to prezygotic late-acting self-incompatibility and early-acting inbreeding depression in Chinese chestnut. *J. Am. Soc. Hortic. Sci.* **2019**, *144*, 172–181. [CrossRef]
12. Hao, Y.-Q.; Zhao, X.-F.; She, D.-Y.; Xu, B.; Zhang, D.-Y.; Liao, W.-J. The role of late-acting self-incompatibility and early-acting inbreeding depression in governing female fertility in monkshood, *Aconitum kusnezoffii*. *PLoS ONE* **2012**, *7*, e47034. [CrossRef]
13. Gibbs, P.E. Late-acting self-incompatibility—The pariah breeding system in flowering plants. *New Phytol.* **2014**, *203*, 717–734. [CrossRef]
14. Seavey, S.R.; Bawa, K.S. Late-acting self-incompatibility in angiosperms. *Bot. Rev.* **1986**, *52*, 196–213. [CrossRef]
15. Sage, T.L.; Sampson, F.B. Evidence for ovarian self-incompatibility as a cause of self-sterility in the relictual woody angiosperm, *Pseudowintera axillaris* (Winteraceae). *Ann. Bot.* **2003**, *91*, 807–816. [CrossRef] [PubMed]
16. Bianchi, M.B.; Meagher, T.R.; Gibbs, P.E. Do s genes or deleterious recessives control late-acting self-incompatibility in *Handroanthus heptaphyllus* (Bignoniaceae)? A diallel study with four full-sib progeny arrays. *Ann. Bot.* **2021**, *127*, 723–736. [CrossRef] [PubMed]
17. Husband, B.C.; Schemske, D.W. Evolution of the magnitude and timing of inbreeding depression in plants. *Evolution* **1996**, *50*, 54–70. [CrossRef] [PubMed]
18. Dag, A.; Degani, C.; Gazit, S. Gene flow in mango orchards and its impact on yield. *Acta Hortic.* **2009**, *820*, 347–350. [CrossRef]
19. Raz, A.; Goldway, M.; Sapir, G.; Stern, R.A. "Hong Long" lychee (*Litchi chinensis* Sonn.) is the optimal pollinizer for the main lychee cultivars in Israel. *Plants* **2022**, *11*, 1996. [CrossRef] [PubMed]
20. Denney, J.O. Xenia includes metaxenia. *HortScience* **1992**, *27*, 722–727. [CrossRef]
21. Yang, Q.; Fu, Y.; Liu, Y.; Zhang, T.; Peng, S.; Deng, J. Novel classification forms for xenia. *HortScience* **2020**, *55*, 980–987. [CrossRef]
22. Wallace, H.M.; Lee, L.S. Pollen source, fruit set and xenia in mandarins. *J. Hortic. Sci. Biotechnol.* **1999**, *74*, 82–86. [CrossRef]
23. Militaru, M.; Butac, M.; Sumedrea, D.; Chiţu, E. Effect of metaxenia on the fruit quality of scab resistant apple varieties. *Agric. Agric. Sci. Procedia* **2015**, *6*, 151–156. [CrossRef]
24. Dung, C.D.; Wallace, H.M.; Bai, S.H.; Ogbourne, S.M.; Trueman, S.J. Cross-pollination affects fruit colour, acidity, firmness and shelf life of self-compatible strawberry. *PLoS ONE* **2021**, *16*, e0256964. [CrossRef] [PubMed]
25. Kämper, W.; Thorp, G.; Wirthensohn, M.; Brooks, P.; Trueman, S.J. Pollen paternity can affect kernel size and nutritional composition of self-incompatible and new self-compatible almond cultivars. *Agronomy* **2021**, *11*, 326. [CrossRef]
26. Trueman, S.J.; Penter, M.G.; Malagodi-Braga, K.S.; Nichols, J.; De Silva, A.L.; Ramos, A.T.M.; Moriya, L.M.; Ogbourne, S.M.; Hawkes, D.; Peters, T.; et al. High outcrossing levels among global macadamia cultivars: Implications for nut quality, orchard designs and pollinator management. *Horticulturae* **2024**, *10*, 203. [CrossRef]
27. Lahav, E.; Zamet, D. Flowers, fruitlets and fruit drop in avocado trees. *Rev. Chapingo Ser. Hortic.* **1999**, *5*, 95–100.
28. Gazit, S.; Degani, C. Reproductive biology. In *The Avocado: Botany, Production and Uses*; Whiley, A.W., Schaffer, B., Wolstenholme, B.N., Eds.; CABI Publishing: New York, NY, USA, 2002; pp. 101–133.
29. Hapuarachchi, N.S.; Kämper, W.; Wallace, H.M.; Bai, S.H.; Ogbourne, S.M.; Nichols, J.; Trueman, S.J. Boron effects on fruit set, yield, quality and paternity of Hass avocado. *Agronomy* **2022**, *12*, 1479. [CrossRef]
30. Alcaraz, M.L.; Hormaza, J.I. Inadequate pollination is a key factor determining low fruit-to-flower ratios in avocado. *Horticulturae* **2024**, *10*, 140. [CrossRef]
31. Toukem, N.K.; Dubois, T.; Mohamed, S.A.; Lattorff, H.M.G.; Jordaens, K.; Yusuf, A.A. The effect of annual flower strips on pollinator visitation and fruit set of avocado (*Persea americana* Mill.) in Kenya. *Arthropod–Plant Interact.* **2023**, *17*, 19–29. [CrossRef]
32. D'Asaro, A.; Reig, C.; Martínez-Fuentes, A.; Mesejo, C.; Farina, V.; Agustí, M. Hormonal and carbohydrate control of fruit set in avocado 'Lamb Hass'. A question of the type of inflorescence? *Sci Hortic.* **2021**, *282*, 110046. [CrossRef]

33. Alcaraz, M.L.; Hormaza, J.I. Fruit set in avocado: Pollen limitation, pollen load size, and selective fruit abortion. *Agronomy* **2021**, *11*, 1603. [CrossRef]

34. Solares, E.; Morales-Cruz, A.; Balderas, R.F.; Focht, E.; Ashworth, V.E.T.M.; Wyant, S.; Minio, A.; Cantu, D.; Arpaia, M.L.; Gaut, B.S. Insights into the domestication of avocado and potential genetic contributors to heterodichogamy. *G3* **2023**, *13*, jkac323. [CrossRef]

35. Scholefield, P.B. A scanning electron microscope study of flowers of avocado, litchi, macadamia and mango. *Sci. Hortic.* **1982**, *16*, 263–272. [CrossRef]

36. Degani, C.; Goldring, A.; Adato, I.; El-Batsri, R. Pollen-parent effect on outcrossing rate, yield, and fruit characteristics of 'Fuerte' avocado. *HortScience* **1990**, *25*, 471–473. [CrossRef]

37. Trueman, S.J.; Nichols, J.; Farrar, M.B.; Wallace, H.M.; Hosseini Bai, S. Outcrossing rate and fruit yield of Hass avocado trees decline at increasing distance from a polliniser cultivar. *Agronomy* **2024**, *14*, 122. [CrossRef]

38. Stahl, P.; Mirom, Y.L.; Stern, R.A.; Goldway, M. Comparing 'Iriet' and 'Ettinger' avocado cultivars as pollinators of 'Hass' using SNPs for paternal identification. *Sci. Hortic.* **2019**, *248*, 50–57. [CrossRef]

39. Kämper, W.; Ogbourne, S.M.; Hawkes, D.; Trueman, S.J. SNP markers reveal relationships between fruit paternity, fruit quality and distance from a cross-pollen source in avocado orchards. *Sci. Rep.* **2021**, *11*, 20043. [CrossRef]

40. Degani, C.; El-Batsri, R.; Gazit, S. Outcrossing rate, yield, and selective fruit abscission in 'Ettinger' and 'Ardith' avocado plots. *J. Am. Soc. Hortic. Sci.* **1997**, *122*, 813–817. [CrossRef]

41. Degani, C.; Goldring, A.; Gazit, S. Pollen-parent effect on outcrossing rate in 'Hass' and 'Fuerte' avocado plots during fruit development. *J. Am. Soc. Hortic. Sci.* **1989**, *114*, 106–111. [CrossRef]

42. Degani, C.; Goldring, A.; Gazit, S.; Lavi, U. Genetic selection during the abscission of avocado fruitlets. *HortScience* **1986**, *21*, 1187–1192. [CrossRef]

43. Alcaraz, M.L.; Hormaza, J.I. Influence of physical distance between cultivars on yield, outcrossing rate and selective fruit drop in avocado (*Persea americana*, Lauraceae). *Ann. Appl. Biol.* **2011**, *158*, 354–361. [CrossRef]

44. Kämper, W.; Trueman, S.J.; Cooke, J.; Kasinadhuni, N.; Brunton, A.J.; Ogbourne, S.M. Single nucleotide polymorphisms (SNPs) that uniquely identify cultivars of avocado (*Persea americana*). *Appl. Plant Sci.* **2021**, *9*, e11440. [CrossRef]

45. Ivanova, N.V.; Fazekas, A.J.; Hebert, P.D.N. Semi-automated, membrane-based protocol for DNA isolation from plants. *Plant Mol. Biol. Rep.* **2008**, *26*, 186–198. [CrossRef]

46. McGeehan, S.L.; Naylor, D.V. Automated instrumental analysis of carbon and nitrogen in plant and soil samples. *Commun. Soil Sci. Plant Anal.* **1988**, *19*, 493–505. [CrossRef]

47. Rayment, G.E.; Higginson, F.R. *Australian Laboratory Handbook of Soil and Water Chemical Methods*; Inkata: Melbourne, Australia, 1992.

48. Martinie, G.D.; Schilt, A.A. Investigation of the wet oxidation efficiencies of perchloric acid mixtures for various organic substances and the identities of residual matter. *Anal. Chem.* **1976**, *48*, 70–74. [CrossRef]

49. Munter, R.C.; Grande, R.A. Plant tissue and soil extract analysis by ICP-atomic emission spectrometry. In *Developments in Atomic Plasma Spectrochemical Analysis*; Byrnes, R.M., Ed.; Heyden: London, UK, 1981; pp. 653–672.

50. Kämper, W.; Trueman, S.J.; Tahmasbian, I.; Bai, S.H. Rapid determination of nutrient concentrations in Hass avocado fruit by Vis/NIR hyperspectral imaging of flesh or skin. *Remote Sens.* **2020**, *12*, 3409. [CrossRef]

51. Richards, T.E.; Kämper, W.; Trueman, S.J.; Wallace, H.M.; Ogbourne, S.M.; Brooks, P.R.; Nichols, J.; Bai, S.H. Relationships between nut size, kernel quality, nutritional composition and levels of outcrossing in three macadamia cultivars. *Plants* **2020**, *9*, 228. [CrossRef] [PubMed]

52. Tagliavini, M.; Zavalloni, C.; Rombolà, A.D.; Quartieri, M.; Malaguti, D.; Mazzanti, F.; Millard, P.; Marangoni, B. Mineral nutrient partitioning to fruits of deciduous trees. *Acta Hortic.* **2000**, *512*, 131–140. [CrossRef]

53. Karley, A.J.; White, P.J. Moving cationic minerals to edible tissues: Potassium, magnesium, calcium. *Curr. Opin. Plant Biol.* **2009**, *12*, 291–298. [CrossRef]

54. Ruan, Y.-L.; Patrick, J.W.; Bouzayen, M.; Osorio, S.; Fernie, A.R. Molecular regulation of seed and fruit set. *Trends Plant Sci.* **2012**, *17*, 656–665. [CrossRef]

55. Dung, C.D.; Wallace, H.M.; Bai, S.H.; Ogbourne, S.M.; Trueman, S.J. Biomass and mineral nutrient partitioning among self-pollinated and cross-pollinated fruit on the same strawberry plant. *PLoS ONE* **2022**, *17*, e0269485. [CrossRef]

56. Vrecenar-Gadus, M.; Ellstrand, N.C. The effect of planting design on out-crossing rate and yield in the 'Hass' avocado. *Sci. Hortic.* **1985**, *27*, 215–221. [CrossRef]

57. Garner, L.C.; Ashworth, V.E.T.M.; Clegg, M.T.; Lovatt, C.J. The impact of outcrossing on yields of 'Hass' avocado. *J. Am. Soc. Hortic. Sci.* **2008**, *133*, 648–652. [CrossRef]

58. Kobayashi, M.; Lin, J.-Z.; Davis, J.; Francis, L.; Clegg, M.T. Quantitative analysis of avocado outcrossing and yield in California using RAPD markers. *Sci. Hortic.* **2000**, *86*, 135–149. [CrossRef]

59. Bittencourt, N.S. Evidence for post-zygotic self-incompatibility in *Handroanthus impetiginosus* (Bignoniaceae). *Plant Reprod.* **2017**, *30*, 69–79.

60. Whiley, A.W. Crop management. In *The Avocado: Botany, Production and Uses*; Whiley, A.W., Schaffer, B., Wolstenholme, B.N., Eds.; CABI Publishing: New York, NY, USA, 2002; pp. 231–253.

61. Lovatt, C.J. *Hass Avocado Nutrition Research in California*; University of California: Riverside, CA, USA, 2013.

62. Lovatt, C.; Zheng, Y.; Khuong, T.; Campisi-Pinto, S.; Crowley, D.; Rolshausen, P. Yield characteristics of 'Hass' avocado trees under California growing conditions. In *Proceedings of the VIII World Avocado Congress: Management and Techniques of Cultivation*; ProHass: Lima, Peru, 2015; pp. 336–341.
63. Avocados Australia. *Avocados Australia Best Practice Resource—Packaging for Export*; Avocados Australia: Rocklea, Australia, 2018.
64. Fulgoni, V.L.; Dreher, M.; Davenport, A.J. Avocado consumption is associated with better diet quality and nutrient intake, and lower metabolic syndrome risk in US adults: Results from the national health and nutrition examination survey (NHANES) 2001–2008. *Nutr. J.* **2013**, *12*, 1. [CrossRef] [PubMed]
65. Dreher, M.L.; Cheng, F.W.; Ford, N.A. A comprehensive review of Hass avocado clinical trials, observational studies, and biological mechanisms. *Nutrients* **2021**, *13*, 4376. [CrossRef] [PubMed]
66. Stolp, L.J.; Kodali, D.R. Naturally occurring high-oleic oils: Avocado, macadamia, and olive oils. In *High Oleic Oils. Development, Properties, and Uses*; Flider, F.J., Ed.; Elsevier: London, UK, 2021; pp. 7–52.
67. Pacheco, L.S.; Li, Y.; Rimm, E.B.; Manson, J.E.; Sun, Q.; Rexrode, K.; Hu, F.B.; Guasch-Ferré, M. Avocado consumption and risk of cardiovascular disease in US adults. *J. Am. Heart Assoc.* **2022**, *11*, e024014. [CrossRef]
68. Monge, A.; Stern, D.; Cortés-Valencia, A.; Catzín-Kuhlmann, A.; Lajous, M.; Denova-Gutiérrez, E. Avocado consumption is associated with a reduction in hypertension incidence in Mexican women. *Br. J. Nutr.* **2023**, *129*, 1976–1983. [CrossRef]
69. Read, S.F.J.; Howlett, B.G.; Jesson, L.K.; Pattemore, D.E. Insect visitors to avocado flowers in the Bay of Plenty, New Zealand. *N. Z. Plant Prot.* **2017**, *70*, 38–44. [CrossRef]
70. Peña, J.F.; Carabalí, A. Effect of honey bee (*Apis mellifera* L.) density on pollination and fruit set of avocado (*Persea americana* Mill.) cv. Hass. *J. Apic. Sci.* **2018**, *62*, 5–14. [CrossRef]
71. Sagwe, R.N.; Peters, M.K.; Dubois, T.; Steffan-Dewenter, I.; Lattorff, H.M.G. Insect pollination and pollinator supplementation enhances fruit weight, quality, and marketability of avocado (*Persea americana*). *Arthropod–Plant Interact.* **2023**, *17*, 753–763. [CrossRef]
72. Dymond, K.; Celis-Diez, J.L.; Potts, S.G.; Howlett, B.G.; Wilcox, B.K.; Garratt, M.P.D. The role of insect pollinators in avocado production: A global review. *J. Appl. Entomol.* **2021**, *145*, 369–383. [CrossRef] [PubMed]
73. Faber, B.; Frankie, G.; Pawelek, J.; Chase, M.; Witt, S. Native pollinators of avocado as affected by constructed pollinator habitat gardens in southern California. *Acta Hortic.* **2020**, *1299*, 329–332. [CrossRef]
74. Celis-Diez, J.L.; García, C.B.; Armesto, J.J.; Abades, S.; Garratt, M.P.D.; Fontúrbel, F.E. Wild floral visitors are more important than honeybees as pollinators of avocado crops. *Agronomy* **2023**, *13*, 1722. [CrossRef]
75. Muñoz, A.E.; Plantegenest, M.; Amouroux, P.; Zavieso, T. Native flower strips increase visitation by non-bee insects to avocado flowers and promote yield. *Basic Appl. Ecol.* **2021**, *56*, 369–378. [CrossRef]

Article

Transcriptomic and Metabolomic Analyses Provide New Insights into the Response of Strawberry (*Fragaria × ananassa Duch.*) to Drought Stress

Lili Jiang [1,2], Ruimin Song [3], Xiaofang Wang [1,2], Jie Wang [3,*] and Chong Wu [1,2,*]

[1] Shandong Institute of Pomology, Tai'an 271000, China; gssjianglili@shandong.cn (L.J.); xiaofang2015@sdau.edu.cn (X.W.)
[2] Shandong Technology Innovation Center of Modern Protected Fruit, Tai'an 271000, China
[3] Tobacco Research Institute, Chinese Academy of Agricultural Sciences, Qingdao 266101, China; songruimin123@126.com
* Correspondence: wangjie@caas.cn (J.W.); wuchongge@163.com (C.W.)

Abstract: Strawberry plants have shallow roots and large leaves, which are highly sensitive to variations in water levels. To explore the physicochemical and molecular mechanisms of strawberry response to water stress, and provide new ideas for strawberry scientific irrigation, we measured the transpiration rate, fresh weight, biomass gain, and other indicators of potted "Zhangji" strawberry plants under drought and waterlogging treatments using a Plantarray system. Transcriptomic and metabolomic analyses of strawberry leaves following mild drought, moderate drought, severe drought, and rehydration treatments were performed to identify key genes and metabolites involved in the response to drought stress. Below a certain threshold, the transpiration rate of strawberry plants was significantly lower after the deficit irrigation treatment than the conventional water treatment. Transcriptome analysis revealed that genes involved in oxidoreductase activity and in sulfur and nitrogen metabolism were up-regulated, as well as starch and sucrose. Strawberry plants secrete various endogenous growth hormones to maintain their normal growth under drought stress. The syntheses of salicylic acid (SA) and abscisic acid (ABA) were up-regulated in the mild and moderate drought treatments. However, the syntheses of 1-aminocyclopropanecarboxylic acid (ACC) and indole-3-acetic acid (IAA) were down-regulated in severe drought treatment and up-regulated in rehydration after severe drought treatment.

Keywords: strawberry; drought stress; transcriptomics; metabolomics

Citation: Jiang, L.; Song, R.; Wang, X.; Wang, J.; Wu, C. Transcriptomic and Metabolomic Analyses Provide New Insights into the Response of Strawberry (*Fragaria × ananassa Duch.*) to Drought Stress. *Horticulturae* **2024**, *10*, 734. https://doi.org/10.3390/horticulturae10070734

Academic Editor: Xiaohu Zhao

Received: 31 May 2024
Revised: 6 July 2024
Accepted: 8 July 2024
Published: 12 July 2024

1. Introduction

Drought is a global problem that can lead to major losses in yield. Drought and semi-drought conditions affect approximately 36% of the world's land, and drought-related problems are especially severe in China [1]. Drought stress has various effects on the growth and development of plants [2]. Strawberry, a member of the Rosaceae family, is a shallow-rooted plant that requires loose, water-permeable, ventilated soils. Long-term drought and water shortage can have deleterious effects on the transpiration and photosynthesis of strawberry plants, thereby decreasing the quality and yield of strawberries [3]. The development of drought-resistant varieties is essential for enhancing strawberry yield.

Most previous studies on the drought stress resistance of strawberry have focused on developing drought-resistant varieties and clarifying the physiological response of strawberry plants to drought stress [4]. Ünal et al. (2023) reported that mild drought stress, with half the quantity of water given as the control treatment, resulted in an increase in the content of amino acids in strawberry [5]. Guo et al. (2023) reported the different expression patterns of FvPP2C genes under ABA (abscisic acid), salt, and drought treatments [6]. Han et al. (2023) reported that ICE (inducer of CBF expression) transcription factors are crucial in the molecular regulation of strawberry confronted by cold and drought stress [7]. And FvlCE1 was significantly up-regulated after cold, drought, salt, and heat treatments.

Biotic stress and abiotic stress have been hot research topics in recent years. There is increasing evidence that indicates that phytohormones, in addition to controlling plant growth and development under normal conditions, also mediate various environmental stresses, including salt and drought, and thus regulate plant growth adaptation [8]. Cao et al. (2022) revealed the molecular regulation of drought stress in wild strawberry (*Fragaria nilgerrensis*) via integrated transcriptome and methylome analyses [9]. However, for the cultivated strawberry varieties, there is little information about the physiological and molecular mechanisms of their response to different degrees of drought treatment.

Here, we studied the drought resistance of one of the main strawberry varieties grown in greenhouses ("Zhangji") (*Fragaria* × *ananassa Duch.*) to clarify the physiological changes in strawberry during drought response. "Zhangji" is the largest cultivated strawberry variety at present, and the cultivated area of "Zhangji" in some areas accounts for more than 80% of the total area. The "Zhangji" fruit has a milky flavor, low acidity, early fruit, and high yield, which are favored by consumers and growers. We used transcriptomic sequencing and mass spectrometry to clarify the molecular mechanism of strawberry stress resistance and the effects of drought on the physiology of strawberry plants. Our findings have important implications for the breeding of drought-resistant strawberry varieties.

2. Materials and Methods

2.1. Experimental Design

Potted "Zhangji" strawberry (*Fragaria* × *ananassa Duch.*), native to Shizuoka, Japan, was taken as the test material. The "Zhangji" strawberry seedling was made by cutting seedlings at the Jinniushan Base of the Shandong Fruit Research Institute. The cultivation substrate was composed of peat, vermiculite, and perlite with a volume ratio of 2:1:1. The experiment was carried out in an artificial climate room, with temperature (25 ± 2) °C, light cycle 16:8, and relative humidity (60 ± 5)%.

For the strawberry physiological indexes test, three treatments were arranged: conventional water supply (control), deficient water supply (water deficit), and excessive water supply (waterlogging).

For transcriptome and metabolome analyses, drought and conventional water supply treatments were set. Every treatment was repeated 3 times, with 10 pots for each replicate. With the decrease in the substrate water content during the drought treatment, 4 periods were sampled during both the drought and conventional treatments: (1) mild drought (Mild-Dr, 70% of conventional water content), (2) moderate drought (Mod-Dr, 50% of conventional water content), (3) severe drought (Sev-Dr, 20% of conventional water content), and (4) rewatering after drought treatment (RW-Dr).

The 4th functional leaves were sampled. After the surface was washed with deionized water and sucked up with filter paper, the leaves were freeze-dried in liquid nitrogen and stored at −80 °C for subsequence analysis.

2.2. Measurement of Physiological Indexes

The plant fresh weight, soil temperature, soil volume water content, daily transpiration, transpiration rate, and biomass gain (estimated using the leaf dry weight method) of "Zhangji" strawberry plants under conventional water supply (control), deficient water supply (water deficit), and excessive water supply (waterlogging) were measured using a Plantarray system, soil temperature tester (FOCHOU ST-02, Guangdong, China), constant temperature drying box (REALE RVO-0B (6020), Guangdong, China), and other instruments and equipment [10].

Daily transpiration: the daily transpiration of plants was measured using the water balance method [11]. The strawberry leaves were placed in a closed environment with a set relative humidity gradient; the daily transpiration of the leaves was calculated by measuring the change in the relative humidity in the environment before and after the leaves were placed in this environment for 24 h.

Transpiration rate: leafy branches with normal growth were used for measurements of the transpiration rate. Immediately after cutting, the weights of the branches were measured and recorded. The detached leafy branches were then returned to their original mother plant; after 3–5 min under the original environmental conditions, they were weighed again. The transpiration loss within 3–5 min was then calculated. The leaf area was measured using the paper-cut weighing method. White paper with the same thickness was cut into 10 cm $\times$ 10 cm pieces (1 dm^2) and weighed (mg); the tested leaves were spread on the same white paper, and the leaf shape (without petiole and branches) was traced with a pencil; the paper leaves were then cut and weighed (mg).

$$\text{Leaf area (dm}^2) = \text{Leaf weight (mg)}/\text{Paper weight (mg)}$$

$$\text{Transpiration rate} = \text{Transpiration water loss}/\text{unit leaf area} \times \text{time}$$

2.3. Transcriptome Sequencing and Analysis

2.3.1. Extraction and Detection of RNA Samples

The total RNA from the strawberry leaf samples was extracted using the Trizol method (Invitrogen, Carlsbad, CA, USA) [12]; the RNA was then stored at $-80\ °C$. The integrity of the extracted total RNA was detected using 1.2% agarose gel electrophoresis, and the 28SrRNA/18 rRNA ratio of high-quality RNA was 2.0. A Nanodrop 2000 ultraviolet spectrophotometer (Thermo Fisher Scientific, Waltham, MA, USA) was used to determine the concentration and purity of the RNA. RNA was considered high quality if A260/A230 = A260/A280 and >1.8 and if RNA integrity (RIN value) > 9.0, as determined by an Agilent 2100 bioanalyzer (2100 Bioanalyzer, AGILENT, Waldbronn, Germany).

2.3.2. RNA Library Construction

rRNA was removed from the total RNA samples (5 μg) using the rRNA Removal Kit (Ribo-zero™ rRNA Removal Kit, EpiCentre, Manalapan Township, NJ, USA). cDNA libraries were constructed using the TruseqTM RNA sample prep Kit (Illumina, San Diego, CA, USA). dUTP was used instead of dTTP for second-strand synthesis; after the synthesis of double-stranded cDNA, short, spliced fragments were connected, and the UNG enzyme was added to induce the degradation of the second-strand cDNA. The cDNA was amplified via 15 PCR cycles, and the cDNA library was obtained using 2% Low Range Ultra Agarose (Bio-Rad, Hercules, CA, USA). Quantification by TBS380 (Picogreen dsDNA Assay, XYbscience, Yueyang, China) was performed according to the data ratio. The bridge PCR amplification was performed on cBot (cBot Truseq PE Cluster Kit v3 murcBotcopyright HSL, Illumina, San Diego, CA, USA) to generate clusters. The results were sent to the HiSeq sequencing platform for 2 $\times$ 150 bp sequencing [12].

2.3.3. Analysis of Differentially Expressed Genes (DEGs)

DEGs were analyzed using DEseq software (Bioconductor, https://www.bioconductor.org/) [13]. The overall distribution of DEGs in samples was visualized using a volcano map. Genes were considered differentially expressed if padj $\leq$ 0.05. The GO software (Bioconductor, https://www.bioconductor.org/) was used to enrich the selected DEGs (gene ontology), and significant enrichment was defined as $p \leq 0.05$. The KEGG (Kyoto Encyclopedia of Genes and Genomes) pathway of differentially expressed genes were enriched by KOBAS, with a significant threshold of Q $\leq$ 0.05.

GO analysis of the DEGs was performed using Blast2GO software (https://www.blast2go.com/) [14], and GO terms were considered significant if the corrected p-value < 0.05. KEGG pathway enrichment analysis of DEGs was performed using BlastKOALA, and KEGG pathways were considered significant if the Q value $\leq$ 0.05.

2.4. Targeted Metabolomic Analysis

2.4.1. Preparation of the Metabolomic Standards

A targeted metabolomic analysis of plant hormones was used to study the metabolism of strawberry leaves in different treatments. The 24 plant hormones examined (24 plant hormone standards and 7 stable isotope labeling standards were purchased from Shanghai Zhenjun Biotechnology Co., Ltd., Shanghai, China) included Indole-3-acetic acid (IAA), 3-indolebutyric acid (IBA), indole-3-carboxylic acid (ICA), methyl indole-3-acetate (ME-IAA), indole-3-carboxaldehyde (ICA), N6-isopentenyladenine (IP), isopentenyl adenosine (IPA), *trans*-zeatin-riboside (tZR), *trans*-zeatin (tZ), dihydrozeatin (Dh-Z), kinetin (K), methylsalicylate (MESA), brassinolide (BL), methyl jasmonate (MeJA), dihydrojasmonic acid (H$_2$JA), *N*-Jasmonic acid-isoleucine (JA-Ile), ($\pm$)-jasmonic acid (JA), Salicylic acid(SA), abscisic acid (ABA), gibberellin A1 (GA$_1$), Gibberellin A3 (GA$_3$), Gibberellin A4 (GA$_4$), Gibberellin A7 (GA$_7$), and 1-aminocyclopropanecarboxylic acid (ACC). The 24 hormone standard products were weighed, mixed, and used to prepare a mother liquor.

The stock solution of individual phytohormones was mixed and prepared in a phytohormone-free matrix to obtain a series of phytohormone calibrators. Certain concentrations of IAA-D4, JA-D5, N6-isopentenyladenine-D6, dihydrozeatin-D3, gibberellin A1-D4, SA-D4, and ABA-D6 were mixed to make internal standard solutions. The stock solution of these phytohormones and the working solution were stored at $-20\,^{\circ}$C.

LC-MS was used to determine the concentrations of a series of standard solutions. The linearity of the standard solutions was evaluated using concentrations of the standards (abscissa) and the ratio of the internal standard peak area (ordinate).

2.4.2. Metabolite Extraction

After grinding with liquid nitrogen, 100 mg samples containing mixed internal standard were homogenized with 400 µL of acetonitrile (50%) and extracted for 30 min at 4 $^{\circ}$C, then centrifuged at 12,000 rpm for 10 min. The supernatant passed through the HLB sorbent, which was subsequently eluted with 500 µL of acetonitrile (30%). With two fractions combined into the same centrifuge tube and mixed well, these solutions were injected into the LC-MS/MS system for analysis.

2.4.3. Chromatographic and Mass Spectrometry Detection

Metabolic extracts [15] were used for liquid chromatography–mass spectrometry (LC-MS, Thermo-Fisher, Waltham, MA, USA). An ultra-high performance liquid chromatography coupled with a tandem mass spectrometry (UHPLC-MS/MS) system (ExionLC™ AD UHPLC-QTRAP 6500+, AB SCIEX Corp., Boston, MA, USA) was used to determine the concentrations of phytohormones at Novogene Co., Ltd. (Beijing, China). Separation was performed on a Waters XSelect HSS T3 column (2.1 $\times$ 150 mm, 2.5 µm), which was maintained at 45 $^{\circ}$C. The mobile phase, consisting of 0.01% formic acid in water (solvent A) and 0.01% formic acid in acetonitrile (solvent B) (LC-MS, Thermo-Fisher, Waltham, MA, USA), was delivered at a flow rate of 0.30 mL/min. The solvent gradient was as follows: 10% B, 1 min; 10–50% B, 3 min; 50–65% B, 4 min; 65–70% B, 6 min; 70–100% B, 7 min; 100–10% B, 9.1 min; and 10% B, 12 min.

The mass spectrometer was operated in multiple reaction mode. The parameters were as follows: ion spray voltage (negative mode: -4500 V, positive mode: 4500 V), curtain gas (35 psi), ion source temp (550 $^{\circ}$C), and ion source gases of 1 and 2 (60 psi).

2.4.4. Metabolic Data Analysis

MetaboAnalysis 6.0 [16] was used to analyze the targeted metabolic data. Differential metabolites were identified using the following criteria: VIP $\geq$ 1, according to the orthogonal partial least squares–discriminant analysis model and $p < 0.05$, according to a Student's *t*-test. A bubble map of differential metabolite enrichment was made using the ggplot2 package in R software (https://www.r-project.org/).

3. Results

3.1. Effects of Drought on the Physiology of Strawberry

Recent studies have shown that stress exposure induces a series of physiological changes in plants, and many plants have evolved strategies that allow them to tolerate drought. The daily transpiration, transpiration rate, fresh weight, and biomass gain of "Zhangji" strawberry plants under a regular water supply (control), deficient water supply (water deficit), and excessive water supply (waterlogging) were measured using a Plantarray system. We found that the transpiration rate in the deficit irrigation treatment was significantly lower than that of the control (Figure 1).

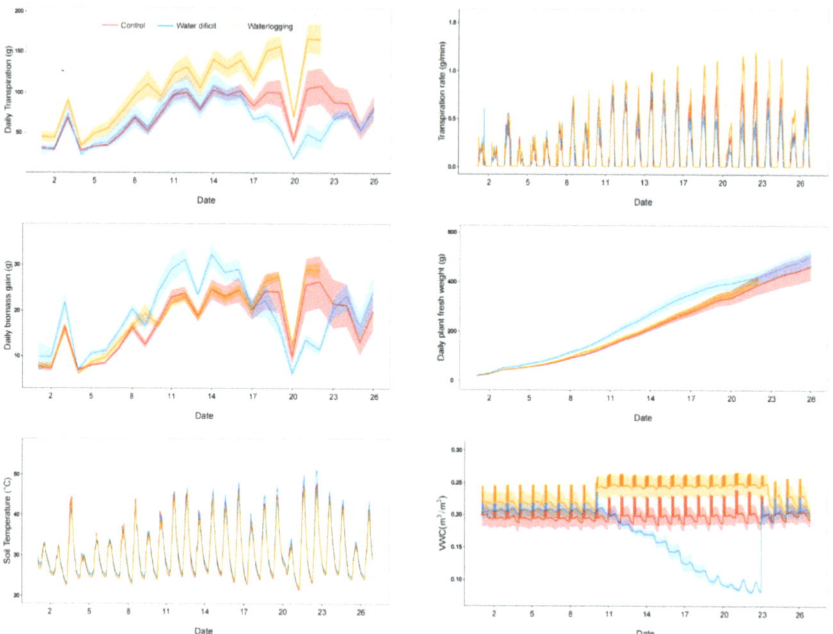

Figure 1. Effects of drought on the physiological indexes of strawberry plants.

The daily transpiration rate of strawberry was consistently lower during the deficit water treatment than the control. The biomass of plants increased continuously, and the rate of increase was initially rapid and then slow. The increase in the biomass of plants was higher in the early stage of the water deficit treatment than in the conventional water treatment, but it was lower in the late stage of the water deficit treatment than in the conventional water treatment. The fresh weight of plants increased gradually over time, and the increase in the fresh weight of plants was significantly higher in the water deficit treatment than in the control. The soil temperature fluctuated over time in the three treatments, and the variation in soil temperature was greater in the water deficit treatment than under control treatment. In the early stage of the experiment, the soil water content was stable and lower in the water deficit treatment than in the conventional water treatment. From day 11 to 23, the soil volume water content gradually decreased in the water deficit treatment; in the excessive water treatment, the soil volume water content gradually increased and was higher than the conventional water treatment at the end of this period.

3.2. Transcriptome Analysis

3.2.1. Transcriptome Sequencing and Identification of DEGs

Differences in the physiological responses of strawberry plants were observed in the Mild-Dr, Mod-Dr, Sev-Dr, RW-Dr, and CK groups, and DEGs in each of these groups were identified.

A total of 428 (209 up-regulated and 219 down-regulated), 5666 (3014 up-regulated and 2652 down-regulated), 3693 (2172 up-regulated and 1521 down-regulated), and 2573 (1079 up-regulated and 1494 down-regulated) DEGs were identified in the Mild-Dr vs. CK, Mod-Dr vs. CK, Sev-Dr vs. CK, and RW-Dr vs. CK comparisons, respectively (Figure 2).

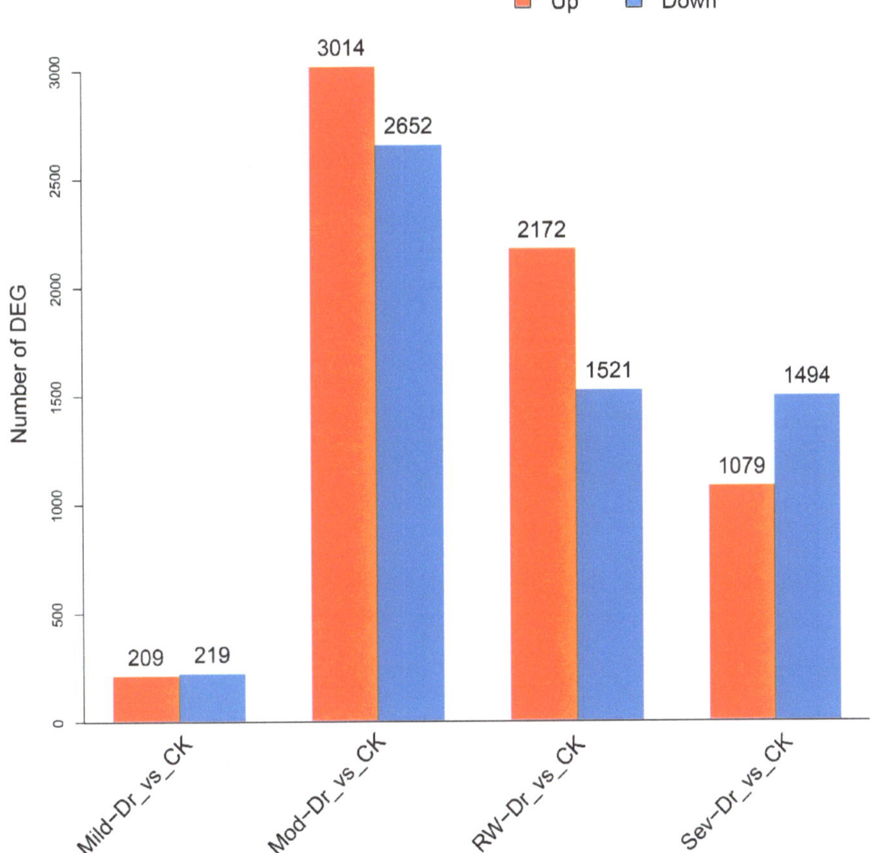

Figure 2. Up-regulated and down-regulated DEGs in four comparison groups.

3.2.2. GO and KEGG Pathway Enrichment Analyses

Gene ontology (GO) and Kyoto Encyclopedia of Genes and Genomes (KEGG) pathway enrichment analyses were conducted on DEGs under normal conditions and drought treatment to clarify their biological functions. GO terms from three different categories—molecular function (MF), cell component (CC), and biological process (BP)—were determined for DEGs in the comparison of different drought (mild, mod, severe and rewatering after drought) degrees with CK; the differential DEGs were all mainly enriched in binding, catalytic activity, metabolic process, and cellular process (Figure 3).

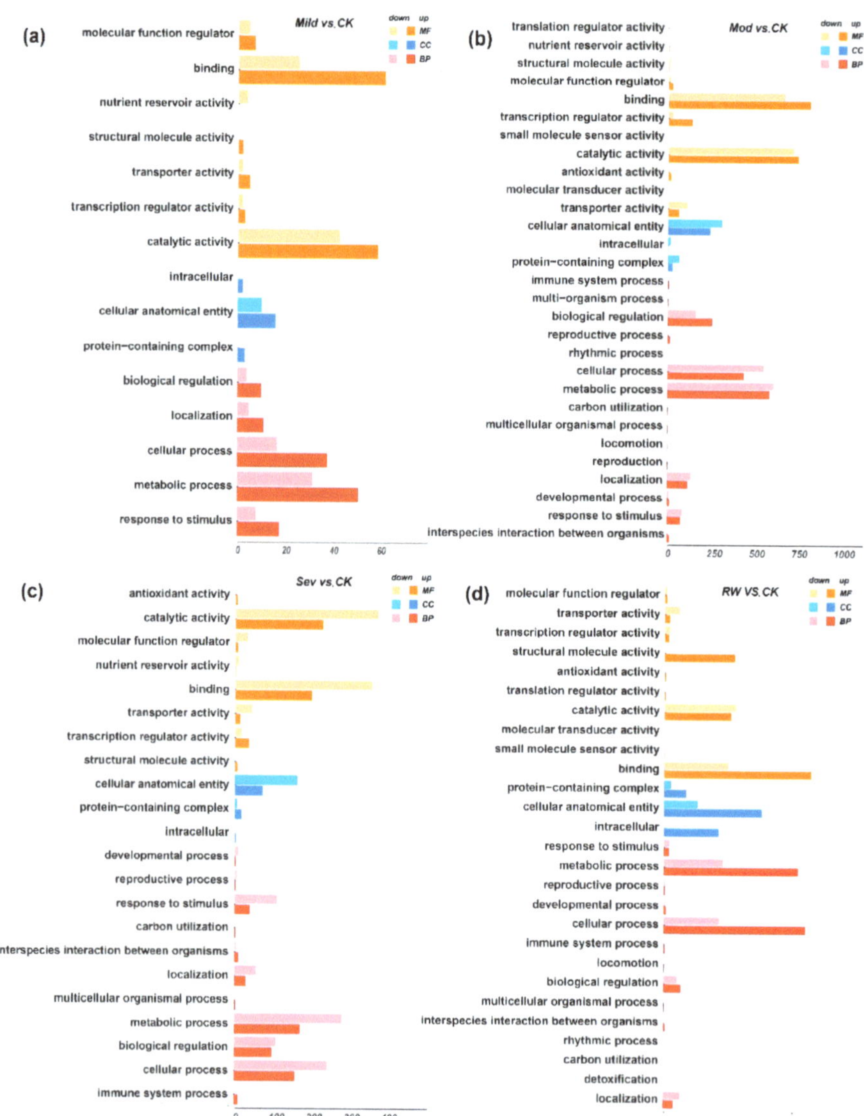

Figure 3. Results of the GO enrichment analysis of DEGs in different comparison groups.

Bubble maps were used to visualize the results of the KEGG pathway enrichment analysis for all DEGs and clarify the effects of drought stress on the enriched pathways (Figure 4). DEGs in the mild-Dr vs. The CK comparison group was highly enriched in "pentose and glucose interconversions" and "sulfur metabolism". DEGs in the Mod-Dr vs. CK comparison group were highly enriched in "biosynthesis of secondary metabolites" and "nitrogen metabolism". DEGs in the Sev-Dr vs. CK comparison group were highly enriched in the "galactose metabolism" and "fatty acid elongation" pathways. DEGs in the RW-Dr vs. CK comparison group were highly enriched in "ribosome", "ribosome biogenesis in eukaryotes", and other pathways. During the moderate drought treatment, the normal growth and development of plants were enhanced through the regulation of sugar, nitrogen, and sulfur metabolism. During the excessive drought treatment, drought resistance was mainly achieved through alterations of the synthesis and metabolism of amino acids and fatty acids. Normal development and growth were restored during the rehydration treatment after drought through changes in the nucleotide metabolism pathway.

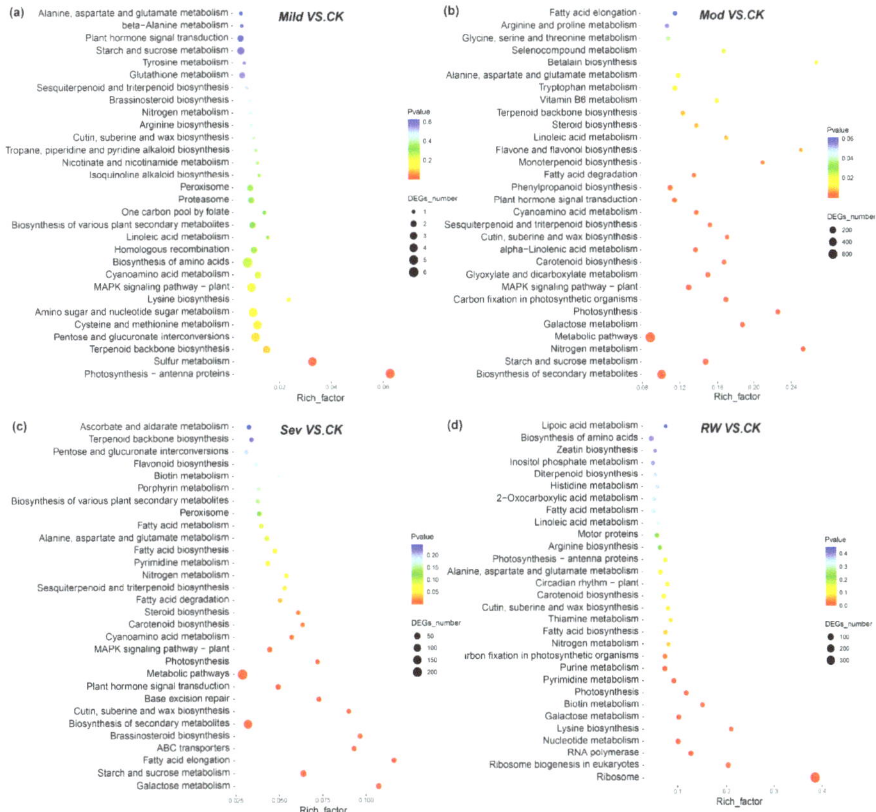

Figure 4. KEGG enrichment analysis of DEGs in different comparison groups.

3.3. Metabolomic Analysis

A targeted metabolomic analysis of "Zhangji" strawberry plants was conducted to identify the metabolites related to drought stress and clarify the mechanism by which "Zhangji" strawberry plants respond to drought stress. As shown in Figure 5, changes in 12 hormones were detected in the mild-Dr, Mod-Dr, Sev-Dr, and RW-Dr treatments compared with CK. Compared with the control, the syntheses of 3-indolebutyric acid (IBA), salicylic acid (SA), and abscisic acid (ABA) were up-regulated in the mild drought treatment. The syntheses of indole-3-carboxyl acid (ICA), salicylic acid (SA), abscisic acid (ABA), gibberellin A1 (GA1), and jasmonic acid (JA) were up-regulated in the moderate drought treatment. The syntheses of salicylic acid (SA), 1-aminocyclopropanecarboxylic acid (ACC), and indole-3-acetic acid (IAA) were down-regulated in severe drought treatment. Although 3-indolebutyric acid (IBA) was down-regulated, 1-aminocyclopropanecarboxylic acid (ACC), indole-3-acetic acid (IAA), indole-3-carboxaldehyde (ICA), and isopentenyl adenosine (IPA) were up-regulated in the rehydration after severe drought treatment compared with the other treatments.

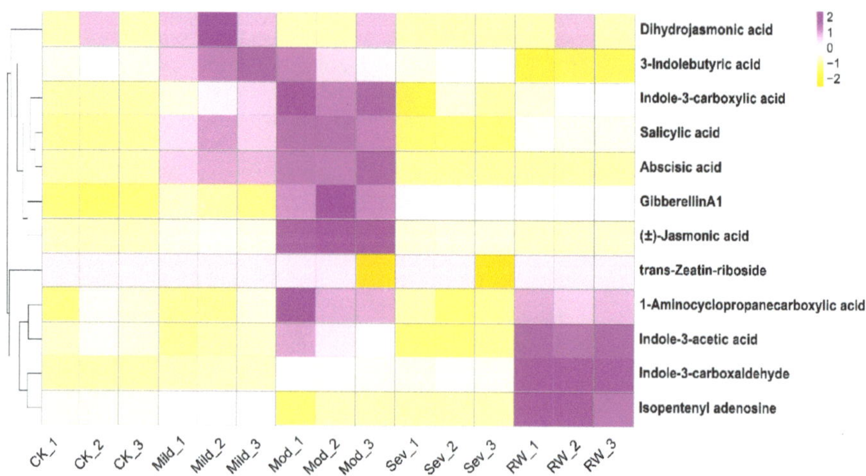

Figure 5. Comparison of heat maps of phytohormones in different treatment groups.

4. Discussion

Strawberry is one of the most well-received fruits because of its sweet smell and rich nutrients [17]. In terms of physiological structure, cultivated strawberry has a large leaf surface and a shallow root system, making it more sensitive to drought [18]. And drought is the most considerable environmental stressor [19], caused by a decrease in water content in plant tissue and resulting in a decrease in somatic osmotic pressure and abnormal metabolism, which has a certain impact on the growth and development of strawberry [20].

The response of strawberry plants to stress is a complex process. Recent studies have shown that plant physiological changes are the primary response mechanism to stress, and many plants have evolved a response mechanism to drought [1]. In this study, the water content of "Zhangji" strawberry plants can be maintained under drought conditions by reducing the transpiration rate and daily transpiration. Under mild drought stress, the physiological processes and hydraulic signals of strawberry plants are altered to mediate adaptation to stress. However, as the stress exposure increases, the expression of genes involved in stress resistance increases, affecting the phenotypic features of plants.

Transcriptomic analyses can be used to characterize quantitative changes in the expression of genes at different times and under different states [21]. In this study, we identified genes that play a key role in the drought tolerance of "Zhangji" strawberries to clarify the mechanism underlying the response to drought. The expression of genes involved in pathways such as "oxidoreductase activity", "cell metabolism", "sulfur metabolism", "sugar metabolism", "pentose and glucose transformation", and "nitrogen metabolism" was up-regulated in the four treatments (mild water deficiency, moderate water deficiency, severe water deficiency, and rehydration after drought), which indicated that strawberry plants could resist the damage induced by drought stress through the accumulation of sugars in leaves. Soluble sugars can reduce the membrane osmotic potential by combining with membrane lipid bilayers, which alleviates changes in osmotic pressure and promotes cell expansion; this might explain why the increase in soluble sugars, such as glucose and sucrose, can enhance the resistance of plants to abiotic stress [22]. In addition, sugars can be used as small molecular signaling substances to mediate resistance to abiotic stress. Previous studies have shown that soluble sugars in roots act as energy storage substances and osmotic regulators in plants, but they also play key roles as signal substances under drought and heavy metal stress [23]. Nitrogen metabolism and carbon metabolism are closely related via a complex network of metabolites, and this has implications for plant growth. The energy and carbon sources needed for nitrogen metabolism are provided by carbon metabolism [24]. The activity of nitrogen metabolism enzymes and the soluble protein content were enhanced in strawberry leaves in the moderate water deficiency treatment. The above results indicate that moderate drought treatment can promote nitrogen metabolism and the conversion of inorganic nitrogen and sulfur into amino acids, which can then be used to assemble proteins. The expression of genes involved in endogenous auxin such as "pyrimidine metabolism", "lysine biosynthesis", "biotin metabolism", and "eukaryotic nuclide generation" was significantly up-regulated in the rehydration treatment after drought, which indicates that the growth and development of the plants were restored after drought stress.

Endogenous hormones are important growth regulators in plants. The relative abundance of endogenous hormones play an important role in regulating the resistance of plants to drought stress, and their responses to drought stress are complex and dynamic [25]. Previous studies have shown that, under drought stress, the synthesis of hormones such as IAA, GA3, and zeatin nucleoside is often suppressed [26], and ABA can be used as a signaling molecule to induce stomatal closure and enhance drought resistance [27]. In this study, under mild water shortage, the concentrations of IBA, SA, and ABA increased. Increases in GA1, ($\pm$)-JA, and ICA were observed in the moderate water deficit treatment. Levels of SA, ACC, and IAA during the severe water deficit treatment were lower than those during other treatments. Increases in ACC, IAA, ICA, and IPA were observed in the rewatering after drought treatment. The above results indicate that drought stress induces an endogenous response in "Zhangji" strawberry plants. The secretion of various endogenous growth hormones maintains the normal growth of plants under drought stress, and the normal growth and development of plants are restored in the rewatering after drought treatment.

5. Conclusions

In this study, various approaches were used to clarify the mechanism by which "Zhangji" strawberry plants respond to different levels of drought stress. At the physiological level, strawberry plants can adapt to drought stress by regulating the water content in their plant body; for example, by reducing their respiration rate. Transcriptome analysis revealed that genes involved in oxidoreductase activity and in sulfur and nitrogen metabolism were up-regulated, as well as starch and sucrose. The results of the metabolomic analysis showed that hormones such as SA, ABA, and ($\pm$)-JA in strawberry plants increased. Strawberry plants secrete various endogenous growth hormones to maintain their normal growth under drought stress. Rehydration after drought restores the normal growth and development of strawberry plants. Our findings provide new insights into the pathways that mediate the response to different levels of drought stress in "Zhangji" strawberry plants and will aid the development of new drought-resistant strawberry varieties.

Author Contributions: Investigation, C.W.; Data curation, X.W.; Writing—original draft, R.S.; Writing—review & editing, L.J. and J.W.; Visualization, C.W.; Supervision, L.J.; Project administration, J.W.; Funding acquisition, J.W. All authors have read and agreed to the published version of the manuscript.

Funding: This study was supported by the Key R&D Project of Shandong Province (2023TZXD057, 2022CXPT017) and the Special Fund Project for Guiding Local Scientific and Technological Development by the Central Government (YDZX2023031).

Data Availability Statement: Data are contained within the article.

Conflicts of Interest: The authors declare that they have no conflicts of interest.

References

1. Zhou, C.C.; Bo, W.H.; El-Kassaby, Y.A.; Li, W. Transcriptome profiles reveal response mechanisms and key role of PsNAC1 in Pinus sylvestris var. mongolica to drought stress. *BMC Plant Biol.* **2024**, *24*, 343. [CrossRef] [PubMed]
2. Arief, M.A.A.; Kim, H.; Kurniawan, H.; Nugroho, A.P.; Kim, T.; Cho, B.K. Chlorophyll fluorescence imaging for early detection of drought and heat stress in strawberry plants. *Plants* **2023**, *12*, 1387. [CrossRef] [PubMed]

3. Gaecía-López, J.V.; Redondo-Gómez, S.; Flores-Duarte, N.J.; Rodríguez-Llorente, I.D.; Pajuelo, E.; Mateos-Naranjo, E. PGPR-based biofertilizer modulates strawberry photosynthetic apparatus tolerance responses by severe drought, soil salinization and short extreme heat event. *Plant Stress* **2024**, *12*, 100448. [CrossRef]

4. Wang, B.; Yang, W.L.; Shan, C.J. Effects of selenomethionine on the antioxidative enzymes, water physiology and fruit quality of strawberry plants under drought stress. *Hortic. Sci.* **2022**, *49*, 10–18. [CrossRef]

5. Űnal, N.; Okatan, V. Effects of drought stress treatment on phytochemical contents of strawberry varieties. *Sci. Hortic.* **2023**, *316*, 112013. [CrossRef]

6. Guo, L.L.; Lu, S.X.; Liu, T.; Nai, G.J.; Ren, J.X.; Gou, H.M.; Chen, B.H.; Mao, J. Genome-wide identification and abiotic stress response analysis of PP2C gene family in woodland and pineapple strawberries. *Int. Mol. Sci.* **2023**, *24*, 4049. [CrossRef] [PubMed]

7. Han, J.X.; Li, X.G.; Li, W.H.; Yang, Q.; Li, Z.H.; Cheng, Z.; Lv, L.; Zhang, L.H.; Han, D.G. Isolation and preliminary functional analysis of FvlCE1, involved in cold and drought tolerance in *Fragaria vesca* through overexpression and CRISPR/Cas9 technologies. *Plant Physiol. Biochem.* **2023**, *196*, 270–280. [CrossRef]

8. Yu, Z.P.; Duan, X.B.; Luo, L.; Dai, S.J.; Ding, Z.J.; Xia, G.M. How plant hormones mediate salt stress responses. *Trends Plant Sci.* **2020**, *25*, 1117–1130. [CrossRef]

9. Cao, Q.; Huang, L.; Li, J.M.; Qu, P.; Tao, P.; Crabbe, M.; James, C.; Zhang, T.C.; Qiao, Q. Integrated transcriptome and methylome analyses reveal the molecular regulation of drought stress in wild strawberry (*Fragaria nilgerrensis*). *BMC Plant Biol.* **2022**, *22*, 613. [CrossRef] [PubMed]

10. Halperin, O.; Gebremedhin, A.; Wallach, R.; Moshelion, M. High-throughput physiological phenotyping and screening system for the characterization of plant-environment interactions. *Plant J. Cell* **2017**, *89*, 839–850. [CrossRef]

11. Yin, M.; Zhang, J.X.; Ding, L.; Liu, X.R. Visualized analysis of the research progress and trend of lysimeter in the measurement of evapotranspiration based on CiteSpace. *Tech. Superv. Water Resour.* **2023**, 241–248. [CrossRef]

12. Wei, S.W.; Zhang, L.; Huo, G.T.; Ge, G.J.; Luo, L.J.; Yang, Q.C.; Yang, X.; Long, P. Comparative transcriptomics and metabolomics analyses provide insights into thermal resistance in lettuce (*Lactuca sativa* L.). *Sci. Hortic.* **2021**, *289*, 110423. [CrossRef]

13. Ma, L.; Ma, S.Y.; Chen, G.P.; Lu, X.; Wei, R.N.; Xu, L.; Feng, X.J.; Yang, X.M.; Chai, Q.; Zhang, X.C.; et al. New insights into the occurrence of continuous cropping obstacles in pea (*Pisum sativum* L.) from soil bacterial communities, root metabolism and gene transcription. *BMC Plant Biol.* **2023**, *23*, 226. [CrossRef] [PubMed]

14. Yu, Y.H.; Jia, X.L.; Wang, W.L.; Jin, Y.M.; Liu, W.Z.; Wang, D.M.; Mao, Y.X.; Xie, C.T.; Liu, T. Floridean Starch and Floridoside Metabolic Pathways of Neoporphyra haitanensis and Their Regulatory Mechanism under Continuous Darkness. *Mar. Drugs* **2021**, *19*, 664. [CrossRef] [PubMed]

15. Yang, X.; Wei, S.W.; Liu, B.; Guo, D.D.; Zheng, B.X.; Feng, L.; Liu, Y.M.; Francisco, A.T.B.; Luo, L.J.; Huang, D.F. A novel integrated non-targeted metabolomic analysis reveals significant metabolite variations between different lettuce (*Lactuca sativa* L) varieties. *Hortic. Res.* **2018**, *5*, 33. [CrossRef] [PubMed]

16. Chong, J.; Wishart, D.S.; Xia, J.G. MetaboAnalyst 4.0 for Comprehensive and Integrative Metabolomics Data Analysis. *Curr. Protoc. Bioinform.* **2019**, *68*, e86. [CrossRef]

17. Galli, V.; Rafael, D.S.M.; Perin, E.C.; Borowski, J.M.; Bamberg, A.L.; Rombaldi, C.V. Mild salt stress improves strawberry fruit quality. *LWT-Food Sci. Technol.* **2016**, *73*, 693–699. [CrossRef]

18. Dong, C.; Xi, Y.; Chen, X.; Cheng, Z.M. Genome-wide identification of AP2/EREBP in *Fragaria vesca* and expression pattern analysis of the FvDREB subfamily under drought stress. *BMC Plant Biol.* **2021**, *21*, 295. [CrossRef]

19. Mozafari, A.A.; Havas, F.; Ghaderi, N. Application of iron nanoparticles and salicylic acid in vitro culture of strawberries (*Fragaria × ananassa* Duch.) to cope with drought stress. *Plant Cell Tissue Organ Cult.* **2018**, *132*, 511–523. [CrossRef]

20. Khan, M.Q.N.; Sevgin, N.; Rizwana, H.; Arif, N. Exogenous melatonin mitigates the adverse effects of drought stress in strawberry by upregulating the antioxidant defense system. *S. Afr. J. Bot.* **2023**, *162*, 658–666. [CrossRef]

21. Anderson, J.T.; Mitchell-Olds, T. Ecological genetics and genomics of plant defenses: Evidence and approaches. *Funct. Ecol.* **2011**, *25*, 312–324. [CrossRef] [PubMed]

22. Anzano, A.; Bonanomi, G.; Mazzoleni, S.; Lanzotti, V. Plant metabolomics in biotic and abiotic stress: A critical overview. *Phytochem. Rev.* **2021**, *21*, 503–524. [CrossRef]

23. Wei, T.L.; Wang, Y.; Xie, Z.Z.; Guo, D.Y.; Chen, C.W.; Fan, Q.J.; Deng, X.D.; Liu, J.H. Enhanced ROS scavenging and sugar accumulation contribute to drought tolerance of naturally occurring autotetraploids in Poncirus trifoliata. *Plant Biotechnol. J.* **2018**, *17*, 1394–1407. [CrossRef] [PubMed]

24. Mao, W.Q.; Xia, Y.H.; Ma, C.; Zhu, G.X.; Wang, Z.C.; Tu, Q.; Chen, X.B.; Wu, J.S.; Su, Y.R. Response of organic carbon mineralization of nitrogen addition in micro-aerobic and anaerobic layers of paddy soil. *Huan Jing Ke Xue* **2023**, *44*, 6248–6256. [CrossRef] [PubMed]

25. Waadt, R.; Seller, C.A.; Hsu, P.K.; Takahashi, Y.; Munemasa, S.; Schroeder, J.I. Plant hormone regulation of abiotic stress responses. *Plant Stress Responses* **2022**, *23*, 680–694. [CrossRef]

26. Carmen, J.M.; Andrea, M.M.P.; Pablo, M.G.; Manuel, R.; Antonio, H.J.; Gregorio, B.E.; Pedro, D.V.; José, M.G.P. Comprehensive study of the hormonal, enzymatic and osmoregulatory response to drought in Prunus species. *Sci. Hortic.* **2024**, *326*, 112786. [CrossRef]
27. Wu, Y.Y.; Sun, Y.; Wang, W.M.; Xie, Z.Z.; Zhan, C.H.; Jin, L.; Huang, J.L. OsJAZ10 negatively modulates the drought tolerance by integrating hormone signaling with systemic electrical activity in rice. *Plant Physiol. Biochem.* **2024**, *211*, 108683. [CrossRef]

Article

Chloroplast Genome Profiling and Phylogenetic Insights of the "Qixiadaxiangshui" Pear (*Pyrus bretschneideri* Rehd.1)

Huijun Jiao [1,†], Qiming Chen [1,†], Chi Xiong [2], Hongwei Wang [1], Kun Ran [1], Ran Dong [1], Xiaochang Dong [1], Qiuzhu Guan [1] and Shuwei Wei [1,3,*]

[1] Shandong Institute of Pomology, Longtan Road No. 66, Tai'an 271000, China; jiaohj_njau@163.com (H.J.); 2018204012@njau.edu.cn (Q.C.)
[2] Huaiyin Institute of Technology, Meicheng East Road No. 1, Huaian 223001, China
[3] National Center of Technology Innovation for Comprehensive Utilization of Saline-Alkali Land, Zhihui Road No. 8, Dongying 257335, China
[*] Correspondence: weisw2007@163.com
[†] These authors contributed equally to this work.

Abstract: The "Qixiadaxiangshui" pear (*Pyrus bretschneideri* Rehd.1) is a highly valued cultivar known for its crisp texture, abundant juice, and rich aroma. In this study, we reported the first complete chloroplast genome sequence of the "Qixiadaxiangshui" pear, which is 159,885 bp in length with a GC content of 36.58%. The genome exhibits a typical circular quadripartite structure, comprising a large single-copy region (LSC), a small single-copy region (SSC), and a pair of inverted repeat regions (IRs). A total of 131 genes were identified, including 84 protein-coding genes, 8 rRNA genes, and 37 tRNA genes. We also identified 209 simple sequence repeats (SSRs) and several mutation hotspots, such as *ndhC-trnM-CAU* and *trnR-UCU-atpA*, which can be applied in molecular identification and phylogenetic studies of *Pyrus*. Comparative genomic analysis showed high conservation among ten pear cultivars. Phylogenetic analysis indicated that the "Qixiadaxiangshui" pear is closely related to germplasm Dangshansuli, Wonwhang, and Yali, suggesting a recent common ancestor. These findings provided valuable insights into the genetic diversity and evolutionary dynamics of the *Pyrus* species and contribute to the conservation and breeding of pear germplasm resources.

Keywords: "Qixiadaxiangshui" pear; chloroplast genome; comparative genome analysis; phylogenetic analysis

Citation: Jiao, H.; Chen, Q.; Xiong, C.; Wang, H.; Ran, K.; Dong, R.; Dong, X.; Guan, Q.; Wei, S. Chloroplast Genome Profiling and Phylogenetic Insights of the "Qixiadaxiangshui" Pear (*Pyrus bretschneideri* Rehd.1). *Horticulturae* **2024**, *10*, 744. https://doi.org/10.3390/horticulturae10070744

Academic Editor: Michailidis Michail

Received: 11 June 2024
Revised: 11 July 2024
Accepted: 11 July 2024
Published: 15 July 2024

1. Introduction

The chloroplast, a crucial organelle in plant cells, is responsible for converting light energy into chemical energy, making it a focal point in botanical studies [1]. The chloroplast genome, with its relatively slow rate of variation, primarily due to insertions, deletions, and point mutations in non-coding regions, serves as an ideal material for phylogenetic analysis [2]. Typically, the chloroplast genome is a double-stranded circular configuration, comprising a large single-copy region (LSC) and a small single-copy region (SSC) separated by two inverted repeat regions (IRa and IRb), forming a quadripartite structure [3].

The chloroplast genome is characterized by the integrity of its replication, transcription, and translation systems, its small molecular weight, the conservatism of gene structure and quantity, and the uniformity of gene arrangement. These features make the chloroplast genome a valuable tool for studies in plant systematics, genetics, and evolutionary biology. Chloroplast genome research has been pivotal in elucidating the phylogenetic relationships within the genus *Pyrus*, including species such as *Pyrus pyrifolia* cultivar Wonwhang [4], *Pyrus ussuriensis* Maxim. [5], and *Pyrus pashia* [6], revealing the complex evolutionary relationships among these species.

Pyrus bretschneideri, commonly known as the Chinese white pear, is one of the most important pear species in China. Notable cultivars of *Pyrus bretschneideri* include "Dangshansuli", "Yali", and the "Qixiadaxiangshui" pear. These cultivars or strains are known for their crisp texture, sweetness, and high juice content, making them highly popular among consumers. Current studies on chloroplast genomes of pear species have primarily focused on a limited number of cultivars, leading to an incomplete understanding of the genetic diversity and evolutionary dynamics within the genus. The limited phylogenetic information from commonly used chloroplast regions, such as *rpl16*, *trnH*, and *matK*, results in phylogenetic trees with low resolution and weak bootstrap support [7]. Therefore, it is necessary to develop high-resolution genetic markers to promote the species identification, germplasm screening, and phylogeny of *Pyrus* to further facilitate the utilization and conservation of pear germplasm resources.

The "Qixiadaxiangshui" pear is a high-quality cultivar of pear native to Qixia City, Shandong Province, China. Known for its crisp texture, abundant juice, and rich aroma, it is highly valued locally but remains relatively unknown outside China [8]. Reports on the chloroplast genome of the "Qixiadaxiangshui" pear are scarce, limiting our understanding of its genetic makeup and evolutionary background. The lack of molecular genetic information on "Qixiadaxiangshui" not only restricts its research on the classification and genetic diversity of the *Pyrus* genus but also impedes the application of this excellent cultivar in pear breeding. To address these gaps, our study undertook an in-depth sequencing of the chloroplast genome of the "Qixiadaxiangshui" pear. We analyzed a series of genomic variation hotspots within the chloroplast genomes of the genus *Pyrus* to thoroughly assess the evolutionary status of the "Qixiadaxiangshui" pear and its phylogenetic relationships with other pear germplasm. Our research aims to provide a scientific basis and new perspectives for genetic breeding and molecular evolution studies of the "Qixiadaxiangshui" pear, contributing to the improvement and optimization of this variety and the conservation and utilization of pear genetic resources.

2. Materials and Methods

2.1. Materials

Fresh leaves of the "Qixiadaxiangshui" pear (*Pyrus bretschneideri* Rehd.1) were collected from ten healthy and mature plants grown at the JinNiu Mountain experimental base of the Shandong Institute of Pomology in Tai'an City, Shandong Province, on 5 April 2021. These plants were selected based on their uniform growth conditions and absence of disease or pest infestation to ensure consistency and reliability of the genomic data. The leaves were washed, dried, and stored at −80 °C. The chloroplast genome sequences of closely related species of "Qixiadaxiangshui" pear were retrieved from the NCBI database. Detailed information on the experimental materials is provided in Table S1.

2.2. DNA Extraction and Sequencing

DNA from fresh leaves of the "Qixiadaxiangshui" pear was extracted using the Plant DNA Extraction Kit (Vazyme Biotech Co., Ltd., Nanjing, China). The quality of the extracted DNA was verified through agarose gel electrophoresis, and its concentration was measured using a NanoDrop 2000 spectrophotometer (Thermo Scientific™, Waltham, MA, USA). After verifying the quality of the sample genomic DNA, the DNA was fragmented using an ultrasonic shearing method. The fragmented DNA was then subjected to fragment purification, end repair, 3′-end adenylation, and ligation with sequencing adapters. Fragment size selection was performed using agarose gel electrophoresis, followed by PCR amplification to construct the sequencing library. The constructed library underwent quality control, and libraries that passed quality control were sequenced using the Illumina NovaSeq 6000 platform with a paired-end read length of 150 bp.

2.3. Chloroplast Genome Assembly and Annotation

High-quality reads were obtained by filtering the raw data with fastp (version 0.20.0). Sequences aligning to a proprietary chloroplast genome database were identified using bowtie2 v2.2.4 in very-sensitive-local mode and were considered as chloroplast DNA (cpDNA) sequences of the sample. The assembly core module utilized SPAdes v3.10.1 [9] with k-mer values of 55, 87, and 121, assembling the chloroplast genome without reference. The assembled genome was quality controlled against the reference sequence Pyrus_bretschneideri_Rehd.1:KY626169.1. Chloroplast genome annotation was performed using two approaches: (1) CDS annotation with Prodigal v2.6.3, rRNA prediction with HMMER v3.1b2, and tRNA prediction with ARAGORN v1.2.38 and (2) gene sequences from closely related species published on NCBI were obtained and aligned against the assembled sequence using BLAST v2.6 to obtain another annotation result. Discrepancies between annotation results were manually reviewed to eliminate erroneous and redundant annotations, define multi-exon boundaries, and finalize the annotation. The chloroplast genome map was generated using OGDRAW v1.3.1 [10].

2.4. Chloroplast Genome Characterization

Codon usage bias (Relative Synonymous Codon Usage, RSCU) was analyzed using CodonW, revealing codon preference by comparing actual to expected frequency ratios. An RSCU value greater than 1 indicates a higher-than-expected usage frequency of that codon. The online software MISA v1.0 [11] recognized single sequence repeats (SSRs) in the chloroplast genome (chloroplast SSR markers, cpSSRs), and cpSSRs were known as microsatellites. The minimum repeat thresholds of 8, 5, 3, 3, 3, and 3 were set for mono-, di-, tri-, tetra-, penta-, and hexanucleotide repeating units, respectively.

2.5. Comparative Analysis of Chloroplast Genomes

Differences in the four boundary regions of chloroplast genomes between "Qixiadaxiangshui" and nine other pears of the Rosaceae family, including the *Pyrus bretschneideri* strain Dangshansuli, the *Pyrus bretschneideri* strain Yali, the *Pyrus communis* strain Bartlett, the *Pyrus calleryana* cultivar OPR125, *Pyrus hopeiensis*, *Pyrus phaeocarpa*, the *Pyrus pyrifolia* cultivar Whangkeumbae, the *Pyrus pyrifolia* cultivar Wonwhang, and *Pyrus ussuriensis*, were analyzed using the IRscope online tool [12]. Multiple sequence alignments of chloroplast genomes from ten pear germplasm were performed using the mVISTA tool [13] in shuffle-LAGAN mode. Nucleotide diversity (Pi) of the chloroplast genomes was calculated with DNAsp, setting the window length to 600 bp and step size to 200 bp.

2.6. Phylogenetic Analysis

Chloroplast genome sequences of 28 Rosaceae species, obtained from the NCBI database and using *Castanea mollissima* from the Fagaceae family as an outgroup (Table S1), were aligned using MAFFT (v7.520) [14]. A phylogenetic tree was constructed using the maximum likelihood method (ML) with RaxML-HPC2 v8.2.12 [15] on the CIPRES Science Gateway server. Bootstrap support (1000 replicates) was evaluated for branch robustness, with results visualized in MEGA11 v11.0.13 [16].

3. Results

3.1. Structure and Composition of the Chloroplast Genome of the "Qixiadaxiangshui" Pear

After removing adapters and low-quality data, a total of 7.62 Gb of raw data were obtained with 25,428,945 original reads, achieving a Q20 of 98.29%. The assembly and annotation of the chloroplast genome of the "Qixiadaxiangshui" pear revealed a typical quadripartite structure, consisting of a large single-copy (LSC) region, a pair of inverted repeats (IRs), and a small single-copy (SSC) region, with lengths of 87,863 bp, 26,392 bp, and 19,238 bp, respectively (Figure 1). The GC content was 36.58%, indicating a higher AT content of 63.42%. Variations in GC values were observed across the IR, LSC, and SSC regions, with the IR region having the highest GC content (42.64%) compared to the LSC (34.28%) and SSC (30.40%) regions (Table 1). The results reveal a highly conserved quadripartite structure with variable GC content across different regions, indicating its complex genetic architecture and adaptation potential.

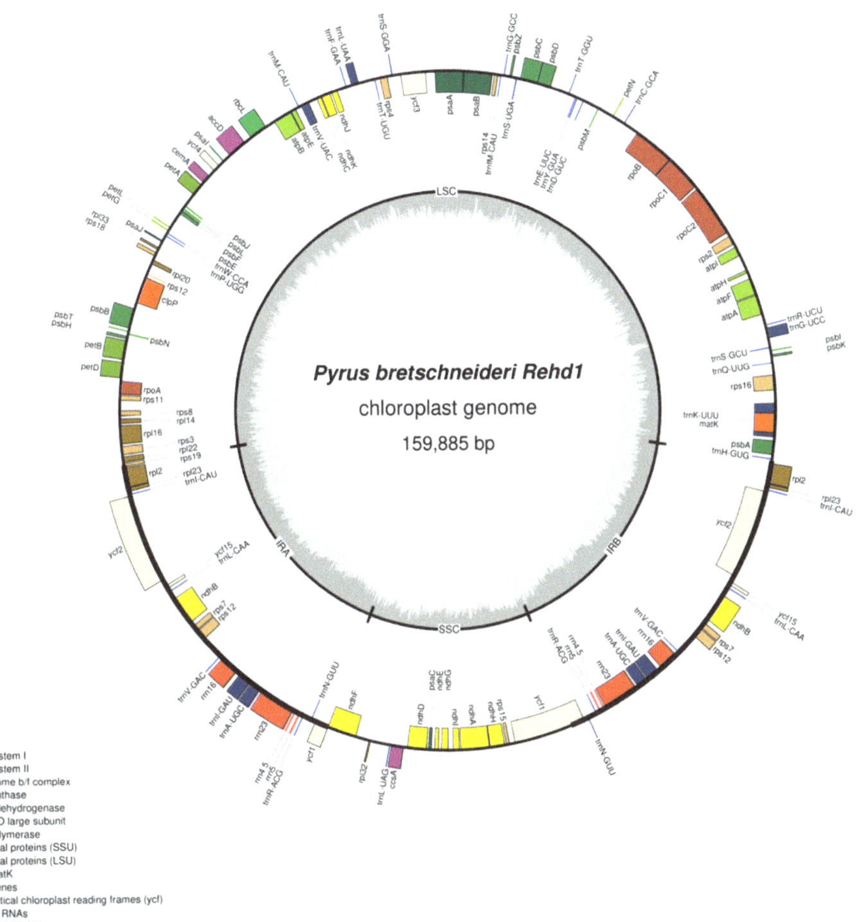

Figure 1. Chloroplast genome map of "Qixiadaxiangshui" Pear. On the bottom left, the genes corresponding to the different functional groups are depicted with distinctive colors. The outer side of the circle represents positive coding genes, the inner side of the circle represents reverse coding genes, and the inner gray circle represents GC content. IRa/IRb, LSC, and SSC regions are shown by black lines surrounding the dark grey area.

Table 1. Base composition of chloroplast genome of "Qixiadaxiangshui" pear.

Region	A (%)	C (%)	G (%)	T (%)	GC (%)	Base Size (bp)
IRa	28.6	22.1	20.54	28.76	42.64	26,392
IRb	28.76	20.54	22.1	28.6	42.64	26,392
LSC	32.12	17.64	16.65	33.6	34.28	87,863
SSC	34.77	15.9	14.5	34.83	30.4	19,238
Total	31.3	18.64	17.93	32.12	36.58	159,885

The chloroplast genome of the "Qixiadaxiangshui" pear contains a total of 131 genes, including 84 protein-coding genes (mRNA), eight ribosomal RNA (rRNA) genes, 37 transfer RNA (tRNA) genes, and 2 pseudo genes. These genes are categorized into three groups: photosynthesis-related genes (45 in total, including those for photosystem I and II), self-replication genes (73, including tRNAs and rRNAs), and other biosynthesis-related genes (13) (Table 2).

Table 2. Classification of chloroplast genome genes of "Qixiadaxiangshui" pear.

Gene Function	Gene Category	Gene Name	Number
Photosynthesis	Subunits of photosystem I	psaA, psaB, psaC, psaI, psaJ	5
	Subunits of photosystem II	psbA, psbB, psbC, psbD, psbE, psbF, psbH, psbI, psbJ, psbK, psbL, psbM, psbN, psbT, psbZ	15
	Subunits of NADH dehydrogenase	ndhA*, ndhB*(2), ndhC, ndhD, ndhE, ndhF, ndhG, ndhH, ndhI, ndhJ, ndhK	12
	Subunits of cytochrome b/f complex	petA, petB*, petD*, petG, petL, petN	6
	Subunits of ATP synthase	atpA, atpB, atpE, atpF*, atpH, atpI	6
	Large subunit of rubis CO	rbcL	1
	Proteins of large ribosomal subunit	#rpl2*, rpl14, rpl16*, rpl2*, rpl20, rpl22, rpl23(2), rpl32, rpl33	10
Self-replication	Proteins of small ribosomal subunit	rps11, rps12**(2), rps14, rps15, rps16*, rps18, rps19, rps2, rps3, rps4, rps7(2), rps8	14
	Subunits of RNA polymerase	rpoA, rpoB, rpoC1*, rpoC2	4
	Ribosomal RNAs	rrn16(2), rrn23(2), rrn4.5(2), rrn5(2)	8
	Transfer RNAs	trnA-UGC*(2), trnC-GCA, trnD-GUC, trnE-UUC, trnF-GAA, trnG-GCC, trnG-UCC*, trnH-GUG, trnI-CAU(2), trnI-GAU*(2), trnK-UUU*, trnL-CAA(2), trnL-UAA*, trnL-UAG, trnM-CAU, trnN-GUU(2), trnP-UGG, trnQ-UUG, trnR-ACG(2), trnR-UCU, trnS-GCU, trnS-GGA, trnS-UGA, trnT-GGU, trnT-UGU, trnV-GAC(2), trnV-UAC*, trnW-CCA, trnY-GUA, trnfM-CAU	37
	Maturase	matK	1
Other genes	Protease	clpP**	1
	Envelope membrane protein	cemA	1
	Acetyl-CoA carboxylase	accD	1
	c-type cytochrome synthesis gene	ccsA	1
Genes of unknown function	Conserved hypothetical chloroplast ORF	#ycf1, ycf1, ycf15(2), ycf2(2), ycf3**, ycf4	8

Notes: Gene*: Gene with one introns; Gene**: Gene with two introns; #Gene: Pseudo gene.

3.2. RSCU Analysis of the "Qixiadaxiangshui" Pear Genome

The Relative Synonymous Codon Usage (RSCU) value, which equals the actual frequency of a codon divided by its expected frequency, highlights the codon usage bias and provides insights into the evolutionary dynamics and gene expression regulation within the "Qixiadaxiangshui" pear chloroplast genome. The RSCU analysis of the chloroplast genome revealed a total of 26,088 codons encoding 20 amino acids (excluding stop codons) (Figure 2 and Table 3). Leucine (Leu) had the highest number of codons (2752), accounting for 10.55% of the total, while cysteine (Cys) had the lowest (299, 1.15%). There are six types of codons encoding Leu, with the UUA codon having the highest RSCU value of 1.9272 and the CUG codon having the lowest at 0.39. Excluding the codon AUG (methionine, Met), the remaining 30 preferred synonymous codons (RSCU > 1) all ended with an A/T(U) base. The preferred terminator used was UAA. The RSCU analysis not only demonstrates a preference for certain codons in the encoding of amino acids but also reflects the genetic and evolutionary intricacies of the "Qixiadaxiangshui" pear, with leucine and cysteine codon frequencies highlighting the nuanced regulation of gene expression.

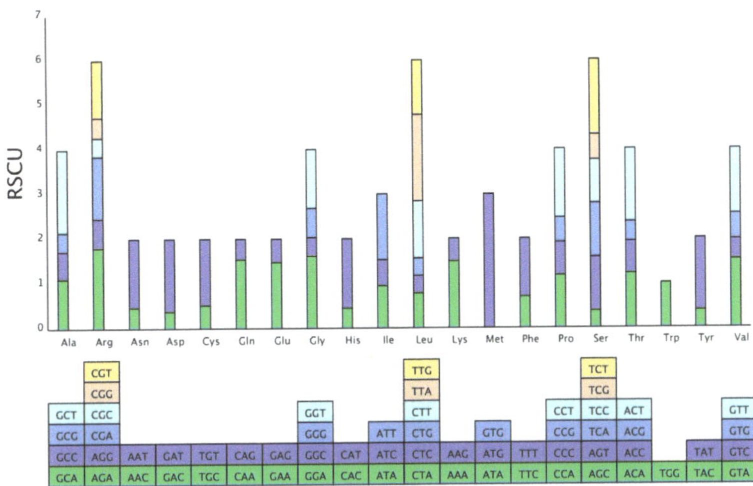

Figure 2. Total RSCU values of amino acids in chloroplast genome of "Qixiadaxiangshui" pear. The numbers on the bar chart represent the proportion of amino acid codons in the total codons (%).

Table 3. RSCU analysis of "Qixiadaxiangshui" pear chloroplast genome.

Amino Acid	Codon	No.	Total	RSCU	Amino Acid	Codon	No.	Total	RSCU
* (Ter)	UAA	47	84	1.6785	M (Met)	AUG	616		2.9904
* (Ter)	UAG	22		0.7857	M (Met)	GUG	1		0.0048
* (Ter)	UGA	15		0.5358	N (Asn)	AAC	298	1269	0.4696
A (Ala)	GCA	384	1384	1.11	N (Asn)	AAU	971		1.5304
A (Ala)	GCC	213		0.6156	P (Pro)	CCA	314	1068	1.176
A (Ala)	GCG	146		0.422	P (Pro)	CCC	197		0.738
A (Ala)	GCU	641		1.8528	P (Pro)	CCG	148		0.5544
C (Cys)	UGC	77	299	0.515	P (Pro)	CCU	409		1.532
C (Cys)	UGU	222		1.485	Q (Gln)	CAA	714	927	1.5404
D (Asp)	GAC	207	1091	0.3794	Q (Gln)	CAG	213		0.4596
D (Asp)	GAU	884		1.6206	R (Arg)	AGA	470	1569	1.7976
E (Glu)	GAA	1010	1367	1.4776	R (Arg)	AGG	174		0.6654
E (Glu)	GAG	357		0.5224	R (Arg)	CGA	363		1.3884
F (Phe)	UUC	518	1483	0.6986	R (Arg)	CGC	109		0.417
F (Phe)	UUU	965		1.3014	R (Arg)	CGG	119		0.4548
G (Gly)	GGA	701	1736	1.6152	R (Arg)	CGU	334		1.2774
G (Gly)	GGC	181		0.4172	S (Ser)	AGC	127	2011	0.3792
G (Gly)	GGG	284		0.6544	S (Ser)	AGU	401		1.1964
G (Gly)	GGU	570		1.3132	S (Ser)	UCA	407		1.2144
H (His)	CAC	140	626	0.4472	S (Ser)	UCC	324		0.9666
H (His)	CAU	486		1.5528	S (Ser)	UCG	190		0.567
I (Ile)	AUA	709	2244	0.948	S (Ser)	UCU	562		1.677
I (Ile)	AUC	433		0.579	T (Thr)	ACA	408	1343	1.2152
I (Ile)	AUU	1102		1.4733	T (Thr)	ACC	242		0.7208
K (Lys)	AAA	1029	1379	1.4924	T (Thr)	ACG	148		0.4408
K (Lys)	AAG	350		0.5076	T (Thr)	ACU	545		1.6232
L (Leu)	CUA	357	2752	0.7782	V (Val)	GUA	539	1415	1.5236
L (Leu)	CUC	181		0.3948	V (Val)	GUC	161		0.4552
L (Leu)	CUG	179		0.39	V (Val)	GUG	202		0.5712
L (Leu)	CUU	588		1.2822	V (Val)	GUU	513		1.45
L (Leu)	UUA	884		1.9272	W (Trp)	UGG	452	452	1
L (Leu)	UUG	563		1.2276	Y (Tyr)	UAC	189	971	0.3892
M (Met)	AUA	1	618	0.0048	Y (Tyr)	UAU	782		1.6108

Note: '*' represents the termination codon.

3.3. cpSSR Analysis of the "Qixiadaxiangshui" Pear Genome

A total of 209 repeat loci were identified in the genome, including 48 pairs of complex repeat sequences (Table 4). Within these, 129 were located in the large single-copy (LSC) region, 38 in the small single-copy (SSC) region, and 42 in the inverted repeat (IR) regions. The analysis revealed 181 mononucleotide simple sequence repeats (SSRs), including 19 dinucleotide SSRs, 64 trinucleotide SSRs, 5 tetranucleotide SSRs, 1 pentanucleotide SSR, and 2 hexanucleotide SSRs. Among the mononucleotide repeats (Figure 3), sequences of A and T were predominant. Dinucleotide repeats were primarily composed of AT and TA sequences. Trinucleotide repeats were the most varied in type, though they occurred less frequently. Tetranucleotide, pentanucleotide, and hexanucleotide repeats were rare in both type and occurrence. The cpSSR analysis uncovers a significant diversity of simple sequence repeats, especially in the mononucleotide category, predominantly composed of A and T sequences, illustrating the complexity and variability within the genome structure.

Table 4. Distribution analysis of SSRs.

Region	Exon	Intron	Intergenic	Total
LSC	31	16	82	129
SSC	22	3	13	38
IR	21	3	18	42

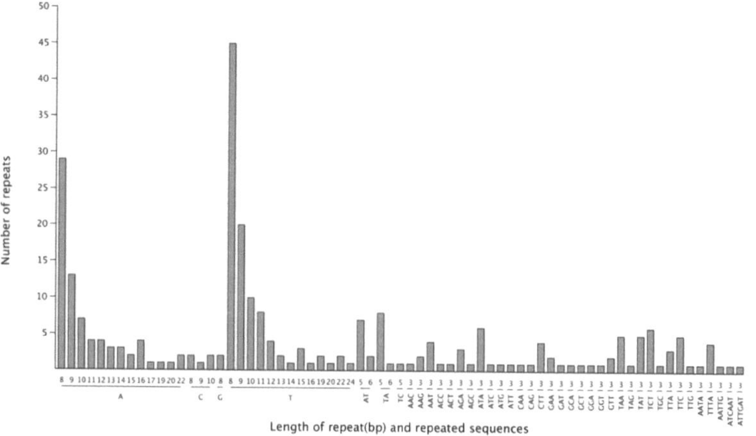

Figure 3. Statistical chart of SSR quantity by type. The horizontal axis represents the number of repetitions, while the vertical axis represents the number of repetitions of the same type.

3.4. Analysis of IR Boundary Features

The comparison of the inverted repeat (IR), large single-copy (LSC), and small single-copy (SSC) boundaries of the chloroplast genomes of ten pear germplasm, including the "Qixiadaxiangshui" pear (*Pyrus bretschneideri* Rehd.1), is illustrated in Figure 4. The length of the LSC region in the chloroplast genomes of these ten pear plants ranges from 87,863 bp to 88,200 bp, the IR region from 26,386 bp to 26,411 bp, and the SSC region from 19,201 bp to 19,251 bp. The sizes of the IR, SSC, and LSC regions are similar across the ten pears, with the genes on both sides of the JLB, JSB, JSA, and JLA boundaries being consistent, indicating a high degree of conservation. At the LSC/IRb boundary (JLB), the *rps19* gene in the LSC region of Bartlett pear (*Pyrus communis* strain Bartlett) extends into the IRb region by 134 bp, while in the other nine pears, the extension is 120 bp. The SSC/IRb boundary (JSB) shows that the boundary is within the *ndhF* gene for all ten germplasm, with the *ndhF* gene of the Bartlett pear extending into the IRb region by 14 bp, which is 2 bp more than the other nine pears. The JSA (SSC/IRa) boundary is located within the *ycf1* gene for all ten germplasm, with the *ycf1* gene of Bartlett extending into the Ira region by 1076 bp, whereas the extension in the other nine pears is 1094 bp. The LSC/Ira (JLA) boundary between the *rpl2* and *trnH* genes shows that the *rpl2* gene is entirely within the Ira region, and the *trnH* gene contracts towards the LSC region by 29 to 104 bp, with little variation in range.

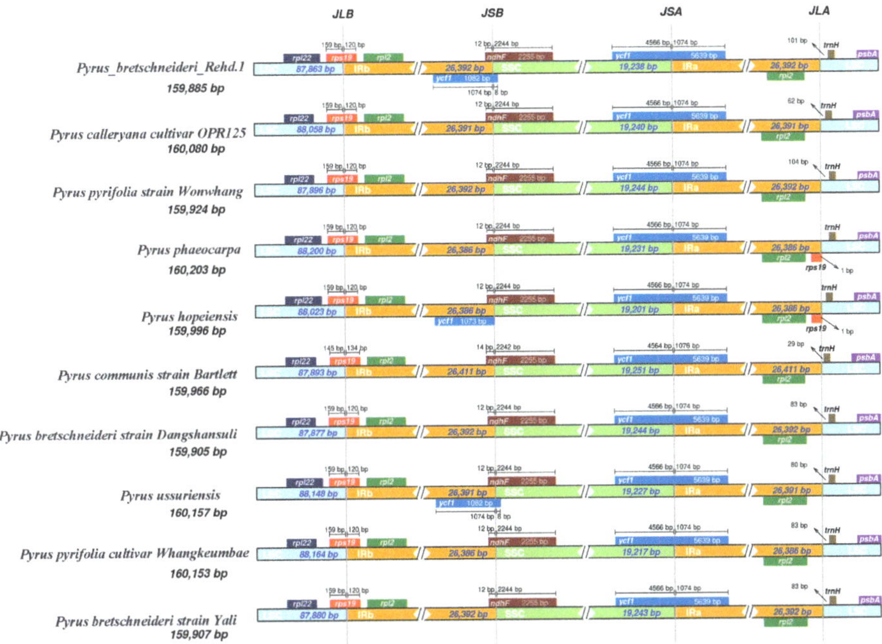

Figure 4. Whole genome boundary map of chloroplast of "Qixiadaxiangshui" pear and nine pear plants. The numbers with the arrows represent the distance in bp between each junction. Note that the distances are not drawn to scale and that the cp genomes follow the typical quadripartite structure with one LSC, one SSC, and two IR regions.

The expansion, contraction, or even complete absence of the IR region is a common evolutionary phenomenon in chloroplast genomes, with the expansion in some species leading to the duplication of the rps19 and ycf1 genes within this region [17]. For the "Qixiadaxiangshui" pear, *Pyrus hopeiensis*, and *Pyrus ussuriensis*, a copy of the *ycf1* gene occurs in the IRb region at the JSB boundary, with an incomplete copy in *Pyrus hopeiensis* and the absence of the *ycf1* gene copy in other pears; similarly, an incomplete copy of the *rps19* gene is found in the Ira region for *Pyrus phaeocarpa* and *Pyrus hopeiensis*, while other species lack this copy. The analysis of IR boundary features demonstrates a remarkable conservation in the structure and size of the LSC, SSC, and IR regions, alongside consistent gene positioning at key boundaries. This conservation reflects the evolutionary stability of these regions, despite the variability in gene extension and contraction at the boundaries, indicative of the dynamic evolutionary processes influencing chloroplast genome architecture.

3.5. Comparative Analysis of Chloroplast Genomes among Pear Germplasm

Using the *Pyrus bretschneideri* strain Yali as a reference, a comprehensive sequence comparison of ten pear germplasm, including the "Qixiadaxiangshui" pear, *Pyrus bretschneideri* strain Yali, *Pyrus pyrifolia* cultivar Whangkeumbae, and *Pyrus ussuriensis*, was conducted using the mVISTA online tool. This comparison aimed to discern the differences in the chloroplast genome sequences among these ten pears. The results indicated a high degree of similarity in the chloroplast genome sequences across the ten pear germplasm. Notably, some variable hotspots were primarily located in non-coding regions (Figure 5), including intergenic spaces such as *rps16-trnQ-UUG*, *trnR-UCU-atpA*, *trnT-GGU-psbD*, *ndhC-trnM-CAU*, *trnM-CAU-atpE*, and *accD-psaI*.

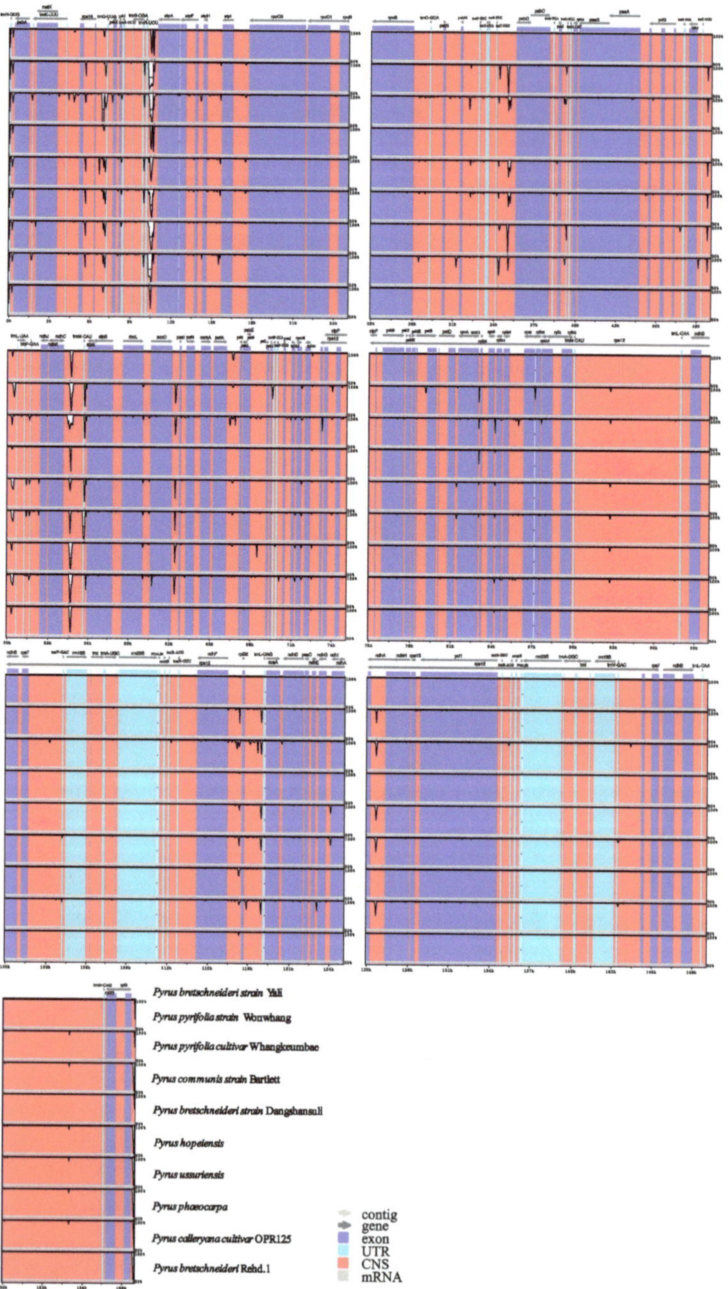

Figure 5. Comparison of chloroplast genes among ten pear genera. The gray scissors above indicate the direction of the gene. The dark blue region represents exons, the light blue region represents untranslated regions (UTRs), and the pink region represents non-coding sequences (CNS). The *y*-axis represents consistency from 50% to 100%.

Additionally, a comparison of the chloroplast genomes of these ten pears was performed using the MAFFT v7.520 software (Figure 6), with the nucleotide diversity (Pi) calculated using DNAsp v6.1 software to analyze the level of genomic variation among the *Pyrus*. The results showed that Pi values ranged from 0.00033 to 0.01426, with an average value of 0.001297. Intergenic regions with Pi values greater than 0.005 included *ndhC-trnM-CAU-atpE* and *trnR-UCU-atpA*. The intergenic region *ndhC-trnM-CAU-atpE* had the highest Pi value of 0.01426, containing 31 variable sites. Following this, the *trnR-UCU-atpA* region's Pi value reached 0.00985, encompassing 17 variable sites.

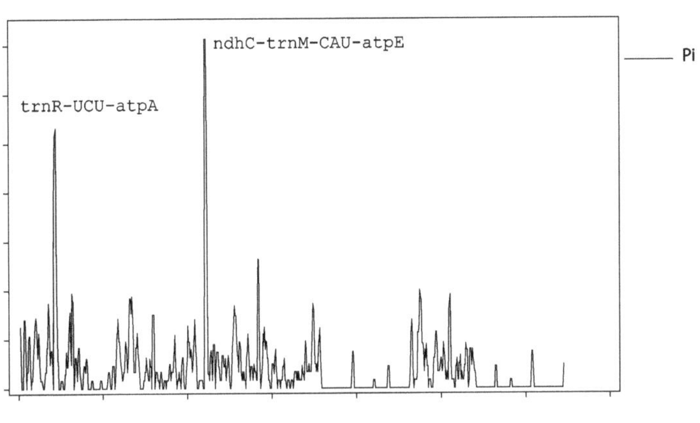

Figure 6. Chloroplast genome nucleotide diversity of ten cultivars in the *Pyrus* genus.

By comparing the results obtained from the mVISTA online tool and MAFFT v7.520 software, several key observations were made. Both methods consistently identified high levels of similarity in the chloroplast genome sequences across the ten pear germplasm and highlighted variable hotspots predominantly located in non-coding regions. While mVISTA identified these variable hotspots, the nucleotide diversity analysis of MAFFT quantified these differences with precise Pi values, revealing the extent of variability within regions like *ndhC-trnM-CAU-atpE* and *trnR-UCU-atpA*. This dual approach underscores the genetic diversity within the genus *Pyrus*, offering both a visual and quantitative understanding of the evolutionary dynamics of these pears.

3.6. Phylogenetic Analysis of Rosaceae Plants

A phylogenetic tree of the complete chloroplast genomes of 28 Rosaceae species and *Castanea mollissima* was constructed using multiple sequence alignments and the maximum likelihood method (Figure 7). The results revealed that the tested Rosaceae species were divided into three clades: one comprising the genera *Rubus*, *Rosa*, and *Fragaria*, which received 100% support; another consisting of the genera *Prunus*, also with 100% support; and a third clade formed by the genera *Malus*, *Crataegus*, and *Pyrus*, again with 100% support. Within the genus *Pyrus*, the "Qixiadaxiangshui" pear clustered closely with the Dangshansuli, Wonwhang, and Yali, all supported with a 100% bootstrap value. The phylogenetic analysis highlights the genetic divergence and common ancestry within the Rosaceae family. The high bootstrap support values confirm the robustness of our phylogenetic tree. The close clustering of the "Qixiadaxiangshui" pear with Dangshansuli, Wonwhang, and Yali suggests these cultivars share a recent common ancestor, indicating conserved genetic traits advantageous for breeding programs.

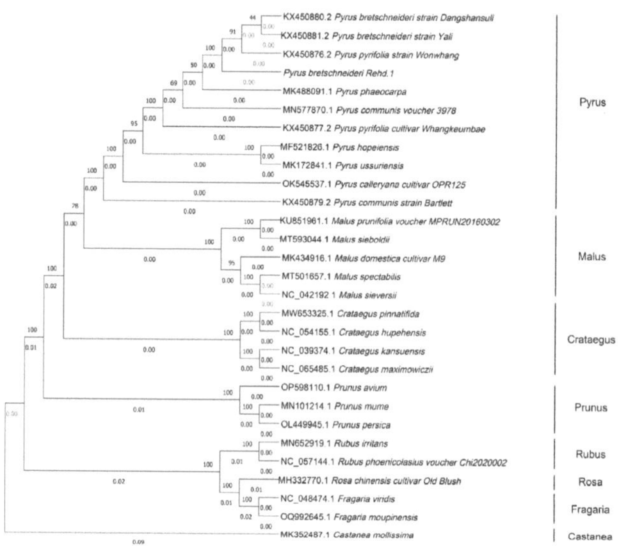

Figure 7. Phylogenetic analysis of chloroplast genomes based on 29 species from nine genera.

4. Discussion

Pears are ranked as the third most significant fruit in China and are revered as the "forebear of all fruits" holding substantial economic value. In this study, we sequenced, assembled, and annotated the chloroplast genome of the "Qixiadaxiangshui" pear (*Pyrus bretschneideri* Rehd.1), employing bioinformatics analyses. Our findings reveal that the chloroplast genome structure of the "Qixiadaxiangshui" pear is a typical circular quadripartite structure, similar to most angiosperms. Previous studies have established that the size of chloroplast genomes in angiosperms generally ranges from 120 to 160 kb [18,19]. In this work, the length of the "Qixiadaxiangshui" pear chloroplast genome is 159,885 bp, which falls within this typical range and is consistent with the sizes reported for other *Pyrus* species [7]. The slight variations in chloroplast genome size are primarily attributed to the expansion and contraction of the IR boundary, a common phenomenon observed in angiosperm chloroplast genomes [20,21]. Comparative analysis of structure boundaries among the "Qixiadaxiangshui" pear and other pear cultivars revealed minor variations in the IR/SC boundary positions, consistent with observations in previous studies on the *Pyrus* genus. Notably, expansions of the IR boundary were observed in the *rps19* gene of the Bartlett pear and the *ndhF* gene of other *Pyrus* germplasm, similar to expansions found in other plant genera such as *Ulmus* [22], *Manglietia* [23], and *Physalis* [24]. These findings suggest a stable chloroplast genome structure within the "Qixiadaxiangshui" pear, although further studies using biological replicates from multiple individuals are necessary to confirm these initial observations and uncover other structural variations.

Repetitive sequences and SSRs are widely distributed in chloroplast genomes and are closely related to genome rearrangement and recombination [25]. These elements serve as important molecular markers and are extensively used in plant population genetics and phylogeny [26–28]. In this study, we identified 209 SSRs in the chloroplast genome of the "Qixiadaxiangshui" pear, predominantly consisting of mononucleotide repeats composed of A/T bases. Most of these SSRs were located in the LSC region. These findings are consistent with SSR analyses of other *Pyrus* chloroplast genomes [4–6], highlighting the conserved nature of these sequences within the genus. The detected long repeats and SSRs provide valuable molecular marker information, which can be utilized to reveal population-level polymorphisms and phylogenetic relationships of the *Pyrus* species in future research.

Mutation, genetic drift, and natural selection are key factors influencing codon preference [7]. RSCU is a crucial evolutionary feature of the genome, significantly impacting gene expression, and can provide insights into the evolutionary processes of organisms [29,30]. The codon analysis for the "Qixiadaxiangshui" pear revealed that 96.77% of the 31 codons with RSCU values greater than 1 ended with A/T(U) bases. This indicates a strong preference for A/T-ending codons in the chloroplast genes of the "Qixiadaxiangshui" pear, consistent with previous observations in the chloroplast genomes of other angiosperms [31,32]. Knight et al. [33] developed a model suggesting that the GC base composition of the genome drives codon usage. However, some researchers argue that genome-wide codon bias is determined by non-random mutations and the selective pressure for protein translation efficiency [34,35]. Therefore, it is speculated that the preference for A/T-ending codons in the chloroplast genome of the "Qixiadaxiangshui" pear is likely due to a combination of mutation biases favoring A/T bases and selection pressure. This finding enhances our understanding of the genetic structure of the "Qixiadaxiangshui" pear and provides a reference for future technical research on chloroplast genetic engineering aimed at improving desirable traits in pear cultivars.

The comparison of the entire chloroplast genome sequences between the "Qixiadaxiangshui" pear and other *Pyrus* cultivars demonstrated high similarity, with coding regions being more conserved than non-coding regions, a pattern consistent with other angiosperms [36,37]. Our nucleotide diversity analysis identified several mutation hotspots with precise Pi values, such as *ndhC-trnM-CAU* and *trnR-UCU-atpA*. These regions are known to undergo faster nucleotide substitution rates at the species level, providing important references for the development of DNA barcodes [38]. The identified mutation hotspots and the sequenced chloroplast genome in our study offer valuable molecular marker tools, which can provide a wealth of informative sites for the phylogeny and molecular identification of the genus *Pyrus*. This information not only enhances our understanding of the genetic diversity within *Pyrus* but also supports the development of efficient strategies for pear breeding and conservation.

To further clarify the phylogenetic position of the "Qixiadaxiangshui" pear within the Rosaceae family, a phylogenetic tree constructed based on the complete chloroplast genomes of the Rosaceae species divided the species into three branches, with the "Qixiadaxiangshui" pear clustering with plants of the genera *Crataegus* and *Malus* on the same branch. The close clustering of the "Qixiadaxiangshui" pear with Dangshansuli, Wonwhang, and Yali suggests these cultivars share a recent common ancestor, indicating conserved genetic traits advantageous for breeding programs. However, the current study is limited by the sample size and the scope of genomic regions analyzed, necessitating further research with larger datasets and additional genomic regions to validate and expand upon these findings.

5. Conclusions

In this study, the complete chloroplast genome of the "Qixiadaxiangshui" pear (*Pyrus bretschneideri* Rehd.1) was first reported, enriching the genetic resources and laying the foundation for exploring its genetic background and resource utilization. Specifically, its structure, gene composition, GC content, and codon bias were similar to those of typical angiosperms. The chloroplast genomes of the "Qixiadaxiangshui" pear shared common characteristics with other pear species, such as size, structure, gene composition, and low sequence variation, demonstrating that the chloroplast genome of *Pyrus* is relatively conservative. Additionally, several mutation hotspots, such as *ndhC-trnM-CAU* and *trnR-UCU-atpA*, were identified, which can be applied in molecular identification and phylogenetic studies of *Pyrus*. The phylogenetic results exhibited the closest genetic relationship between the "Qixiadaxiangshui" pear and other cultivars like Dangshansuli, Wonwhang, and Yali, suggesting a recent common ancestor. In summary, these results contribute to a better understanding of the phylogeny and genetic improvement of *Pyrus* germplasm resources. However, further research with larger datasets and additional genomic regions is necessary to validate and expand upon these findings.

Supplementary Materials: The following supporting information can be downloaded at: https: //www.mdpi.com/article/10.3390/horticulturae10070744/s1, Table S1: Data source of tested species.

Author Contributions: H.J. and Q.C.: contributing to the data collection, data analysis, preparation of figures, and manuscript drafting. C.X., H.W., K.R., R.D., X.D. and Q.G.: investigations and formal analysis. Q.C. and S.W.: writing and editing of the manuscript. All authors have read and agreed to the published version of the manuscript.

Funding: This research was supported by the National Center of Technology Innovation for Comprehensive Utilization of Saline-Alkali Land (Innovation Team for the Evaluation of Salt-Alkali Tolerant Germplasm and Breeding of New Varieties of Fruits and Vegetables), the National Center of Technology Innovation for Comprehensive Utilization of Saline-Alkali Land (GYJ2023004), the earmarked fund for China Agriculture Research System (CARS-28-37), the Natural Science Foundation of Shandong Province, China (ZR2023MC061, ZR2021MC177, ZR2023MC108), the Agricultural Science and Technology Innovation Project of Shandong Academy of Agricultural Sciences, China (CXGC2024F21), the Youth Foundation of Shandong Institute of Pomology, China (GSS2022QN11), and the Science and Technology Innovation Project of Shandong Land Group Dongying Co., Ltd.

Data Availability Statement: The raw sequencing data are available at NCBI, accession ID number: 2831255.

Conflicts of Interest: The authors have declared that no competing interests exist.

References

1. Van Dingenen, J.; Blomme, J.; Gonzalez, N.; Inzé, D. Plants grow with a little help from their organelle friends. *J. Exp. Bot.* **2016**, *67*, 6267–6281. [CrossRef] [PubMed]
2. Hamsher, S.E.; Keepers, K.G.; Pogoda, C.S.; Stepanek, J.G.; Kane, N.C.; Kociolek, J.P. Extensive chloroplast genome rearrangement amongst three closely related *Halamphora* spp. (Bacillariophyceae), and evidence for rapid evolution as compared to land plants. *PLoS ONE* **2019**, *14*, e0217824. [CrossRef] [PubMed]
3. Cheng, H.; Li, J.; Zhang, H.; Cai, B.; Gao, Z.; Qiao, Y.; Mi, L. The complete chloroplast genome sequence of strawberry (*Fragaria* × *ananassa* Duch.) and comparison with related species of Rosaceae. *PeerJ* **2017**, *5*, e3919. [CrossRef] [PubMed]
4. Chung, H.Y.; Lee, T.H.; Kim, Y.K.; Kim, J.S. Complete chloroplast genome sequences of Wonwhang (*Pyrus pyrifolia*) and its phylogenetic analysis. *Mitochondrial DNA Part B* **2017**, *2*, 325–326. [CrossRef] [PubMed]
5. Gil, H.Y.; Kim, Y.; Kim, S.H.; Jeon, J.H.; Kwon, Y.; Kim, S.C.; Park, J. The complete chloroplast genome of *Pyrus ussuriensis* Maxim. (Rosaceae). *Mitochondrial DNA Part B* **2019**, *4*, 1000–1001. [CrossRef]
6. Bao, L.; Li, K.; Teng, Y.; Zhang, D. Characterization of the complete chloroplast genome of the wild Himalayan pear Pyrus pashia (Rosales: Rosaceae: Maloideae). *Conserv. Genet. Resour.* **2017**, *9*, 569–571. [CrossRef]
7. Xu, Y.; Liu, Y.; Yu, Z.; Jia, X. Complete chloroplast genome sequence of the long blooming cultivar Camellia 'Xiari Qixin': Genome features, comparative and phylogenetic analysis. *Genes* **2023**, *14*, 460. [CrossRef] [PubMed]
8. Wei, S.; Wang, S.M. Effects of Bagging on Aroma of 'Qixiadaxiangshui' Pear Fruit. *Agric. Sci. Technol.* **2015**, *31*, 1676. [CrossRef]
9. Bankevich, A.; Nurk, S.; Antipov, D.; Gurevich, A.A.; Dvorkin, M.; Kulikov, A.S.; Pevzner, P.A. SPAdes: A new genome assembly algorithm and its applications to single-cell sequencing. *J. Comput. Biol.* **2012**, *19*, 455–477. [CrossRef] [PubMed]
10. Greiner, S.; Lehwark, P.; Bock, R. OrganellarGenomeDRAW (OGDRAW) version 1.3.1: Expanded toolkit for the graphical visualization of organellar genomes. *Nucleic Acids Res.* **2019**, *47*, W59–W64. [CrossRef] [PubMed]
11. Beier, S.; Thiel, T.; Minch, T.; Scholz, U.; Mascher, M. MISA-web: A web server for microsatellite prediction. *Bioinformatics* **2017**, *33*, 2583–2585. [CrossRef] [PubMed]
12. Amiryousefi, A.; Hyvönen, J.; Poczai, P. IRscope: An online program to visualize the junction sites of chloroplast genomes. *Bioinformatics* **2018**, *34*, 3030–3031. [CrossRef] [PubMed]
13. Frazer, K.A.; Pachter, L.; Poliakov, A.; Rubin, E.M.; Dubchak, I. VISTA: Computational tools for comparative genomics. *Nucleic. Acids Res.* **2004**, *32*, W273–W279. [CrossRef] [PubMed]
14. Katoh, K.; Standley, D.M. MAFFT multiple sequence alignment software version 7: Improvements in performance and usability. *Mol. Biol. Evol.* **2013**, *30*, 772–780. [CrossRef] [PubMed]
15. Stamatakis, A. Raxml-vi-hpc: Maximum likelihood-based phylogenetic analyses with thousands of taxa and mixed models. *Bioinformatics* **2006**, *22*, 2688–2690. [CrossRef] [PubMed]
16. Kumar, S.; Stecher, G.; Li, M.; Knyaz, C.; Tamura, K. MEGA X: Molecular evolutionary genetics analysis across computing platforms. *Mol. Biol. Evol.* **2018**, *35*, 1547–1549. [CrossRef] [PubMed]
17. Lu, L.; Li, X.; Hao, Z.D.; Yang, L.M.; Zhang, J.B.; Peng, Y.; Xu, H.; Lu, Y.; Zhang, J.; Shi, J.S.; et al. Phylogenetic studies and comparative chloroplast genome analyses elucidate the basal position of halophyte *Nitraria sibirica* (Nitrariaceae) in the Sapindales. *Mitochondrial DNA Part A* **2018**, *29*, 745–755. [CrossRef] [PubMed]

18. Raubeson, L.A.; Peery, R.; Chumley, T.W.; Dziubek, C.M.; Fourcade, H.M.; Boore, J.L.; Jansen, R.K. Comparative chloroplast genomics: Analyses including new sequences from the angiosperms *Nuphar advena* and *Ranunculus macranthus*. *BMC Genom.* **2007**, *8*, 174. [CrossRef] [PubMed]
19. Palmer, J.D. Comparative organization of chloroplast genomes. *Annu. Rev. Genet.* **1985**, *19*, 325–354. [CrossRef]
20. Dugas, D.V.; Hernandez, D.; Koenen, E.J.M.; Schwarz, E.; Straub, S.; Hughes, C.E.; Jansen, R.; Nageswara-Rao, M.; Staats, M.; Trujillo, J.T.; et al. Mimosoid legume plastome evolution: IR expansion, tandem repeat expansions, and accelerated rate of evolution in *clpP*. *Sci. Rep.* **2015**, *5*, 16958. [CrossRef] [PubMed]
21. Li, W.; Zhang, C.; Guo, X.; Liu, Q.; Wang, K. Complete chloroplast genome of *Camellia japonica* genome structures, comparative and phylogenetic analysis. *PLoS ONE* **2019**, *14*, e216645. [CrossRef] [PubMed]
22. Zuo, L.; Shang, A.; Zhang, S.; Yu, X.; Ren, Y.; Yang, M.; Wang, J. The first complete chloroplast genome sequences of *Ulmus* species by *de novo* sequencing: Genome comparative and taxonomic position analysis. *PLoS ONE* **2017**, *12*, e171264. [CrossRef] [PubMed]
23. Yang, L.; Tian, J.; Xu, L.; Zhao, X.; Song, Y.; Wang, D. Comparative chloroplast genomes of six Magnoliaceae species provide new insights into intergeneric relationships and phylogeny. *Biology* **2022**, *11*, 1279. [CrossRef] [PubMed]
24. Feng, S.; Zheng, K.; Jiao, K.; Cai, Y.; Chen, C.; Mao, Y.; Wang, L.; Zhan, X.; Ying, Q.; Wang, H. Complete chloroplast genomes of four *Physalis* species (Solanaceae): Lights into genome structure, comparative analysis, and phylogenetic relationships. *BMC Plant Biol.* **2020**, *20*, 242. [CrossRef] [PubMed]
25. Amiteye, S. Basic concepts and methodologies of DNA marker systems in plant molecular breeding. *Heliyon* **2021**, *7*, e08093. [CrossRef] [PubMed]
26. Li, L.; Fang, Z.; Zhou, J.; Chen, H.; Hu, Z.; Gao, L.; Chen, L.; Ren, S.; Ma, H.; Zhang, L.; et al. An accurate and efficient method for large-scale SSR genotyping and applications. *Nucleic. Acids Res.* **2017**, *45*, e88. [CrossRef] [PubMed]
27. Zhao, Y.; Qu, D.; Ma, Y. Characterization of the chloroplast genome of *Argyranthemum frutescens* and a comparison with other species in Anthemideae. *Genes* **2022**, *13*, 1720. [CrossRef] [PubMed]
28. Yang, X.; Zhou, T.; Su, X.; Wang, G.; Zhang, X.; Guo, Q.; Cao, F. Structural characterization and comparative analysis of the chloroplast genome of *Ginkgo biloba* and other gymnosperms. *J. For. Res.* **2021**, *32*, 765–778. [CrossRef]
29. Liu, S.X.; Xue, D.Y.; Cheng, R.; Han, H.X. The complete mitogenome of *Apocheima cinerarius* (Lepidoptera: Geometridae: Ennominae) and comparison with that of other lepidopteran insects. *Gene* **2014**, *547*, 136–144. [CrossRef] [PubMed]
30. Wang, W.; Yu, H.; Wang, J.; Lei, W.; Gao, J.; Qiu, X.; Wang, J. The complete chloroplast genome sequences of the medicinal plant *Forsythia suspensa* (Oleaceae). *Int. J. Mol. Sci.* **2017**, *18*, 2288. [CrossRef] [PubMed]
31. Wang, Y.; Zhan, D.F.; Jia, X.; Mei, W.L.; Dai, H.F.; Chen, X.T.; Peng, S.Q. Complete chloroplast genome sequence of *Aquilaria sinensis* (Lour.) Gilg and evolution analysis within the Malvales order. *Front. Plant Sci.* **2016**, *7*, 280. [CrossRef] [PubMed]
32. Li, B.; Lin, F.; Huang, P.; Guo, W.; Zheng, Y. Complete chloroplast genome sequence of Decaisnea insignis: Genome organization, genomic resources and comparative analysis. *Sci. Rep.* **2017**, *7*, 10073. [CrossRef]
33. Knight, R.D.; Freeland, S.J.; Landweber, L.F. A simple model based on mutation and selection explains trends in codon and amino-acid usage and GC composition within and across genomes. *Genome Biol.* **2001**, *2*, H10. [CrossRef] [PubMed]
34. Chen, S.L.; Lee, W.; Hottes, A.K.; Shapiro, L.; Mcadams, H.H. Codon usage between genomes is constrained by genome-wide mutational processes. *Proc. Natl. Acad. Sci. USA* **2004**, *101*, 3480–3485. [CrossRef] [PubMed]
35. Quax, T.E.F.; Claassens, N.J.; Söll, D.; van der Oost, J. Codon Bias as a Means to Fine-Tune Gene Expression. *Mol. Cell* **2015**, *59*, 149–161. [CrossRef] [PubMed]
36. Wang, Y.; Wang, S.; Liu, Y.; Yuan, Q.; Sun, J.; Guo, L. Chloroplast genome variation and phylogenetic relationships of *Atractylodes* species. *BMC Genom.* **2021**, *22*, 103. [CrossRef] [PubMed]
37. Jeon, J.; Kim, S. Comparative Analysis of the complete chloroplast genome sequences of three closely related East-Asian wild roses (*Rosa* sect. Synstylae; Rosaceae). *Genes* **2019**, *10*, 23. [CrossRef] [PubMed]
38. Katayama, H.; Uematsu, C. Comparative analysis of chloroplast DNA in *Pyrus* species: Physical map and gene localization. *Theor. Appl. Genet.* **2003**, *106*, 303–310. [CrossRef] [PubMed]

Article

Integrated Metabolome and Transcriptome Analyses Reveal the Mechanisms Regulating Flavonoid Biosynthesis in Blueberry Leaves under Salt Stress

Bin Ma [1], Yan Song [1], Xinghua Feng [1], Pu Guo [1], Lianxia Zhou [1], Sijin Jia [1], Qingxun Guo [1,2,*] and Chunyu Zhang [1,2,*]

[1] Department of Horticulture, College of Plant Science, Jilin University, Changchun 130062, China
[2] Jilin Engineering Research Center for Crop Biotechnology Breeding, College of Plant Science, Jilin University, Changchun 130062, China
* Correspondence: guoqx@jlu.edu.cn (Q.G.); cy_zhang@jlu.edu.cn (C.Z.)

Abstract: The flavonoids play important roles in plant salt tolerance. Blueberries (*Vaccinium* spp.) are extremely sensitive to soil salt increases. Therefore, improving the salt resistance of blueberries by increasing the flavonoid content is crucial for the development of the blueberry industry. To explore the underlying molecular mechanism, we performed an integrated analysis of the metabolome and transcriptome of blueberry leaves under salt stress. We identified 525 differentially accumulated metabolites (DAMs) under salt stress vs. control treatment, primarily including members of the flavonoid class. We also identified 20,920 differentially expressed genes (DEGs) based on transcriptome data; of these, 568 differentially expressed transcription factors (TFs) were annotated, and bHLH123, OsHSP20, and HSP20 TFs might be responsible for blueberry leaf salt tolerance. DEGs involved in the flavonoid biosynthesis pathway were significantly enriched at almost all stages of salt stress. Salt treatment upregulated the expression of most flavonoid biosynthetic pathway genes and promoted the accumulation of flavonols, flavonol glycosides, flavans, proanthocyanidins, and anthocyanins. Correlation analysis suggested that 4-coumarate CoA ligases (*4CL5* and *4CL1*) play important roles in the accumulation of flavonols (quercetin and pinoquercetin) and flavan-3-ol (epicatechin and prodelphinidin C2) under salt stress, respectively. The flavonoid 3′5′-hydroxylases (*F3′5′H*) regulate anthocyanin (cyanidin 3-O-beta-D-sambubioside and delphinidin-3-O-glucoside chloride) biosynthesis, and leucoanthocyanidin reductases (*LAR*) are crucial for the biosynthesis of epicatechin and prodelphinidin C2 during salt stress. Taken together, it is one of the future breeding goals to cultivate salt-resistant blueberry varieties by increasing the expression of flavonoid biosynthetic genes, especially *4CL*, *F3′5′H*, and *LAR* genes, to promote flavonoid content in blueberry leaves.

Keywords: blueberry; metabolome; transcriptome; salt stress; flavonoids

Citation: Ma, B.; Song, Y.; Feng, X.; Guo, P.; Zhou, L.; Jia, S.; Guo, Q.; Zhang, C. Integrated Metabolome and Transcriptome Analyses Reveal the Mechanisms Regulating Flavonoid Biosynthesis in Blueberry Leaves under Salt Stress. *Horticulturae* 2024, 10, 1084. https://doi.org/10.3390/horticulturae10101084

Academic Editor: Haijun Gong

Received: 12 August 2024
Revised: 25 September 2024
Accepted: 7 October 2024
Published: 9 October 2024

1. Introduction

Soil salinity is a major abiotic stress that adversely affects the growth and development of horticultural crops [1]. According to the FAO, about 90 million hectares of land worldwide are affected by salinization, accounting for approximately 6% of the global arable land area [2]. Salt stress is commonly caused by high concentrations of sodium ions (Na cations) and chloride ions (Cl anions) in the soil [3]. Plants have developed a variety of physiological, biochemical, and molecular adaptive mechanisms in response to salt stress [4]. Systematic studies of salt tolerance mechanisms have been performed in many plant species [5]. In general, the excessive accumulation of reactive oxygen species (ROS) can damage DNA, proteins, and lipids and inhibit plant growth [6]. Secondary metabolites, especially phenylpropanoid metabolites including cinnamic acids, coumarins, and flavonoids, serve as nonenzymatic systems to eliminate ROS and thus improve plant acclimation to salt stress [6,7] and tolerance of other stresses [8]. Some studies also showed that flavonoid metabotites, including quercetin, catechincan, anthocyanin, and proanthocyanidin, enhanced tolerance to salinity in plants [9–12].

The phenylpropanoid biosynthetic pathway has been extensively studied in several plant species. The precursor of all phenylpropanoid metabolites, p-coumaric acid, is biosynthesized via general phenylpropanoid pathway enzymes including phenylalanine ammonia-lyase (PAL), cinnamate 4-hydroxylase (C4H), and 4-coumarate CoA ligase (4CL). P-coumaric acid is then converted into various phenolic acids by shikimate O-hydroxycinnamoyltransferase (HCT), caffeoyl shikimate esterase (CSE), or caffeic acid 3-O-methyltransferase (COMT) and into flavonoids by chalcone synthase (CHS), chalcone isomerase (CHI), flavanone 3-hydroxylase (F3H), flavonoid 3′-hydroxylase (F3′H), and flavonoid 3′5′-hydroxylase (F3′5′H). Finally, flavonol synthase (FLS) catalyzes flavonol biosynthesis, while dihydroflavonol 4-reductase (DFR) and anthocyanidin synthase (ANS) catalyze anthocyanin and flavan-3-ol (flavans and proanthocyanidin) biosynthesis, and UDP-glucose flavonoid 3-O-glucosyl transferase (UFGT) and UDP-glycosyltransferase (UGT) are responsible for anthocyanin biosynthesis. Finally, leucoanthocyanidin reductase (LAR) and anthocyanidin reductase (ANR) are crucial for flavan-3-ol biosynthesis [8,13].

Some studies reported that salt tress promoted the expression of the most genes encoding these proteins [14,15]. For example, 1.2% NaCl promoted the expression of *PAL, C4H, 4CL, CHS, CHI,* and *ANR* for Sophora alopecuroides roots [15]. At the same time, heterologous expression of *AvFLS* from Apocynum venetum increased the total flavonoid content and enhanced plant salinity tolerance in tobacco [16]. The *NtCHS1* RNAi-silenced transgenic tobacco plants reduced flavonoid accumulation and weakened ROS-scavenging ability under salt stress; in contrast, *NtCHS1* overexpressing plants had higher tolerance to salinity [17]. Overexpression of a tea (Camellia sinensis) CsF3H gene confers tolerance to salt stress in transgenic tobacco [18]. Thus, most genes from the flavonoid biosynthetic pathway can enhance plant salinity tolerance.

Blueberries, *Vaccinium* sp., native to North America, are a popular fruit worldwide known for its health benefits due to its abundant secondary metabolites, particularly flavonoids [19,20]. According to data reported in 2022 by FAOSTAT, currently, the global area used for blueberries is 256,701 hectares. Of the continents, the largest blueberry producer is China (77,641 hectares), followed by the USA (46,539 hectares) and Canada (42,216 hectares). In recent years, blueberries have seen an increase not only in fresh consumption but also in the processing industry. The rising demand for blueberry plants has led to the global expansion of blueberry cultivation [21]. In China, *Vaccinium corymbosum*, the most important blueberry species, is widely cultivated in northern China. Like most perennial fruit crops, blueberry has low salt tolerance and is extremely sensitive to soil salt increases [22–24]. For the blueberry cultivar 'Tifblue', the plants' shoot and root dry weight increase in plants subjected to 100 mM NaCl for 76 d was only 38.51% and 42.64%, respectively, for the unsalinized controls [25]. With the increase in blueberry planting area and decrease in suitable soil conditions, cultivating salt-tolerant varieties to expand blueberry growing conditions are important measures to promote the development of the blueberry industry [24]. However, molecular regulatory mechanisms of blueberry responses to salt stress are unclear. Therefore, it is important to explore the molecular regulatory mechanisms of the response of blueberry to salt stress by combining metabolome and transcriptome data to further improve salt tolerance in blueberry.

In this study, we identified differentially accumulated metabolites (DAMs) and differentially expressed genes (DEGs) in blueberry leaves in response to salt stress by performing metabolome and transcriptome analysis, respectively. To elucidate the mechanism driving flavonoid biosynthesis in blueberry leaves, we mapped the flavonoid biosynthetic Kyoto Encyclopedia of Genes and Genomes (KEGG) pathways in blueberry leaves under salt stress and analyzed the correlation between differentially accumulated flavonoid metabolites and DEGs from the flavonoid biosynthesis pathway by integrating transcriptome and metabolome data. The results of this study broaden our understanding of flavonoid biosynthesis in response to salt stress and provide a basis for breeding salt-resistant blueberry varieties.

2. Materials and Methods

2.1. Plant Materials and Salt Stress Treatments

The plants of the blueberry (*Vaccinium corymbosum*) cultivar 'Northland' were used as experimental materials. The in vitro-grown plants were kept in laboratory conditions in the Department of Horticulture at Jilin University, China. We obtained them from an institutional laboratory.

About two hundred in vitro-grown blueberry plants were transferred to 7-cm pots containing soil and grown in a growth chamber at 25 °C with 70% relative humidity under a 16 h light/8 h dark photoperiod for six months. Then the potted plants were irrigated with 1/2 Hoagland solution containing 200 mM NaCl [26,27]. The first to third fully expanded leaves were collected from randomly selected plants at 6 h (S6), 12 h (S12), 24 h (S24), and 48 h (S48) of NaCl treatment, using samples collected at time 0 h (S0) with 1/2 Hoagland solution (without NaCl) as control. The samples were flash-frozen in liquid nitrogen and stored at −80 °C for transcriptome deep sequencing (RNA-seq) analysis and metabolomic profiling. RNA-seq analysis was performed in three replicates, while the metabolomic profiling experiment was performed in five replicates.

2.2. Metabolomics Analysis by UHPLC-MS/MS

Frozen samples of blueberry leaves from NaCl treated for 6 h (S6), 12 h (S12), 24 h (S24), and 48 h (S48) and NaCl untreated for 0 h (S0) control were ground to a powder. The 80 mg of powder were resuspended in 1 mL methanol:acetonitrile:H_2O (2:2:1, $v/v/v$) solution for metabolite extraction. The mixture was centrifuged for 20 min (14,000× g, 4 °C), and the supernatant was dried in a vacuum centrifuge. The sample was dissolved in 0.1 mL acetonitrile: water (1:1, v/v) and centrifuged for 15 min (14,000× g, 4 °C), and 2 µL of the supernatant was used for LC–MS analysis. The ultra-high-performance liquid chromatography-quadrupole time-of-flight mass spectrometry (UHPLC-Q-TOF-MS) was performed using an Agilent 1290 infinity LC ultra-performance liquid chromatography (UHPLC) system with a C-18 column (ACQUITY UPLC BEH C-18 1.7 µm, 2.1 mm × 100 mm; Waters, Ireland) coupled to a quadrupole time-of-flight instrument (AB Sciex Triple TOF 6600). The column temperature was 40 °C, and the flow rate was set at 0.4 mL/min. Mobile phase A consisted of 25 mM ammonium acetate and 0.5% formic acid in water, and mobile phase B was methanol. The gradient elution procedure was as follows: 5% B (0–0.5 min); then B changed linearly to 100% from 0.5 to 10 min; 100% B (10–12 min); B changed linearly from 100% to 5% from 12.0 to 12.1 min; 5% B (12.1–16 min). Throughout the analysis, each sample was placed in an automatic sampler at 4 °C. Quality control (QC) samples were inserted into the sample queue to monitor and evaluate the stability and reliability of the data. The electrospray ionization (ESI) source conditions were described previously. During MS data acquisition, the instrument was set to acquire data over a m/z range of 60–1000 Da, and the accumulation time for TOF MS scans was set at 0.20 s/spectra. For automatic MS/MS acquisition, the instrument was set to acquire data over a m/z range of 25–1000 Da, and the accumulation time for the product ion scan was set at 0.05 s/spectra. The product ion scan was acquired using information-dependent acquisition (IDA) at high sensitivity mode.

To analyze the metabolome data, the raw data were converted to the final data format by employing ProteoWizard MSConvert (https://sourceforge.net/projects/proteowizard/, accessed on 1 December 2023), and the matched peak data and peak area data were obtained using MS-DIAL software (ver. 4.60) for normalization. Collection of Algorithms of MEtabolite pRofile Annotation (CAMERA) was used to annotate isotopes and adducts. Among the extracted ion features, only variables having more than 50% of the nonzero measurement values in at least one group were retained. Compound identification of metabolites was performed by comparing the accuracy of m/z values (<10 ppm) and MS/MS spectra using an in-house database established with available authentic standards.

All metabolites detected in positive and negative ion mode were analyzed for differences according to FC > 1.5 or FC < 0.67 and p-value < 0.05 based on univariate statistical analysis. Pareto-scaled principal component analysis (PCA) and orthogonal partial least-squares discriminant analysis (OPLS-DA) were performed via multivariate statistical analysis using the ropls R package. Seven-fold cross-validation and response permutation testing were used to evaluate the robustness of the model.

2.3. Transcriptomic Analysis by RNA Sequencing

Total RNA was extracted from blueberry leaves using TRIzol reagent (Invitrogen, CA, USA), and paired-end libraries were prepared using an ABclonal mRNA-seq Lib Prep Kit (ABclonal, Wuhan, China) according to the manufacturer's instructions. The libraries were sequenced on an Illumina NovaSeq 6000 (or MGISEQ-T7) instrument (San Diego, CA, USA) and 150-bp paired-end reads were generated. Raw data in fastq format were processed using in-house Perl scripts. The clean reads were used for subsequent analysis by removing the adapter sequences and filtering out low-quality reads and reads with N ratios > 5%. The GC content of clean reads and quality score of Q20 and Q30 were calculated to evaluate base quality. The clean reads were separately aligned to the reference *Vaccinium corymbosum* cv. Draper V1.0 genome sequence (https://www.vaccinium.org/genomes, accessed on 10 March 2024) using HISAT2 software (http://daehwankimlab.github.io/hisat2/, accessed on 10 March 2024) to obtain mapped reads.

FeatureCounts (http://subread.sourceforge.net/, accessed on 20 March 2024) was used to count the number of reads mapped to each gene. The fragments per kilobase of transcript per million fragments mapped (FPKM) of each gene was calculated based on the length of the gene and the number of reads mapped to this gene. DEGs were identified based on the criteria absolute $\log_2$(fold change) $\geq$ 1 and p-value < 0.05 using DESeq2 (http://bioconductor.org/packages/release/bioc/html/DESeq2.html, accessed on 20 March 2024). The clusterProfiler R software package (Version 4.12.6) was used for gene ontology (GO) functional enrichment and KEGG pathway enrichment analysis. A GO or KEGG function was considered to be significantly enriched when $p < 0.05$.

2.4. RNA Sequencing Data Validation

Total RNA was extracted using an RNA Extraction Kit (Sangon Biotech, Shanghai, China), and RT-qPCR was performed on an ABI 7900HT real-time PCR system. Eight genes of flavonoid biosynthetic genes (*PAL-1, 4CL1, C4H, DFR, F3'5'H-1, F3H, UFGT-1,* and *LAR-3*) were selected for RT-qPCR analysis, and *GAPDH* (AY123769) was used as the reference transcript. Primer sequences are shown in Table S1. The relative expression levels of each gene were calculated using the $2^{-\Delta\Delta Ct}$ method [28].

2.5. Integrative Analysis of Transcriptomic and Metabolomic Data

The phenylpropanoid and flavonoid biosynthetic pathways were mapped against the phenylpropanoid, flavonoid, flavonol, and anthocyanin KEGG pathways using the target DEGs and DAMs identified above. To perform integrative analysis of DEGs and DAMs in both the metabolomic and transcriptomic data from flavonoid biosynthetic pathways, Pearson correlation coefficients were calculated to determine the correlation between DEGs and DEMs using SPSS-V 29.0 software. Heatmaps of gene expression levels were constructed using $\log_{10}$ (FPKM), and heatmaps of metabolites were constructed using scale-normalized $\log_{10}$ (peak intensity) values with Tbtools (v1.098761) software [29].

3. Results

3.1. Metabolome Analysis for Blueberry Leaves under Salt Stress

Using non-targeted metabolomic profiling, we detected 1174 metabolomic substances in positive (745) and negative (429) ion modes in blueberry leaves in response to salt stress

(Figure 1a and Table S2). Most of these metabolomic substances belonged to 10 superclasses, including lipids and lipid-like molecules (24%), phenylpropanoids and polyketides (23%), organic oxygen compounds (9%), benzenoids (9%), organoheterocyclic compounds (8%), organic acids and derivatives (6%), and so on (Figure 1b and Table S3). Thus, lipids and lipid-like molecules, phenylpropanoids, and polyketides are the main metabolites for blueberry leaves under salt stress.

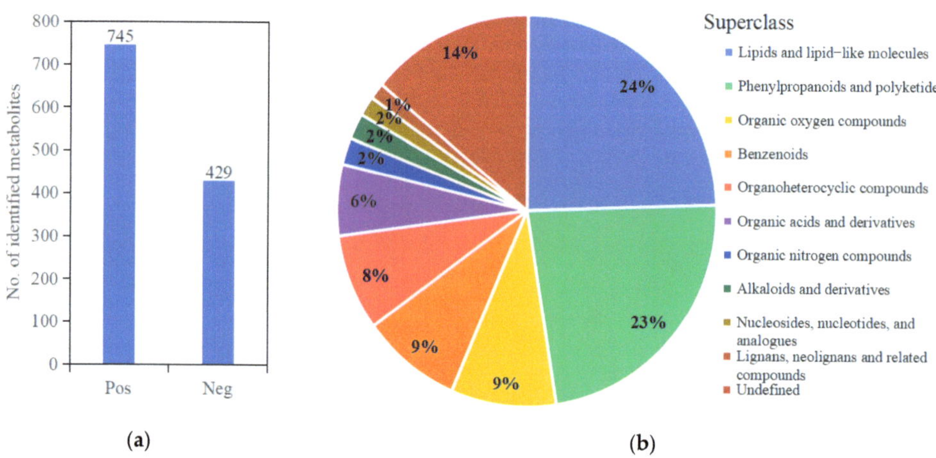

Figure 1. Metabolomic substances in blueberry leaves in response to salt stress identified by non-targeted metabolomic profiling in positive and negative ion mode. (**a**) Number of metabolomic substances in positive (Pos) and negative (Neg) ion modes. (**b**) Classification of the detected metabolomic substances.

We analyzed the peaks obtained from all experimental samples and quality control (QC) samples by principal component analysis (PCA). The QC samples in positive and negative ion mode were closely clustered together (Figure S1). At the same time, Pearson correlation coefficient analysis showed that the correlations between QC samples were >0.992 for negative ion mode and 0.973 for positive ion mode, indicating good repeatability of the experiment (Figure S2a,b).

3.2. Analysis of DAMs for Blueberry Leaves under Salt Stress

To explore the effects of salt stress on metabolism in blueberry, we constructed volcano plots of all metabolites detected in positive and negative ion mode under salt treatment for 6 h (S6), 12 h (S12), 24 h (S24), and 48 h (S48) compared to 0 h (S0) control, based on FC > 1.5 or FC < 0.67 and p-value < 0.05 by univariate statistical analysis (Figure S3). A total of 525 DAMs (333 positive ion mode; 192 negative ion mode) were significantly upregulated or downregulated under salt stress, with more DAMs identified in positive ion mode than negative ion mode. The number of DAMs increased during salt treatment. For example, 121 DAMs were identified in S6_vs_S0, 272 in S12_vs_S0, 328 in S24_vs_S0, and 383 in S48_vs_S0. A line regression analysis also showed that the number of DAMs increased gradually with increasing salt treatment duration (Figure 2a and Table S4). A total of 31 DAMs were shared by all four pairwise comparisons; 7 DAMs were identified only in S6_vs_S0, 27 in S12_vs_S0, 40 in S24_vs_S0, and 57 in S48_vs_S0, indicating that the number of specific DAMs also increased with increasing duration of salt treatment (Figure 2b). Most of these DAMs belonged to the lipid and lipid-like molecules superclass (137 DAMs) and the phenylpropanoid and polyketide superclass (133 DAMs). Flavonoid metabolites (79 DAMs) accounted for more than 50% of DAMs in the phenylpropanoid and polyketide superclass, representing the largest group of major DAMs in blueberry leaves under salt stress (Figure 2c). These results indicated that flavonoid metabolites are the main DAMs for blueberry leaves under salt stress.

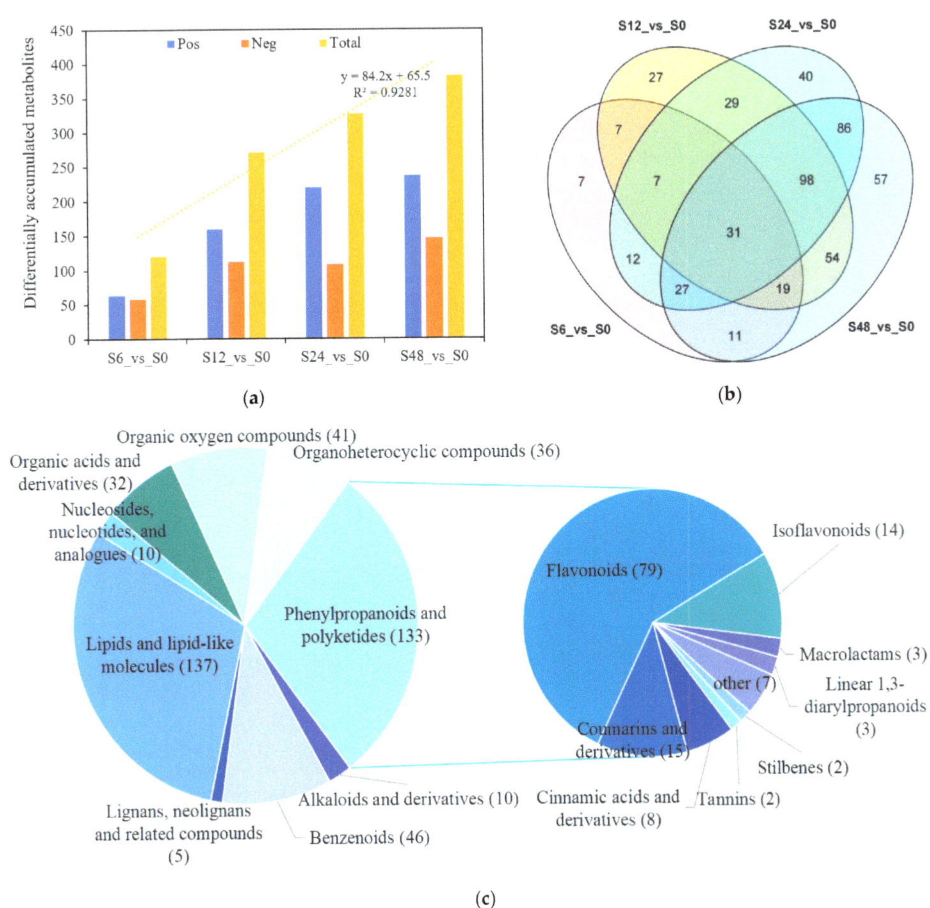

Figure 2. Differentially accumulated metabolites (DAMs) in blueberry leaves in response to salt stress identified by non-targeted metabolomic profiling in positive and negative ion mode. (**a**) Number of DAMs detected in positive (Pos) and negative (Neg) ion mode and total number of DAMs. (**b**) Venn diagram showing the extent of overlap between DAMs across pairwise comparisons. Pink, DAMs for 6 h vs. 0 h; yellow, DAMs for 12 h vs. 0 h; green, DAMs for 24 h vs. 0 h; blue, DAMs for 48 h vs. 0 h. (**c**) Classification of the DAMs.

3.3. Enrichment Analysis of the KEGG Pathways of the DAMs

All DAMs were assigned to 89 significantly enriched KEGG pathways at p-value < 0.05 under salt stress, and 59 significantly enriched KEGG pathways were shared by three pairwise comparisons, as shown in a Venn diagram (Figure 3a and Table S5). Of these, 61 significantly enriched KEGG pathways were found in S6_vs_S0, 24 in S12_vs_S0, 66 in S24_vs_S0, and 60 in S48_vs_S0 (Figure 3b). The 272 DAMs in S12_vs_S0 were only annotated to 24 enriched KEGG pathways, mainly because more DAMs were annotated to the same enriched KEGG pathway. For example, five DAMs were annotated to the flavonoid biosynthesis pathway (ko00941) in S12_vs_S0; however, only two DAMs were annotated to the flavonoid biosynthesis pathway in S24_vs_S0. The top 20 significantly enriched KEGG pathways were the hedgehog signaling pathway (ko04340), cell cycle (ko04111), longevity-regulating pathway (ko04213), circadian rhythm (ko04710), and vasopressin-regulated water reabsorption (ko04962) for DAMs after 6, 24, and 48 h of salt stress compared to the 0 h control samples. In addition, flavonoid biosynthesis was significantly enriched at 12 h and 24 h of salt stress compared to the 0 h control samples (Figure 3c).

3.4. Transcriptome Analysis and DEGs for Blueberry Leaves under Salt Stress

To investigate the responses of blueberries to salt stress, we sequenced the same materials used in the metabolite analysis. In total, 6.5–8.61 Gb of clean bases were obtained for each sample, with a Q20 ranging from 95.32 to 97.2% and a Q30 ranging from 87.62 to 92.12%. The GC content was greater than 45.7%. The clean reads were successfully mapped to the blueberry reference genome with an average mapping efficiency of 90.13% (Table S6). These results implied that the sequencing quality was high.

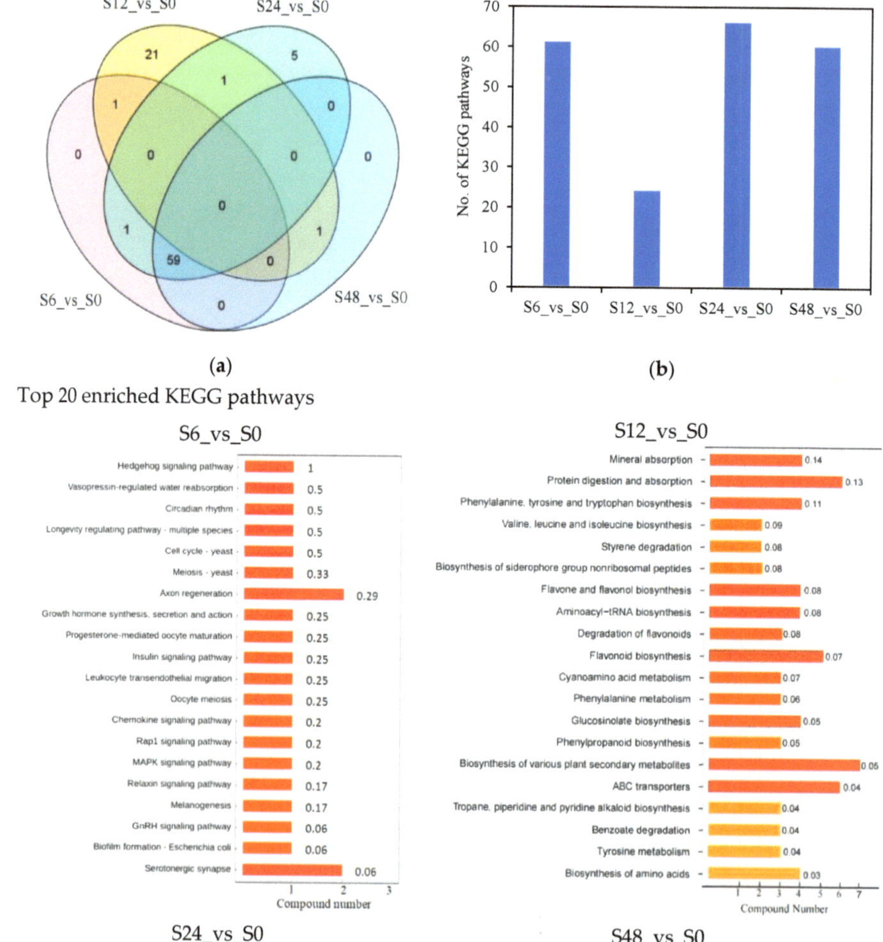

Figure 3. *Cont.*

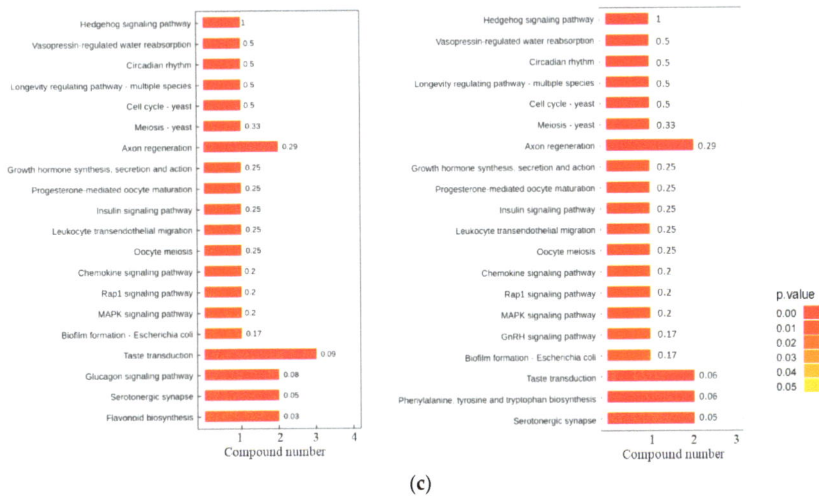

(c)

Figure 3. Enriched KEGG pathways of differentially accumulated metabolites (DAMs) in blueberry leaves in response to salt stress revealed by non-targeted metabolomic profiling. (**a**) Venn diagram of significantly enriched KEGG pathways of the DAMs. Pink, DAMs for 0 h vs. 6 h; yellow, DAMs for 0 h vs. 12 h; green, DAMs for 0 h vs. 24 h; blue, DAMs for 0 h vs. 48 h. (**b**) The number of significantly enriched KEGG pathways among the DAMs. (**c**) The top 20 enriched KEGG pathways among the DAMs, and rich factors in different comparison groups.

To examine changes in gene expression under salt stress, we identified DEGs during NaCl treatment (Table S7). A total of 20,920 DEGs were detected, including 4468 DEGs in S6_vs_S0 (2150 upregulated and 2318 downregulated), 9814 in S12_vs_S0 (5353 upregulated and 4461 downregulated), 10,333 in S24_vs_S0 (4859 upregulated and 5474 downregulated), and 14,195 in S48_vs_S0 (6570 upregulated and 7625 downregulated). The line regression analysis also showed that the number of DEGs increased gradually with increasing salt treatment duration (Figure 4a). Moreover, 1326 DEGs were shared among the 6, 12, 24, and 48 h of salt stress groups relative to the 0 h control, as revealed in a Venn diagram (Figure 4b).

3.5. Functional Annotation and Enrichment Analysis of DEGs

We annotated the functions of the DEGs under salt stress using the gene ontology (GO) and KEGG databases (Tables S8 and S9). The DEGs were assigned to 3195 GO terms and 290 KEGG pathways, including 1146 GO terms and 167 KEGG pathways that were shared in all four pairwise comparisons (Figure 5a). We also analyzed the top 30 GO terms and top 20 KEGG pathways. Among the top 30 GO terms, photosystem was the most highly enriched in the cellular components category, photosynthesis in the biological process category, and tetrapyrrole binding and oxidoreductase activity in the molecular functions category. In addition, phenylalanine ammonia-lyase activity was significantly enriched at the later stages of salt stress (24 h and 48 h) compared to 0 h control samples (Figure 5b). Among the top 20 KEGG pathways, biosynthesis of secondary metabolites, metabolic pathways, porphyrin and chlorophyll metabolism, photosynthesis, and photosynthesis—antenna proteins were significantly enriched at almost all stages of salt treatment (Figure 5c). Moreover, many DEGs were involved in the biosynthesis of phenylpropanoid-derived compounds, including the phenylpropanoid and flavonoid biosynthesis pathways, which were significantly enriched at almost all stages of salt stress. Likewise, DEGs were significantly enriched in the anthocyanin biosynthesis pathway at 6, 12, and 48 h of salt stress compared to 0 h control samples (Table S9). Therefore, DEGs involved in the biosynthesis of phenylpropanoid-derived compounds were significantly enriched under salt stress.

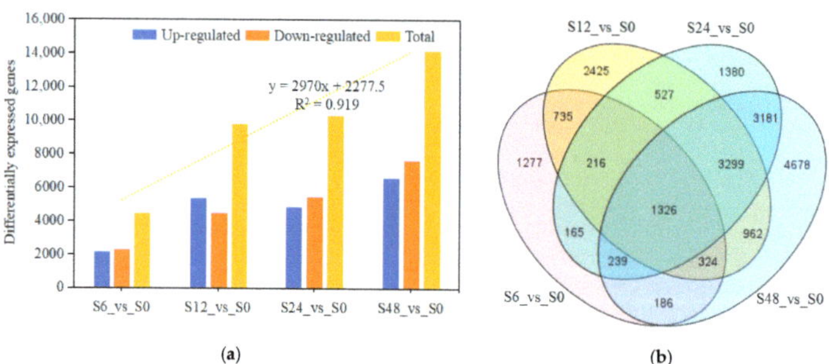

Figure 4. Differentially expressed genes (DEGs) in blueberry leaves in response to salt stress identified by transcriptome deep sequencing (RNA-seq). (**a**) Number of DEGs in response to salt stress. Upregulated, upregulated DEGs; Down-regulated, downregulated DEGs; Total, total DEGs. (**b**) Venn diagram showing the extent of overlap between DEGs across pairwise comparisons. Pink, DEGs for 0 h vs. 6 h; yellow, DEGs for 0 h vs. 12 h; green, DEGs for 0 h vs. 24 h; blue, DEGs for 0 h vs. 48 h.

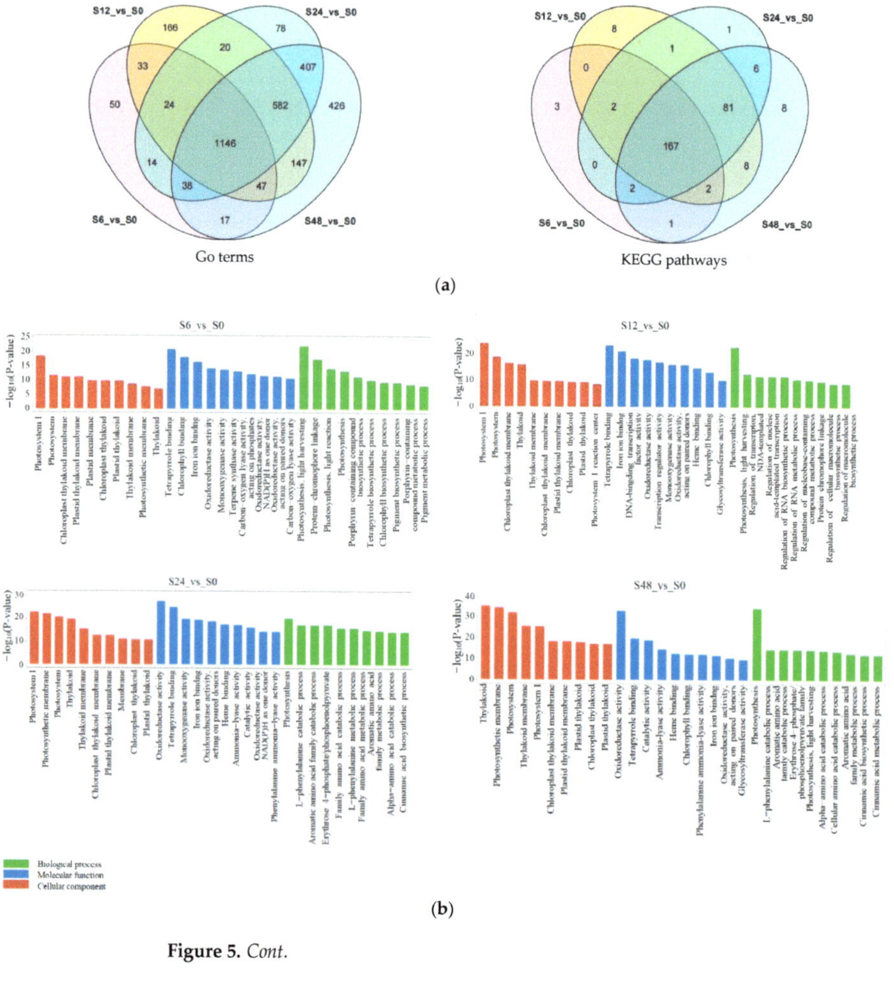

Figure 5. *Cont.*

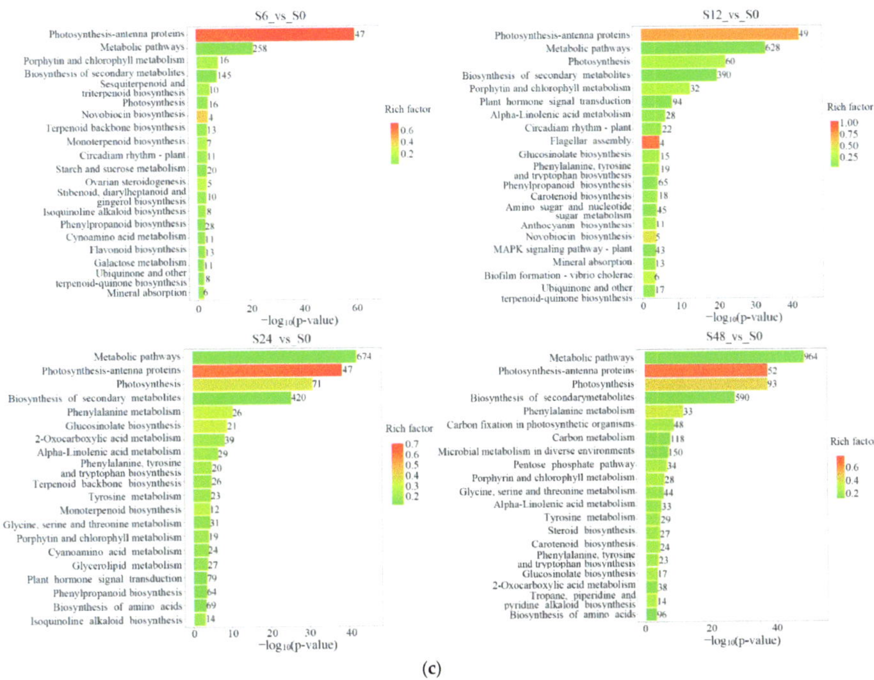

(c)

Figure 5. GO and KEGG pathway enrichment analysis of differentially expressed genes (DEGs) in blueberry leaves in response to salt stress identified by transcriptome deep sequencing (RNA-seq). (**a**) Venn diagrams of GO terms and significantly enriched KEGG pathways among the DEGs. Pink, DEGs for 0 h vs. 6 h; yellow, DEGs for 0 h vs. 12 h; green, DEGs for 0 h vs. 24 h; blue, DEGs for 0 h vs. 48 h. (**b**) GO terms of DEGs in different comparison groups. Green bar, biological process; blue bar, molecular function; red bar, cellular component. (**c**) The top 20 enriched KEGG pathways of DEGs in different comparison groups.

3.6. Identification of Differentially Expressed Transcription Factors (TFs) under Salt Stress

To explore the responses of TFs to salt stress in blueberry, we further screened the differentially expressed TFs (Table S10). The total of 568 differentially expressed transcription factor genes were identified by merging the redundant sequences under salt stress. According to the different functional domains, these TFs were annotated to 12 major TF families, including zinc finger (133 members), AP2 (70 members), MYB (57 members), bHLH (50 members), and HSP (46 members), WRKY (40 members), NAC (31 members), Homeobox (31 members), AUX/IAA (15 members), GATA (11 members), bZip (10 members), and B3 (10 members). For zinc finger proteins, CCCH-type (18 members) was the most, followed by RING-type (14 members), CCT-type (9 members), C2H2-type (9 members), and so on (Figure 6). Among them, 13 TFs were significantly upregulated more than 4.190 times at 6, 12, 24, and 48 h of salt stress compared to 0 h control (Table 1). These TFs were from the AP2 family (ERF114, ABR1, ERF4, ERF110, and ERF011), HSP family (HSP20), bHLH family (bHLH35, bHLH123, and bHLH162), MYB family (MYB102, MYB13, and MYB14), and zinc finger family (ZAT12), and may play an important role under salt stress.

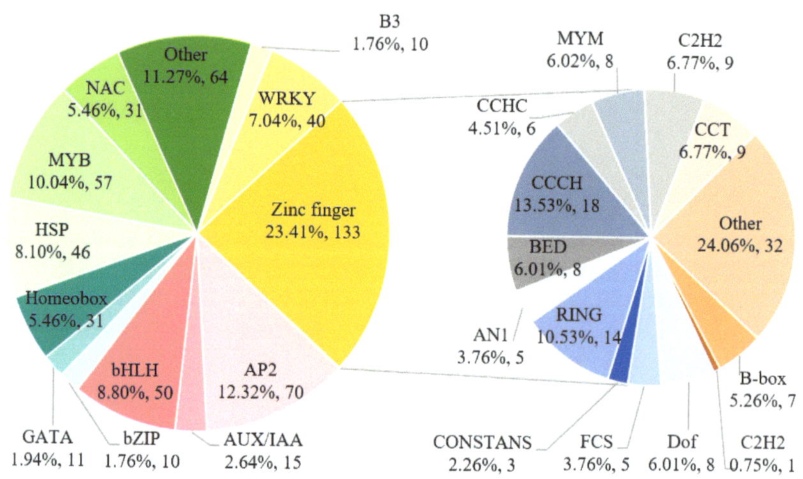

Figure 6. Differentially expressed transcription factors (TF) in blueberry leaves in response to salt stress identified by transcriptome deep sequencing (RNA-seq).

Table 1. The upregulated transcription factors according to fold change ≥4 and *p*-value <0.05 during salt stress compared to the 0 h control samples.

Gene ID	Fold Change				
	Gene Name	S6_vs_S0	S12_vs_S0	S24_vs_S0	S48_vs_S0
VaccDscaff3-augustus-gene-84.42	ERF114	4.924	9.832	9.964	11.758
VaccDscaff13-augustus-gene-7.21	ABR1	5.842	7.966	9.182	7.754
VaccDscaff39-processed-gene-249.7	ERF4	6.942	11.268	9.92	10.268
VaccDscaff38-augustus-gene-106.21	ERF110	6.986	9.476	10.994	10.43
VaccDscaff21-processed-gene-16.2	ERF011	5.714	9.802	5.024	6.278
VaccDscaff30-processed-gene-33.28	HSP20	6.736	9.258	6.182	8.018
VaccDscaff99-augustus-gene-4.57	bHLH35	4.930	7.946	4.868	4.190
VaccDscaff10-augustus-gene-14.13	bHLH123	9.756	8.986	7.578	10.622
VaccDscaff9-augustus-gene-346.25	bHLH162	5.108	9.73	6.768	7.444
VaccDscaff9-augustus-gene-136.29	MYB102	6.452	7.546	6.276	7.056
VaccDscaff3-augustus-gene-219.29	MYB13	6.262	10.038	8.412	6.076
VaccDscaff43-augustus-gene-40.46	MYB14	5.066	5.700	7.13	6.102
VaccDscaff21-processed-gene-155.6	ZAT12	7.062	9.660	6.632	8.080

3.7. DAMs and DEGs in Phenylpropanoid and Flavonoid Biosynthetic KEGG Pathways under Salt Stress

Metabolomic and transcriptomic analyses indicated that the DEGs and DAMs in response to salt stress were highly enriched in phenylpropanoid and flavonoid biosynthesis pathways. We identified 46 DEGs encoding 16 types of enzymes and 103 DAMs from four classes (phenylpropionic acids, cinnamic acids and derivatives, coumarins and derivatives, and flavonoids) of the phenylpropanoid and flavonoid biosynthesis pathways (Tables S11 and S12). Most DAMs belonging to the cinnamic acids and derivatives class (2-hydroxycinnamic acid, 4-hydroxycinnamic acid, calceolarioside, rosmarinic acid, caffeic acid, and caffeoylshikimic acid) were upregulated under salt stress. DAMs in flavonoid biosynthesis pathways were upregulated or downregulated under salt stress. Specifically, quercetin and pinoquercetin (flavonols), epicatechin (flavan), prodelphinidin C2 (proanthocyanidin), cyanidin 3-O-beta-D-sambubioside, cyanidin-3-O-rhamnoside, and delphinidin-3-O-glucoside chloride (anthocyanins), and kaempferol 3,7-diglucoside, kaempferol 3-O-arabinoside, kaempferol 3-O-beta-D-xyloside, myricetin-3-o-galactoside, and isorhamnetin-3-O-glucoside (flavonol glycosides) were upregulated under salt stress. By contrast, cyanidin (anthocyanidin) and cyanidin 3-arabinoside, cyanidin 3-O-glucoside, cyanidin-3-O-alpha-arabinoside, and delphinidin 3-rutinoside (anthocyanins), and quercetin-3-neohesperidoside-7-rhamnoside and syringetin-3-O-galactoside (flavonol glycosides) were downregulated under salt stress (Table S11 and Figure 7). In general, most DAMs involved in the phenylpropanoid and flavonoid biosynthesis pathways were upregulated under salt stress.

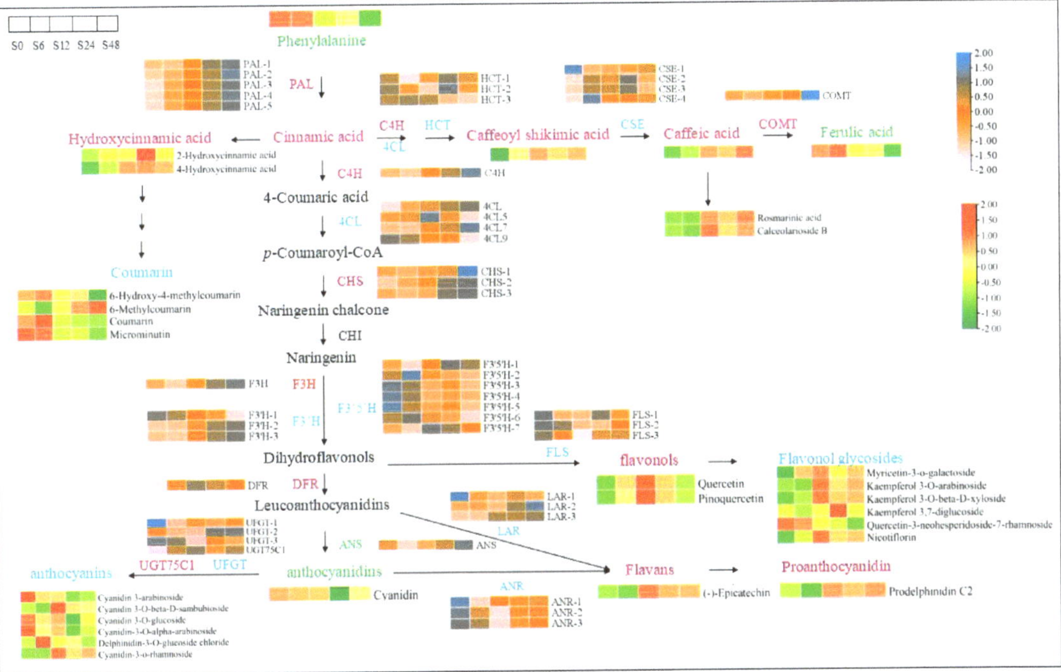

Figure 7. Phenylpropanoid and flavonoid KEGG biosynthetic pathways in blueberry leaves under salt stress. Red and magenta fonts indicate upregulated and downregulated metabolites or genes, respectively; blue fonts indicate both upregulated and downregulated metabolites or genes; and black fonts indicate no significant changes in response to salt stress. Heatmaps show the expression of DEGs or the accumulation of DAMs. Colored bars on the upper right indicate low expression (pink) or high expression (blue) of differentially accumulated metabolites (DAM) based on $\log_{10}$ (peak intensity). Colored bars on the right indicate low expression (green) or high expression (red) of differentially expressed genes (DEGs) based on scale normalized $\log_{10}$ (FPKM). PAL, phenylalanine ammonia-lyase; C4H, cinnamate 4-hydroxylase; 4CL, 4-coumarate CoA ligase; HCT, shikimate O-hydroxycinnamoyltransferase; CSE, caffeoyl shikimate esterase; COMT, caffeic acid 3-O-methyltransferase; CHS, chalcone synthase; CHI, chalcone isomerase; F3H, flavanone 3-hydroxylase; F3′H, flavonoid 3′-hydroxylase; F3′5′H, flavonoid 3′5′-hydroxylase; FLS, flavonol synthase; DFR, dihydroflavonol 4-reductase; ANS, anthocyanidin synthase; ANR, anthocyanidin reductase; LAR, leucoanthocyanidin reductase; UFGT, UDP-glucose flavonoid 3-O-glucosyl transferase; UGT, UDP-glycosyltransferase.

Most DEGs were upregulated under salt stress, including *PAL*, *C4H*, *COMT*, *4CL1*, *4CL5*, *4CL7*, *CHS*, *DFR*, and *UGT75C1*, and their expression reached their highest levels at 48 h. In addition, four *ANS* genes were downregulated at 6 h of salt stress. Most *HCT* genes were downregulated, and most *CSE* genes in the phenylpropanoid biosynthesis pathway were upregulated under salt stress. Most *F3H* and *F3′H* genes, which are involved in the flavonoid biosynthetic pathway, were upregulated under salt stress, especially at

48 h of salt treatment. By contrast, most *F3'5'H* and *FLS* genes (flavonol biosynthesis genes) were downregulated under salt stress. Both *ANR* and *LAR* genes are involved in flavan and proanthocyanin biosynthesis. Almost all *ANR* genes were downregulated under salt stress, especially at 6 h of treatment, whereas most *LAR* genes were upregulated, especially at 24 and 48 h of salt stress. *UFGT* genes, which are involved in anthocyanin biosynthesis, were upregulated or downregulated under salt stress. These results indicate that most genes involved in phenylpropanoid and flavonoid biosynthetic pathways were induced under salt stress, and most of these genes were rapidly upregulated at 48 h of salt treatment. We mapped the 46 DEGs and 29 DAMs onto KEGG phenylpropanoid and flavonoid biosynthetic pathways (Figure 7). These genes play an important role in phenylpropanoid and flavonoid biosynthesis for blueberry leaves under salt stress.

To validate the accuracy and reliability of the RNA-seq data, we selected eight flavonoid biosynthetic genes for RT-qPCR analysis (Figure S4). A linear regression analysis showed significant correlation between the RNA-seq and RT-qPCR results, with correlation coefficients (R^2) greater than 0.8312 for all the compared groups. This result confirmed that the RNA-seq data are accurate and reliable in this study.

3.8. Combined Metabolome and Transcriptome Analysis of the Flavonoid Biosynthetic Pathway under Salt Stress

To investigate the functions of DEGs in the flavonoid pathway, we calculated the Pearson's correlation coefficients (r) between the FPKM values of the DEGs and peak areas of DAMs involved in flavonoid biosynthesis during salt stress (Table 2). The levels of the flavonol metabolites quercetin and pinoquercetin were positively correlated with *4CL5* and negatively correlated with *FLS* (*FLS-3*) expression, suggesting that *4CL5* and *FLS-3* genes might be the key genes involved in flavonol accumulation in blueberries under salt stress. Among the five flavonol glycosides, the levels of myricetin-3-o-galactoside, kaempferol 3-O-beta-D-xyloside, and nicotiflorin were negatively correlated with *FLS* and *F3'5'H* expression; the level of kaempferol 3-O-arabinoside was negatively correlated with *FLS* expression; and the level of quercetin-3-neohesperidoside-7-rhamnoside was negatively correlated with *PAL*, *C4H*, *4CL1*, and *CHS* expression and positively correlated with *4CL9*, *F3'5'H*, and *FLS* expression. These results indicate that *FLS* and *F3'5'H* genes play important roles in flavonol glycoside biosynthesis.

Epicatechin (flavan) and prodelphinidin C2 (proanthocyanidin) levels were positively correlated with *4CL1* and *LAR-3* expression and negatively correlated with ANR expression, indicating that *4CL1* and *LAR-3* are the key genes for epicatechin and prodelphinidin C2 biosynthesis under salt stress. Cyanidin, an anthocyanidin metabolite, was negatively correlated with *C4H*, *4CL7*, *CHS-2*, and *F3'5'H-1* expression. Among the six anthocyanin metabolites, *UFGT* was negatively correlated with cyanidin 3-O-beta-D-sambubioside and cyanidin-3-o-rhamnoside and positively correlated with cyanidin-3-O-alpha-arabinoside. Cyanidin 3-arabinoside was significantly correlated with most flavonoid biosynthesis genes, including *PAL*, *C4H*, *4CL1*, *4CL7*, *CHS*, *F3'H*, *F3'5'H*, and *ANR*. Cyanidin 3-O-glucoside was negatively correlated with *C4H*, *4CL7*, and *F3'5'H* (Table 2). These results indicate that salt stress regulates anthocyanin biosynthesis in blueberry leaves by upregulating or downregulating the expression of genes in the anthocyanin biosynthesis pathway.

Table 2. Pearson's correlation coefficients (r) between differentially accumulated metabolites (DAMs) and differentially expressed genes (DEGs) from the flavonoid metabolite pathway in response to salt stress.

	Flavonol			Flavonol Glycosides				Flavan	Proanthocyanidin Anthocyanindin			Anthocyanin				
	Que	Pin	Myr-Gal	Kae-Ara	Kae-Xyl	Que-Neo-Rha	Nic	Epi	Pro C2	Cya	Cya-Ara	Cya-Sam	Cya-Glu	Cya-a-Ara	Del-Glu-Chl	Cya-Rha
PAL-3	--	--	--	--	--	-0.895 *	--	--	--	--	--	--	--	--	--	--
PAL-4	--	--	--	--	--	-0.902 *	--	--	--	--	-0.899 *	--	--	--	--	--
PAL-5	--	--	--	--	--	-0.886 *	--	--	--	--	-0.894 *	--	--	--	--	--
C4H	--	--	--	--	--	-0.878 *	--	--	--	-0.912 *	-0.967 **	--	0.944 *	--	--	--
4CL1	--	--	--	--	--	-0.989 **	--	0.909 *	0.907 *	--	-0.928 *	--	--	--	--	--
4CL5	0.932 *	0.930 *	--	--	--	--	--	--	--	--	--	--	--	--	--	--
4CL7	--	--	--	--	--	--	--	--	--	-0.949 *	-0.903 *	--	-0.942 *	--	--	--
4CL9	--	--	--	--	--	0.954 *	--	--	-0.893 *	--	--	--	--	--	--	--
CHS-2	--	--	--	--	--	-0.911 *	--	--	--	-0.885 *	-0.898 *	--	--	--	--	--
F3'5'H-1	--	--	--	--	--	--	--	--	--	-0.898 *	--	--	0.879 *	--	--	--
F3'5'H-2	--	--	--	--	--	--	--	--	--	--	--	--	--	--	0.890 *	--
F3'5'H-3	--	--	--	--	-0.947 *	0.982 **	-0.947 *	-0.894 *	--	--	--	--	--	--	--	-0.892 *
F3'5'H-4	--	--	-0.919 *	--	-0.928 *	0.965 **	-0.941 *	--	--	--	0.893 *	--	--	--	--	--
F3'5'H-5	--	--	--	--	-0.892 *	0.942 *	-0.921 *	--	--	--	0.900 *	0.936 *	--	--	--	--
F3'5'H-7	--	--	--	--	--	--	--	0.884 *	--	--	--	--	--	--	--	--
FLS-1	--	--	-0.942 *	--	--	--	--	--	--	--	--	--	--	--	--	-0.909 *
FLS-2	--	--	--	-0.892 *	-0.944 *	0.907 *	-0.921 *	-0.946 *	-0.881 *	--	--	--	--	--	--	--
FLS-3	-0.923 *	-0.929 *	--	--	-0.895 *	0.906 *	-0.938 *	--	--	--	--	--	--	--	--	--
ANR-1	--	--	-0.904 *	--	--	--	--	--	--	--	0.900 *	--	--	--	--	--
ANR-2	--	-0.903 *	--	--	-0.960 **	0.909 *	-0.985 **	-0.885 *	--	--	--	--	--	--	--	-0.899 *
ANR-3	--	--	--	-0.934 *	-0.979 **	--	-0.983 **	-0.943 *	--	--	--	--	-0.915 *	--	--	-0.949 *
LAR-1	--	--	-0.973 **	--	--	--	-0.883 *	--	--	--	--	--	--	--	--	--
LAR-3	--	--	--	0.971 **	0.984 *	-0.946 *	0.946 *	0.990 **	0.969 **	--	--	--	0.882 *	--	--	0.978 *
DFR	--	0.954 *	--	--	--	--	0.895 *	--	--	--	--	--	--	--	--	--
UFGT-1	--	--	--	--	--	--	--	--	--	--	--	--	--	0.879 *	--	--
UFGT-3	--	--	--	-0.977 *	-0.996 **	0.936 *	0.979 **	-0.968 **	-0.945 *	--	--	--	-0.880 *	--	--	-0.985 **
UGT75C1	--	--	0.990 **	--	--	--	--	--	--	--	--	--	--	--	--	--

Que, quercetin; Pin, pinoquercetin; Myr-Gal, myricetin-3-o-galactoside; Kae-Ara, kaempferol 3-O-arabinoside; Kae-Xyl, kaempferol 3-O -beta-D -xyloside; Que-Neo-Rha, quercetin-3-neohesperidoside-7-rhamnoside; Nic, nicotiflorin; Epi, epicatechin; Pro C2, prodelphinidin C2; Cya, cyanidin; Cya-Ara, Cyanidin 3-arabinoside; Cya-Sam, cyanidin 3-O-beta-D-sambubioside; Cya-Glu, cyanidin 3-O-glucoside; Cya-a-Ara, cyanidin-3-O-alpha-arabinoside; Del-Glu-Chl, delphinidin-3-O-glucoside chloride, Cya-Rha, and cyanidin-3-o-rhamnoside. * Correlation significant at the 0.05 level; ** Correlation significant at the 0.01 level; -- The correlation is not significant.

4. Discussion

4.1. Salt Stress Regulates the Accumulation of Phenylpropanoid and Flavonoid Metabolites

Phenylpropanoid metabolites are beneficial to human health, and they also contribute to abiotic stress tolerance in plants [30–32]. Cinnamic acid and its derivatives are important phenylpropanoid metabolites that can improve the adaptabilities of plants to salt stress [33,34]. In the current study, the levels of cinnamic acid and its derivatives, including 2-hydroxycinnamic acid, 4-hydroxycinnamic acid, calceolarioside, rosmarinic acid, caffeic acid, and caffeoylshikimic acid, increased in blueberry leaves in response to salt stress (Table S11 and Figure 7). These results suggest that blueberry leaves may mitigate damage from salt stress by promoting the accumulation of cinnamic acid and its derivatives.

Flavonoids are a large group of phenylpropanoid metabolites that are widespread among plants. Flavonols, flavans, anthocyanins, and proanthocyanidins are the main flavonoids in the leaves of *Vaccinium* species [35,36]. Many studies have shown that flavonoid accumulation is strongly influenced by salt stress [37–39]. For example, salt stress promoted the accumulation of total flavonoids, quercetin-3-O-glucuronide, quercetin-3-O-galactoside, isoquercitrin, kaempferol-3-O-glucuronide, quercetin-3-O-rhamnoside, and kaempferol-3-O-rhamnoside and affected the accumulation of pelargonidin, dihydrokaempferol, kaempferol, astragalin, quercetin, quercitrin, isoquercitrin, rutin, epicatechin, and cyanidin in *Cyclocarya paliurus* leaves [6]. Salt stress also upregulated the levels of quercetin in the root tissues of S. alopecuroides [15]. Saline-alkali stress increased the accumulation of dihydroquercetin, dihydromyricetin, delphinidin, cyanidin, and cyanidin-3-O-malonylhexoside and decreased the accumulation of kaempferol and rutin in sorghum (*Sorghum bicolor*) leaves [40].

Here, we determined that salt stress significantly promoted the accumulation of flavonoid metabolites in blueberry leaves (Table S11 and Figure 7), in which quercetin increased 1.85-fold after 12 h of salt treatment and pinoquercetin increased 2-fold and 1.52-fold after 12 h and 24 h of salt treatment, respectively, relative to the 0 h control. Kaempferol 3,7-diglucoside and kaempferol 3-O-beta-D-xyloside increased 2.03-fold and 1.56-fold after 12 h of salt treatment relative to the 0 h control, respectively. Kaempferol 3-O-arabinoside increased 1.71-fold at 12 h and 1.52-fold at 48 h; myricetin-3-o-galactoside increased 1.60-fold at 6 h, 1.17-fold at 12 h, and 1.58-fold at 48 h under salt stress compared to 0 h control. Epicatechin increased 2.99-fold, 2.44-fold, and 2.44-fold after 12 h, 24 h, and 48 h, respectively, and prodelphinidin C2 increased 2.23-fold after 24 h of salt treatment relative to the 0 h control. These results indicated that most flavonoid metabolites of blueberry leaves were upregulated during salt stress, especially at 24 h of salt stress compared to 0 h control.

Flavonols include quercetin, myricetin, kaempferol, rutin, and isorhamnetin as well as their derivatives (primarily glycosides); quercetin is the predominant flavonol in the fruits and leaves of *Vaccinium* species [33–35]. Proanthocyanidin, catechin, and epicatechin are the predominant flavan-3-ols in blueberry [41,42]. Some reports suggest that flavonoids can improve salt tolerance by removing excess ROS in soybeans (*Glycine max*), and the accumulation of flavonoids also enhanced salt tolerance in Arabidopsis [43,44]. In addition, quercetin can minimize salt-induced toxicity in tomato [9]. Thus, quercetin, pinoquercetin, kaempferol glycosides, epicatechin, and prodelphinidin C2 may play important roles in alleviating salt stress in blueberry leaves.

The most common anthocyanidins in blueberries are cyanidin, delphinidin, petunidin, peonidin, and malvidin. Most anthocyanins are glycosylated with glucopyranose, galactopyranose, or arabinopyranose at position 3 of the C-ring [45–47]. In the current study, the contents of cyanidin and some anthocyanins (cyanidin 3-arabinoside, cyanidin 3-O-glucoside, cyanidin-3-O-alpha-arabinoside, and delphinidin 3-rutinoside) were downregulated, and some anthocyanins (cyanidin 3-O-beta-D-sambubioside, cyanidin-3-O-rhamnoside, and delphinidin-3-O-glucoside chloride) were upregulated under salt stress (Table S11 and Figure 7). High anthocyanin content in plants is potentially a crucial physiological characteristic that enhances plant tolerance to salt stress [48]. For example, high anthocyanin accumulation enhances tolerance to high salinity environments in Arabidopsis [11,49]. The increasing of anthocyanin content in transgenic *Brassica napus* plants also enhanced the tolerance of transgenic plants to salt stress [50]. Thus, various anthocyanins may play different roles in the response of blueberry to salt stress, and cyanidin 3-O-beta-D-sambubioside, cyanidin-3-O-rhamnoside, and delphinidin-3-O-glucoside chloride may be able to alleviate the damage of salt stress on blueberry leaves.

4.2. Salt Stress Regulates the Expression of Genes from the Phenylpropanoid and Flavonoid Pathways

Plant responses to salt stress are accompanied by changes in the expression patterns of numerous genes. Many studies have been performed on the responses of the phenylpropanoid and flavonoid biosynthetic pathways to salt stress in plants using transcriptomic approaches [15,26,51]. Here, we determined that the phenylpropanoid biosynthesis and flavonoid biosynthesis pathways were significantly enriched in blueberry leaves under salt stress (Table S9 and Figure 5c). Genes from the phenylpropanoid biosynthesis pathway and the flavonoid biosynthesis pathway also contribute to plant tolerance to drought and cold stress in blueberries [52,53].

PAL, the first enzyme of the phenylpropanoid biosynthesis pathway, produces cinnamic acid, and C4H and 4CL subsequently produce 4-coumaric acid. These enzymes lead to the production of downstream phenylpropanoid metabolites, including cinnamic acids and derivatives, coumarins and derivatives, and flavonoids. Several studies have shown that *PAL* and *C4H* are upregulated in response to salt stress, leading to a switch to the biosynthesis of other secondary metabolites in various plants [15,39,54]. Indeed, in the present study, *PAL* and *C4H* expression was also promoted by salt stress in blueberries, indicating that PAL and C4H play important roles in the accumulation of phenylpropanoid metabolites (Table S12 and Figure 7). The 4CL enzymes are present in multiple isoforms, and genes encoding 4CL isoforms are upregulated or downregulated in response to abiotic stress [55–57]. In the leaves of Persian walnut (Juglans regia) and Manchurian walnut (*Juglans mandshurica*), 75% of *Jr4CLs* and 78.95% of *Jm4CLs* showed increased expression, and 25% of *Jr4CLs* and 21.05% of *Jm4CLs* showed decreased expression in response to salt treatment [58]. Similarly, in the current study, most *4CL* genes (*4CL1*, *4CL5*, and *4CL7*) were upregulated and only *4CL9* was downregulated in response to salt stress; similar results were also found in other plants [59]. Thus, *4CLs* might be positively or negatively regulated in response to salt stress, and positive regulation appears to be predominant.

CHS, F3H, F3'H, F3'5'H, and DFR function in the salt stress response and flavonoid biosynthesis [60–63]. Several studies have indicated that salt stress inhibits or promotes the expression of these genes [64–66]. In the current study, all *CHS*, *F3H*, and *DFR* genes, most *F3'H* genes, and some *F3'5'H* genes were upregulated by salt stress in blueberry leaves, suggesting that these genes play important roles in the plant response to salt stress by promoting flavonoid biosynthesis (Table S12 and Figure 7). Some studies showed that the *NtCHS1* promoted flavonoid accumulation and increased salt tolerance, the *CsF3H* gene confers tolerance to salt stress in transgenic tobacco, and the *AtDFR* gene can be effectively manipulated to modulate salinity stress tolerance by directing to high accumulation of anthocyanins in oilseed plants [17,18,50]. FLS is the key enzyme of flavonol biosynthesis, and overexpression of the *FLS* gene increases the salt tolerance in plants [16]. Here, we found that salt stress promoted or inhibited the expression of *FLS* genes (Table S12 and Figure 7). These data suggested that *CHS*, *F3H*, *DFR*, and *FLS* genes play important roles in enhancing salt tolerance by promoting the accumulation of flavonoids in blueberry leaves.

Both *ANR* and *LAR*, the key genes of flavan and proanthocyanidin biosynthesis, are upregulated or downregulated under salt stress [15,67,68]. The expression of *ANS*, *UFGT*, and *UGT*, the key genes of anthocyanin biosynthesis, is also regulated by salt stress in many plants [6,69]. Our study showed that *ANR*, *LAR*, and *UFGT* were upregulated or downregulated, *ANS* was downregulated, and *UGT* was upregulated in blueberry leaves under salt stress (Table S12 and Figure 7). In Arabidopsis, UGT79B2 and UGT79B3 contribute to salt stress tolerance via modulating anthocyanin accumulation [70]. These results indicate that salt stress affects flavonoid biosynthesis by modulating the expression of flavonoid biosynthesis genes, and the *UGT* gene may play an important role in promoting anthocyanin accumulation of blueberry leaves under salt stress.

4.3. The Differentially Expressed Transcription Factors (TFs) in Response to Salt Stress

The transcription factors play a crucial role in stress adaptation [71]. It has been reported that many TFs, including AP2, MYB, bHLH, WRKY, NAC, bZip, and GATA, are involved in the regulation of the salt stress response in plants [72,73]. In this study, we identified 568 TFs that mainly belonged to zinc finger, AP2, MYB, bHLH, and HSP families under salt stress in blueberry leaves (Figure 6). APETALA2/ethylene-responsive factor (AP2/ERF) family transcription factors are well-documented in plant responses to a wide range of biotic and abiotic stresses. Overexpression of tomato AP2/ERF transcription factor *SlERF.B1* increases sensitivity to salt and drought stresses [74]. Overexpression of *SlERF5* in transgenic tomato plants resulted in high tolerance to drought and salt stress and increased levels of relative water content compared with wild-type plants [75]. In rice (*Oryza sativa*), *OsSERF1* gene overexpression improved salinity tolerance [76]. Overexpression of the *Zoysia japonica ZjABR1/ERF10* enhanced tolerances to salt stress in transgenic rice [77]. Here, AP2/ERF TFs, *ERF114*, *ABR1*, *ERF4*, *ERF110*, and *ERF011* were significantly upregulated by 4.924 to 11.758 times during salt stress compared to the 0 h control (Table 1). Therefore, we predict that these TFs may be involved in the blueberry leaf salt tolerance.

The plant bHLHs and MYBs play crucial roles in abiotic stresses, especially salt stress [78–80]. In our study, *bHLH35*, *bHLH123*, *bHLH162*, *MYB102*, *MYB13*, and *MYB14* were significantly upregulated during salt stress. The heterologous expression of the *Anthurium andraeanum AabHLH35* gene in Arabidopsis improved tolerance to cold and drought stresses, and the overexpression of tobacco *NtbHLH123* resulted in a greater resistance to salt stress, while *NtbHLH123*-silenced plants had reduced resistance to salt stress but improved salt tolerance [81,82]. In peanut (*Arachis hypogaea*), the overexpression of *AhbHLH121* improved salt resistance, whereas silencing *AhbHLH121* resulted in the inverse correlation [83]. The heterologous overexpression of the *Vitis amurensis VaMyb14* gene in Arabidospis improves cold and drought tolerance, and *Atmyb102* was upregulated in Arabidospis upon treatment with ABA, JA, or a combined treatment of osmotic stress and wounding [84,85]. Thus, bHLH35, bHLH123, bHLH162, MYB102, MYB13, and MYB14 may be involved in the salt tolerance of blueberry leaves, especially, bHLH123 (NtbHLH123 homolog) transcription factor.

The HSP are a group of proteins found in living organisms and are called stress proteins. The zinc finger proteins are one of the most abundant groups of proteins and have a wide range of molecular functions [86,87]. The heterologous expression of rice *OsHSP20* in *Escherichia coli* or *Pichia pastoris* cells enhanced heat and salt stress tolerance when compared with the control cultures, and Zat12 is thought to be involved in cold and oxidative stress signaling in Arabidopsis [88–90]. In Arabidospis, knockout *AtZat12* plants were more sensitive than wild-type plants to salinity stress [91]. Here, *HSP20* (*OsHSP20* homolog) and *ZAT12* (*AtZAT12* homolog) were induced significant upregulation under salt stress in blueberry leaves. Thus, HSP20 and ZAT12 might be responsible for blueberry leaf salt tolerance.

4.4. Salt Stress Promotes Flavonoid Biosynthesis by Activating the Expression of Flavonoid Biosynthesis Pathway Genes

The flavonoid biosynthesis pathway was enriched among DAMs and DEGs in blueberry leaves under salt stress, as revealed by transcriptome and metabolome analysis. PAL, C4H, and 4CL in the general phenylpropanoid pathway provide the precursors for flavonoid biosynthesis, and the expression of their underlying genes affects the accumulation of flavonoid metabolites [6]. Here, we showed that the expression levels of *4CL* genes were positively correlated with the accumulation of flavonoids, including *4CL-3* with epicatechin and prodelphinidin C2, *4CL5* with flavonols (quercetin and pinoquercetin), and *4CL9* with quercetin-3-neohesperidoside-7-rhamnoside. Epicatechin, prodelphinidin C2, quercetin, and pinoquercetin levels increased under salt stress, and 4CL family members may play key roles in flavonol, flavan, and proanthocyanidin biosynthesis in blueberries under salt stress (Table 2).

The "early" flavonoid biosynthetic genes *CHS, F3H, F3'H,* and *F3'5'H* promote flavonoid biosynthesis. For example, *BdCHS* from *Brachypodium distachyon, VcF3H* from blueberry, *GmF3'H* from soybean, and *ScF3'5'H* from *Senecio cruentus* promote anthocyanin biosynthesis [60,61,92,93]. We determined that the expression levels of *CHS, F3'H,* and *F3'5'H* were significantly correlated with quercetin-3-neohesperidoside-7-rhamnoside and cyanidin 3-arabinoside levels, suggesting that these genes are coordinately regulated to function in the early steps of flavonoid biosynthesis in response to salt stress in blueberry leaves. Moreover, *F3'5'H* expression was positively correlated with quercetin-3-neohesperidoside-7-rhamnoside, epicatechin, cyanidin 3-arabinoside, cyanidin 3-O-beta-D-sambubioside, and delphinidin-3-O-glucoside chloride accumulation and negatively correlated with myricetin-3-o-galactoside, kaempferol 3-O-beta-D-xyloside, nicotiflorin, epicatechin, cyanidin, cyanidin 3-O-glucoside, and cyanidin-3-o-rhamnoside accumulation (Table 2). Thus, these *F3'5'H* genes might be responsive to salt stress and positively or negatively regulate the accumulation of flavonol glycosides, flavans, anthocyanidins, and anthocyanins. Salt stress promoted the accumulation of cyanidin 3-O-beta-D-sambubioside and delphinidin-3-O-glucoside chloride and upregulated the expression of *F3'5'H* genes (*F3'5'H-7* and *F3'5'H-2*), indicating that *F3'5'H* genes might be the key genes of cyanidin 3-O-beta-D-sambubioside and delphinidin-3-O-glucoside chloride biosynthesis in blueberry leaves under salt stress.

The *DFR, ANS,* and *UFGT* genes, which are "late" structural genes in the anthocyanin biosynthetic pathway, are key genes for anthocyanin biosynthesis [94–96]. *FLS* is the key gene responsible for the accumulation of flavonols and flavonol glycosides, and both *LAR* and *ANR* are key genes for flavan and proanthocyanidin biosynthesis [97–99]. In this study, transcriptome and metabolome analyses showed that *UFGT-1* expression was positively correlated with cyanidin-3-O-alpha-arabinoside accumulation and *UFGT-3* expression was negatively correlated with cyanidin 3-O-beta-D-sambubioside and cyanidin-3-o-rhamnoside accumulation in blueberry under salt stress. Most flavonols or flavonol glycosides were negatively correlated with FLS family genes, and only quercetin-3-neohesperidoside-7-rhamnoside was positively correlated with *FLS. ANRs* were negatively correlated with epicatechin, and *LAR-3* were positively correlated with epicatechin and prodelphinidin C2 (Table 2). Thus, these genes positively or negatively regulate anthocyanin, flavonol, flavan, or proanthocyanidin biosynthesis under salt stress. We also found that salt stress promoted the accumulation of epicatechin prodelphinidin C2 and increased the expression of *LAR-3* genes, suggesting that *LAR-3* are the key genes for the biosynthesis of epicatechin prodelphinidin C2 in response to salt stress (Tables S11 and S12 and Figure 7).

5. Conclusions

In this study, we conducted metabolomic and transcriptomic analyses for six-month-old blueberry cultivar 'Northland' via UHPLC-MS/MS and RNA-seq, respectively, to explore the underlying molecular mechanism of flavonoid biosynthesis in blueberry leaves in response to salt stress. Metabolomics analysis showed that salt stress significantly promoted the accumulation of flavonols, flavonol glycosides, flavan, proanthocyanidin, and anthocyanins. Transcriptomic analysis indicated that salt stress upregulated the expression of *PALs, C4Hs, 4CL1, 4CL5, 4CL7s, CHSs, F3H,* and *DFRs* and most *F3'5'H, UFGT,* and *ANR* genes. Integrative metabolomic and transcriptomic analysis suggested that *4CL5* and *F3'5'H* are key genes for flavonol and anthocyanin biosynthesis, respectively, and that *4CL1* and *LAR-3* are key genes for flavan-3-ol biosynthesis in the flavonoid biosynthetic pathway in blueberry leaves under salt stress. The bHLH123, OsHSP20, and HSP20 TFs might be responsible for blueberry leaf salt tolerance. These findings reveal the possible molecular mechanism of flavonoid accumulation in blueberries during salt stress, providing a basis for future studies aimed at cultivating salt-resistant blueberry varieties.

Supplementary Materials: The following supporting information can be downloaded at: https://www.mdpi.com/article/10.3390/horticulturae10101084/s1, Table S1: Primers used for qRT-PCR; Table S2: The identified metabolites under salt stress from blueberry leaves. Table S3: The number of metabolites identified in each superclass under salt stress from blueberry leaves; Table S4: Identification of differential metabolites based on FC > 1.5 or FC < 0.67 and *p* value < 0.05 under salt stress; Table S5: The enrichment analysis of KEGG pathways for differential metabolites under salt stress; Table S6: Summary of RNA-seq database; Table S7: Identification of differential expressed genes under salt stress; Table S8: The enrichment analysis of GO terms for differential expressed genes under salt stress; Table S9: The enrichment analysis of KEGG pathways for differential expressed genes under salt stress; Table S10: The differential expressed transcription factors under salt stress; Table S11: Identification of differential metabolites from phenylpropanoid metabolite pathway based on FC > 1.5 or FC < 0.67 and *p*-value < 0.05 under salt stress; Table S12: The differential expressed genes from phenylpropanoid metabolite pathway under salt stress; Figure S1: Principal component analysis (PCA) of the peaks obtained from all experimental samples and quality control (QC) samples. S0, S6, S12, S24, and S48 represent that 200 mM NaCl treated the samples for 0, 6, 12, 24, and 48 h, respectively; Figure S2: Pearson correlation coefficients between quality control samples (QC) samples detected in negative ion mode and positive ion mode. (a) MultiScatter negative ion mode; (b) MultiScatter positive ion mode. The points in each small grid represent the ion peaks (metabolites) extracted from QC samples; Figure S3: Volcano plots of all metabolites detected in positive and negative ion mode (including unidentified metabolites). Metabolites with FC > 1.5 or FC < 0.67 and a *p*-value < 0.05 were considered to be differentially accumulated. S0, S6, S12, S24, and S48 represent the 200 mM NaCl-treated samples for 0, 6, 12, 24, and 48 h, respectively; Figure S4: Validation of transcript expression changes by qRT-qPCR. The figure is based on $\log_2(2^{-\Delta\Delta Ct})$ data from qRT-PCR and $\log_2$(fold change) data from RNA-seq. The linear trend line and the R^2 are shown.

Author Contributions: Conceptualization, Q.G., C.Z. and B.M.; methodology, Y.S.; software, B.M., Y.S. and X.F.; validation, Y.S. and P.G.; formal analysis, B.M. and X.F.; investigation, S.J.; resources, Q.G.; data curation, L.Z.; writing—original draft preparation, B.M.; writing—review and editing, Q.G. and C.Z.; visualization, L.Z.; supervision, Q.G. and C.Z.; project administration, C.Z. and Q.G.; funding acquisition, C.Z. All authors have read and agreed to the published version of the manuscript.

Funding: This study was supported by the National Natural Science Foundation of China (grant number 31700260).

Data Availability Statement: Data are contained within the article and Supplementary Materials. We have uploaded the RNA-Seq data generated in this study to the BioProject in the NCBI repository with the following accession number: PRJNA1128395.

Conflicts of Interest: The authors declare no conflicts of interest.

References

1. Hao, S.; Wang, Y.; Yan, Y.; Liu, Y.; Wang, J.; Chen, S. A Review on Plant Responses to Salt Stress and Their Mechanisms of Salt Resistance. *Horticulturae* **2021**, *7*, 132. [CrossRef]
2. Parihar, P.; Singh, S.; Singh, R.; Singh, V.P.; Prasad, S.M. Effect of salinity stress on plants and its tolerance strategies: A review. *Environ. Sci. Pollut. R.* **2015**, *22*, 4056–4075. [CrossRef]
3. Guo, M.; Wang, X.; Guo, H.; Bai, S.; Khan, A.; Wang, X.; Gao, Y.; Li, J. Tomato salt tolerance mechanisms and their potential applications for fighting salinity: A review. *Front. Plant Sci.* **2022**, *13*, 949541. [CrossRef]
4. Gupta, B.; Huang, B. Mechanism of salinity tolerance in plants: Physiological, biochemical, and molecular characterization. *Int. J. Genom.* **2014**, *2014*, 701596. [CrossRef]
5. Van Zelm, E.; Zhang, Y.; Testerink, C. Salt Tolerance Mechanisms of Plants. *Annu. Rev. Plant Biol.* **2020**, *71*, 403–433. [CrossRef]
6. Zhang, L.; Zhang, Z.; Fang, S.; Liu, Y.; Shang, X. Integrative analysis of metabolome and transcriptome reveals molecular regulatory mechanism of flavonoid biosynthesis in *Cyclocarya paliurus* under salt stress. *Ind. Crops Prod.* **2021**, *170*, 113823. [CrossRef]
7. Lin, S.; Zeng, S.; Biao, A.; Yang, X.; Yang, T.; Zheng, G.; Mao, G.; Wang, Y. Integrative analysis of transcriptome and metabolome reveals salt stress orchestrating the accumulation of specialized metabolites in *Lycium barbarum* L. Fruit. *Int. J. Mol. Sci.* **2021**, *22*, 4414. [CrossRef]

8.   Vogt, T. Phenylpropanoid biosynthesis. *Mol. Plant* **2010**, *3*, 2–20. [CrossRef]
9.   Parvin, K.; Hasanuzzaman, M.; Bhuyan, M.; Mohsin, S.M.; Fujita, A.M. Quercetin mediated salt tolerance in tomato through the enhancement of plant antioxidant defense and glyoxalase systems. *Plants* **2019**, *8*, 247. [CrossRef]
10.   Yiu, J.; Tseng, M.; Liu, C.; Kuo, C. Modulation of NaCl stress in *Capsicum annuum* L. seedlings by catechin. *Sci. Hortic.* **2012**, *134*, 200–209. [CrossRef]
11.   Truong, H.A.; Lee, W.J.; Jeong, C.Y.; Trịnh, C.S.; Lee, S.; Kang, C.S.; Cheong, Y.K.; Hong, S.W.; Lee, H. Enhanced anthocyanin accumulation confers increased growth performance in plants under low nitrate and high salt stress conditions owing to active modulation of nitrate metabolism. *J. Plant Physiol.* **2018**, *231*, 41–48. [CrossRef] [PubMed]
12.   Guo, B.; Chen, F.; Liu, G.; Li, W.; Li, W.; Zhuang, J.; Zhang, X.; Wang, L.; Lei, B.; Hu, C.; et al. Effects and mechanisms of proanthocyanidins-derived carbon dots on alleviating salt stress in rice by muti-omics analysis. *Food Chem. X* **2024**, *22*, 101422. [CrossRef] [PubMed]
13.   Jaakola, L.; Määttä, K.; Pirttilä, A.M.; Törrönen, R.; Kärenlampi, S.; Hohtola, A. Expression of genes involved in anthocyanin biosynthesis in relation to anthocyanin, proanthocyanidin, and flavonol levels during bilberry fruit development. *Plant Physiol.* **2002**, *130*, 729–739. [CrossRef] [PubMed]
14.   Zhang, Q.; Wang, S.; Qin, B.; Sun, H.; Yuan, X.; Wang, Q.; Xu, J.; Yin, Z.; Du, Y.; Du, J.; et al. Analysis of the transcriptome and metabolome reveals phenylpropanoid mechanism in common bean (*Phaseolus vulgaris*) responding to salt stress at sprout stage. *Food Energy Secur.* **2023**, *12*, e481. [CrossRef]
15.   Zhu, Y.; Wang, Q.; Wang, Y.; Xu, Y.; Li, J.; Zhao, S.; Wang, D.; Ma, Z.; Yan, F.; Liu, Y. Combined transcriptomic and metabolomic analysis reveals the role of phenylpropanoid biosynthesis pathway in the salt tolerance process of *Sophora alopecuroides*. *Int. J. Mol. Sci.* **2021**, *22*, 2399. [CrossRef]
16.   Wang, M.; Ren, T.; Huang, R.; Li, Y.; Zhang, C.; Xu, Z. Overexpression of an Apocynum venetum flavonols synthetase gene confers salinity stress tolerance to transgenic tobacco plants. *Plant Physiol. Biochem.* **2021**, *162*, 667–676. [CrossRef]
17.   Chen, S.; Wu, F.; Li, Y.; Qian, Y.; Pan, X.; Li, F.; Wang, Y.; Wu, Z.; Fu, C.; Lin, H.; et al. NtMYB4 and NtCHS1 are critical factors in the regulation of flavonoid biosynthesis and are involved in salinity responsiveness. *Front. Plant Sci.* **2019**, *10*, 178. [CrossRef]
18.   Mahajan, M.; Yadav, S.K. Overexpression of a tea flavanone 3-hydroxylase gene confers tolerance to salt stress and alternaria solani in transgenic tobacco. *Plant Mol. Biol.* **2014**, *85*, 551–573. [CrossRef]
19.   Norberto, S.; Silva, S.; Meireles, M.; Faria, A.; Pintado, M.; Calhau, C. Blueberry anthocyanins in health promotion: A metabolic overview. *J. Funct. Foods.* **2013**, *5*, 1518–1528. [CrossRef]
20.   Ribera, A.E.; Reyes-Diaz, M.; Alberdi, M.; Zuñiga, G.E.; Mora, M.L. Antioxidant compounds in skin and pulp of fruits change among genotypes and maturity stages in highbush blueberry (*Vaccinium corymbosum* L.) grown in southern Chile. *J. Soil Sci. Plant Nutr.* **2010**, *10*, 509–536. [CrossRef]
21.   Rodriguez-Saona, C.; Vincent, C.; Isaacs, R. Blueberry IPM: Past successes and future challenges. *Annu. Rev. Entomol.* **2019**, *64*, 95–114. [CrossRef] [PubMed]
22.   Muralitharan, M.S.; Chandler, S.; Van, S.R. Effects of NaCl and Na$_2$SO$_4$ on growth and solute composition of highbush blueberry (*Vaccinium corymbosum*). *Funct. Plant Biol.* **1992**, *19*, 155–164. [CrossRef]
23.   Bryla, D.R.; Scagel, C.F.; Lukas, S.B.; Sullivan, D.M. Ion-specific limitations of sodium chloride and calcium chloride on growth, nutrient uptake, and mycorrhizal colonization in northern and southern highbush blueberry. *J. Am. Soc. Hort. Sci.* **2021**, *146*, 399–410. [CrossRef]
24.   Molnar, S.; Clapa, D.; Pop, V.C.; Hârța, M.; Andrecan, F.A.; Bunea, C.I. Investigation of salinity tolerance to different cultivars of highbush blueberry (*Vaccinium corymbosum* L.) grown in vitro. *Not. Bot. Horti Agrobot.* **2024**, *52*, 13691. [CrossRef]
25.   Wright, G.C.; Patten, K.D.; Drew, M.C. Salinity and supplemental calcium influence growth of rabbiteye and southern highbush blueberry. *J. Am. Soc. Hortic. Sci.* **1992**, *117*, 749–756. [CrossRef]
26.   Gan, T.; Lin, Z.; Bao, L.; Hui, T.; Cui, X.; Huang, Y.; Wang, H.; Su, C.; Jiao, F.; Zhang, M.; et al. Comparative proteomic analysis of tolerant and sensitive varieties reveals that phenylpropanoid biosynthesis contributes to salt tolerance in mulberry. *Int. J. Mol. Sci.* **2021**, *22*, 9402. [CrossRef]
27.   Wu, Q.; Bai, X.; Zhao, W.; Xiang, D.; Wan, Y.; Yan, J.; Zou, L.; Zhao, G. De novo assembly and analysis of tartary buckwheat (*Fagopyrum tataricum* Garetn.) transcriptome discloses key regulators involved in salt-stress response. *Genes* **2017**, *8*, 255. [CrossRef]
28.   Pfaffl, M.W. A new mathematical model for relative quantification in real-time RT-PCR. *Nucl. Acids Res.* **2001**, *29*, e45. [CrossRef]
29.   Chen, C.; Chen, H.; Zhang, Y.; Thomas, H.R.; Frank, M.H.; He, Y.; Xia, R. TBtools: An integrative toolkit developed for interactive analyses of big biological data. *Mol. Plant* **2020**, *13*, 1194–1202. [CrossRef]
30.   Sellappan, S.; Akoh, C.C.; Krewer, G. Phenolic compounds and antioxidant capacity of Georgia-grown blueberries and blackberries. *J. Agric. Food Chem.* **2002**, *50*, 2432–2438. [CrossRef]
31.   Giovanelli, G.; Buratti, S. Comparison of polyphenolic composition and antioxidant activity of wild Italian blueberries and some cultivated varieties. *Food Chem.* **2009**, *112*, 903–908. [CrossRef]
32.   Zhang, H.; Han, B.; Wang, T.; Chen, S.; Li, H.; Zhang, Y.; Dai, S. Mechanisms of plant salt response: Insights from proteomics. *J. Proteome Res.* **2012**, *11*, 49–67. [CrossRef] [PubMed]
33.   Häkkinen, S.H.; Törrönen, A.R. Content of flavonols and selected phenolic acids in strawberries and Vaccinium species: Influence of cultivar, cultivation site and technique. *Food Res. Int.* **2000**, *33*, 517–524. [CrossRef]

34. Riihinen, K.; Jaakola, L.; Kärenlampi, S.; Hohtola, A. Organ-specific distribution of phenolic compounds in bilberry (*Vaccinium myrtillus*) and 'northblue' blueberry (*Vaccinium corymbosum* × *V. angustifolium*). *Food Chem.* **2008**, *110*, 156–160. [CrossRef]

35. Raudone, L.; Vilkickyte, G.; Pitkauskaite, L.; Raudonis, R.; Vainoriene, R.; Motiekaityte, V. Antioxidant actvities of *Vaccinium vitis-idaea* L. leaves within cultivars and their phenolic compounds. *Molecules* **2019**, *24*, 844. [CrossRef]

36. Vyas, P.; Kalidindi, S.; Chibrikova, L.; Igamberdiev, A.U.; Weber, J.T. Chemical analysis and effect of blueberry and lingonberry fruits and leaves against glutamate-mediated excitotoxicity. *J. Agric. Food Chem.* **2013**, *61*, 7769–7776. [CrossRef]

37. Ren, G.; Yang, P.; Cui, J.; Gao, Y.; Yin, C.; Bai, Y.; Zhao, D.; Chang, J. Multiomics analyses of two sorghum cultivars reveal the molecular mechanism of salt tolerance. *Front. Plant Sci.* **2022**, *13*, 886805. [CrossRef]

38. Ma, W.; Kim, J.K.; Jia, C.; Yin, F.; Kim, H.J.; Akram, W.; Hu, X.; Li, X. Comparative transcriptome and metabolic profiling analysis of buckwheat (*Fagopyrum tataricum* (L.) Gaertn.) under salinity stress. *Metabolites* **2019**, *9*, 225. [CrossRef]

39. Abdallah, S.B.; Aung, B.; Amyot, L.; Lalin, I.; Lachal, M.; Karray-Bouraou, N.; Hannoufa, A. Salt stress (NaCl) affects plant growth and branch pathways of carotenoid and flavonoid biosyntheses in *Solanum nigrum*. *Acta Physiol. Plant* **2016**, *38*, 72. [CrossRef]

40. Ma, S.; Lv, L.; Meng, C.; Zhang, C.; Li, Y. Integrative analysis of the metabolome and transcriptome of *Sorghum bicolor* reveals dynamic changes in flavonoids accumulation under saline-alkali stress. *J. Agric. Food Chem.* **2020**, *68*, 14781–14789. [CrossRef]

41. Li, X.; Jin, L.; Pan, X.; Yang, L.; Guo, W. Proteins expression and metabolite profile insight into phenolic biosynthesis during highbush blueberry fruit maturation. *Food Chem.* **2019**, *290*, 216–228. [CrossRef] [PubMed]

42. Yang, B.; Song, Y.; Li, Y.; Wang, X.; Guo, Q.; Zhou, L.; Zhang, Y.; Zhang, C. Key genes for phenylpropanoid metabolite biosynthesis during half-highbush blueberry (*Vaccinium angustifolium* × *Vaccinium corymbosum*) fruit development. *J. Berry Res.* **2022**, *12*, 297–311. [CrossRef]

43. Pi, E.; Zhu, C.; Fan, W.; Huang, Y.; Qu, L.; Li, Y.; Zhao, Q.; Ding, F.; Qiu, L.; Wang, H.; et al. Quantitative phosphoproteomic and metabolomic analyses reveal GmMYB173 optimizes flavonoid metabolism in soybean under salt stress. *Mol. Cell. Proteom.* **2018**, *17*, 1209–1224. [CrossRef] [PubMed]

44. Wang, F.; Zhu, H.; Kong, W.; Peng, R.; Liu, Q.; Yao, Q. The *Antirrhinum* AmDEL gene enhances flavonoids accumulation and salt and drought tolerance in transgenic Arabidopsis. *Planta* **2016**, *244*, 59–73. [CrossRef] [PubMed]

45. Bunea, A.; Rugină, D.; Sconța, Z.; Pop, R.M.; Pintea, A.; Socaciu, C.; Tăbăran, F.; Grootaert, C.; Struijs, K.; VanCamp, J. Anthocyanin determination in blueberry extracts from various cultivars and their antiproliferative and apoptotic properties in B16-F10 metastatic murine melanoma cells. *Phytochemistry* **2013**, *95*, 436–444. [CrossRef]

46. Müller, D.; Schantz, M.; Richling, E. High performance liquid chromatography analysis of anthocyanins in bilberries (*Vaccinium myrtillus* L.), blueberries (*Vaccinium corymbosum* L.), and corresponding juices. *J. Food Sci.* **2012**, *77*, C340–C345. [CrossRef]

47. Gavrilova, V.; Kajdzanoska, M.; Gjamovski, V.; Stefova, M. Separation, characterization and quantification of phenolic compounds in blueberries and red and black currants by HPLC-DAD-ESI-MSn. *J. Agric. Food Chem.* **2011**, *59*, 4009–4018. [CrossRef]

48. Naing, A.H.; Kim, C.K. Abiotic stress-induced anthocyanins in plants: Their role in tolerance to abiotic stresses. *Physiol. Plant* **2021**, *172*, 1711–1723. [CrossRef]

49. Sun, M.; Feng, X.; Gao, J.; Peng, R.; Yao, Q.; Wang, L. VvMYBA6 in the promotion of anthocyanin biosynthesis and salt tolerance in transgenic Arabidopsis. *Plant Biotechnol. Rep.* **2017**, *11*, 299–314. [CrossRef]

50. Kim, J.; Lee, W.J.; Vu, T.T.; Jeong, C.Y.; Hong, S.W.; Lee, H. High accumulation of anthocyanins via the ectopic expression of AtDFR confers significant salt stress tolerance in *Brassica napus* L. *Plant Cell Rep.* **2017**, *36*, 1215–1224. [CrossRef]

51. Jia, C.; Guo, B.; Wang, B.; Li, X.; Yang, T.; Li, N.; Wang, J.; Yu, Q. Integrated metabolomic and transcriptomic analysis reveals the role of phenylpropanoid biosynthesis pathway in tomato roots during salt stress. *Front. Plant Sci.* **2022**, *13*, 1023696. [CrossRef] [PubMed]

52. Wang, A.; Wang, L.; Liu, K.; Liang, K.; Yang, S.; Cao, Y.; Zhang, L. Comparative transcriptome profiling reveals the defense pathways and mechanisms in the leaves and roots of blueberry to drought stress. *Fruit Res.* **2022**, *2*, 18. [CrossRef]

53. Zhang, F.; Ji, S.; Wei, B.; Cheng, S.; Wang, Y.; Hao, J.; Wang, S.; Zhou, Q. Transcriptome analysis of postharvest blueberries (*Vaccinium corymbosum* 'Duke') in response to cold stress. *BMC Plant Biol.* **2020**, *20*, 80. [CrossRef] [PubMed]

54. Li, C.; Qi, Y.; Zhao, C.; Wang, X.; Zhang, Q. Transcriptome profiling of the salt stress response in the leaves and roots of halophytic *Eutrema salsugineum*. *Front. Genet.* **2021**, *12*, 770742. [CrossRef]

55. Lavhale, S.G.; Kalunke, R.M.; Giri, A.P. Structural, functional and evolutionary diversity of 4-coumarate-CoA ligase in plants. *Planta.* **2018**, *248*, 1063–1078. [CrossRef]

56. Lavhale, S.G.; Joshi, R.S.; Kumar, Y.; Giri, A.P. Functional insights into two *Ocimum kilimandscharicum* 4-coumarate-CoA ligases involved in phenylpropanoid biosynthesis. *Int. J. Biol. Macromol.* **2021**, *181*, 202–210. [CrossRef]

57. Gao, S.; Liu, X.; Ni, R.; Fu, J.; Tan, H.; Cheng, A.; Lou, H. Molecular cloning and functional analysis of 4-coumarate: CoA ligases from *Marchantia paleacea* and their roles in lignin and flavanone biosynthesis. *PLoS ONE* **2024**, *19*, e0296079. [CrossRef]

58. Ma, J.; Zuo, D.; Zhang, X.; Li, H.; Ye, H.; Zhang, N.; Li, M.; Dang, M.; Geng, F.; Zhou, H.; et al. Genome-wide identification analysis of the 4-Coumarate: CoA ligase (*4CL*) gene family expression profiles in *Juglans regia* and its wild relatives *J. Mandshurica* resistance and salt stress. *BMC Plant Biol.* **2024**, *24*, 211. [CrossRef]

59. Zhang, C.; Ma, T.; Luo, W.; Xu, J.; Liu, J.; Wan, D. Identification of *4CL* genes in desert poplars and their changes in expression in response to salt stress. *Genes* **2015**, *6*, 901–917. [CrossRef]

60. Wang, J.; Wang, X.; Zhao, S.; Xi, X.; Feng, J.; Han, R. *Brachypodium* BdCHS is a homolog of *Arabidopsis* AtCHS involved in the synthesis of flavonoids and lateral root development. *Protoplasma* **2023**, *260*, 999–1003. [CrossRef]

61. Huang, H.; Hu, K.; Han, K.; Xiang, Q.; Dai, S. Flower colour modification of chrysanthemum by suppression of *F3'H* and overexpression of the exogenous *Senecio cruentus F3'5'H* gene. *PLoS ONE* **2013**, *8*, e74395. [CrossRef]
62. Baba, S.A.; Ashraf, N. Functional characterization of flavonoid 3'-hydroxylase, CsF3'H, from *Crocus sativus* L: Insights into substrate specificity and role in abiotic stress. *Arch. Biochem. Biophys.* **2019**, *667*, 70–78. [CrossRef] [PubMed]
63. Ma, L.; Jia, W.; Duan, Q.; Du, W.; Li, X.; Cui, G.; Wang, X.; Wang, J. Heterologous expression of *Platycodon grandiflorus PgF3'5'H* modifies flower color pigmentation in tobacco. *Genes* **2023**, *14*, 1920. [CrossRef] [PubMed]
64. Isıyel, M.; İlhan, E.; Kasapoğlu, A.G.; Muslu, S.; Öner, B.M.; Aygören, A.S.; Yiğider, E.; Aydın, M.; Yıldırım, E. Identification and characterization of *Phaseolus vulgaris CHS* genes in response to salt and drought stress. *Genet. Resour. Crop Evol.* **2024**, *9*, 189. [CrossRef]
65. Wang, J.; Lan, Z.; Wang, H.; Xu, C.; Zhou, Z.; Cao, J.; Liu, Y.; Sun, Z.; Mu, D.; Han, J.; et al. Genome-wide identification of chalcone synthase (CHS) family members and their expression patterns at the sprouting stage of common bean (*Phaseolus vulgaris*) under abiotic stress. *Sci. Hortic.* **2024**, *334*, 113309. [CrossRef]
66. Martinez, V.; Mestre, T.C.; Rubio, F.; Girones-Vilaplana, A.; Moreno, D.A.; Mittler, R.; Rivero, R.M. Accumulation of flavonols over hydroxycinnamic acids favors oxidative damage protection under abiotic stress. *Front. Plant Sci.* **2016**, *7*, 838. [CrossRef]
67. Wang, Y.; Jiang, W.; Li, C.; Wang, Z.; Lu, C.; Cheng, J.; Wei, S.; Yang, J.; Yang, Q. Integrated transcriptomic and metabolomic analyses elucidate the mechanism of flavonoid biosynthesis in the regulation of mulberry seed germination under salt stress. *BMC Plant Biol.* **2024**, *24*, 132. [CrossRef]
68. Sui, D.; Wang, B.; El-Kassaby, Y.A.; Wang, L. Integration of physiological, transcriptomic, and metabolomic analyses reveal molecular mechanisms of salt stress in *Maclura tricuspidata*. *Plants* **2024**, *13*, 397. [CrossRef]
69. Saad, K.R.; Kumar, G.; Mudliar, S.N.; Giridhar, P.; Shetty, N.P. Salt stress-induced anthocyanin biosynthesis genes and MATE transporter involved in anthocyanin accumulation in *Daucus carota* cell culture. *ACS Omega* **2021**, *6*, 24502–24514. [CrossRef]
70. Li, P.; Li, Y.; Zhang, F.; Zhang, G.; Jiang, X.; Yu, H.; Hou, B. The Arabidopsis UDP-glycosyltransferases UGT79B2 and UGT79B3, contribute to cold, salt and drought stress tolerance via modulating anthocyanin accumulation. *Plant J.* **2017**, *89*, 85–103. [CrossRef]
71. Singh, K.B.; Foley, R.C.; Oñate-Sánchez, L. Transcription factors in plant defense and stress responses. *Curr. Opin. Plant Biol.* **2002**, *5*, 430–436. [CrossRef] [PubMed]
72. Bokolia, M.; Kumar, A.; Singh, B. Plant tolerance to salinity stress: Regulating transcription factors and their functional role in the cellular transcriptional network. *Gene Rep.* **2024**, *34*, 101873. [CrossRef]
73. Golldack, D.; Lüking, I.; Yang, O. Plant tolerance to drought and salinity: Stress regulating transcription factors and their functional significance in the cellular transcriptional network. *Plant Cell Rep.* **2021**, *30*, 1383–1391. [CrossRef] [PubMed]
74. Wang, Y.; Xia, D.; Li, W.; Cao, X.; Ma, F.; Wang, Q.; Zhan, X.; Hu, T. Overexpression of a tomato AP2/ERF transcription factor *SlERF.B1* increases sensitivity to salt and drought stresses. *Sci. Hortic.* **2022**, *304*, 111332. [CrossRef]
75. Pan, Y.; Seymour, G.B.; Lu, C.; Hu, Z.; Chen, X.; Chen, G. An ethylene response factor (ERF5) promoting adaptation to drought and salt tolerance in tomato. *Plant Cell Rep.* **2012**, *31*, 349–360. [CrossRef]
76. Schmidt, R.; Mieulet, D.; Hubberten, H.M.; Obata, T.; Hoefgen, R.; Fernie, A.R.; Fisahn, J.; San Segundo, B.; Guiderdoni, E.; Schippers, J.H.; et al. Salt-responsive ERF1 regulates reactive oxygen species-dependent signaling during the initial response to salt stress in rice. *Plant Cell* **2013**, *25*, 2115–2131. [CrossRef]
77. Guo, T.; Wang, S.; Fan, B.; Zou, S.; Chen, S.; Liu, W.; Wang, S.; Ai, L.; Han, L. Overexpression of the *Zoysia japonica ZjABR1/ERF10* regulates plant growth and salt tolerance in transgenic *Oryza sativa*. *Environ. Exp. Bot.* **2023**, *206*, 105117. [CrossRef]
78. Zhang, Z.; Fang, J.; Zhang, L.; Jin, H.; Fang, S. Genome-wide identification of bHLH transcription factors and their response to salt stress in *Cyclocarya paliurus*. *Front. Plant Sci.* **2023**, *14*, 1117246. [CrossRef]
79. Ye, D.; Liu, J.; Tian, X.; Wen, X.; Zhang, Y.; Zhang, X.; Sun, G.; Xia, H.; Liang, D. Genome-wide identification of bHLH gene family and screening of candidate gene in response to salt stress in kiwifruit. *Environ. Exp. Bot.* **2024**, *222*, 105774. [CrossRef]
80. Wang, X.; Niu, Y.; Zheng, Y. Multiple functions of MYB transcription factors in abiotic stress responses. *Int. J. Mol. Sci.* **2021**, *22*, 6125. [CrossRef]
81. Jiang, L.; Tian, X.; Li, S.; Fu, Y.; Xu, J.; Wang, G. The AabHLH35 transcription factor identified from *Anthurium andraeanum* is involved in cold and drought tolerance. *Plants* **2019**, *8*, 216. [CrossRef] [PubMed]
82. Liu, D.; Li, Y.; Zhou, Z.; Xiang, X.; Liu, X.; Wang, J.; Hu, Z.; Xiang, S.; Li, W.; Xiao, Q.; et al. Tobacco transcription factor bHLH123 improves salt tolerance by activating NADPH oxidase *NtRbohE* expression. *Plant Physiol.* **2021**, *186*, 1706–1720. [CrossRef] [PubMed]
83. Zhao, X.; Wang, Q.; Yan, C.; Sun, Q.; Wang, J.; Li, C.; Yuan, C.; Mou, Y.; Shan, S. The bHLH transcription factor *AhbHLH121* improves salt tolerance in peanut. *Int. J. Biol. Macromol.* **2024**, *256*, 128492. [CrossRef]
84. Fang, L.; Wang, Z.; Su, L.; Gong, L.; Xin, H. *Vitis* Myb14 confer cold and drought tolerance by activating lipid transfer protein genes expression and reactive oxygen species scavenge. *Gene* **2024**, *890*, 147792. [CrossRef] [PubMed]
85. Denekamp, M.; Smeekens, S.C. Integration of wounding and osmotic stress signals determines the expression of the *AtMYB102* transcription factor gene. *Plant Physiol.* **2003**, *132*, 1415–1423. [CrossRef] [PubMed]
86. Laity, J.H.; Lee, B.M.; Wright, P.E. Zinc finger proteins: New insights into structural and functional diversity. *Curr. Opin. Struct. Biol.* **2001**, *11*, 39–46. [CrossRef]
87. Hagymasi, A.T.; Dempsey, J.P.; Srivastava, P.K. Heat-shock proteins. *Curr. Protoc.* **2022**, *2*, e592. [CrossRef]

88. Guo, L.; Li, J.; He, J.; Liu, H.; Zhang, H. A class I cytosolic HSP20 of rice enhances heat and salt tolerance in different organisms. *Sci. Rep.* **2020**, *10*, 1383. [CrossRef]

89. Vogel, J.T.; Zarka, D.G.; Van Buskirk, H.A.; Fowler, S.G.; Thomashow, M.F. Roles of the CBF2 and ZAT12 transcription factors in configuring the low temperature transcriptome of Arabidopsis. *Plant J.* **2005**, *41*, 195–211. [CrossRef]

90. Rizhsky, L.; Davletova, S.; Liang, H.; Mittler, R. The zinc-finger protein Zat12 is required for cytosolic ascorbate peroxidase 1 expression during oxidative stress in Arabidopsis. *J. Biol. Chem.* **2004**, *279*, 11736–11743. [CrossRef]

91. Davletova, S.; Schlauch, K.; Coutu, J.; Mittler, R. The zinc-finger Protein Zat12 plays a central role in reactive oxygen and abiotic stress signaling in *Arabidopsis*. *Plant Physiol.* **2005**, *139*, 847–856. [CrossRef] [PubMed]

92. Toda, K.; Yang, D.; Yamanaka, N.; Watanabe, S.; Harada, K.; Takahashi, R. A single-base deletion in soybean flavonoid 3′-hydroxylase gene is associated with gray pubescence color. *Plant Mol. Biol.* **2002**, *50*, 187–196. [CrossRef] [PubMed]

93. Zhang, C.; Guo, Q.; Liu, Y.; Liu, H.; Wang, F.; Jia, C. Molecular cloning and functional analysis of a flavanone 3-hydroxylase gene from blueberry. *J. Hortic. Sci. Biotechnol.* **2017**, *92*, 57–64. [CrossRef]

94. Sun, W.; Zhou, N.; Feng, C.; Sun, S.; Tang, M.; Tang, X.; Ju, Z.; Yi, Y. Functional analysis of a dihydroflavonol 4-reductase gene in *Ophiorrhiza japonica* (OjDFR1) reveals its role in the regulation of anthocyanin. *PeerJ* **2021**, *9*, e12323. [CrossRef] [PubMed]

95. Vainio, J.; Mattila, S.; Abdou, S.M.; Sipari, N.; Teeri, T.H. Petunia dihydroflavonol 4-reductase is only a few amino acids away from producing orange pelargonidin-based anthocyanins. *Front. Plant Sci.* **2023**, *14*, 1227219. [CrossRef] [PubMed]

96. Nguyen, H.M.; Putterill, J.; Dare, A.P.; Plunkett, B.J.; Cooney, J.; Peng, Y.; Souleyre, E.J.F.; Albert, N.W.; Espley, R.V.; Günther, C.S. Two genes, ANS and UFGT2, from *Vaccinium* spp. are key steps for modulating anthocyanin production. *Front. Plant Sci.* **2023**, *14*, 1082246. [CrossRef]

97. Zhang, C.; Liu, H.; Jia, C.; Liu, Y.; Wang, F.; Wang, J. Cloning, characterization and functional analysis of a flavonol synthase from *Vaccinium corymbosum*. *Trees-Struct. Funct.* **2016**, *30*, 1595–1605. [CrossRef]

98. Vu, T.T.; Jeong, C.Y.; Nguyen, H.N.; Lee, D.; Lee, S.A.; Kim, J.H.; Hong, S.W.; Lee, H. Characterization of *Brassica napus* Flavonol Synthase Involved in Flavonol Biosynthesis in *Brassica napus* L. *J. Agric. Food Chem.* **2015**, *63*, 7819–7829. [CrossRef]

99. Wang, P.; Zhang, L.; Jiang, X.; Dai, X.; Xu, L.; Li, T.; Xing, D.; Li, Y.; Li, M.; Gao, L.; et al. Evolutionary and functional characterization of leucoanthocyanidin reductases from *Camellia sinensis*. *Planta* **2018**, *247*, 139–154. [CrossRef]

Article

Altered Photoprotective Mechanisms and Pigment Synthesis in *Torreya grandis* with Leaf Color Mutations: An Integrated Transcriptome and Photosynthesis Analysis

Yujia Chen [1,2], Lei Wang [1], Jing Zhang [1], Yilu Chen [1] and Songheng Jin [1,2,*]

[1]   Jiyang College, Zhejiang A&F University, Zhuji 311800, China; chenyujia414@stu.zafu.edu.cn (Y.C.); leiwang33@zafu.edu.cn (L.W.); zj2534015733@163.com (J.Z.); chenyl@stu.zafu.edu.cn (Y.C.)

[2]   School of Forestry and Biotechnology, Zhejiang A&F University, Hangzhou 311300, China

*   Correspondence: shjin@zafu.edu.cn

Citation: Chen, Y.; Wang, L.; Zhang, J.; Chen, Y.; Jin, S. Altered Photoprotective Mechanisms and Pigment Synthesis in *Torreya grandis* with Leaf Color Mutations: An Integrated Transcriptome and Photosynthesis Analysis. *Horticulturae* **2024**, *10*, 1211. https://doi.org/10.3390/horticulturae10111211

Academic Editor: Gunārs Lācis

Received: 20 October 2024
Revised: 14 November 2024
Accepted: 15 November 2024
Published: 17 November 2024

Abstract: *Torreya grandis* is a widely cultivated fruit species in China that is valued for its significant economic and agricultural importance. The molecular mechanisms underlying pigment formation and photosynthetic performance in *Torreya* leaf color mutants remain to be fully elucidated. In this study, we performed transcriptome sequencing and measured photosynthetic performance indicators to compare mutant and normal green leaves. The research results indicate that the identified *Torreya* mutant differs from previously reported mutants, exhibiting a weakened photoprotection mechanism and a significant reduction in carotenoid content of approximately 33%. Photosynthetic indicators, including the potential maximum photosynthetic capacity (F_v/F_m) and electron transport efficiency (Ψ_o, φ_{Eo}), decreased significantly by 32%, 52%, and 49%, respectively. While the quantum yield for energy dissipation (φ_{Do}) increased by 31%, this increase was not statistically significant, which may further reduce PSII activity. A transcriptome analysis revealed that the up-regulation of chlorophyll degradation-related genes—*HCAR* and *NOL*—accelerates chlorophyll breakdown in the *Torreya* mutant. The down-regulation of carotenoid biosynthesis genes, such as *LCY1* and *ZEP*, is strongly associated with compromised photoprotective mechanisms and the reduced stability of Photosystem II. Additionally, the reduced expression of the photoprotective gene *psbS* weakened the mutant's tolerance to photoinhibition, increasing its susceptibility to photodamage. These changes in gene expression accelerate chlorophyll degradation and reduce carotenoid synthesis, which may be the primary cause of the yellowing in *Torreya*. Meanwhile, the weakening of photoprotective mechanisms further impairs photosynthetic efficiency, limiting the growth and adaptability of the mutants. This study emphasizes the crucial roles of photosynthetic pigments and photosystem structures in regulating the yellowing phenotype and the environmental adaptability of *Torreya*. It also provides important insights into the genetic regulation of leaf color in relation to photosynthesis and breeding.

Keywords: photosynthetic performance; comparative transcriptome; leaf color mutants; *Torreya grandis*

1. Introduction

Renowned for its economic value, *Torreya grandis* is a highly prized tree species. Its popularity stems from its versatility, with applications in various fields such as food, medicine, timber, oil production, and ecological restoration [1]. Leaf color mutation is a frequently observed dominant phenotypic trait in higher plants. Previous research has identified a yellow-leafed *Torreya* mutant in Zhejiang Province [2]. Recently, we identified a novel yellow-leafed mutant of *Torreya* in the natural environment of Zhuji County, Zhejiang Province. Different mutants may display variations in photosynthetic performance. The photosynthesis of plants mainly occurs in leaves, so leaf color is an important phenotypic characteristic that affects plant traits. Hence, various leaf color mutants provide the optimal materials for studying the regulation of a plant's photosynthetic performance, photosynthetic pigment biosynthesis, and chlorophyll structure development in relation to the cultivation of efficient photosynthetic plants [2–5].

Some leaf color mutants have become valuable genetic resources and are widely used in genetic breeding studies due to their unique traits. For example, the formation of red leaf mutants is often attributed to anthocyanin accumulation—a phenomenon that has been extensively studied [6]. In contrast, the mechanisms underlying yellow leaf mutant development in woody plants are significantly more complex and hold substantial research value, particularly in economically important tree species. In contrast, the formation mechanism of yellow leaves in native plants, especially in economic tree species, is more complex and remains insufficiently studied, offering substantial research value. The molecular mechanisms underlying leaf yellowing can be classified into several categories, such as changes in the expression levels of genes related to chlorophyll synthesis or degradation, changes in the expression of genes related to chloroplast development and function, changes in the expression of genes in pathways related to photosynthesis, and mutations affecting genes related to carotenoid biosynthesis [2,7,8]. Changes in the expression levels of genes associated with chlorophyll synthesis and degradation, genes related to chloroplast development and function, genes involved in photosynthesis, and genes linked to carotenoid biosynthesis constitute the molecular mechanisms that lead to the yellowing phenotype of leaves. Thus far, the inclusion of a comprehensive analysis of photosynthetic performance to explore the mechanisms of the yellowing of leaves in woody plants has been limited to a few species.

Chloroplasts contain a variety of pigments, of which chlorophyll and carotenoids are particularly prominent. Carotenoids function as antenna pigments in photosynthesis, capturing light energy and facilitating its transfer to chlorophyll, as well as functioning to protect chlorophyll from light damage [9]. Chlorophyll forms a photosynthetic complex by binding to specific proteins that capture light energy and facilitate electron transport to reaction centers [10]. The electron transport in photosynthesis occurs sequentially through three major pigment–protein complexes that are situated on the thylakoid membrane—Photosystem II (PSII), cytochrome b_6f (cytb$_6$f), and Photosystem I (PSI) [11]. Chlorophyll fluorescence has become an indispensable tool for studying plant photosynthesis; it enables the rapid, accurate, and non-invasive detection of plant photosystems and photosynthetic electron transport processes that contain delayed chlorophyll a fluorescence (DF) and fast chlorophyll a fluorescence (PF) [12–14]. PF kinetics has emerged as a critical tool for studying the structural stability of PSII, offering a comprehensive understanding of the energy flow among its various components [15]. By introducing JIP test parameters, it is possible to effectively characterize the changes in energy flux within the photosynthetic electron transport chain, enabling a quantitative assessment of photosynthetic efficiency [16]. DF reveals the charge recombination reaction in PSII, which is markedly influenced by the energy state of the thylakoid membrane [17,18]. Plant survival is dependent on the occurrence of photosynthesis within chloroplasts, which furnish the vital materials and energy required for plant activities. Thus, investigating the photosynthetic mechanisms and regulatory pathways of mutant yellow leaves represents a significant area of focus in biological research.

Next-generation sequencing (NGS) technology has been widely applied to elucidate the molecular mechanisms underlying the formation of plant leaf color mutants. This technology is essential for identifying key genes and exploring biological processes at the transcriptional level [4,5,8,19].

The objective of this study was to compare the differences in photosynthetic performance and molecular mechanisms between the *Torreya* mutant and its wild type, with a focus on elucidating how leaf color regulation affects photosynthetic efficiency. These findings provide theoretical foundations for a deeper understanding of the genetic regulatory mechanisms of plant photosynthesis and contribute to the cultivation of plants with enhanced photosynthetic performance.

2. Materials and Methods

2.1. Plant Material

Torreya seedlings were sourced from the Chinese *Torreya* Museum, with one-year-old *Torreya* scions grafted onto one-year-old *Torreya* Fort. ex Lindl. rootstocks. The experimental materials were obtained from normal plants and natural mutants displaying yellow leaves, all cultivated under the same environmental conditions. The rootstock was trimmed approximately 5–6 cm above the root collar before being grafted onto the scion. Each grafted tree was individually cultivated in liter plastic pots filled with 7 kg of loam. These were watered to saturation when the soil dried naturally and were fertilized monthly to maintain nutrient levels. While wild-type plants display dark green leaves, the mutant features yellow leaves. Reduced pigmentation makes the veins more visible, adding contrast against the deeper leaf background (Figure 1). In August 2023, three replicates of green wild-type *Torreya* and yellow *Torreya* natural mutant leaf tissues were randomly selected as research materials.

Figure 1. Leaf appearance of wild type and mutant *Torreya*.

2.2. Measurement of Photosynthetic Pigment

The chlorophyll (Chl) and total carotenoid contents were assessed using the ethanol extraction method, while calculations were carried out following Lichetenthaler's methodology [20].

2.3. Measurements of Chlorophyll a Fluorescence

Using a Multi-Function Plant Efficiency Analyzer M-PEA (Hansatech, Norfolk, UK), the simultaneous determination of the kinetic profiles of PF and DF was carried out following 30 min of the leaves being subjected to complete dark adaptation. The measurement method was detailed by Strasser et al. [21]. The M-PEA emitted red light at an intensity of $5000~\mu mol \cdot m^{-2} \cdot s^{-1}$ and a wavelength of 627 ± 10 nm, and the light–dark conversion began after 300 μs of exposure. The PF light reflection signal was recorded under illumination, and the DF signal was recorded in the dark [21]. Based on the methodologies described in prior studies, the PF and DF curves were generated [21,22].

The parameters that could be directly obtained in the kinetics curve included F_o at 20 μs, the maximum fluorescence intensity F_m, F_J at 3 ms, and F_I at 30 ms. The standardization calculated from the O-P phase is represented by the following equation: $V_t = (F_t - F_o)/(F_m - F_o)$. The JIP test parameters are defined as follows: the potential maximum photosynthetic capacity (Fv/Fm), the ratio of exciton-driven electron transfer captured by the active reaction center of PSII (Ψ_o), the quantum yield of electron transport (φ_{Eo}), the quantum yield for the reduction of the end electron acceptors on the PSI acceptor side (φ_{Ro}), the probability with which an electron from the intersystem electron carriers moves to reduce the end electron acceptors on the PSI acceptor side (δ_{Ro}), and the quantum efficiency of energy dissipation (φ_{Do}). The specific fluxes (per active reaction center in PSII) quantify the absorbed energy flux (ABS/RC), the trapped energy flux (TR$_o$/RC), the electron transport flux (ET$_o$/RC), the electron flux reducing the end electron acceptors at the PSI acceptor side (RE$_o$/RC), and the energy dissipation flux (DI$_o$/RC). The phenomenological fluxes (per cross-sectional area unit) include the flux absorbed (ABS/CS$_m$), the flux trapped (TR$_o$/CS$_m$), the flux dissipated as heat (DI$_o$/CS$_m$), the flux transferred through electron transport (ET$_o$/CS$_m$), and the reduction in the specific electron flux to the terminal electron acceptor of PSI (RE$_o$/CS$_m$).

2.4. Transcriptome Sequencing

All gathered leaf samples were promptly frozen in liquid nitrogen and maintained at $-80\ ^\circ$C for later use. Total RNA from the samples was extracted using the TRIzol reagent (Invitrogen, Waltham, MA, USA). After performing RNA extraction, assessing RNA integrity, and constructing the DNA library, sequencing was conducted using NGS technology on the Illumina platform. This process was facilitated by Huanran Biotechnology Co., Ltd., (Hangzhou, China). Prior to the bioinformatics analysis, the sequencing data underwent filtering to eliminate low-quality reads and those containing adaptor sequences.

Clean reads derived from the raw data were mapped to the *Torreya* reference genome (https://doi.org/10.6084/m9.figshare.21089869) (accessed on 12 December 2023) using HISAT v2.0.4 software. A differential expression pattern analysis was performed using DESeq [23]. Differentially expressed genes (DEGs) were identified based on the criteria of | log2FoldChange | > 1 and Padj < 0.05. Gene function annotation was conducted using multiple public databases, including NCBI non-redundant (NR), Gene Ontology (GO), the Kyoto Encyclopedia of Genes and Genomes (KEGG), Swiss-Prot, and the Evolutionary Genealogy of Genes: Non-supervised Orthologous Groups (eggNOGs).

2.5. Validation by RT-qPCR of Some DEGs

Total RNA was used as a template for cDNA synthesis, utilizing a HiScript III All-in-One RT SuperMix Perfect for qPCR (Vazyme, Nanjing, China) according to the manufacturer's instructions. The gene-specific primer sequences were designed by PrimerPremier 5.0 software (Table S1). The gene Actin from *Torreya* was used as the internal reference gene. qRT-PCR was conducted using a Taq Pro Universal SYBR qPCR Master Mix Kit (Vazyme, Nanjing, China) on a LightCycler$^\circledR$ 480 II (Roche Diagnostics, Mannheim, Germany). The cycling program consisted of an initial denaturation step at 95 $^\circ$C for 30 s, followed by 40 cycles comprising 95 $^\circ$C for 10 s, 60 $^\circ$C for 15 s, and 72 $^\circ$C for 10 s. Gene expression levels were quantified using the $2^{-\Delta\Delta Ct}$ method.

2.6. Statistical Analysis

Physiological data were analyzed using a one-way analysis of variance (ANOVA) with SPSS version 25.0 (SPSS Inc., Chicago, IL, USA). Results are expressed as mean $\pm$ standard deviation (SD). When $p < 0.05$, there was a statistically significant difference, which is denoted by different letters.

3. Results

3.1. Determination of Photosynthetic Pigment

Table 1 demonstrates that the contents of Chl a and b in the mutant significantly decreased by 59% and 55%, respectively, while carotenoid levels also decreased substantially by 33%. Similarly, the ratio of Chl a to b displayed a comparablea trend.

Table 1. Comparison of pigment concentrations between mutant and wild-type *Torreya* leaves.

Torreya grandis	Chl (a + b) (mg/g FW)	Chl a (mg/g FW)	Chl b (mg/g FW)	Chl a/b (mg/g FW)	Carotenoid (mg/g FW)
Wild type	0.42 ± 0.01 a	0.32 ± 0.02 a	0.09 ± 0.01 a	3.47 ± 0.08 a	0.12 ± 0.00 a
Mutant	016 ± 0.00 b	0.12 ± 0.01 b	0.04 ± 0.01 b	2.82 ± 0.20 b	0.08 ± 0.01 b

The values are presented as mean $\pm$ SD ($n = 3$); significant differences are indicated by different letters; $p < 0.05$.

3.2. Assessment of Fluorescence Kinetic Parameters

The PF curve of the mutant leaves exhibited significant changes, including a marked decrease in the maximum fluorescence (F_m). Additionally, the transitions between the J-I and I-p phases disappeared, and the J-step approached the p level (Figure 2A). To facilitate a comparison, the PF curve was standardized from step O (20 μs) to step p (300 ms) (Figure 2B). In the mutant, V_t was elevated as a whole, with V_J and V_I showing particularly significant increases. At the same time, the I_2 peak was not obvious, and the DF decay between the I_1 and I_2 peaks nearly disappeared; this observation aligns with the OJIP transient phase of PF (Figure 2C). Additionally, the decay kinetics of DF at the I_1 peak showed a significant decrease in the mutant (Figure 2D).

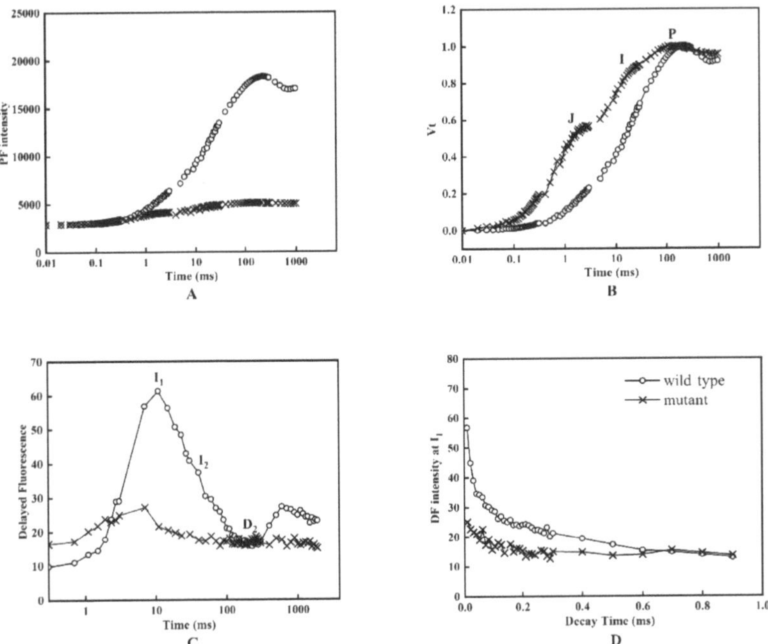

Figure 2. (**A**) Prompt chlorophyll a fluorescence (PF). (**B**) Normalized curve; $V_t = [(F_t - F_O)/(F_M - F_o)]$, J reflects early electron transport blockage; I reflects the size of the PQ pool and the efficiency of electron flow; P represents the maximum PSII photochemical efficiency. (**C**) Delayed chlorophyll a fluorescence (DF). I_1 represents the redox state of Q_A and PSII functionality; I_2 represents the reduction of the PQ pool and the efficiency of electron transfer; D_2 represents the charge separation stability and recombination dynamics. (**D**) The decay kinetics of DF at the characteristic maxima I_1 (7 ms). Each curve represents the mean value derived from three replicate measurements.

There was a significant reduction in the parameters Ψ_o, φ_{Eo}, φ_{Ro}, δ_{Ro}, and F_v/F_m. Conversely, φ_{Do} increased in the mutant leaves, though with minimal variations in magnitude (Table 2).

Table 2. JIP test parameters in the leaves of mutant and wild-type *Torreya grandis*.

Torreya grandis	F_v/F_m	φ_{Eo}	φ_{Ro}	φ_{Do}	δ_{Ro}	Ψ_o
Wild type	0.84 ± 0.01 a	0.70 ± 0.02 a	0.26 ± 0.01 a	0.16 ± 0.01 b	0.38 ± 0.01 a	0.83 ± 0.02 a
Mutant	0.57 ± 0.12 b	0.36 ± 0.16 b	0.18 ± 0.03 b	0.21 ± 0.14 b	0.27 ± 0.05 b	0.40 ± 0.06 b

F_v/F_m: the potential maximum photosynthetic capacity; Ψ_o: the ratio of exciton-driven electron transfer captured by the active reaction center of PSII at t = 0; φ_{Eo}: quantum efficiency of electron transfer at t = 0; φ_{Ro}: quantum yield for the reduction of the terminal electron acceptors on the PSI acceptor side; φ_{Do}: quantum yield for energy dissipation at t = 0; δ_{Ro}: quantum yield for the reduction of the terminal electron acceptors on the PSI acceptor side. The values are presented as mean ± SD (*n* = 3); significant differences are indicated by different letters; *p* < 0.05.

The energy flux absorption is illustrated in Figure 3. In the mutant leaves, ET_0/RC and RE_0/RC were lower, while DI_0/RC increased significantly. In the mutant, the phenomenological energy fluxes, including ABS/CS_m, TR_0/CS_m, ET_0/CS_m, and RE_0/CS_m, were significantly reduced. Only DI_0/CS_m increased slightly, although this change was not significant (Figure 3).

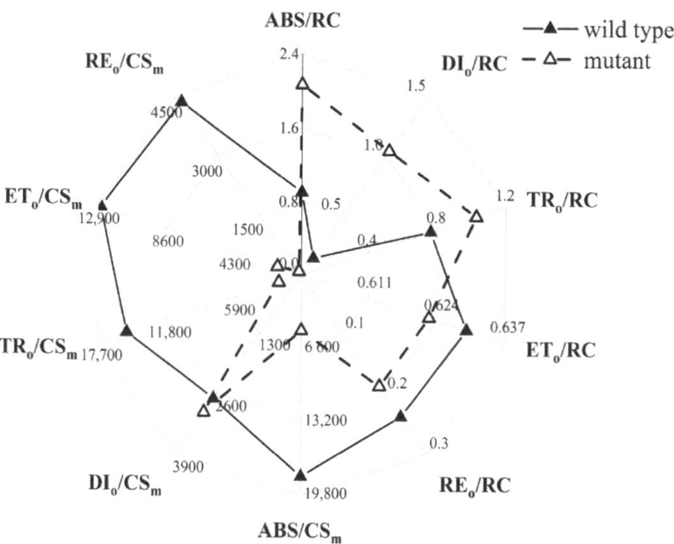

Figure 3. Radar plot of energy fluxes in mutant and wild type *Torreya*. The radar plot reflected specific activity values at individual PSII reaction centers (RCs) and cross-sections (CSs). Each data point represents the mean value derived from three replicate measurements.

3.3. Transcriptomic Analysis of Mutant and Wild-Type Torreya

Transcriptome sequencing was conducted on both wild-type and mutant samples; a total of 300,334,628 raw reads were generated. After removing low-quality sequences, filtering adapters, and eliminating ambiguous reads, we obtained 294,982,568 clean reads. In the filtered RNA data, the average Q20 and Q30 scores exceeded 97.69% and 95.85%, respectively, with the GC content ranging from 43.55% to 43.72%. The raw sequence data and assemblies were deposited in the NCBI BioProject under the accession number PRJNA1150738. Subsequently, the clean read sequences were aligned with the published *Torreya* genome, revealing that over 94.97% of the sequences successfully mapped the reference genome (Table S2). The Pearson correlation coefficient (R^2) among the biological replicates exceeded 0.77. The above results indicate a good within-group reproducibility of the samples (Figure S1).

A total of 47,089 genes were annotated using the NR, eggNOG, Swiss-Pro, KEGG, and GO databases to provide annotation information for transcripts. The highest annotation success rate was achieved with the NR database (82.55%) and the lowest was achieved with the KEGG database (32.09%) (Table S3).

3.4. Functional Annotation and Classification of the DEGs

In this study, differentially expressed genes (DEGs) were identified using the criteria |log2FoldChange| > 1 and a corrected Padj value < 0.05. A total of 4607 DEGs were detected between wild-type and mutant samples, with 2433 being up-regulated and 2174 being down-regulated. The overall distribution of mRNA is illustrated in a volcano plot (Figure 4A). Hierarchical clustering of the DEGs was performed, with leaves of the same type being clustered together. This clustering suggests that genes within the same cluster may have similar biological functions. Additionally, the abundance of DEGs was visually represented using FPKM values and color coding (Figure 4B). GO annotation of these DEGs was conducted based on the categories of Cellular Components (CCs), Molecular Functions (MFs), and Biological Processes (BPs) to evaluate their potential functions. The results indicated that genes associated with the 'cell periphery' and the 'plasma membrane' were the most highly enriched; these genes were classified under the CC category (Figure 4A). Following this, 'response to stimulus' was the most enriched category within Biological Processes (BPs). In the Molecular Function (MF) category, 'glucosyltransferase activity' exhibited the highest level of enrichment (Figure 5A). Based on the KEGG analysis, a total of 1235 DEGs were identified and enriched in 119 pathways. The top 20 enrichment pathways, based on the false discovery rate, are shown in Figure 4B. The most significantly enriched KEGG pathways included 'plant–pathogen interaction', 'amino sugar and nucleotide sugar metabolism', and 'galactose metabolism' (Figure 5B).

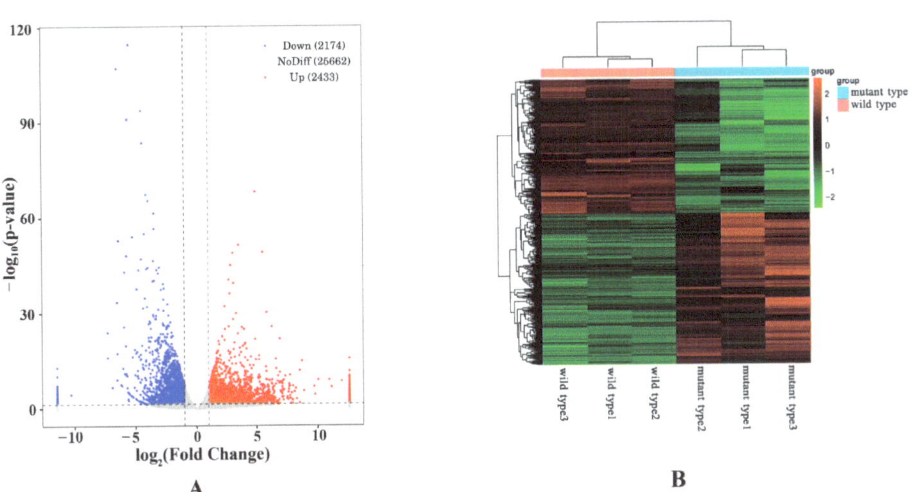

Figure 4. Differentially expressed genes (DEGs) in different leaves of *Torreya*. (**A**) DEGs were displayed in the form of a volcano plot. The red dots denote up-regulated genes, while the blue dots denote down-regulated genes. (**B**) Hierarchical cluster analysis of DEGs. Each row represents a gene, with red indicating a more pronounced up-regulation and green indicating a more pronounced down-regulation.

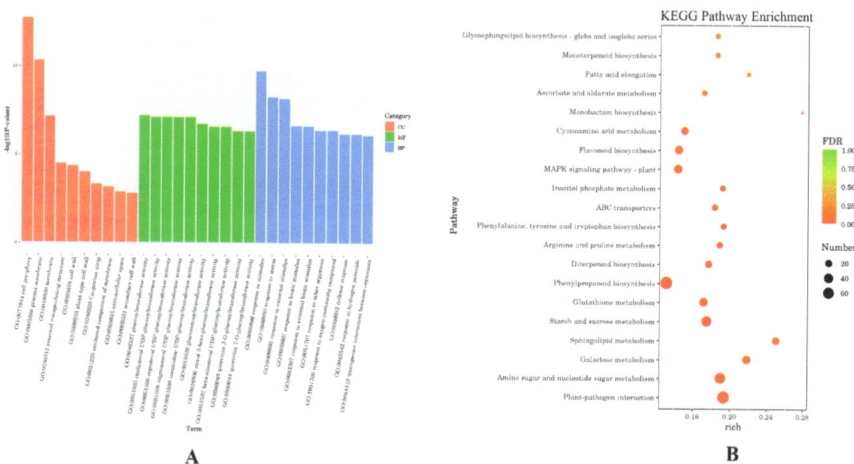

Figure 5. Functional annotation of DEGs. (**A**) GO classification of differentially expressed genes—the top 30 enriched GO terms. (**B**) KEGG enrichment of differentially expressed genes; the larger the bubble, the more DEGs that are enriched.

3.5. Identification of the DEGs Associated with Leaf Coloration

Using the KEGG database, we analyzed the metabolic pathways directly associated with leaf color, specifically focusing on carotenoid biosynthesis, porphyrin metabolism, and photosynthesis. We identified the DEGs annotated within these pathways. In this study, five DEGs were identified as being the core genes involved in encoding carotenoid biosynthetic enzymes.

These included carotenoid cleavage dioxygenase 4 (*CCD4*), zeaxanthin epoxidase (*ZEP*), 9-cis-epoxy carotenoid dioxygenase (*NCED1*), and lycopene beta-cyclase (*LCY1*). *CCD4* and *NCED1* are involved in carotenoid degradation, whereas *ZEP* and *LCY1* are involved in carotenoid synthesis. The expression levels of *CCD4* genes (*TG3g00961* and TG3g00950) showed different degrees of regulation, with some genes being up-regulated and some being down-regulated compared to the wild type. In contrast, the expression levels of the other three genes—*NCED1* (TG3g00946), *ZEP* (TG3g01751), and *LCY1* (TG5g03735)—were uniformly down-regulated in the mutant (Figure 6A).

Four DEGs were identified within the photosynthesis pathway, encoding enzymes associated with photosynthesis—PSI P700 chlorophyll a apoprotein A2 (*psaB*), ATP synthase subunit alpha (*atpA*), PSI P700 chlorophyll a apoprotein A1 (psaA), and PSII 22 kDa protein (*psbS*). The expression levels of *psaB* (TG11g00862), *psaA* (TG8g03353), and *atpA* (TG5g03088) were significantly higher in the mutant compared to the wild type. Conversely, the expression of psbS (TG4g01167) was significantly down-regulated in the mutant relative to the wild type (Figure 6C).

In the porphyrin metabolic pathway, we concentrated on analyzing the coding genes of the key enzymes involved in chlorophyll metabolism. A total of six DEGs were identified, with each encoding an enzyme involved in chlorophyll metabolism. These enzymes included chlorophyllide a oxygenase (*CAO*), Chl b reductase (*NOL*), 7-hydroxymethyl chlorophyll a reductase (HCAR), uroporphyrinogen decarboxylase (*hemE1*), chlorophyllase-2 (*CLH2*), and light-independent protochlorophyllide reductase (*chlN*) (Figure 6B).

Among these genes, chlN, CAO, and hemE1 are involved in chlorophyll synthesis, while HCAR, NOL, and CLH2 are associated with chlorophyll degradation. The expression levels of *CAO* (TG8g02191), *NOL* (TG9g00978), and *HCAR* (TG11g01233) were significantly up-regulated in the mutant compared to the wild type. Conversely, the expression levels of *chlN* (TG7g03643), *hemE1* (TG7g00721), and *CLH2* (TG8g02048) were down-regulated in the mutant relative to the wild type (Figure 5B). The expression levels of these genes are further detailed in Figure 6C.

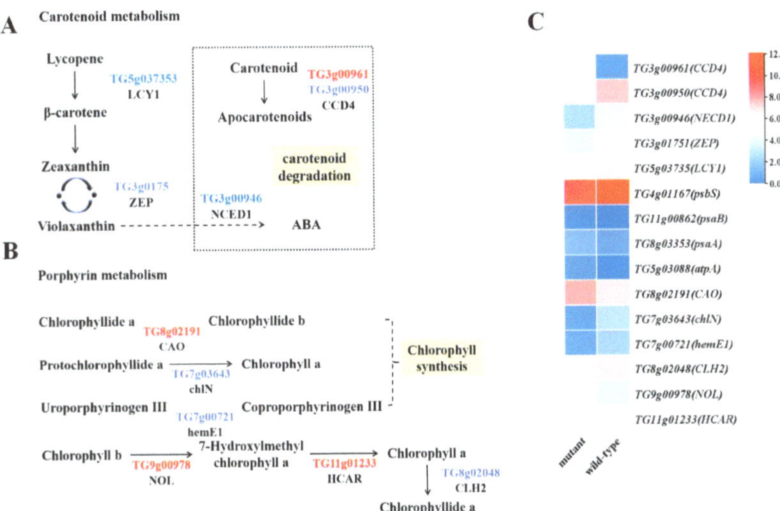

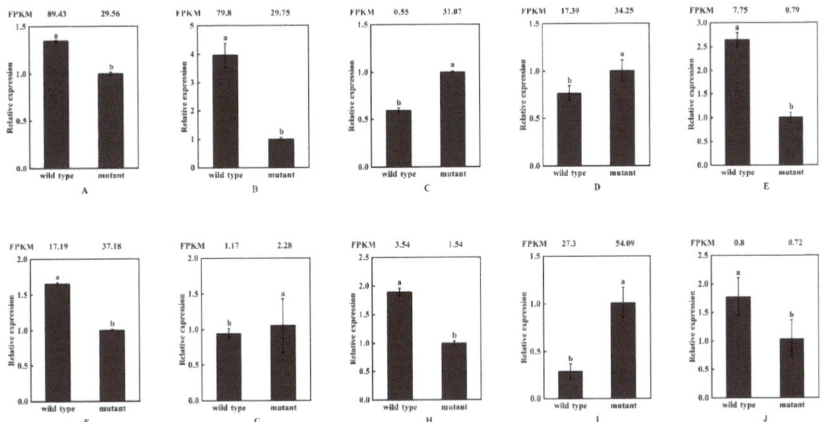

Figure 6. Regulation of gene expression and metabolic pathways associated with leaf color at the transcriptional level. (**A**) Analysis of differentially expressed genes related to chlorophyll biosynthesis and degradation pathways. (**B**) Differential expressions of genes related to carotenoid biosynthesis and degradation pathways. (**C**) Transcriptome data (FPKM) were used for heat mapping.

3.6. Validation of Transcription Data Accuracy

Several DEGs associated with carotenoid biosynthesis, photosynthesis, and porphyrin metabolism pathways were validated using qRT-PCR to further assess the confidence of the RNA-seq data. The DEGs from qRT-PCR are consistent with the expression patterns observed in the RNA-seq data, indicating that the RNA-seq data were highly accurate (Figure 7).

Figure 7. The expression levels of DEGs in wild-type and mutant *Torreya*. (**A**) Gene ID: TG8G02048 (CHL2). (**B**) Gene ID: TG5G03735 (LCY1). (**C**) Gene ID: TG3G00961 (CCD4). (**D**) Gene ID: TG9G00978 (NOL). (**E**) Gene ID: TG7G03643 (chlN). (**F**) Gene ID: TG3G0175 (ZEP). (**G**) Gene ID: TG8G003353 (psaA). (**H**) Gene ID: TG4G00167 (psbS). (**I**) Gene ID: TG11G01233 (HCAR). (**J**) Gene ID: TG3G00946 (NECD1). Each error bar represents the SD calculated from three biological replicates, each of which includes three technical replicates.

4. Discussion

4.1. Photosynthesis Performance in the Mutant Torreya

Previous research has demonstrated that Chl-deficient mutants consistently display reduced levels of chlorophyll and carotenoids. The concentration and composition of these pigments are closely linked to the development of yellowing leaves in plants [4,5]. The changes in photosynthetic pigment content observed in a previous mutant of *Torreya* differ from those found in this research, as the earlier study reported an increase in carotenoid content [2]. We observed a significant reduction in chlorophyll and carotenoid content in the mutant leaves of *Torreya*, which altered the normal leaf coloration and resulted in a Chl-deficient phenotype (Table 1). In this study, the induction kinetics of PF and DF were analyzed in both wild-type and mutant leaves. This method, which involves simultaneous measurements, provides complementary information on electron transport through optical coupling [22]. The increase in PF kinetics from the lowest point (O) to the highest point (p) primarily indicates alterations in the original photochemical processes of PSII [21,24]. The results indicated that the p point in the mutant *Torreya* was significantly lower than that in the wild type (Figure 2A). Previous studies suggest that the reduction in fluorescence at p may result from several different factors, i.e., the denaturation of chlorophyll–protein complexes [25], the inhibition of the PSI acceptor side [26], a reduction in the number of active reaction centers available to support electron transport to PSI [27], changes in the properties of PSII electron acceptors [28], and the development of radiation-free dissipation of the chlorophyll excited state within PSII antennas [29]. From the standardized Vt curve obtained after the O-p phase, it can be observed that the J and I steps were elevated in the mutant compared to the wild type, while the O and P steps remained unchanged (Figure 2B). The apparent enhancement of V_J and V_I reflected the kinetic inhibition of the electron transport chain in the subsequent processes. The significant enhancement of V_J and V_I indicated kinetic inhibition in the electron transfer chain during later stages. The observed inhibition may be related to a decreased electron transfer efficiency between QA and QB on the acceptor side of PSII, as well as a limited reoxidation of plastoquinol (PQH_2) [30,31]. DF is directly generated from the reaction centers of PSII, making it a reliable indicator of PSII function [32]. The kinetic values of DF were altered by the reduced electron transfer rate in PSII (Figure 2C,D). This reduction is manifested as a decrease in DF yield, which subsequently affects charge recombination and the regeneration of antenna pigments [21]. The loss of DF observed between peaks I_1 and I_2 was associated with the absence of the J-I process in the PF curve. The significant reduction in I_1 indicated a potential decrease in P680, which could result in decreased electron transfer efficiency between the donor and acceptor sides of PSII [28]. Additionally, changes in I_2 may have been correlated with the prolonged reopening of PSII reaction centers [33], which is consistent with the PF results mentioned above. F_v/F_m is regarded as a crucial indicator for assessing the extent of photoinhibition in plants [16]. Previous studies have indicated that the F_v/F_m in green leaves is significantly higher than that in mutant leaves [2], which is consistent with the results of this study. The elevated ABS/RC values observed in the mutant samples compared to the wild-type samples are primarily due to the accelerated degradation of pigments within PSII [34]. Furthermore, the significant increase in the ABS/RC and TR_o/RC values is believed to result from the enlargement of the antenna complex relative to the size of P_{680} [35]. The down-regulation of Ψ_o and φ_{Ro} has been observed in the mutant, which is consistent with previous findings [2]. The significantly lower φ_{Eo} in the mutant suggests the utilization of trapped photons for electron transfer from Q_A to Q_B, and beyond Q_B, in the electron transport chain [16].

The variations in φ_{Ro} and δ_{Ro} reflect changes in PSI electron transfer efficiency, with their decreased values in mutant leaves being interpreted as an impairment of PSI activity. DI_o/RC and φ_{Do} are not significantly enhanced. Moreover, the rise in DI_o/CSm values indicates the intensified thermal inactivation of RCs, which consequently lowers the values of RE_o/CS_m and ET_o/CS_m [36]. The chlorophyll and carotenoid content in the mutants was significantly reduced, leading to impaired photosynthetic efficiency, which was evidenced by decreased photosynthetic electron transfer efficiency and intensified photoinhibition. Unlike previous studies that reported enhanced PSI activity in *Torreya* mutants [2], the mutant in this study lacked additional photoprotective mechanisms. This was reflected not only in the reduced photosynthetic efficiency, but also in the decreased stability and tolerance of the photosystem. Such inefficiency and instability may place the mutant plants at a disadvantage in terms of growth and development, potentially impairing their adaptability and competitiveness in natural environments.

4.2. Integrated Analysis of DEGs in Carotenoid, Chlorophyll, and Photosynthesis Pathways

The molecular mechanisms underlying yellow leaf formation vary among different plant species. The yellowing of plants is typically attributed to a deficiency in chlorophyll, which is the primary pigment that is responsible for imparting green coloration to leaves [11]. To date, numerous mutants with defects in chlorophyll biosynthesis and degradation have been discovered among different plant species, including rice [37], Arabidopsis [38], and pakchoi [39]. In this study, we applied a combined criterion of |log2FoldChange| > 1 and a corrected Padj value < 0.05 to identify DEGs associated with porphyrin metabolism in both wild-type and mutant samples. Among the selected DEGs, *NOL* and *HCAR*, which are involved in chlorophyll degradation, exhibited up-regulation, whereas *CLH2* showed down-regulation. The expression level of the chlN gene, which is associated with the chlorophyll biosynthesis pathway, was down-regulated. Six enzymes involved in chlorophyll metabolism have been identified in both Arabidopsis and rice, including *HCAR* and *NOL*, which were annotated in this study [37,40]. Chlorophyll degradation plays a crucial role in the disassembly of PSII and the reduction of chlorophyll–protein complexes, which directly impacts the efficiency of photosynthesis [41]. In a study on tobacco plants overexpressing cucumber *HCAR*, accelerated chlorophyll degradation was observed, which was accompanied by a reduction in photosynthetic proteins, ultimately leading to a decline in quantum yield and net photosynthetic rate [42]. In Arabidopsis, plants overexpressing *HCAR* exhibited accelerated leaf yellowing [38]. The up-regulation of the *ZjNOL* gene, cloned from Zoysia japonica, can impair the structure and function of the photosystem, thereby reducing photosynthetic efficiency [43]. These findings suggest that the up-regulation of *NOL* and *HCAR* may reduce the photosynthetic capacity of *Torreya* mutants by accelerating chlorophyll degradation and impairing the photosystem, thereby limiting resource accumulation, growth, and development. The expression of *CLH2* was inversely correlated with the yellowing of broccoli [44]. However, in our study, the expression level of the *CLH2* gene decreased, which typically leads to chlorophyll accumulation or preservation, suggesting that *CLH2* may have a more complex regulatory role in the color modulation of *Torreya* leaves. The reduction of protochlorophyllide (Pchlide) to chlorophyllide (Chlide) is a critical step in the chlorophyll biosynthesis pathway; this process is mediated by the light-dependent protochlorophyllide oxidoreductase (LPOR) and the dark-active Pchlide oxidoreductase (DPOR) [45]. The dark treatment of Norway spruce resulted in etiolation, which was attributed to the inhibition of the transcription of genes encoding DPOR [46]. The chlorophyll gene *chlN*, which encodes DPOR, was down-regulated in the mutant of *Torreya*. The alterations in these gene expression levels may serve as the primary factor contributing to the reduction in the chlorophyll content in leaves, subsequently facilitating the development of yellowing phenotypes. At the same time, the DEGs associated with the carotenoid metabolic pathway in the *Torreya* mutant also exhibited changes in expression. The *LCY1* gene encodes a carotenoid cyclase involved in the conversion of lycopene to α-carotene [47]. This process is crucial for a plant's photosynthetic efficiency, particularly in enhancing leaf photoprotection and antioxidant responses [48]. The overexpression of *Ntβ-LCY1* in tobacco has been proven to result in a strong accumulation of carotenoids [49]. The expression level of the *LCY1* gene identified in this study was significantly lower in the mutant compared to that in the wild-type. Carotenoids not only capture light energy as photosynthetic pigments, but also regulate the stability of PSII in the light protection mechanism [50]. We speculate that in the yellowing mutant of Torreya, the down-regulation of the *LCY1* gene may reduce carotenoid synthesis, resulting in a decreased photosynthetic capacity and yellowing. Zeaxanthin epoxidase (*ZEP*) serves as a crucial enzyme in the xanthophyll cycle, catalyzing the downstream reactions within the carotenoid biosynthesis pathway [4,47]. The xanthophyll cycle mechanism has been shown to protect plants from photodamage [49], particularly in PSII [51], where excess light energy is dissipated primarily through thermal dissipation and photochemical quenching mechanisms [52]. In addition, the accumulation of zeaxanthin plays a crucial role in mitigating light stress, thereby supporting normal plant growth [11]. Previous studies have demonstrated that the elevated abundance of zeaxanthin in striped leaves forms an adaptive protective mechanism, which enables rice plants to alleviate the negative impact of surrounding white leaf stripes on overall plant growth and development; this mechanism is primarily driven by the up-regulation of *OsZEP* expression [53]. The down-regulation of *ZEP* expression in the mutant leaves of *Torreya* may impair the function of the xanthophyll cycle, making the leaves more susceptible to photo-oxidative damage, thereby compromising the stability of PSII. Abscisic acid (ABA) synthesis is mainly achieved by the oxidative cleavage of carotenoids, and 9-cis-epoxycarotenoid dioxygenase (NCED1) is a key regulatory gene in ABA biosynthesis [54,55]. In the *Torreya* mutant, the expression levels of the *NCED1* and *ZEP* genes were found to be down-regulated, which aligns with the findings observed in the poplar golden leaf mutant [4].

The expression of the *psbB* and *psbD* genes was blocked in the yellow leaves of common wheat, and a further chloroplast ultrastructure analysis suggested that the formation of yellow leaves was closely associated with an abnormal chloroplast structure [56]. According to the KEGG database, it was revealed that a total of 37 unigenes were identified within the photosynthesis pathway of the mutant yellow leaves of ginkgo [5]. In the mutant, only the *psbS* gene was down-regulated in the photosynthetic pathway compared to that in the wild type. The level of *psbS* plays a regulatory role in cyclic electron flow, and *psbS* absence increases susceptibility to damage to PSI and PSII [57]. Thus, the *Torreya* mutants examined in this study exhibited significant deficiencies in their ability to cope with chlorophyll deficiency and the decline in PSII linear electron transport [2]. Related studies have shown that transgenic tobacco lines overexpressing *SaPsbS* exhibit higher F_v/F_m values [58], while the photochemical efficiency of Photosystem II is significantly reduced in Arabidopsis mutants lacking PsbS [59]. This finding aligns with the chlorophyll fluorescence results obtained in this study. Studies have shown that Chl-deficient mutants attempt to increase non-photochemical quenching (NPQ), which enhances heat dissipation and improves the photoprotection provided by the xanthophyll cycle pool [60]. The PsbS protein, an essential component of the PSII core complex, has a critical function in NPQ [61,62], while *ZEP* serves a critical function in the lutein cycle [19]. The expression levels of *ZEP* and *psbS* were significantly down-regulated in *Torreya* mutants, suggesting a deficiency in photoprotection mechanisms. This may increase the mutant's susceptibility to light-induced damage, leading to reduced photosystem efficiency.

In order to further advance the understanding of the pigment formation and photosynthetic efficiency in Torreya mutants, several future directions of study could be pursued. CRISPR technology could complement the findings of this study on chlorophyll and carotenoid biosynthesis. Recent studies have shown that CRISPR/Cas9 can be used to knock out or overexpress genes in plant systems, enabling the targeted modifications of the pathways involved in pigment synthesis. For example, in the case of carotenoid biosynthesis, CRISPR has been used to successfully knock out genes like lycopene ε-cyclase (LCYE) in Chlamydomonas reinhardtii, which led to altered carotenoid profiles, enhancing the production of valuable pigments like astaxanthin [63]. This technology can be employed for the selective knockout or overexpression of annotated genes, such as HCAR, NOL, and ZEP, in the context of this study. By generating precise genetic modifications, we can explore the molecular mechanisms behind the observed phenotype and their impact on the plant's photosynthetic performance. Additionally, combining CRISPR with transcriptomic analyses could enable the identification of novel regulators or signaling pathways that influence pigment biosynthesis, opening up new avenues for crop improvement strategies under changing environmental conditions.

5. Conclusions

In this study, the content of Chl and carotenoids was found to be reduced in *Torreya* mutants, which correlated with impaired photosynthetic activity compared to the wild type. This impairment led to a significant decrease in parameters associated with electron transfer (ET_o/CS_m, ET_o/RC, and φ_{Eo}). The genes involved in chlorophyll biosynthesis (*chlN*, *hemE1*, and *CLH2*) were down-regulated; those associated with carotenoid biosynthesis (*NCED1*, *LCY1*, and *ZEP*) also exhibited a downward trend. In contrast, genes related to chlorophyll degradation (*NOL* and *HCAR*) showed up-regulated expression. These findings highlight significant differences in the regulation and alteration of photosynthetic pigment biosynthesis among the mutant varieties of *Torreya*. The weakening of photoprotective mechanisms results in a marked decrease in electron transfer, which may restrict the growth and competitiveness of these mutants in natural environments. Overall, this study provides valuable insights for further investigation into leaf color changes and their molecular mechanisms in other plant species, while also offering important information for the cultivation and breeding of *Torreya*.

Supplementary Materials: The following supporting information can be downloaded at: https://www.mdpi.com/article/10.3390/horticulturae10111211/s1, Figure S1: Pearson's correlation coefficients between two biological replicates in the six samples. Table S1: The gene-specific primer sequences used in qRT-PCR. Table S2: Statistics of sequencing data quality from *Torreya grandis* samples. Table S3: Reference genome information.

Author Contributions: Conceptualization, Y.C. (Yujia Chen), S.J. and L.W.; methodology, Y.C. (Yujia Chen) and J.Z.; software, Y.C. (Yilu Chen) and J.Z.; validation, Y.C. (Yujia Chen) and J.Z.; formal analysis, Y.C. (Yujia Chen); resources, S.J.; writing—original draft preparation, Y.C. (Yujia Chen); writing—review and editing, Y.C. (Yujia Chen), L.W. and S.J.; funding acquisition, S.J. All authors have read and agreed to the published version of the manuscript.

Funding: This research was funded by the National Key Research and Development Project (2019YFE0118900), the National Natural Science Foundation of China (31971641), and the Zhejiang Provincial Team Technology Commissioner Project (Horticulture Team in Wencheng).

Data Availability Statement: Raw sequence data and assemblies were deposited under NCBI BioProject PRJNA1150738.

Conflicts of Interest: The authors declare no competing interests.

References

1. Chen, X.; Jin, H. Review of cultivation and development of Chinese torreya in China. *For. Trees Livelihoods* **2019**, *28*, 68–78. [CrossRef]
2. Shen, J.; Li, X.; Zhu, X.; Ding, Z.; Huang, X.; Chen, X.; Jin, S. Molecular and photosynthetic performance in the yellow leaf mutant of *Torreya grandis* according to transcriptome sequencing, chlorophyll a fluorescence, and modulated 820 nm reflection. *Cells* **2022**, *11*, 431. [CrossRef] [PubMed]
3. Yang, Y.; Chen, X.; Xu, B.; Li, Y.; Ma, Y.; Wang, G. Phenotype and transcriptome analysis reveals chloroplast development and pigment biosynthesis together influenced the leaf color formation in mutants of *Anthurium andraeanum* 'Sonate'. *Front. Plant Sci.* **2015**, *6*, 139. [CrossRef]
4. Tian, Y.; Rao, S.; Li, Q.; Xu, M.; Wang, A.; Zhang, H.; Chen, J. The coloring mechanism of a novel golden variety in *Populus deltoides* based on the RGB color mode. *For. Res.* **2021**, *1*, 5. [CrossRef]
5. Wu, Y.; Guo, J.; Wang, T.; Cao, F.; Wang, G. Metabolomic and transcriptomic analyses of mutant yellow leaves provide insights into pigment synthesis and metabolism in Ginkgo biloba. *BMC Genom.* **2020**, *21*, 858. [CrossRef] [PubMed]
6. Wang, H.; Wang, X.; Song, W.; Bao, Y.; Zhang, H. PdMYB118, isolated from a red leaf mutant of Populus deltoids, is a new transcription factor regulating anthocyanin biosynthesis in poplar. *Plant Cell Rep.* **2019**, *38*, 927–936. [CrossRef]
7. Fan, L.; Hou, Y.; Zheng, L.; Shi, H.; Liu, Z.; Wang, Y.; Li, S.; Liu, L.; Guo, M.; Yang, Z. Characterization and fine mapping of a yellow leaf gene regulating chlorophyll biosynthesis and chloroplast development in cotton (*Gossypium arboreum*). *Gene* **2023**, *885*, 147712. [CrossRef]

8. Qin, H.; Guo, J.; Jin, Y.; Li, Z.; Chen, J.; Bie, Z.; Luo, C.; Peng, F.; Yan, D.; Kong, Q. Integrative analysis of transcriptome and metabolome provides insights into the mechanisms of leaf variegation in *Heliopsis helianthoides*. *BMC Plant Biol.* **2024**, *24*, 731. [CrossRef]

9. Yuan, H.; Zhang, J.; Nageswaran, D.; Li, L. Carotenoid metabolism and regulation in horticultural crops. *Hortic. Res.* **2015**, *2*, 15036. [CrossRef]

10. Levin, G.; Schuster, G. LHC-like proteins: The guardians of photosynthesis. *Int. J. Mol. Sci.* **2023**, *24*, 2503. [CrossRef]

11. Su, X.; Cao, D.; Pan, X.; Shi, L.; Liu, Z.; Dall'Osto, L.; Bassi, R.; Zhang, X.; Li, M. Supramolecular assembly of chloroplast NADH dehydrogenase-like complex with photosystem I from *Arabidopsis thaliana*. *Mol. Plant* **2022**, *15*, 454–467. [CrossRef] [PubMed]

12. Goltsev, V.; Zaharieva, I.; Chernev, P.; Kouzmanova, M.; Kalaji, H.M.; Yordanov, I.; Krasteva, V.; Alexandrov, V.; Stefanov, D.; Allakhverdiev, S.I. Drought-induced modifications of photosynthetic electron transport in intact leaves: Analysis and use of neural networks as a tool for a rapid non-invasive estimation. *Biochim. Biophys. Acta Bioenerg.* **2012**, *1817*, 1490–1498. [CrossRef] [PubMed]

13. Kan, X.; Ren, J.; Chen, T.; Cui, M.; Li, C.; Zhou, R.; Zhang, Y.; Liu, H.; Deng, D.; Yin, Z.; et al. Effects of salinity on photosynthesis in maize probed by prompt fluorescence, delayed fluorescence and P700 signals. *Environmental* **2017**, *140*, 56–64. [CrossRef]

14. Chen, W.; Jia, B.; Chen, J.; Feng, Y.; Li, Y.; Chen, M.; Liu, H.; Yin, Z. Effects of different planting densities on photosynthesis in maize determined via prompt fluorescence, delayed fluorescence and P700 signals. *Plants* **2021**, *10*, 276. [CrossRef]

15. Andrzejowska, A.; Hájek, J.; Puhovkin, A.; Harańczyk, H.; Barták, M. Freezing temperature effects on photosystem II in Antarctic lichens evaluated by chlorophyll fluorescence. *J. Plant Physiol.* **2024**, *294*, 154192. [CrossRef]

16. Rastogi, A.; Kovar, M.; He, X.; Zivcak, M.; Kataria, S.; Kalaji, H.; Skalicky, M.; Ibrahimova, U.; Hussain, S.; Mbarki, S. JIP-test as a tool to identify salinity tolerance in sweet sorghum genotypes. *Photosynthetica* **2020**, *58*, 333–343. [CrossRef]

17. Evans, E.H.; Crofts, A.R. The relationship between delayed fluorescence and the H+ gradient in chloroplasts. *Biochim. Biophys. Acta Bioenerg.* **1973**, *292*, 130–139. [CrossRef]

18. Jursinic, P.; Govindjee; Wraight, C. Membrane potential and microsecond to millisecond delayed light emission after a single excitation flash in isolated chloroplasts. *Photochem. Photobiol.* **1978**, *27*, 61–71. [CrossRef]

19. Gang, H.; Liu, G.; Chen, S.; Jiang, J.J.F. Physiological and transcriptome analysis of a yellow-green leaf mutant in Birch (*Betula platyphylla* × *B. pendula*). *Forests* **2019**, *10*, 120. [CrossRef]

20. Lichtenthaler, H.K. Chlorophylls and carotenoids: Pigments of photosynthetic biomembranes. In *Methods in Enzymology*; Elsevier: Amsterdam, The Netherlands, 1987; Volume 148, pp. 350–382.

21. Strasser, R.J.; Tsimilli-Michael, M.; Qiang, S.; Goltsev, V. Simultaneous in vivo recording of prompt and delayed fluorescence and 820-nm reflection changes during drying and after rehydration of the resurrection plant *Haberlea rhodopensis*. *Biochim. Biophys. Acta Bioenerg.* **2010**, *1797*, 1313–1326. [CrossRef]

22. Gao, J.; Li, P.; Ma, F.; Goltsev, V. Photosynthetic performance during leaf expansion in Malus micromalus probed by chlorophyll a fluorescence and modulated 820 nm reflection. *J. Photochem.* **2014**, *137*, 144–150.

23. Love, M.I.; Huber, W.; Anders, S. Moderated estimation of fold change and dispersion for RNA-seq data with DESeq2. *Genome Biol.* **2014**, *15*, 550. [CrossRef] [PubMed]

24. Zhang, R.; Zhang, X.; Camberato, J.; Xue, J. Photosynthetic performance of maize hybrids to drought stress. *Russ. J. Plant Physiol.* **2015**, *62*, 788–796. [CrossRef]

25. Ducruet, J.-M.; Lemoine, Y. Increased heat sensitivity of the photosynthetic apparatus in triazine-resistant biotypes from different plant species. *Plant Cell Physiol.* **1985**, *26*, 419–429. [CrossRef]

26. Oukarroum, A.; Goltsev, V.; Strasser, R.J. Temperature effects on pea plants probed by simultaneous measurements of the kinetics of prompt fluorescence, delayed fluorescence and modulated 820 nm reflection. *PLoS ONE* **2013**, *8*, e59433. [CrossRef]

27. Paunov, M.; Koleva, L.; Vassilev, A.; Vangronsveld, J.; Goltsev, V. Effects of different metals on photosynthesis: Cadmium and zinc affect chlorophyll fluorescence in durum wheat. *Int. J. Mol. Sci.* **2018**, *19*, 787. [CrossRef]

28. Kalaji, H.M.; Schansker, G.; Ladle, R.J.; Goltsev, V.; Bosa, K.; Allakhverdiev, S.I.; Brestic, M.; Bussotti, F.; Calatayud, A.; Dąbrowski, P. Frequently asked questions about in vivo chlorophyll fluorescence: Practical issues. *Photosynth. Res.* **2014**, *122*, 121–158. [CrossRef]

29. Oukarroum, A.; Bussotti, F.; Goltsev, V.; Kalaji, H.M.; Botany, E. Correlation between reactive oxygen species production and photochemistry of photosystems I and II in *Lemna gibba* L. plants under salt stress. *Environmental* **2015**, *109*, 80–88. [CrossRef]

30. Mihaljević, I.; Viljevac Vuletić, M.; Tomaš, V.; Horvat, D.; Zdunić, Z.; Vuković, D. PSII photochemistry responses to drought stress in autochthonous and modern sweet cherry cultivars. *Photosynthetica* **2021**, *59*, 517–528. [CrossRef]

31. Jiang, D.; Chu, X.; Li, M.; Hou, J.; Tong, X.; Gao, Z.P.; Chen, G.X. Exogenous spermidine enhances salt-stressed rice photosynthetic performance by stabilizing structure and function of chloroplast and thylakoid membranes. *Photosynthetica* **2020**, *58*, 61–71. [CrossRef]

32. Goltsev, V.; Zaharieva, I.; Chernev, P.; Strasser, R.J. Delayed fluorescence in photosynthesis. *Photosynth. Res.* **2009**, *101*, 217–232. [CrossRef] [PubMed]

33. Kalaji, H.M.; Goltsev, V.; Bosa, K.; Allakhverdiev, S.I.; Strasser, R.J.; Govindjee. Experimental in vivo measurements of light emission in plants: A perspective dedicated to David Walker. *Photosynth. Res.* **2012**, *114*, 69–96. [CrossRef] [PubMed]

34. Zhang, K.; Chen, B.-H.; Yan, H.; Rui, Y.; Wang, Y.-A. Effects of short-term heat stress on PSII and subsequent recovery for senescent leaves of *Vitis vinifera* L. cv. Red Globe. *J. Integr. Agric.* **2018**, *17*, 2683–2693. [CrossRef]

35. Zhou, R.; Xu, J.; Li, L.; Yin, Y.; Xue, B.; Li, J.; Sun, F. Exploration of the Effects of Cadmium Stress on Photosynthesis in *Oenanthe javanica* (Blume) DC. *Toxics* **2024**, *12*, 307. [CrossRef]
36. Chen, S.; Yang, J.; Zhang, M.; Strasser, R.J.; Qiang, S.; Botany, E. Classification and characteristics of heat tolerance in *Ageratina adenophora* populations using fast chlorophyll a fluorescence rise OJIP. *Environmental* **2016**, *122*, 126–140.
37. Sato, Y.; Morita, R.; Katsuma, S.; Nishimura, M.; Tanaka, A.; Kusaba, M. Two short-chain dehydrogenase/reductases, NON-YELLOW COLORING 1 and NYC1-LIKE, are required for chlorophyll b and light-harvesting complex II degradation during senescence in rice. *Plant J.* **2009**, *57*, 120–131. [CrossRef]
38. Sakuraba, Y.; Kim, Y.-S.; Yoo, S.-C.; Hörtensteiner, S.; Paek, N.-C.J.B.; Communications, B.R. 7-Hydroxymethyl chlorophyll a reductase functions in metabolic channeling of chlorophyll breakdown intermediates during leaf senescence. *Biochemical* **2013**, *430*, 32–37. [CrossRef]
39. Zhang, K.; Liu, Z.; Shan, X.; Li, C.; Tang, X.; Chi, M.; Feng, H. Physiological properties and chlorophyll biosynthesis in a Pak-choi (*Brassica rapa* L. ssp. *chinensis*) yellow leaf mutant, pylm. *Acta Physiol. Plant.* **2017**, *39*, 22. [CrossRef]
40. Meguro, M.; Ito, H.; Takabayashi, A.; Tanaka, R.; Tanaka, A. Identification of the 7-hydroxymethyl chlorophyll a reductase of the chlorophyll cycle in Arabidopsis. *Plant Cell* **2011**, *23*, 3442–3453. [CrossRef]
41. Tanaka, A.; Ito, H.; Physiology, C. Chlorophyll degradation and its physiological function. *Plant* **2024**, pcae093. [CrossRef]
42. Liu, W.; Chen, G.; Chen, J.; Jahan, M.S.; Guo, S.; Wang, Y.; Sun, J.J.P. Overexpression of 7-hydroxymethyl chlorophyll a reductase from cucumber in tobacco accelerates dark-induced chlorophyll degradation. *Plants* **2021**, *10*, 1820. [CrossRef] [PubMed]
43. Guan, J.; Teng, K.; Yue, Y.; Guo, Y.; Liu, L.; Yin, S.; Han, L. *Zoysia japonica* chlorophyll b reductase gene NOL participates in chlorophyll degradation and photosynthesis. *Front. Plant Sci.* **2022**, *13*, 906018. [CrossRef] [PubMed]
44. Wang, H.-T.; Ou, L.-Y.; Chen, T.-A.; Kuan, Y.-C. Refrigeration, forchlorfenuron, and gibberellic acid treatments differentially regulate chlorophyll catabolic pathway to delay yellowing of broccoli. *Postharvest Biol. Technol.* **2023**, *197*, 112221. [CrossRef]
45. Armstrong, G.A. Greening in the dark: Light-independent chlorophyll biosynthesis from anoxygenic photosynthetic bacteria to gymnosperms. *J. Photochem. Photobiol. B Biol.* **1998**, *43*, 87–100. [CrossRef]
46. Wang, Q.; Liu, S.; Li, X.; Wu, H.; Shan, X.; Wan, Y. Expression of Genes in New Sprouts of *Cunninghamia lanceolata* grown under dark and light conditions. *J. Plant Growth Regul.* **2020**, *39*, 481–491. [CrossRef]
47. Weber, A.P.; Schneidereit, J.; Voll, L.M. Using mutants to probe the in vivo function of plastid envelope membrane metabolite transporters. *J. Exp. Bot.* **2004**, *55*, 1231–1244. [CrossRef]
48. Kössler, S.; Armarego-Marriott, T.; Tarkowská, D.; Turečková, V.; Agrawal, S.; Mi, J.; de Souza, L.P.; Schöttler, M.A.; Schadach, A.; Fröhlich, A.J.; et al. Lycopene β-cyclase expression influences plant physiology, development, and metabolism in tobacco plants. *J. Exp. Bot.* **2021**, *72*, 2544–2569. [CrossRef]
49. Nisar, N.; Li, L.; Lu, S.; Khin, N.C.; Pogson, B.J. Carotenoid metabolism in plants. *Mol. Plant* **2015**, *8*, 68–82. [CrossRef]
50. Zakar, T.; Laczko-Dobos, H.; Toth, T.N.; Gombos, Z. Carotenoids assist in cyanobacterial photosystem II assembly and function. *Front. Plant Sci.* **2016**, *7*, 295. [CrossRef]
51. Dong, L.; Tu, W.; Liu, K.; Sun, R.; Liu, C.; Wang, K.; Yang, C. The PsbS protein plays important roles in photosystem II supercomplex remodeling under elevated light conditions. *J. Plant Physiol.* **2015**, *172*, 33–41. [CrossRef]
52. Fernández-Marín, B.; Neuner, G.; Kuprian, E.; Laza, J.M.; García-Plazaola, J.I.; Verhoeven, A. First evidence of freezing tolerance in a resurrection plant: Insights into molecular mobility and zeaxanthin synthesis in the dark. *Physiol. Plant.* **2018**, *163*, 472–489. [CrossRef] [PubMed]
53. Hao, Q.; Zhang, G.; Zuo, X.; He, Y.; Zeng, H. Cia zeaxanthin biosynthesis, OsZEP and OsVDE regulate striped leaves occurring in response to deep transplanting of rice. *Int. J. Mol. Sci.* **2022**, *23*, 8340. [CrossRef] [PubMed]
54. Shi, Y.; Guo, J.; Zhang, W.; Jin, L.; Liu, P.; Chen, X.; Li, F.; Wei, P.; Li, Z.; Li, W. Cloning of the lycopene β-cyclase gene in Nicotiana tabacum and its overexpression confers salt and drought tolerance. *Int. J. Mol. Sci.* **2015**, *16*, 30438–30457. [CrossRef] [PubMed]
55. Alquezar, B.; Rodrigo, M.J.; Lado, J.; Zacarías, L. A comparative physiological and transcriptional study of carotenoid biosynthesis in white and red grapefruit (*Citrus paradisi* Macf.). *Tree Genet. Genomes* **2013**, *9*, 1257–1269. [CrossRef]
56. Wu, H.; Shi, N.; An, X.; Liu, C.; Fu, H.; Cao, L.; Feng, Y.; Sun, D.; Zhang, L. Candidate genes for yellow leaf color in common wheat (*Triticum aestivum* L.) and major related metabolic pathways according to transcriptome profiling. *Int. J. Mol. Sci.* **2018**, *19*, 1594. [CrossRef]
57. Roach, T.; Krieger-Liszkay, A. The role of the PsbS protein in the protection of photosystems I and II against high light in *Arabidopsis thaliana*. *Biochim. Biophys. Acta Bioenerg.* **2012**, *1817*, 2158–2165. [CrossRef]
58. Zhang, M.; Yang, X. PsbS expression analysis in two ecotypes of *Sedum alfredii* and the role of SaPsbS in Cd tolerance. *Russ. J. Plant Physiol.* **2014**, *61*, 427–433. [CrossRef]
59. Yang, Y.N.; Le, T.T.L.; Hwang, J.-H.; Zulfugarov, I.S.; Kim, E.-H.; Kim, H.U.; Jeon, J.-S.; Lee, D.-H.; Lee, C.-H. High light acclimation mechanisms deficient in a PsbS-knockout *Arabidopsis mutant*. *Int. J. Mol. Sci.* **2022**, *23*, 2695. [CrossRef]
60. Štroch, M.; Čajánek, M.; Kalina, J.; Špunda, V. Regulation of the excitation energy utilization in the photosynthetic apparatus of chlorina f2 barley mutant grown under different irradiances. *J. Photochem. Photobiol. B Biol.* **2004**, *75*, 41–50. [CrossRef]
61. Kim, Y.-K.; Lee, J.-Y.; Cho, H.S.; Lee, S.S.; Ha, H.J.; Kim, S.; Choi, D.; Pai, H.-S. Inactivation of organellar glutamyl-and seryl-tRNA synthetases leads to developmental arrest of chloroplasts and mitochondria in higher plants. *J. Biol. Chem.* **2005**, *280*, 37098–37106. [CrossRef]

62. Valencia, W.M.; Pandit, A. Photosystem II subunit S (PsbS): A nano regulator of plant photosynthesis. *J. Mol. Biol.* **2023**, *436*, 168407.
63. Kang, S.; Jeon, S.; Kim, S.; Chang, Y.K.; Kim, Y.C. Development of a pVEC peptide-based ribonucleoprotein (RNP) delivery system for genome editing using CRISPR/Cas9 in *Chlamydomonas reinhardtii*. *Sci. Rep.* **2020**, *10*, 22158. [CrossRef] [PubMed]

 *horticulturae* 

Article

Comprehensive Identification and Expression Profiling of Lipoxygenase Genes in Sweet Cherry During Fruit Development

Quanjuan Fu [1,2], Di Xu [2], Sen Hou [1,2], Rui Gao [2], Jie Zhou [2], Chen Chen [2], Shengnan Zhu [3], Guoqin Wei [1,2,*] and Yugang Sun [2,*]

[1] National Center of Technology Innovation for Comprehensive Utilization of Saline-Alkali Land, Dongying 257000, China; yantaifqj@163.com (Q.F.); housenmail@163.com (S.H.)
[2] Shandong Institution of Pomology, Taian 271000, China; xudi20020120@163.com (D.X.); gaorui368@163.com (R.G.); zhoujie17870386870@163.com (J.Z.); chenchen4065@163.com (C.C.)
[3] Tai'an Forestry Protection and Development Center, Taian 271000, China; judyhappy84@126.com
* Correspondence: guoqinw1983@126.com (G.W.); sds129@126.com (Y.S.)

Abstract: Lipoxygenase (LOX) is involved in the oxidation of fatty acids in plants and is a ubiquitous oxygenase that plays an important role in the process of plant resistance to adversity. In this study, the LOX gene family in the sweet cherry genome was identified by bioinformatics methods, the chromosomal mapping of different LOX genes was carried out, and the homology alignment and functional domain analysis of the encoded proteins were performed. The results showed that there were nine LOX gene sequences in the sweet cherry LOX gene family, and the subcellular localization was mainly located in the cytoplasm, chloroplast, or plasma membrane, and was concentrated on chromosomes 1, 2, 3, 4, 6, and 8. During the ripening process of sweet cherry fruits, the LOX gene family showed five different expression patterns, the expression peak of different LOX genes reached the peak of expression at a specific development period, all LOX genes jointly promoted the growth and development of fruits, the enzyme activities of LOX in different varieties of early and late ripening cherries exhibited great differences during the development process, and the results of volatile content in the later stages also showed that different varieties of cherries had their specificity. The results of this study provide a theoretical basis for further revealing the specific functions of LOX gene family members in sweet cherry.

Keywords: sweet cherry; LOX gene family; bioinformatics; expression patterns; volatile analysis

Citation: Fu, Q.; Xu, D.; Hou, S.; Gao, R.; Zhou, J.; Chen, C.; Zhu, S.; Wei, G.; Sun, Y. Comprehensive Identification and Expression Profiling of Lipoxygenase Genes in Sweet Cherry During Fruit Development. *Horticulturae* **2024**, *10*, 1361. https://doi.org/10.3390/horticulturae10121361

Academic Editor: Christopher M. Menzel

Received: 30 October 2024
Revised: 11 December 2024
Accepted: 11 December 2024
Published: 18 December 2024

1. Introduction

Sweet cherry is a subgenus of *Prunus* (L.), which is widely cultivated worldwide, mainly in Europe, Asia, and North America. Sweet cherry plants have developed rapidly in China, with a cultivation area of 140,000 hm^2 and an output of about 500,000 tons, which can be eaten fresh or processed into cherry wine, cherry juice, canned cherries, preserved fruits, etc., with high cultivation efficiency. The ripe fruit of each fruit tree contains its own unique aroma compounds, and sweet cherries are no exception. At present, more than 60 volatile compounds have been isolated and identified in sweet cherry fruits, including aldehydes, alcohols, terpenes, lipids, and ketones [1]. Studies have shown that C6 aldehydes and alcohols such as hexanal, (E)-2-hexenal, and (E)-2-hexenol are characteristic components of sweet cherry fruit aroma, accounting for more than 50% of the volatile content [2]. The C6 aldehydes and alcohols with "green-flavored" aroma are mainly derived from the fatty acid metabolism pathway [3].

Lipoxygenase (LOX) is a key enzyme in the fatty acid metabolism pathway, which is a class of double oxygenases containing non-heme iron ions widely present in aerobic organisms, which mainly catalyzes the oxygenation reaction of unsaturated fatty acids such as linolenic acid and linoleic acid to generate hydroperoxides [4,5]; this participates in the synthesis of aromatic substances such as aldehydes and alcohols in plants [6]. LOX has always been a research hotspot and has a variety of physiological functions in plants, for example, seed germination, tuber growth, root development, the signal transduction of insect and disease resistance, maturity, and the senescence of pollen, leaves, fruits, and fruit aroma formation [7].

LOX is a polygenic family, and studies have shown that there are 21 LOX genes in poplars [8], 18 LOX genes in grapes [9,10], and 6 LOX genes in *Arabidopsis thaliana* [11,12]. Different forms of LOX proteins may be localized to different subcellular organelles, such as chloroplasts, microsomes, and liposomes [13], while LOX genes exhibit different expression patterns at different stages of development in plants or under different stress conditions [14–16]. For example, the expression of LOX9 gene was significantly up-regulated when tobacco was infected by pathogenic bacteria *Phytophthora Parasitica* Var. *nicotianae* [17], and the specific expression of LOX13 gene and induced SA biosynthesis in tomato and rice when it was mechanically damaged [18–20].

Based on the successful sequencing of the sweet cherry genome, it is possible to study the LOX gene of sweet cherry at the genome-wide level. In this study, the LOX gene family of cherries was identified and classified by bioinformatics methods, the expression level of LOX gene during the development of sweet cherry fruit was analyzed by qRT-PCR, the enzyme activity of LOX at different development stages was determined, and the aroma components of C6 aldehyde and alcohol in the fruit were analyzed, which provided a theoretical basis for further study of the LOX gene function of sweet cherry.

2. Materials and Methods

2.1. Experimental Materials

The test materials were two sweet cherry varieties, the early-maturing sweet cherry variety 'Brooks' and the late-maturing sweet cherry variety 'Lu Yu'. The materials were collected from the cherry orchard of Tianpinghu Experimental Base of Shandong Provincial Fruit Research Institute in Tai'an Prefecture, and the rootstock was 'Gisela 5'. Consistently managed 7-year-old plants with good growth were selected and sampled every 5 days during the fruit development period of sweet cherry (10–25 May), and the sampling period was at 9:00~10:00 a.m. each time. The sampling site was 1~2 m outside the canopy, using random sampling method, 100 fruits of each variety are collected each time, and the samples are immediately stored with liquid nitrogen and brought back to the laboratory for later use.

2.2. LOX Gene Family Identification and Structural Analysis

The sequencing data of the sweet cherry genome, Coding Sequence (CDS) regions, protein sequence, and annotation information are all from the NCBI website. By consulting the domestic and foreign literature, the LOX amino acid sequences of five species of plants, namely *Arabidopsis thaliana*, rice, tomato, grape, and poplar, were obtained. The domain information of LOX gene was downloaded from the Pfam database, the Profile HMM model was established by using Hmmer3.0 software, and the protein sequences in the sweet cherry genome data were retrieved and downloaded by using this model; in addition, a new HMM model was established according to the obtained sweet cherry LOX gene family sequence, the sweet cherry protein sequences were retrieved again with the new model, the obtained sequences were considered to be candidate protein sequences, and all protein domains were identified using SMART Validation (http://smart.embl-Heidelberg.de/, accessed on 20 September 2023).

The GSDS online tool (http://gsds.cbi.pku.edu.cn/, accessed on 25 September 2023) was used to create a gene exon-intron structure diagram. The chromosomal mapping of the LOX gene in sweet cherry was carried out, and the subcellular localization of the LOX gene was predicted by using the Plant-Ploc online tool, WOLFPSORT, and Softberry.

2.3. Prediction of LOX Protein Structure in Sweet Cherry

Online tools such as ProtParam (http://web.expasy.org/protparam/, accessed on 30 September 2023) were used to assess the theoretical isoelectric point, molecular weight, and other physicochemical properties of Sakura LOX protein, and the online analytics tools HNN Secondary Structure Prediction were taken advantage of for forecasting the secondary structures, such as α-helix, extension chains, etc., of LOX proteins.

2.4. Phylogenetic Tree Construction of LOX Amino Acid Sequence in Sweet Cherry

DNAMAN software (version 6) was used to compare the amino acid sequences of different LOX genes in *Arabidopsis thaliana* and sweet cherry, and the results of the comparison were constructed by software MEGA6.0.

2.5. Analysis of the Expression Pattern of the LOX Gene Family in Sweet Cherry

The samples were collected at the five stages of sweet cherry fruit development (T1, T2, T3, T4, and T5). Total RNA was extracted by using the RNA Extraction Kit (Tiangen, DP504, Beijing, China) after the cryopreservation of liquid nitrogen, and the specific primers (Supplement file-PaLOX Primer) for different LOX genes were designed by Primer 5.0 software (Sangon Biotech (Shanghai) Co., Ltd., Shanghai, China). The expression levels of different genes in the sweet cherry LOX gene family were detected by qRT-PCR, and the data were processed and graphed by Microsoft EXCEL 2010 software.

2.6. LOX Enzyme Activity Assay

Lipoxygenase (LOX) activity was determined according to the method of Chen Kunsong et al. [21]: 2.0 g of pulp tissue was taken and fully ground into powder form with liquid nitrogen and 10 mL of 50 mmol$\cdot$L^{-1} phosphate buffer (pH 7.0); the mixture was pre-cooled at 4 °C, mixed well, and extracted; 15,000 r$\cdot$min^{-1} (4 °C) was centrifuged for 15 min; and the supernatant was used for LOX activity assay. The 3 mL reaction system contained 25 µL of sodium linoleate mother liquor, 2.775 mL of 100 mmol$\cdot$L^{-1} acetic acid buffer (pH 5.5), crude enzyme solution 0.2 mL, reaction temperature 30 °C, and detection wavelength 234 nm. After adding the crude enzyme solution, the change in OD value within 15 s to 1 min was recorded, and the enzyme activity was measured in ΔOD_{234} g$^{-1}\cdot$FW. Each experiment was repeated three times.

2.7. Headspace Solid-Phase Microextraction and Gas Chromatography–Mass Spectrometry Analysis

The fruit aroma compounds were determined by headspace solid-phase microextraction (HS-SPME) and gas chromatography–mass spectrometry (QP 2010Plus, Shimadzu, Tokyo, Japan) according to the method of Zhang Xu et al. [1]. Chromatographic conditions: injection port temperature 250 °C, splitless injection, programmed temperature increase, 36 °C for 5 min, 2 °C$\cdot$min^{-1} to 50 °C, then 5 °C$\cdot$min^{-1} to 100 °C, then 8 °C$\cdot$min^{-1} to 180 °C, 15 °C$\cdot$min^{-1} to 250 °C, and hold for 3 min; carrier gas, He gas; column flow rate, 50 mL$\cdot$min^{-1}. Mass spectrometry conditions: ionization, EI; electron energy, 70 eV. Full scanning method: scanning range, 35~500 m$\cdot$z^{-1}; ion source temperature, 250 °C. Dioctanone was used as the internal standard, and the content of each component was calculated by calculating the peak area ratio. Each component was searched and analyzed using the computer spectral library (NBS/WILEY) and NIST14 mass spectrometry library.

3. Results and Analysis

3.1. Identification of the LOX Gene Family in Sweet Cherry

The conserved domain of the LOX gene was analyzed by SMART (version 9) and CD-Search (CDD version 3.21) software, and it was found that the protein domain numbered Pav_sc0001305.1_g630.1.mk was quite different from other proteins, excluding this protein; the rest of the proteins all exhibited the LH2 domain and most of the proteins also possessed a lipoxygenase conserved domain (Figure 1). A total of nine LOX gene family members were identified in the sweet cherry genome (Table 1), with PaLOX5b and PaLOX6 located on chromosome 1, PaLOX2b and PaLOX2c on chromosome 2, PaLOX3 on chromosome 3, PaLOX4 on chromosome 4, PaLOX1 and PaLOX2a on chromosome 6, and PaLOX5a on chromosome 8 (Figure 2).

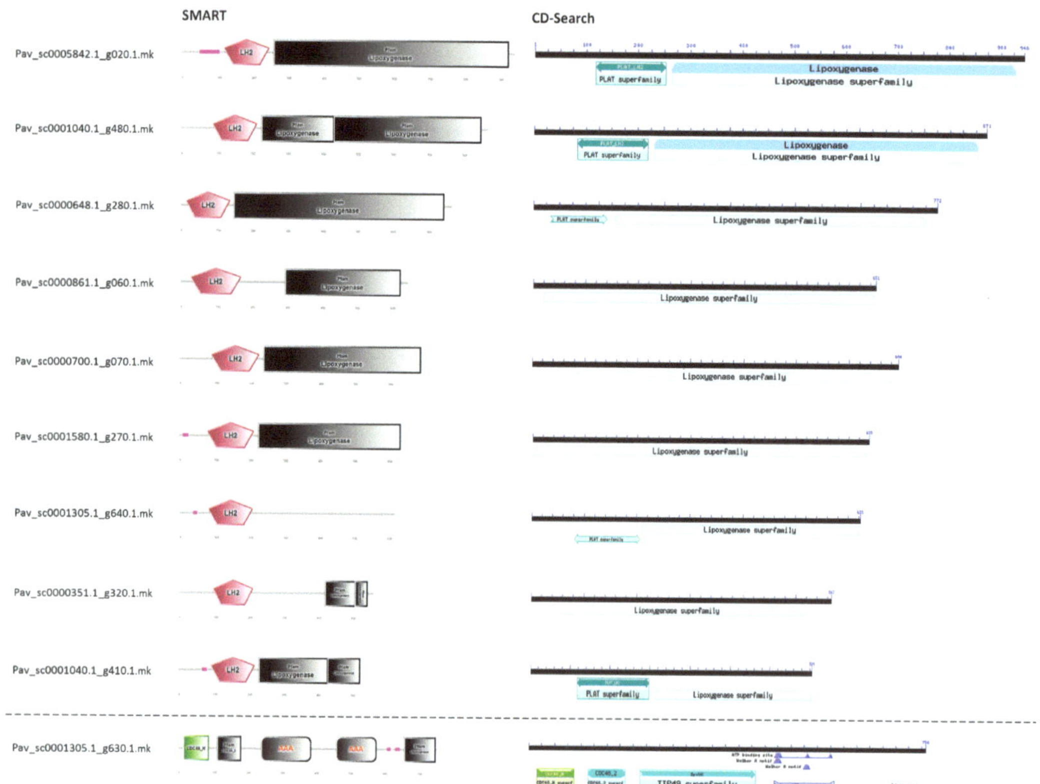

Figure 1. Analysis of conserved domains of sweet cherry LOX genes. Pink blocks: low complexity regions. LH2: lipoxygenase homology 2 (beta barrel) domain. CDC48_N: cell division protein 48 (CDC48) N-terminal domain. CDC48_2: Pfam domain (PF02933), cell division protein 48 (CDC48), domain 2. AAA: ATPases associated with a variety of cellular activities.

Table 1. LOX gene family information of sweet cherry types.

Query (ID)	Gene Name	Chromosome	Genomic Location	Gene Size (bp)
Pav_sc0000700.1_g070.1.mk	PaLOX1	Chr6	22272340-22275934	2085
Pav_sc0005842.1_g020.1.mk	PaLOX2a	Chr6	3557941-3563600	2841
Pav_sc0001040.1_g480.1.mk	PaLOX2b	Chr2	6171098-6176380	2616
Pav_sc0001040.1_g410.1.mk	PaLOX2c	Chr2	6226304-6228884	1590
Pav_sc0001580.1_g270.1.mk	PaLOX3	Chr3	1958879-1963407	1920
Pav_sc0001305.1_g640.1.mk	PaLOX4	Chr4	2866292-2869838	1872
Pav_sc0000648.1_g280.1.mk	PaLOX5a	Chr8	15960601-15964259	2319
Pav_sc0000861.1_g060.1.mk	PaLOX5b	Chr1	932547-937116	1956
Pav_sc0000351.1_g320.1.mk	PaLOX6	Chr1	18734255-18738288	1704

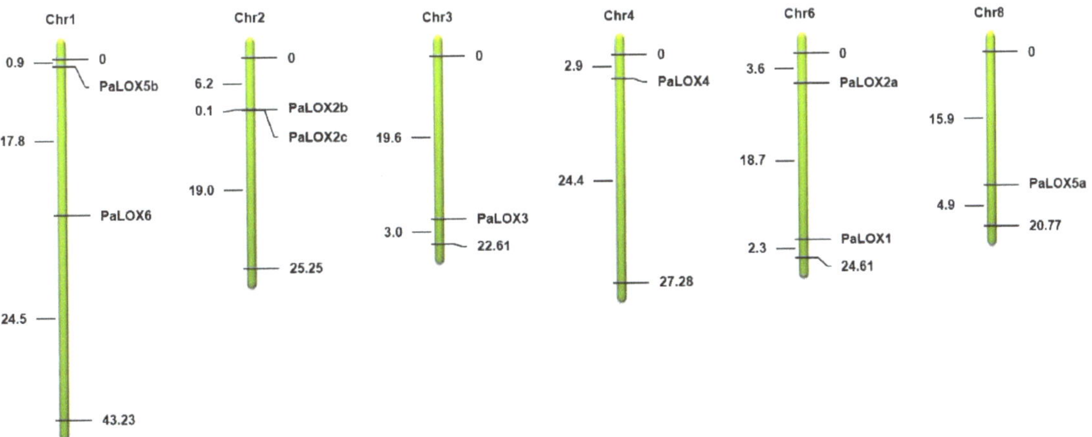

Figure 2. Chromosome location of sweet cherry LOX genes.

3.2. Analysis of Physicochemical Properties of LOX Gene Family in Sweet Cherry

The physicochemical properties of LOX proteins were analyzed using the Expasy online website. Table 2 showed that among the nine LOX gene families in sweet cherry, the maximum number of amino acids encoded by PaLOX2a was 946aa and the number of amino acids encoded by PaLOX6 was at least 567aa; the results showed that the number of amino acids encoded by the LOX gene family was positively correlated with its molecular weight, and the amino acid quantities and molecular weights of the nine LOX genes were different. According to the theoretical isoelectric point amino acids, they can be divided into acidic amino acids and basic amino acids; PaLOX2a, PaLOX2b, PaLOX4, PaLOX5a, and PaLOX5b belong to the sweet cherry LOX gene family, and PaLOX1, PaLOX2c, PaLOX3, and PaLOX6 belong to basic amino acids. The prediction of subcellular localization showed that PaLOX1, PaLOX2a, PaLOX2b, PaLOX2c, PaLOX5a, and PaLOX5b were localized in the cytoplasm; PaLOX3 and PaLOX4 were localized in chloroplasts; and PaLOX6 was localized in the plasma membrane.

Table 2. Analysis of physicochemical properties of LOX gene family in sweet cherry.

Gene Name	Amino Acid (aa)	Mass Weight (Da)	pI	Protein Subcellular Location
PaLOX1	694	78.7	7.26	cytoplasm
PaLOX2a	946	107.1	6.38	cytoplasm
PaLOX2b	871	98.2	6	cytoplasm
PaLOX2c	529	59.4	7.79	cytoplasm
PaLOX3	639	72.3	9.45	chloroplast
PaLOX4	623	70.4	6.95	chloroplast
PaLOX5a	772	88.3	6.29	cytoplasm
PaLOX5b	651	74.6	6.77	cytoplasm
PaLOX6	567	63	8.71	plasma membrane

3.3. Cluster Analysis of Sweet Cherry LOX Amino Acid Sequences

According to the length and sequence structure characteristics of sweet cherry LOX family amino acids, the phylogenetic tree was constructed by using MEGA to construct the LOX amino acid sequence of sweet cherry and the LOX amino acid sequence of *Arabidopsis thaliana*, which was classified and functional. The phylogenetic tree showed that PaLOX1 protein was most closely related to AtLOX1 protein, and the other PaLOX2, PaLOX3, PaLOX4, PaLOX5, and PaLOX6 proteins in sweet cherry were closely related to AtLOX2, AtLOX3, AtLOX4, AtLOX5, and AtLOX6 proteins in *Arabidopsis thaliana*, respectively. The distribution range of introns in sweet cherry LOX gene was 5~8, including three genes with 6 introns, three genes with 7 introns, two genes with 8 introns, and one gene with 5 introns (Figure 3).

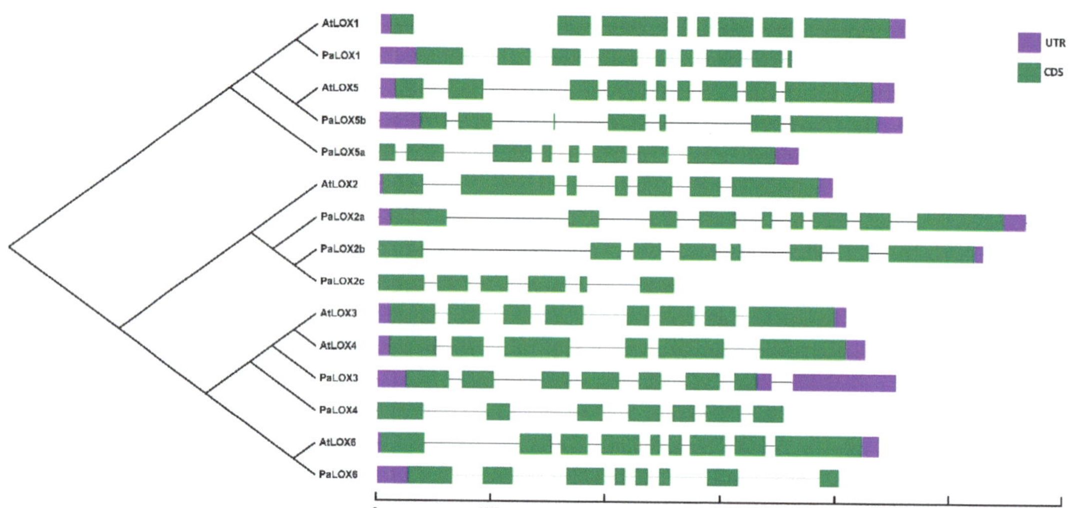

Figure 3. Phylogenetic tree of the amino acid sequences of LOX in sweet cherry and *Arabidopsis*. The purple and green blocks represent the untranslated region (UTR) and coding sequence (CDS) regions, respectively, and the introns are represented by black lines.

3.4. Quantitative Analysis of the Expression Pattern of the LOX Gene Family in Sweet Cherry

The sweet cherry ripening process was divided into five stages, namely T1, T2, T3, T4, and T5 (Figure 4A), and the expression levels of different LOX genes in sweet cherry fruit development were detected by qRT-PCR. The quantitative results showed that different LOX genes showed different expression patterns, which could be divided into five categories; the expression pattern of the first type of gene increased first and then decreased, and while LOX1, LOX3, and LOX5b all belonged to this type, the peak expression of different genes appeared in different periods and the expression levels were highest in the T2 or T3 periods, respectively. The second type of gene expression decreased first and then increased—LOX2a and LOX2c belonged to this category—and the expression levels of LOX2a and LOX2c were lowest at the T3 stage. The expression level of the LOX2b gene was basically stable from T1 to T3, and the expression level of the LOX4 gene was gradually down-regulated after T3, while the down-regulated expression of the LOX4 gene showed a trend of first being fast and then slowing. The fourth type of gene showed an up-regulated expression pattern; for example, the expression level of the LOX6 gene was gradually up-regulated. The expression of the fifth type of gene was relatively stable, such as the LOX5a gene, and there was no significant difference in the expression level of the LOX5a gene at different stages of fruit ripening (Figure 4B).

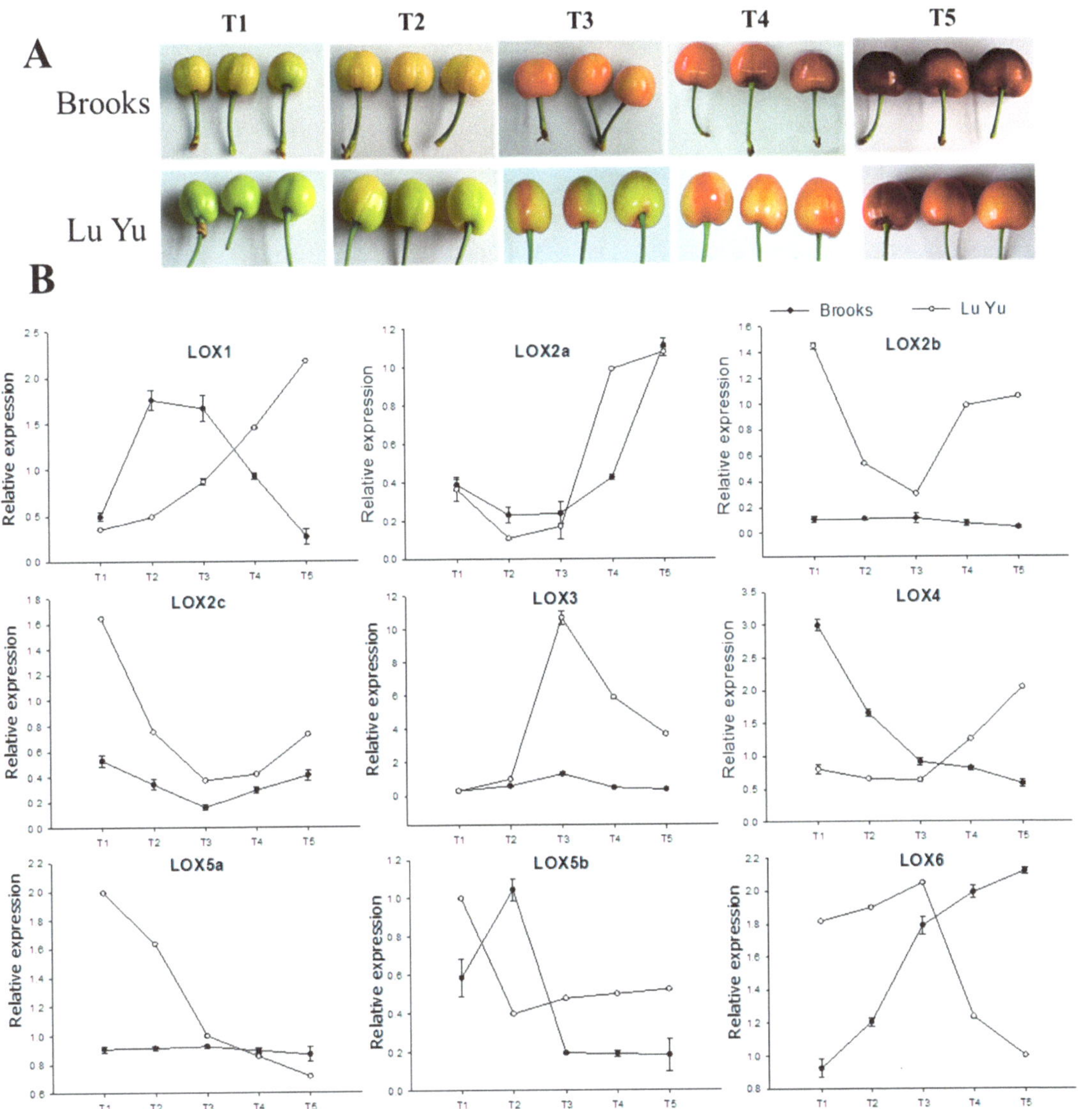

Figure 4. Analysis of LOX gene expression pattern in different developmental stages of sweet cherries. (**A**) different development stages of sweet cherry fruit, (**B**) LOX gene expression pattern in different developmental stages of sweet cherries.

3.5. Changes in LOX Enzyme Activity During Sweet Cherry Fruit Development

According to the different ripening time of sweet cherry fruits, this study selected the early-maturing sweet cherry variety 'Brooks' and the late-maturing sweet cherry variety 'Lu Yu', respectively, to determine the LOX enzyme activity (Figure 5). The results showed that the enzyme activity of LOX decreased first and then increased during the ripening process of different varieties of sweet cherry, reaching the lowest level around the T2 stage and then gradually increasing. However, there were differences in the enzyme activity of the early and late varieties of LOX; in the T1 period, the LOX enzyme activity of the late-maturing cultivar 'Lu Yu' was significantly higher than that of the early-maturing cultivar 'Brooks', the enzyme activity of the late-maturing cultivar 'Brooks' decreased rapidly, and the lowest level of LOX enzyme activity of the early-maturing cultivar 'Brooks' was still higher than that of the late-maturing cultivar 'Brooks' at about the T2 period. In addition, the LOX enzyme activity of the two varieties increased, but the LOX enzyme activity of the early-maturing cultivar 'Brooks' was higher than that of the late-maturing cultivar 'Lu Yu' in each period.

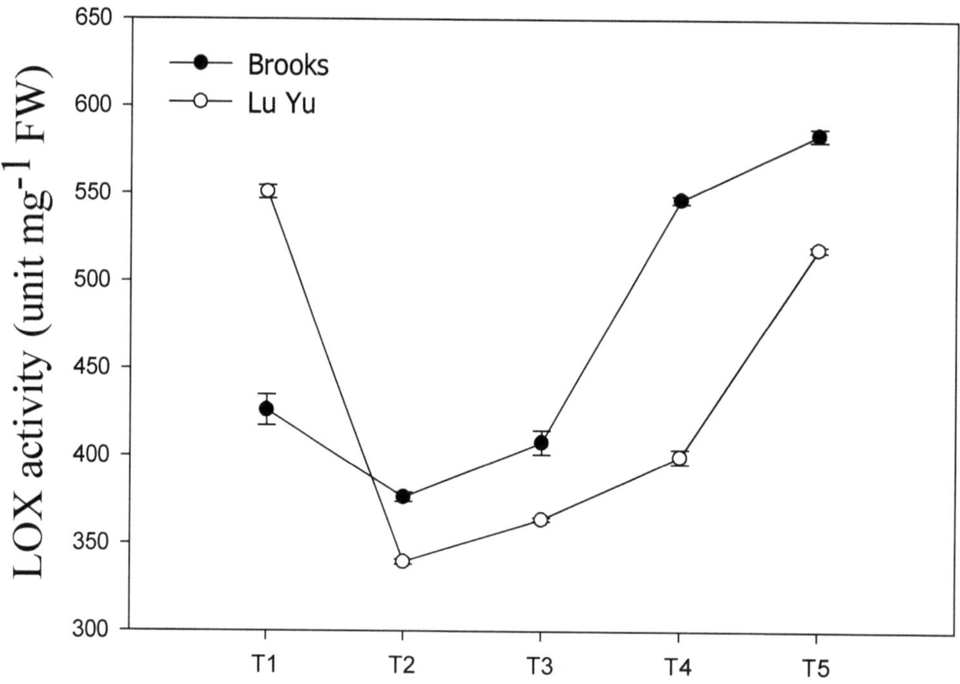

Figure 5. Analysis of the LOX enzymatic activity in different developmental stages of sweet cherries.

3.6. Changes in Aroma Compounds During Sweet Cherry Fruit Ripening

A total of 27 aroma substances were detected in sweet cherry 'Tieton' fruits, with a total content of 1615 μg·kg⁻¹, including five kinds of esters with a content of 265.24 μg·kg⁻¹ and eight kinds of alcohols with a content of 1675.42 μg·kg⁻¹, six kinds of aldehydes with a content of 648.19 μg·kg⁻¹, one kind of ketone with a content of 5.67 μg·kg⁻¹, five kinds of hydrocarbons with a content of 95.47 μg·kg⁻¹, and two acetic acids and other substances with a content of 17.16 μg·kg⁻¹ [22].

This study selected several aroma components with a high content, mainly n-hexanal, (E)-2-hexenal, and (E)-2-hexene-1-ol. Table 3 showed that in the early-maturing sweet cherry 'Brooks', the total contents accounted for 80.64%, 79.36%, 72.18%, 82.78%, and 71.93% of the total aroma components in the five stages of sweet cherry ripening, respectively. Table 4 showed that in the late-maturing sweet cherry 'Lu Yu', the total contents of these aroma components accounted for 84%, 52.25%, 66.18%, 66.7%, and 66.79% of the total aroma components, respectively. There were no significant differences in the contents of the three main aroma compounds, such as n-hexanal at the T1 stage, but with the ripening of the fruit, the content of the main aroma compounds in the early-maturing sweet cherry 'Brooks' first decreased, then increased and then decreased, while the content of the main aroma compounds in the late-maturing sweet cherry 'Lu Yu' showed a trend of first decreasing and then increasing, and was basically stable from T3 to T5.

Table 3. Content of main aroma components of the sweet cherry 'Brooks' (%).

	T1	**T2**	**T3**	**T4**	**T5**
n-hexanal	26.99 b	21.72 b	14.49 b	18.29 b	15.11 b
(E)-2-hexenal	15.07 c	11.70 c	4.21 c	5.44 c	5.22 c
(E)-2-hexen-1-ol	38.58 a	45.93 a	53.48 a	59.06 a	51.59 a
Relative total content	80.64	79.36	72.18	82.78	71.93

Values followed by different letters are signifcantly different at $p < 0.05$.

Table 4. Content of main aroma components of the sweet cherry 'Lu Yu' (%).

	T1	**T2**	**T3**	**T4**	**T5**
n-hexanal	25.25 b	14.29 b	9.38 b	12.03 b	6.42 b
(E)-2-hexenal	25.89 b	12.76 c	3.72 c	5.81 c	3.45 c
(E)-2-hexen-1-ol	32.87 a	25.20 a	53.08 a	48.87 a	56.92 a
Relative total content	84.00	52.25	66.18	66.70	66.79

Values followed by different letters are signifcantly different at $p < 0.05$.

4. Discussion

Through genome-wide analysis, nine LOX genes were identified from cherries, and the results of chromosomal mapping showed that the localization of nine LOX genes on chromosomes was random. However, the functional domain analysis of LOX gene showed that most of the proteins had a conserved domain of lipoxygenase, which was consistent with the study of the conserved domain of LOX gene in poplar [8]. The results of LOX amino acid sequence alignment between sweet cherry and *Arabidopsis thaliana* showed that the LOX gene in the two plants had a certain evolutionary homology relationship. The greater number of LOX genes in cherries compared to *Arabidopsis thaliana* suggests that in the process of evolution, through the duplication or tandemization of genes, introns and exons changed, resulting in new genes, and the functional subfunctionalization and new functionalization of genes occurred [23].

Some studies have shown that lipoxygenase is involved in the synthesis of jasmonic acid, aldehydes, and alcohols, and plays a role in resisting stress and plant growth and development [24–27]. Chauvin et al. [28] studied the *Arabidopsis thaliana* 13-LOX gene and demonstrated that this gene plays a role in the process of defending against mechanical damage by promoting JA biosynthesis. The lipoxygenase of persimmon plays an important role in fruit ripening senescence and stress response by regulating the accumulation of early reactive oxygen species and the synthesis of SA [29].

Fruit ripening undergoes a series of complex metabolic changes, and the LOX gene in tomato has also been shown to be involved in fruit ripening [30]. The results of these studies imply that the LOX gene of sweet cherry is involved in the process of resisting biotic and abiotic stresses and fruit ripening, so this study studied the changes in the expression of the LOX gene in sweet cherry fruit during ripening. The expression patterns of sweet cherry LOX genes in different fruit development stages were diverse, different types of LOX genes were expressed at high levels at specific developmental stages, and all LOX genes may play a role in sweet cherry fruit ripening process through functional complementarity. The expression level of the LOX5a gene was basically stable throughout the fruit development period, and it was speculated that this gene may play the role of "housekeeping gene", which is particularly important for the synthesis of certain essential substances and exerts its function throughout the development period.

In the study of tomato LOX genes, TomloxC was expressed at the veraison and ripening stages of fruits and was involved in the synthesis of related volatile compounds [31]. During the development of sweet cherry fruits, the enzyme activity of LOX decreased first and then increased, and the enzyme activity was lowest in the T2 stage, which was the key period when the sweet cherry fruit was about to be colored. Therefore, LOX plays a certain role in the process of sweet cherry fruit coloration and may be involved in the synthesis of aroma compounds during fruit ripening. In the four stages of T2, T3, T4, and T5, the

LOX enzyme activity of the early-maturing sweet cherry 'Brooks' was higher than that of the late-maturing sweet cherry 'Lu Yu', which indicated that higher LOX enzyme activity had a promoting effect on the ripening of sweet cherry fruits. Ju YL et al. [32] found that exogenous ABA and MeJA not only promoted the accumulation of total anthocyanins in grape fruits but also affected the increase in lipoxygenase (LOX) activity, indicating a certain correlation between the lipoxygenase metabolic pathway and anthocyanin synthesis. However, further research on the mechanism on the synthesis of anthocyanins and volatile aromas formed in the LOX pathway in cherry is still necessary.

Zhang Xu et al. [1] suggested that hexanal, (E)-2-hexenal, benzaldehyde, (E)-2-hexelol, ethyl acetate, and ethyl caproate are characteristic aroma components of ripe sweet cherry fruits. Girard et al. [33] considered hexanal, (E)-2-hexenal, (E)-2-hexel, and benzaldehyde to be the main important aroma components of sweet cherries, accounting for more than 50% of the volatile content. The results of the determination of the content of aroma compounds in the ripening process of sweet cherry fruits showed that the content of the main aroma compounds in the early-maturing cherries accounted for 71.93%~82.78% of the total volatile substance content, the content of the main aroma compounds in the late-maturing cherries accounted for 52.25~84% of the total volatile substance content, and the values were higher than 50%, which was consistent with the previous research results. In early-ripening cherries, the contents of the main aroma compounds such as n-hexanal, (E)-2-hexenal, and (E)-2-hexen-1-ol remained at a high level, while the proportion of these aroma compounds in late-ripening cherries decreased significantly, indicating that the increase in the proportion of other volatile compounds in late-ripening cherries may produce more kinds of aroma components. Combined with the results of LOX enzyme activity, the results showed that the high level of LOX gene expression in cherries increased the LOX enzyme activity and contributed to the ripening of sweet cherry fruits. In this study, the transcripts of several LOX genes, such as LOX2b, LOX2c, LOX3, and LOX5b, were higher in the 'Lu Yu' cherry cultivar at T3, T4, and T5. However, this was not reflected in a greater increase in enzyme activity compared to the 'Brooks' cultivar. We hypothesized that although these genes are all LOX enzyme-encoding genes, only certain members of this family, such as LOX6 and LOX5a, are decisive for the enzyme activity in the fruit, resulting in lower enzyme activity in the 'Lu Yu' fruit than in the 'Brooks' fruit in the later stages of fruit development. In future studies, we will perform further gene cloning and transgenic functional characterization. More in-depth research is needed on the functions of specific genes in the LOX gene family of cherries.

5. Conclusions

This study reveals that the sweet cherry LOX gene family comprises nine LOX gene sequences, primarily located on chromosomes 1, 2, 3, 4, 6, and 8. During the ripening process of sweet cherry fruit, the LOX gene family demonstrates five distinct expression patterns. The expression peaks of various LOX genes occur at specific developmental stages, and collectively, all LOX genes facilitate the growth and development of the fruit. Significant variations in LOX enzyme activity were observed among early and late maturing cherries of different varieties during the development process.

Supplementary Materials: The following supporting information can be downloaded at: https://www.mdpi.com/article/10.3390/horticulturae10121361/s1, Supplement file-PaLOX Primer.

Author Contributions: Conceptualization, G.W. and Y.S.; methodology, S.H.; software, R.G.; validation, J.Z. and C.C.; formal analysis, S.Z.; writing—original draft preparation, Q.F.; writing—review and editing, D.X. All authors have read and agreed to the published version of the manuscript.

Funding: The project supported by the National Center of Technology Innovation for Comprehensive Utilization of Saline-Alkali Land, Taian, Shandong, China. This work was supported by the Key R&D Program of Shandong Province, China (2023TZXD059); the Shandong Province 'High Talent Plan' on Foreign Experts (WRS2023076); the International (Regional) Science and Technology Cooperation Project of Shandong Academy of Agricultural Sciences (CXGC2024G25); the Shanxi Province

Key Research and Development Plan Project (2022ZDLNY03-05); and the Tai'an Agricultural Seed Improvement Project (2022NYLZ04).

Data Availability Statement: The datasets presented in this article are not readily available because the data are part of an ongoing study. Requests to access the datasets should be directed tocorresponding author.

Conflicts of Interest: All authors disclosed no relevant relationships.

References

1. Zhang, X.; Jiang, Y.; Peng, F.; He, N.; Li, Y.; Zhao, D. Changes of Aroma Components in 'Hongdeng' Sweet cherry During Fruit Development. *Sci. Agric. Sin.* **2007**, *40*, 1222–1228. [CrossRef]
2. Qin, L.; Cai, A.; Zhang, Z.; Qi, Y.; Geng, L. Analysis of Volatile Components in Sweet cherry Fruit by HS-SPME-GC/MS. *J. Chin. Mass. Spectrom. Soc.* **2010**, *31*, 228–234+251.
3. Schwab, W.; Davidovich-Rikanati, R.; Lewinsohn, E. Biosynthesis of plant-derived flavor compounds. *Plant J.* **2008**, *54*, 712–732. [CrossRef] [PubMed]
4. Jiang, X.; Yang, J.; Han, M.; Xie, D.; Li, X.; Zhang, C. Cloning and Expression Analysis of LuLOX1 Encoding a Lipoxygenase Gene in Flax (*Linum usitatissimum* L.). *Acta Laser Biol. Sin.* **2019**, *28*, 431–438.
5. Zhu, J.; Chen, H.; Liu, L. JA-mediated MYC2/LOX/AOS feedback loop regulates osmotic stress response in tea plant. *Hortic. Plant J.* **2024**, *10*, 931–946. [CrossRef]
6. Xu, Q.; Cheng, L.; Mei, Y.; Huang, L.; Zhu, J.; Mi, X.; Yu, Y.; Wei, C. Alternative Splicing of Key Genes in LOX Pathway Involves Biosynthesis of Volatile Fatty Acid Derivatives in Tea Plant (*Camellia sinensis*). *J. Agric. Food Chem.* **2019**, *67*, 13021–13032. [CrossRef] [PubMed]
7. González-Gordo, S.; Cañas, A.; Muñoz-Vargas, M.A.; Palma, J.M.; Corpas, F.J. Lipoxygenase (LOX) in Sweet and Hot Pepper (*Capsicum annuum* L.) Fruits during Ripening and under an Enriched Nitric Oxide (NO) Gas Atmosphere. *Int. J. Mol. Sci.* **2022**, *23*, 15211. [CrossRef]
8. Chen, Z. *Genome-Wide Identification of Lipoxygenase Gene Family in Poplar and Function Analysis of PtLOX11*; Anhui Agricultural University: Hefei, China, 2017.
9. Liu, M.; Ju, Y.; Min, Z.; Zhao, X.; Chen, T.; Fang, Y. Cloning and Expression Analysis of Lipoxygenase Gene in Grape (*Vitis vinifera*). *J. Agric. Biotechnol.* **2017**, *25*, 1809–1819.
10. Pilati, S.; Wild, K.; Gumiero, A.; Holdermann, I.; Hackmann, Y.; Dalla Serra, M.; Guella, G.; Moser, C.; Sinning, I. Vitis vinifera lipoxygenase LoxA is an allosteric dimer activated by lipidic surfaces. *J. Mol. Biol.* **2024**, *16*, 168821. [CrossRef]
11. Sun, Y.; Lv, D.; Wang, L.; Miao, C.; Lin, H. Analysis of the role of 13-Lipoxygenases in Arabidopsis leaf senescence. *J. Sichuan Univ. (Nat. Sci. Ed.)* **2013**, *50*, 1391–1396.
12. Ozawa, R.; Shiojiri, K.; Kishimoto, K.; Matsui, K.; Arimura, G.-I.; Urashimo, S.; Takabayashi, J. Cytosolic LOX overexpression in Arabidopsis enhances the attractiveness of parasitic wasps in response to herbivory and incidences of parasitism. *J. Plant Interact.* **2013**, *8*, 207–215. [CrossRef]
13. Zhu, L.; Qing, J.; Du, Q.; He, F.; Du, H. Genome-wide Identification and Expression Characteristics of LOX Gene Family in Eucommia ulmoides. *Bull. Bot. Res.* **2019**, *39*, 927–934.
14. Zhang, P. *Effect of Crop Development on the Emission of Volatile from Tomato Leaf and Their Response to Oligosaccharide*; Shandong University: Jinan, China, 2008.
15. González-Gordo, S.; López-Jaramillo, J.; Palma, J.M.; Corpas, F.J. Soybean (*Glycine max* L.) Lipoxygenase 1 (LOX 1) Is Modulated by Nitric Oxide and Hydrogen Sulfide: An In Vitro Approach. *Int. J. Mol. Sci.* **2023**, *24*, 8001. [CrossRef] [PubMed]
16. Mou, Y.; Sun, Q.; Yuan, C.; Zhao, X.; Wang, J.; Yan, C.; Li, C.; Shan, S. Identification of the LOX Gene Family in Peanut and Functional Characterization of AhLOX29 in Drought Tolerance. *Front. Plant Sci.* **2022**, *13*, 832785. [CrossRef] [PubMed]
17. Li, J.; Ma, C. Recent advances on study of plant lipoxygenases. *J. Biol.* **2007**, *6*, 5–8+29.
18. Zhou, G. *Transcriptional Analysis of Rice Plants Infested by Rice Stripped Stem Borer Chilo Suppressalis (Walker) and the Functional Characterization of a Defense-Related Gene OsHI-LOX*; Zhejiang University: Hangzhou, China, 2009.
19. Marla, S.S.; Singh, V.K. LOX genes in blast fungus (*Magnaporthe grisea*) resistance in rice. *Funct. Integr. Genom.* **2012**, *12*, 265–275. [CrossRef]
20. Camargo, P.O.; Calzado, N.F.; Budzinski, I.G.F.; Domingues, D.S. Genome-Wide Analysis of Lipoxygenase (LOX) Genes in Angiosperms. *Plants* **2023**, *12*, 398. [CrossRef]
21. Zhang, Y.; Chen, K.; Zhang, S.; Ferguson, I. The role of salicylic acid in postharvest ripening of kiwifruit. *Postharvest Biol. Technol.* **2003**, *28*, 67–74. [CrossRef]
22. Wu, P.; Shan, C.; Cui, T.; Zhao, Z.; Wang, C.; Zhou, T.; Sun, Y. Analysis of aroma components of "Meizao" cherry in two rootstock. *Food Ferment. Ind.* **2017**, *43*, 231–236.
23. Pei, Y.; Yu, Q.; Zhao, X. Research progress on the association between polyploidy and new phenotypes in plants. *Acta Hortic. Sin.* **2023**, *50*, 1854–1866. [CrossRef]
24. Ju, L. *An Oriental Melon Gene CmLOX09 Responded to Stress, Hormones and Signal Substances and Transformed into Tomato*; Shenyang Agricultural University: Shenyang, China, 2017.

25. Liang, X.; Guo, X.; Qi, H. The role of ethylene and lipoxygenase in aroma synthesis of tomato fruits. *Acta Hortic. Sin.* **2017**, *44*, 2117–2125. [CrossRef]
26. Chen, X.; Shi, X.; Ai, Q. Transcriptomic and metabolomic analyses reveal that exogenous strigolactones alleviate the response of melon root to cadmium stress. *Hortic. Plant J.* **2022**, *8*, 637–649. [CrossRef]
27. Bai, C.; Wu, C.; Ma, L. Transcriptomics and metabolomics analyses provide insights into postharvest ripening and senescence of tomato fruit under low temperature. *Hortic. Plant J.* **2023**, *9*, 109–121. [CrossRef]
28. Chauvin, A.; Caldelari, D.; Wolfender, J.L.; Farmer, E.E. Four 13-lipoxygenases contribute to rapid jasmonate synthesis in wounded Arabidopsis thaliana leaves: A role for lipoxygenase 6 in responses to long-distance wound signals. *New Phytol.* **2013**, *197*, 566–575. [CrossRef]
29. Hou, L. *Functional Verification of Persimmon 9-Lipoxygenase in Plant Senescence and Defense Response*; Northwest A&F University: Xianyang, China, 2016.
30. Liang, X. *Role of Lipoxygenase and Ethylene in Tomato Fruit Ripening and Aroma Synthesis*; Shenyang Agricultural University: Shenyang, China, 2017.
31. Heitz, T.; Bergey, D.R.; Ryan, C.A. A gene encoding a chloroplast-targeted lipoxygenase in tomato leaves is transiently inced by wounding, systemin and methyl jasmonate. *Plant Physiol.* **1997**, *114*, 1085–1093. [CrossRef]
32. Ju, Y.L.; Liu, M.; Zhao, H.; Meng, J.F.; Fang, Y.L. Effect of Exogenous Abscisic Acid and Methyl Jasmonate on Anthocyanin Composition, Fatty Acids, and Volatile Compounds of Cabernet Sauvignon (*Vitis vinifera* L.) Grape Berries. *Molecules* **2016**, *21*, 1354. [CrossRef]
33. Girard, B.; Kopp, T.G. Physicochemical characteristics of selected sweet cherry cultivars. *J. Agric. Food Chem.* **2000**, *46*, 471–476. [CrossRef]

 *horticulturae*

Review

Why Olive Produces Many More Flowers than Fruit—A Critical Analysis

Julián Cuevas

Department of Agronomy, ceiA3, CIAIMBITAL, University of Almería, 04120 Almería, Spain; jcuevas@ual.es

Abstract: Olive (*Olea europaea* L.) trees produce many more flowers than fruit. In an "on" year, an adult olive tree may produce as many as 500,000 flowers, but 98% of them will drop soon after bloom as unfertilized flowers or juvenile fruit. This waste of resources that could be better invested in fruit reaching maturation requires an explanation. Several, not mutually exclusive, hypotheses explaining the possible significance of heavy flowering followed by massive and premature flower and fruit abscission are analyzed and compared based on previously published works and recent observations on olive reproductive biology. The results suggest that olive trees selectively abort fruits to enhance the quality of the seeds in the surviving fruits. Additionally, a considerable proportion of flowers appears to contribute to the male fitness of the plant by increasing pollen export. Conversely, the hypotheses attributing to resource limitation, pollination deficits, pollinator attraction, or extra flowers functioning as an ovary reserve, must be rejected for explaining the ultimate functions of massive flower production. Implications for olive orchard management are discussed.

Keywords: *Olea europaea*; massive flowering; fruit abortion; pollinator attraction; pollination deficits; male function of flowers; resource limitation; bet hedging; sexual selection; selective abortion

Academic Editors: Zhaohe Yuan, Bo Li and Yujie Zhao

Received: 15 November 2024
Revised: 26 December 2024
Accepted: 31 December 2024
Published: 2 January 2025

Citation: Cuevas, J. Why Olive Produces Many More Flowers than Fruit—A Critical Analysis. *Horticulturae* **2025**, *11*, 26. https://doi.org/10.3390/horticulturae11010026

1. Introduction

Many plant species regularly abort a large portion of their flowers and juvenile fruits that could otherwise develop to maturity. The overproduction of flowers that do not reach maturation is a widespread feature among many Angiosperms, but it is especially common and particularly pronounced in woody, outcrossed, hermaphrodite plants [1,2], including many fruit crops. In these species, surplus flowers and fruit are abscised early. This drop of juvenile fruit is known in pomology as "June drop" because, in the Northern Hemisphere, most fruitlet abortion in apple, and in other temperate fruit crops, mostly occur in June. This waste of energy is surprising, as it seems that the resources invested in surplus flowers and aborting fruitlets could be better allocated to increasing the number and size of fruit that reach maturation [3,4]. For this reason, significant effort has been devoted to determining the causes and significance of this massive flower and fruit abortion in cultivated fruit trees, as well as in wild species. In this context, several hypotheses aim to explain the benefits of producing these massive numbers of flowers. Since the cost of over-initiating flowers and fruit is substantial [5], the benefits of their formation must outweigh the expenditures incurred, in order to reasonably explain why this trait has been fixed and widespread during the evolution of flowering plants.

In olive, different research teams have explored the mechanisms underlying massive flower and fruit abscission, aiming to suggest crop management practices that might increase the low fruit set characteristic of this crop. An olive tree may produce as many as 500,000 flowers in its "on" year, but only 1–2% of them reach harvest as ripened fruit [6]. The remaining 98–99% abscise as flowers or young fruit, mostly within the first two months after bloom [7]. In this study, I critically analyze current hypotheses formulated in the literature to explain the significance of premature flower and fruit abscission in domesticated and wild species to check its appropriateness to explain the olive massive flowering and low flower/fruit ratio. These hypotheses are evaluated using results from previous published investigations, as well as new knowledge about the reproductive biology of *Olea europaea*. The aim of this work is to determine the ultimate reasons why olive trees produce many more flowers than fruit. Stephenson [1], in his seminal review on this topic, distinguished between proximate causes, as the mechanisms responsible for the failure of flowers to set fruit, from the ultimate reasons, referring to the functions that surplus flowers that consistently fail to reach maturation may play in the reproduction of these plants. This distinction is particularly pertinent to the objectives of this review.

2. The Plant

The domesticated olive tree (*Olea europaea* subsp. *europaea* L.) is a woody wind-pollinated, preferentially allogamous crop of immense importance in the Mediterranean region. Its longevity and historical significance have attained the status of myth in several monotheistic religions of the Mediterranean, being, frequently mentioned in the sacred texts of Christianity, Judaism, and Islam. The olive tree is andromonoecious, producing both hermaphrodite and staminate flowers in panicles of between 15 and 30 flowers on average, depending on the cultivar and growing conditions [8]. The panicles develop in high numbers (often more than twenty), clustered in the axil of leaves on 1-year-old shoots (Figure 1).

Figure 1. Massive bloom in an olive tree during its "on" year. Panicles develop clustered on 1-year-old shoots. Source: own author.

Olive panicles are ramified and, in addition to the apical flower ("king flower"), they have primary, secondary, and, occasionally, tertiary branches [9]. Hermaphrodite flowers are mostly located on primary branches, while staminate flowers are more frequent on secondary and higher order branches [10,11] (Figure 2).

Figure 2. Olive panicles showing hermaphrodite and staminate flowers. (**Left**): Intact panicle. (**Right**): petals removed to better expose the well-developed pistil (white arrow) or the rudimentary pistil (black arrow). Source: Seifi et al. [8]; with permission.

The olive tree exhibits a strong alternative bearing habit, with "on" years of massive flowering followed by "off" years with few or no flowers. In its "on" year, an adult olive tree blooms profusely, producing as many as 500,000 flowers [6]. However, not all these flowers reach harvest because up to 98–99% of them drop, either as flowers or developing fruitlets, within the first two months following bloom [7]. Three main waves of abscission can be distinguished in olive. The first wave affects staminate flowers, which drop a few days after bloom, once they have accomplished their mission of exporting their pollen grains [10]. The second wave impacts unfertilized, yet fertile, hermaphrodite flowers, which drop in the second week after bloom. Despite the magnitude of these two phases of flower drop, the olive tree still initially set a much larger proportion of fruit than those ultimately reaching harvest. Thus, after fertilization of the flowers, many developing fruitlets begin to grow, triggering an intense fruitlet competition resolved with the abscission of many of them. This process stabilizes the fruit population after the third wave of fruitlet abscission, occurring five to seven weeks after flowering [12,13]. This final drop wave may partially overlap with the abscission of unfertilized flowers [14]. Regardless of the reasons behind the failure of flowers and fruit to reach maturity, sound explanations are required for explaining the formation of such a large number of flowers and the over-initiation of fruit that will ultimately abscise. Below, I explore several current hypotheses trying to explain the benefits of producing this extraordinary number of surplus flowers in olive.

3. Hypotheses to Explain Massive Fruitlet Abortion

Six valid, not mutually exclusive, hypotheses attempt to explain the massive flower and fruitlet abscission observed in many hermaphroditic angiosperms. The hypotheses explaining the production of surplus flowers are: (1) pollinator attraction, where flowers serve to attract a greater number of biotic pollination vectors; (2) abscission due to pollination deficits, either from the scarcity or the inadequacy of the pollen grains landing on the stigma, or because pollinators fail to visit all the flowers; (3) excess flowers serving solely a male function, exporting their pollen grains and thus enhancing male and, consequently, the total fitness of the progenitor plant; (4) flower drop caused by resource or nutrient limitations, including here water deficits; (5) the bet-hedging hypothesis, which suggests that there are no surplus flowers, but rather that they are formed in case unexpectedly favorable seasons might allow a full fruit set; and (6) sexual selection and selective abortion among developing fruitlets based on fruit and seed sink strength (genotype based).

The validity of these hypotheses in explaining massive flower and fruitlet abscission in olive is evaluated based on both my own research and previously published studies by other colleagues on olive reproductive biology. These findings are critically analyzed to determine the ultimate reasons behind the production of surplus flowers in olive.

3.1. Pollinator Attraction

The first hypothesis postulates that an excess number of flowers are formed, especially when bloom occurs in inflorescences [15] to attract biotic pollinating agents (pollinators) more effectively by offering floral rewards at a lower cost. In this regard, the basic unit of attraction for pollinators is the inflorescence. However, olive is wind-pollinated. Indeed, olive belong to the *Oleaceae* family where many species are insect-pollinated species, and olive flowers still possess a conspicuous white corolla and emit a sweet aroma that attracts insects, primarily honeybees [9]. This ambophily has been identified in the closely related species *Olea ferruginea* [16]. Nonetheless, it is evident that olive has evolved toward anemophily, and during an "on" year, adult plants produce billions of pollen grains (approximately 50,000 million pollen grains per tree, extrapolating data from Cuevas and Polito [10] and Rojas-Gómez et al. [17]). This pollen is readily transported by the wind, as evidenced by the seasonal allergies experienced by many people in Mediterranean countries (Figure 3).

Figure 3. Olive pollen clouds commonly seen during blooming season in Andalusia (Spain). Source: José Angel García, Ideal 6 May 2020.

Although a ancient function of surplus flowers attracting insects cannot be entirely dismissed, if this were their primary role in olive, then a large number of flowers formed during "on" years would result in greater pollinator attraction and, consequently, in a proportionally higher reproductive success [18]. However, olive trees with more flowers set proportionally fewer fruits (Figure 4). On the contrary, trees in their "off" year set proportionally more fruits on average [19]. Consequently, surplus flowers do not function to attract pollinators in olive.

Figure 4. Fruitful shoots from an "on" and an "off" olive tree. Despite "off" trees setting a proportionally higher number of flowers, the total number of fruits is still much larger in "on" trees due to the much higher number of inflorescences. Note the different shoot lengths. Fruit ripening occurs earlier in "off" trees. Source: own author.

3.2. Pollination Deficits

A different hypothesis suggests that fruit set is limited by pollination deficits. These deficits can be due to a shortage or inadequacy of the pollen grains or from the absence of pollinators [20,21]. As stated earlier, olive is wind-pollinated, and a single adult tree can produce billions of pollen grains. Although wind is a random pollination vector, many field experiments have shown that most, if not all, hermaphrodite flowers receive enough pollen grains on their stigmas to achieve fertilization. This has been demonstrated in multiple studies where pollen loads on stigmas were quantified in flowers left exposed under conditions of open free-pollination [22–27]. Cases of pollen scarcity on stigmas have been rarely documented [13], and when observed, they likely reflect the late abortion of a small portion of flowers with non-functional stigmas (Figure 5D) rather than a problem with pollen transport.

An alternative approach here is to suggest that it is not the quantity, but the quality of the pollen that might lead to a lack of fertilization and, consequently, to flower abscission in olive orchards. Self-incompatibility in olive has recently been confirmed to be of the sporophytic type and is known to severely affect many olive cultivars [28,29]. While olive has repeatedly been acknowledged as a self-incompatible species [22,28–30], massive flower and fruitlet abscission still occurs even under optimal cross-pollination conditions. In fact, highly successful fertilization achieved by cross-pollen anticipates fruitlet abscission, as earlier and higher levels of flower fertilization promote earlier fruitlet growth that triggers the abscission of less competitive fruitlets [13].

Fertilization levels of 40–50% of the hermaphrodite flowers have been reported for different olive cultivars under cross-pollination conditions [22,23,31,32]. Furthermore,

Rapoport and Rallo [7] found that more than 50% of the abscised pistils of Manzanillo (syn. Manzanilla de Sevilla) cultivar were fertilized, ruling out pollination deficits as a cause of their drop. Unfortunately, high fertilization rates in olive do not guarantee similar high levels of final fruit set. Even under the best conditions, fruit set does not exceed 10% of the hermaphrodite flowers produced [13,22]. Certainly, the benefits of cross-pollination are undeniable, as it significantly increases fruit set and yield in olive orchards [31,33–35]. However, pollination deficits, either in quantity or quality, are insufficient to explain the cost of producing significantly more flowers than fruits in olive. Therefore, although cross-pollination markedly enhances fruit set in most cultivars, and although some level of pollination deficits might appear in large monovarietal orchards due to the strong olive self-incompatibility, even under those optimal pollination conditions, massive June drop persists. Therefore, the pollination deficits theory does not adequately explain the surplus flower production in cultivated olive.

3.3. Male Function

This hypothesis posits that surplus flowers that later drop are in fact accomplishing only a male function; that is, they serve for exporting their pollen grains to achieve fertilization in flowers of other plants [36–39]. In other words, according to this theory, despite being morphologically hermaphrodite, most or all abscised flowers are, in fact, functionally male, and the surplus flower production is explained by enhancing the male competition component of sexual selection [40]. Extensive analyses across numerous species have shown that pistillate flower abortion rates are lower in monoecious and dioecious plants compared to hermaphrodite plants [2]. This pattern is explained because monoecious and dioecious plants can separate better the investments in male and female organs, adjusting therefore more tightly the resource allocation for pistillate flowers to those more likely to reach ripening. If this were the case for olive, it is not easily understandable why olive has not transitioned to monoecy to form some of the flowers just as staminate, saving resources otherwise spent on surplus pistil formation. In hermaphrodite plants, reproductive resource allocation is divided between male (pollen) and female (fruit and seeds) organs [1,41], with the latter being significantly costlier to produce. Evolution towards monoecy or dioecy would allow plants to allocate resources more strategically based on environmental constraints and available resources. However, in olive, andromonoecy appears to be a fixed trait [10], and the overproduction of hermaphrodite flowers in excess cannot be fully explained by the male function hypothesis. Additionally, the production of an important number of fruitlets that later abscise during the June drop period also denies the notion that these flowers were formed solely to accomplish the male function by exporting their pollen grains.

Staminate flowers in olive result from pistil abortion (Figure 5B,D). Pistil abortion can occur at various stages of flower development, but in most cases, pistil growth is arrested early, leading to flowers that either lack pistils or have rudimentary ones [42,43]. In some cases, however, pistil formation progresses almost to completion, although the stigma papillae fail to develop fully (Figure 5D), so the flowers are not able to adhere pollen grains and, as a result, they are unable to set fruit [43]. The occurrence of pistil abortion at different developmental stages suggests that olive trees are continuously balancing the amount of the available resources [44], including water, to invest in pistil development with suboptimal conditions leading to higher rates of pistil abortion. In this regard, the production of staminate flowers is extremely variable among cultivars, years, trees, branches and even among panicles on the same shoot. However, it tends to increase during dry seasons and under nutritional deficiencies [42,44].

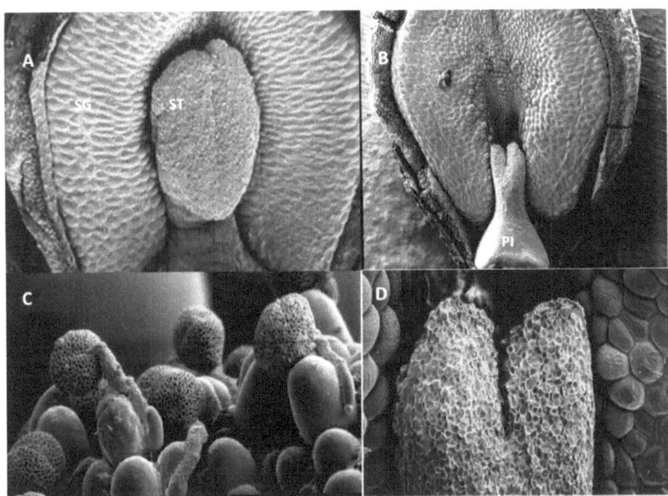

Figure 5. Scanning electron microscopy images of hermaphrodite (**left**) and staminate (**right**) olive flowers. (**A**). Detail of a well-developed stigma (SG) adjacent to one stamen (ST). (**B**). Pistil-aborted flower showing a rudimentary pistil (PI). (**C**). Pollen germination on fully developed stigma papillae. (**D**). Functionally male, late pistil-aborted flower displaying stigma with non-functional papillae. Source: own author [42].

Staminate flowers also formed in less favorable sites within panicles, suggesting again resource limitation as the cause of their formation. Their selective placement on poorly nurtured sites, such as secondary and tertiary branches of the panicle [10,11] (Figure 2), underscores the role of resource competition. This architectural negative effect on their fate reinforces the idea that competition for resources plays a role in their formation [11,44]. Plant architecture, particularly pedicel thickness, is closely associated with critical components of female reproductive success, including fruit and seed set, seed number, and size [45,46]. In an apparent contradiction, the resources saved by not developing a pistil in the staminate flowers of olive are not allocated to other organs within the same flower. This is evidenced by the smaller size of petals and sepals of staminate flower compared to hermaphrodite ones [10]. Nonetheless, compared to hermaphrodite flowers, the staminate flowers produce an equal number of pollen grains per anther, pollen grains that are of the same size and viability, and with the same capacity to perform ovule fertilization in hermaphrodite flowers of other cultivars [10]. These findings indicate that staminate flowers enhance male and, consequently, total plant fitness at a lower cost, since the staminate flowers can sire the same number of embryos in other plants than the hermaphrodite flowers while being less costly to produce [10,47].

Consequently, an important proportion of olive flowers are staminate and hence do not have the capacity to develop into fruit. Their effective achievement of the male function explains the formation of those staminate flowers, but this cannot explain why many extra hermaphrodite flowers are still formed.

3.4. Resource Limitation

If the formation of staminate flowers in olive is at least partially attributable to resource limitation, we may think that, perhaps, the reduced fruit set in hermaphrodite flowers is also due to the lack of resources, being this lack of resources circumstantial rather than structural [48,49]. According to this hypothesis, under certain conditions, all or most hermaphrodite flowers could potentially mature into fruit. However, fruit set in olive is always very low, and there is no significant seasonal or spatial (orchard) variation allowing all flowers to reach ripening. While proper orchard management practices, particularly irrigation and fertilization, can modestly increase fruit set in olive, unfortunately improvements are limited at the most to 5–6% of the flowers formed [5].

It is clear, however, that competition for resources among developing fruitlets is the proximate cause, the mechanism, triggering the intensive post-anthesis fruitlet abscission in olive. This competition occurs initially among fruitlet within the same panicle and later between fruit on neighboring panicles [11]. This conclusion is supported by results of different thinning experiments and by the observation that early fruit growth immediately precedes the abscission of both fertilized and unfertilized flowers [13,14]. In this regard, thinning experiments have demonstrated that removing more than 50% of panicles increases the fruit set of the remaining flowers without reducing the final number of fruit at harvest [50]. This result confirms the surplus nature of many flowers, with up to 60% deemed excessive in certain experiments. Similar findings are obtained when entire inflorescences are removed or when selective thinning is applied by eliminating some flowers within a panicle. Notably, the result remains consistent regardless of whether flower removal is performed before or during the first days after bloom [12,50,51].

It is worth noting that small-fruited cultivars exhibit higher fruit set than large-fruited cultivars [47], although overall productivity per tree does not differ significantly. Additionally, small-fruited cultivars have smaller ovaries at bloom compared to large-fruited cultivars [52,53], so they are less competitive, or less "selfish" we might say. Similar results were observed in experiments with 'Galego' olive trees grown in pots under low temperatures (20/14 °C). At these low temperatures, reduced fruit growth rates decrease fruitlet competition and allowed fruit set to double compared to trees grown at 25/20 °C, although the final fruit size at harvest halved under the lower temperatures [54]. Cool springs also have a positive effect on olive fruit set, likely by reducing fruit growth rate and, therefore, fruitlet competition by resources.

Thus, while good orchard management can and should increase fruit set and yield in olive orchards, it will never do so to the extent of completely avoiding fruitlet drop. On the other hand, flower thinning experiments confirm that competition for resources is the proximate cause of the abscission of fruitlets, but not the ultimate reason for the formation of so many flowers in olive. Therefore, an alternative explanation for the production of surplus hermaphrodite flowers remains necessary.

3.5. Bet Hedging-Ovary Reserve Hypothesis

The bet hedging hypothesis shares some common ground with resource limitation theory, as it also justifies the formation of extra flowers to maximize fruit set during favorable seasons. According to the bet hedging hypothesis, surplus flowers are produced to exploit unexpected prodigious seasons with abundant resources (for wild *Olea europaea*, we will say rain) [1,55], allowing full fruit set in favorable environments. However, in domesticated olive, there is neither spatial nor temporal variation in fruit set that allow all fruit to complete maturation. As mentioned before, this holds true under optimal orchard management or in exceptionally favorable years. Moreover, the fact that many olive flowers drop after fertilization and at various stages of embryo development [14] further suggests that olive trees cannot develop all the hermaphrodite flowers they produce, making it unrealistic to expect all of them to reach maturity. On the other hand, the activity of pollinating insects plays no significant role in olive reproduction, as olive is predominantly wind-pollinated. Therefore, the idea that optimal pollination in exceptional seasons might justify the production of surplus flowers is not a suitable explanation for olive massive flower production. Consequently, bet hedging hypothesis fails to justify the formation of surplus hermaphrodite flowers in cultivated olive trees.

A variant of bet hedging theory, the ovary reserve hypothesis, posits that many flowers are produced to counteract unpredictable, externally induced mortality that is not maternally regulated [56], especially, but not only, herbivory [4]. According to this hypothesis, surplus flowers serve as an insurance against flower losses [57], or even as a bait for herbivores, achieving seed predators' satiation by producing more flowers (and seeds) than necessary [48,55]. If this were the case, the sacrifice of some pistils for the benefit of others would take place throughout the entire reproductive season, providing protection to sibling fruits from seed predators from bloom to fruit maturation and seed dispersal. However, in olive, most juvenile fruit drop occurs very early in the season, leaving sibling fruits exposed to predators for the majority of the reproductive period. It remains possible that the plant's strategy is to safeguard the bloom period, with resource competition among fruitlets driving the abscission of weaker juvenile fruits thereafter. Ehrlen [56] proposed that the number of surplus flowers increases with increasing fruit-cost/flower-cost ratio and ovary mortality. As a result, species with costly fruit and small flowers, such as olive, would exhibit higher rates of fruit abortion.

Guitián et al. [58] found that when flower mortality is high in *Cornus sanguinea*, subsequent fruit abortion is reduced, suggesting that, in this species, surplus flowers act, at least partially, as an insurance against flower predators. A similar conclusion was reached for *Prunus mahaleb* [59]. Certainly, in olive, larvae of certain pests, such as the second generation of *Prays oleae*, prey on hermaphrodite flowers (eating the ovary), while the third generation damages small fruitlets. However, the extent of this predation seems not sufficient to pose a significant threat to olive fruit production. On the other hand, if ovary reserve hypothesis were applicable to olive, we would expect substantial seasonal variations in fruit set depending on herbivory incidence. However, fruit set in olive consistently remains low, and massive flower and fruitlet abscission occurs even in the absence of pest damage. We could argue that domestic olive cultivation protects the crop against extensive pest damage, but this is not always the case. Furthermore, wild olives (*Olea europaea* subsp. *sylvestris*) and close relatives, such as *Olea europaea* subsp. *cuspidata*, are not subjected to farmer protection against pests, and they also produce few fruits per panicle and exhibit similar levels of flower and fruitlet abscission [60,61].

Thus, the conclusion is that surplus flowers are indeed formed in olive, and that their elimination, whether by herbivores, by thinning, or pruning, is partially compensated by increasing fruit set among the survival flowers. However, this compensation does not suffice to achieve a full fruit set. Consequently, circumstantial high levels of herbivory do not justify the significant expenditure incurred by olive trees in producing up to fifty times more flowers than the number of fruit that reach harvest.

3.6. Sexual Selection, Sibling Competition and Selective Abortion

Previous hypotheses fail to explain fully the ultimate reasons for, and the significance of, surplus flower production and massive fruit abortion in olive. One last hypothesis suggests that olive produces more flowers and fruit than needed to facilitate the selection of the best of them based on the quality of the next generation ("the survival of the fittest"; [62]). Selective abortion has been documented in many species with multi-seeded fruits. In such species, maternal investment in fruit development is maximized by aborting fruit with fewer or weaker seeds [1,63,64]. In olive, which produce drupes, selective fruit abortion is primarily determined by the size and strength of the single seed that typically forms within the endocarp. Drupes rarely contain more than one seed, although some olive cultivars show a tendency to produce bi-seeded fruits, especially under cross-pollination conditions [65]. Since the endocarp functions as the dispersal unit, the presence of two seeds within a single stone is detrimental for the size of each individual seed [65], and likely reduces seedlings survival in nature due to intensified sibling competition in the same ground site after endocarp dispersal by birds.

It is plausible that mother olive trees produce extra flowers and fruitlets for letting many of them drop. As discussed earlier, this is not a behavior exclusive to olive trees; as many woody plants, including other fruit crops, produce significantly more flowers than fruits [1]. The right question itself extends to why many plants produce many more ovules than seeds [66], considering that seeds represent the next generation of sporophytes, with fruits serving as dispersal vehicles for seeds. Regardless of how the question is framed, one plausible answer is that olive form surplus flowers and ovules to enable the selection of the fittest seedlings, promoting the abortion of weaker fruitlets and the selection of the best embryos. Lee [67] elegantly developed the gametophyte competition theory, proposing that plants selectively mature fruits from ovaries where pollen tube competition has been most intense. He hypothesized that the embryos resulting from the fertilization of ovules by faster-growing pollen tubes would produce sporophytes with higher fitness. Resource limitation and differences in the quality of pollen achieving fertilization will thus open the possibility for sexual selection during fruit development [63]. Experimental evidence supporting the theory of selective abortion, sexual selection, and sibling competition and how it operates in olive is presented below.

4. How Sexual Selection Operates in Olive

Olive trees appear to select the best embryos through a series of mechanisms: selecting the fastest pollen tubes (carrying the male gametes), the most attractive ovules (enclosing the embryo sac with the egg cell and polar nuclei) for double fertilization, and the strongest siblings after fertilization. The first step of sexual selection occurs in the transmitting tissue of the recipient pistil. A single olive pistil may receive over 1000 pollen grains on its stigma [26,68], many of which germinate and grow forming pollen tubes carrying the male gametes at their tips. The initial selection of the male gametes is based on the adequacy of the genotype of the pollen grain, since most olive cultivars are predominantly allogamous and discriminate against self-pollen through self-incompatibility reactions [28–30]. Nevertheless, under open- and cross-pollination conditions, dozens of pollen tubes often grow within the transmitting tissue of the pistil, although usually only one pollen tube reaches the base of the style to enter the ovary [22,33,68,69] (Figure 6). Intense pollen tube growth attrition ensures that only the fastest pollen tube access the ovary, where now, the "winner" fastest pollen tube encounters four viable ovules.

Intense competition within the transmitting tissue of the pistil is not limited to the male gametes; female gametes contained in the ovules may also be subjected to selective pressures, since only one of the four ovules of the ovary will normally become the seed. This raises an important question: Why do olive flowers contain four ovules when only one typically develops into a seed? In olive, the four ovules are equally sized at bloom, occupy symmetric positions within the ovary, and do not seem to differ in fertility or longevity. Secondary ovules exist in the flowers of stone fruit crops, where secondary ovules usually lag in development, being the primary ovule normally the only one fertilized. However, in olive, there are no obvious reasons why the selected pollen tube would choose one ovule over another. Nevertheless, this selection process is unlikely to be random. Cuevas et al. [68] proposed that this scenario enables ovule competition for the sole pollen tube accessing the ovary, representing a rare case of female competition and male choice in flowering plants.

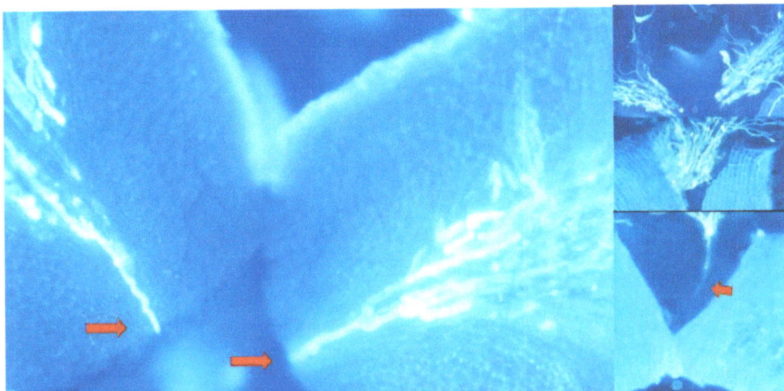

Figure 6. Composition of pollen tube attrition in cross-pollinated olive flowers. (**Left**): The flower has been dissected, softened with NaOH 1N, squashed, and stained with aniline blue, then observed under fluorescence microscopy. (**Right**): Different pistil sections showing progressive pollen tube attrition. (**Up**): stigma; (**down**): style. Red arrows indicate the "winner" pollen tube. Source: own author.

Upon exiting the transmitting tissue, the pollen tube responds to attractant signals from the ovules, growing along the funiculus surface and entering the ovule's micropyle to perform double fertilization, a process exclusive to angiosperms. In this regard, we know that synergids cells play a crucial role in attracting the pollen tube to the filiform apparatus [70,71]. Calcium ions (Ca^{2+}) serve as critical signaling elements for pollen tube guidance and fertilization. As the pollen tube nears the filiform apparatus, attractants secreted by the synergid cells modulate Ca^{2+} concentration at the pollen tube tip, facilitating the interaction between the pollen tube and the receiving synergid cell [72]. The synergid cells also provide resources in this last step of pollen tube growth, culminating in the discharge of two male gametes into the embryo sac [70,73]. A differential capacity of the synergid cells of different ovules to attract the pollen tube might explain in olive the preference of the "winner" pollen tube for a specific ovule. This, along with the formation of four ovules per flower could represent the first reported case of ovule (female) selection and male choice in flowering plants [68], even though this circumstance may be present in many other multi-ovulated ovaries sharing the same characteristics of olive flowers, and producing single-seeded fruit. It is noteworthy that the genes of synergid cells are not transmitted to the next generation, unlike the genes of the egg cell after double fertilization. Thus, the role of the synergids, sisters to the egg cell, may represent a unique example of altruism.

If the strong selection of male and female gametes were not enough, olive trees further promote sibling competition after fertilization by displaying numerous flowers together in a single panicle. This arrangement intensifies competition for limited resources, first within the panicle and then among developing fruit on nearby panicles [12]. Large inflorescences and simultaneous flower opening reinforce effective sexual selection and sibling competition by favoring the abortion of weaker fruitlets that grow at a lower rate [36]. Similar observations have been reported in *Prunus mahaleb*, where the inflorescence also functions as the fructification unit [59]. In this context, fruitlet abscission primarily affects the smaller fruitlets, which may result either from delayed fertilization achieved by a slower pollen tube or from weaker seeds that grow more slowly. Both factors reduce the competitiveness of these fruitlets for survival, leading ultimately to their drop. In olive, it is still unclear which flowers are more likely to become fruit, whether those with pistils fertilized first or those that grow faster [13]. In contrast, Hiei and Ohara [74] observed in *Melampyrum roseum* that flower better positioned in the inflorescence and those that open earlier exhibit higher fruit set. Medrano et al. [75] suggest that resource sequestration by the earliest developing fruit in *Pancratium maritimum* is also the cause of the abscission of the smaller fruitlets.

Simultaneous and massive full bloom is a prerequisite for making sexual selection more effective, as scattered flowering diminishes sibling competition and sexual selection opportunities. In olive, there are only slight phenological differences within a panicle and among nearby panicles, with bloom mostly occurring simultaneously. In some species, it has been shown that the percentage of fruitlets resulting from self-fertilization decreases during the fruitlet abscission period, as maternal plants preferentially select fruitlets produced through cross-fertilization [1]. In avocado, this selection favoring stronger seeds leads to the production of larger fruit, which benefits farmers economically [76,77]. In olive, demonstrating this phenomenon presents challenges due to the difficulty extracting DNA from abscised fruitlets and dead seeds. Sexual selection in domesticated olive might somehow benefit table olive farmers, as studies have shown that, in olive, seed weight is linearly related to fruit weight, reflecting seed's sink strength in attracting photo assimilates for fruit growth [65]. Seed size, in turn, is related to seedling survival and might be a critical factor in harsh Mediterranean environments. Ultimately, the goal of this strong selection among seedlings in wild olives is to produce seeds of superior quality capable of surviving after germination in an environment characterized by prolonged drought and nutrient-poor soils.

5. Implications for Olive Cultivation, Future Prospects

The consequences of strong seedling selection extends to different aspects of orchard management in domestic olive. In this regard, farmers must understand that massive flower and fruitlet abscission are inherent to the reproductive system of the olive tree and cannot be eliminated by increasing fertilization or by over-irrigation. The lack of resources is not the ultimate reason why olive produce so many flowers and regularly abort many fruit. Certainly, flower thinning in olive can increase fruit/flower ratio, but it does not raise the number of fruits per panicle. Therefore, surplus flower production is not explained by resource limitation, although the competition for resources is the mechanism through which olive selectively abort less competitive fruitlets. On the contrary, these extra flowers are primarily formed to facilitate the selection of the best embryos by means of the selective abortion of the weaker fruitlets. Therefore, neither irrigation, despite being a key factor for olive production, nor heavy orchard fertilization can prevent fruitlet competition and drop. However, these management practices wisely carried out can enhance available resources, increasing thus fruit set and the proportion of flowers that reach harvest.

Flower thinning and pruning reduce the flowers population, thereby lowering fruitlet competition. However, as competition starts within the panicle and later extends to nearby panicles, these techniques cannot completely eliminate fruitlet abscission in olive. The fact that fruit load at harvest remains unchanged after removing a large proportion of flowers at bloom indicates that fruit set is proportionally increased by reducing the population of flowers. Nonetheless, heavy flowering during an "on" year usually leads to high yields, which in turn inhibit flower induction for the following season [78], resulting in a reduced crop during the "off" year, a characteristic behavior of alternate-bearing cultivars. This suggests that selecting cultivars with lower flowering levels could minimize the wastage of resources due to heavy flowering and, at the same time, reduce olive alternate bearing habit [5,6]. Such selection should focus on genotypes with fewer flower per panicle and panicles uniformly distributed around the canopy to minimize fruitlets competition. In this context, chemical fruit thinning in olive oil cultivars warrants some more research. As previously mentioned, pruning and flower thinning reduce sibling competition by limiting the number of flowers. In the case of flower thinning, the production of photo assimilates remains mostly unaffected, as this technique does not remove leaves. These practices are widely recognized for increasing fruit size in many fruit crops. However, reducing competition by limiting fruitlet numbers may result in a less effective selection compared to scenarios where many siblings compete. On the other hand, cross-pollination, when other factors are not limiting, enables olive trees to reach their full production potential by increasing initial fruit set and seed size. However, larger fruits may lead to higher rates of fruit drop if the available resources for fruiting remain constant. In this context, small-fruited olive cultivars exhibit a higher proportion of flowers setting fruit, as they have reduced fruit abscission [52]. Additionally, fruit with greater genetic similarity tends to exhibit reduced sibling rivalry and, then, lower rates of fruit abortion [48]. Conversely, genetically diverse seeds intensify competition among fruit, leading to greater variability in seed and fruit size. Seed size, a critical trait, significantly influences germination, seedling survival, and growth [79–81]. In any case, cross-pollination benefits in self-incompatible olive cultivars are undeniable. In order to maximize fruit set, cross-pollination must occur extensively across all panicles, which requires the implementation of an effective pollination design and the careful selection of pollinizers, including their quantity and placement within the orchard. Pollinizer selection should take into account their inter-compatibility relationships and overlapping blooming periods. Other factors to consider when selecting pollinizers include the destination of the olive fruit (oil versus table), regular bearing habits, and similar vigor [27]. Contrary to this, no evidence of xenia has been reported in olive [34], although seeds (and fruit) tend to be slightly heavier under compatible cross-pollination treatments [25].

The final result is that from approximately 500,000 flowers produced in "on" olive trees, only 1–2% of them reach harvest as ripened fruit, a level still regarded as a good commercial yield [6,34]. Orchard management strategies must aim to maintain this level of fructification. This analysis is also relevant to other fruit crops that bloom in panicles, such as mango, avocado, and loquat. However, the applicability of these hypotheses must be assessed in light of the specific reproductive strategies, pollination requirements, and seeding patterns characteristic of each species.

Funding: Julián Cuevas received partial support by the University of Almeria's programme for research and knowledge transfer years 2023 and 2024.

Data Availability Statement: No new data were created in this study. Data sharing is not applicable to this article.

Acknowledgments: To the student Ramón Rodríguez who helped me a lot to improve the manuscript's readability. To all who inspired this review, in particular to the colleagues of the Department of Agronomy at the University of Córdoba (Spain) who performed many of the experiments cited here.

Conflicts of Interest: The author declares no conflicts of interest.

References

1. Stephenson, A.G. Flower and fruit abortion: Proximate causes and ultimate functions. *Ann. Rev. Ecol. Syst.* **1981**, *12*, 253–279. [CrossRef]
2. Sutherland, S. Patterns of fruit-set: What controls fruit-flower ratios in plants? *Evolution* **1986**, *40*, 117–128. [CrossRef]
3. Lloyd, D.G. Sexual strategies in plants. I. An hypothesis of serial adjustment of maternal investment during one reproductive season. *New Phytol.* **1980**, *86*, 69–80. [CrossRef]

4. Stephenson, A.G. Fruit set, herbivory, fruit reduction, and the fruiting strategy of *Catalpa speciosa* (Bignoniaceae). *Ecology* **1980**, *61*, 57–64. [CrossRef]

5. Famiani, F.; Farinelli, D.; Gardi, T.; Rosati, A. The cost of flowering in olive (*Olea europaea* L.). *Sci. Hortic.* **2019**, *252*, 268–273. [CrossRef]

6. Martin, G.C. Olive flower and fruit population dynamics. *Acta Hortic.* **1990**, *286*, 141–154. [CrossRef]

7. Rapoport, H.F.; Rallo, L. Postanthesis flower and fruit abscission in 'Manzanillo' olive. *J. Amer. Soc. Hort. Sci.* **1991**, *116*, 720–723. [CrossRef]

8. Seifi, E.; Guerin, J.; Kaiser, B.; Sedgley, M. Flowering and fruit set in olive: A review. *Iran. J. Plant Physiol.* **2015**, *5*, 1263–1272.

9. Lavee, S. Olea europaea. In *Handbook of Flowering*; Halevy, A.H., Ed.; CRC Press: Boca Raton, FL, USA, 1985; Volume 6, pp. 423–434.

10. Cuevas, J.; Polito, V.S. The role of staminate flowers in the breeding system of *Olea europaea* (*Oleaceae*): An andromonoecious, wind-pollinated taxon. *Ann. Bot.* **2004**, *93*, 547–553. [CrossRef]

11. Seifi, E.; Guerin, J.; Kaiser, B.; Sedgley, M. Inflorescence architecture of olive. *Sci. Hortic.* **2008**, *116*, 273–279. [CrossRef]

12. Rallo, L.; Fernández-Escobar, R. Influence of cultivar and flower thinning within the inflorescence on competition among olive fruit. *J. Amer. Soc. Hort. Sci.* **1985**, *110*, 303–308. [CrossRef]

13. Cuevas, J.; Rapoport, H.F.; Rallo, L. Relationship among reproductive processes and fruitlet abscission in 'Arbequina' olive. *Adv. Hortic. Sci.* **1995**, *9*, 92–96.

14. Rapoport, H.F.; Rallo, L. Fruit set and enlargement in fertilized and unfertilized olive ovaries. *HortSci* **1991**, *29*, 890–898. [CrossRef]

15. Willson, M.F.; Price, P.W. The evolution of inflorescence size in *Asclepias* (Asclepiadaceae). *Evolution* **1977**, *31*, 495–511. [CrossRef] [PubMed]

16. Khan, S.; Kumari, P.; Verma, S. Ambophily in *Olea ferruginea*: A transitional state in the pollination syndrome. *Plant Biosyst.* **2022**, *157*, 221–232. [CrossRef]

17. Rojas-Gómez, M.; Moral, J.; López-Orozco, R.; Cabello, D.; Oteros, J.; Barranco, D.; Galán, C.; Díez, C.M. Pollen production in olive cultivars and its interannual variability. *Ann. Bot.* **2023**, *132*, 1145–1158. [CrossRef]

18. Guitian, J. Why *Prunus mahaleb* (Rosaceae) produces more flowers than fruits. *Am. J. Bot.* **1993**, *80*, 1305–1309. [CrossRef]

19. Cuevas, J.; Rallo, L.; Rapoport, H.F. Crop load effects on floral quality in olive. *Sci. Hortic.* **1994**, *59*, 123–130. [CrossRef]

20. Willson, M.F.; Schemske, D.W. Pollinator limitation, fruit production, and floral display in pawpaw (*Asimina triloba*). *Bull. Torrey Bot. Club* **1980**, *107*, 401–408. [CrossRef]

21. Bawa, K.S.; Webb, C.J. Flower, fruit and seed abortion in tropical forest trees: Implications for the evolution of paternal and maternal reproductive patterns. *Am. J. Bot.* **1984**, *71*, 736–751. [CrossRef]

22. Bradley, M.V.; Griggs, W.H. Morphological evidence of incompatibility in *Olea europaea* L. *Phytomorphology* **1963**, *13*, 141–156.

23. Ateyyeh, A.F.; Stösser, R.; Qrunfleh, M. Reproductive biology of the olive (*Olea europaea* L.) cultivar 'Nabali Baladi'. *J. Appl. Bot.* **2000**, *74*, 255–270.

24. Seifi, E.; Guerin, J.; Kaiser, B.; Sedgley, M. Sexual compatibility and floral biology of some olive cultivars. *N. Z. J. Crop Hortic. Sci.* **2011**, *39*, 141–151. [CrossRef]

25. Sánchez-Estrada, A.; Cuevas, J. 'Arbequina' olive is self-incompatible. *Sci. Hortic.* **2018**, *230*, 50–55. [CrossRef]

26. Sánchez-Estrada, A.; Cuevas, J. Pollen-pistil interaction in 'Manzanillo' olive (*Olea europaea* L.) under self-, free- and cross-pollination. *Rev. Chapingo Ser. Hortic.* **2019**, *25*, 141–150. [CrossRef]

27. Cuevas, J.; Chiamolera, F.M.; Pinillos, V.; Rodríguez, F.; Salinas, I.; Cabello, D.; Arbeiter, A.B.; Bandelj, D.; Raboteg Božiković, M.; Vuletin Selak, G. Arbosana olive is self-incompatible, but inter-compatible with some other low-vigor olive cultivars. *Horticulturae* **2024**, *10*, 739. [CrossRef]

28. Breton, C.M.; Bervillé, A. New hypothesis elucidates self-incompatibility in the olive tree regarding S-alleles dominance relationships as in the sporophytic model. *Comptes Rendus Biol.* **2012**, *335*, 563–572. [CrossRef]

29. Saumitou-Laprade, P.; Vernet, P.; Vekemans, X.; Billiard, S.; Gallina, S.; Essalouh, L.; Mhaïs, A.; Moukhli, A.; El Bakkali, A.; Barcaccia, G.; et al. Elucidation of the genetic architecture of self-incompatibility in olive: Evolutionary consequences and perspectives for orchard management. *Evol. Appl.* **2017**, *10*, 867–880. [CrossRef]

30. Cuevas, J. Incompatibilidad polen-pistilo. In *Variedades de Olivo Cultivadas en España*, 1st ed.; Barranco, D., Caballero, J.M., Martín, A., Rallo, L., Del Río, C., Tous, J., Trujillo, I., Eds.; Junta de Andalucía; Mundi-Prensa and COI: Seville, Spain, 2004; pp. 303–308.

31. Cuevas, J.; Polito, V.S. Compatibility relationships in 'Manzanillo' olive. *HortSci.* **1997**, *32*, 1056–1058. [CrossRef]

32. Vuletin Selak, G.; Cuevas, J.; Goreta Ban, S.; Perica, S. Pollen tube performance in assessment of compatibility in olive (*Olea europaea* L.) cultivars. *Sci. Hortic.* **2014**, *165*, 36–43. [CrossRef]

33. Bradley, M.V.; Griggs, W.H.; Hartmann, H.T. Studies on self- and cross-pollination of olives under varying temperature conditions. *Calif. Agric.* **1961**, *15*, 4–5. [CrossRef]

34. Griggs, W.H.; Hartmann, H.T.; Bradley, M.V.; Iwakiri, B.T.; Whisler, J.E. *Olive Pollination in California, Bulletin 869*; University of California—Agricultural Experiment Station: Berkeley, CA, USA, 1975; pp. 1–49.

35. Lavee, S.; Datt, A.C. The necessity of cross-pollination for fruit set of Manzanillo olives. *J. Horticult. Sci.* **1978**, *53*, 261–266. [CrossRef]
36. Willson, M.F.; Rathcke, B.J. Adaptive design of the floral display in *Asclepias syriaca* L. *Am. Midl. Nat.* **1974**, *92*, 47–57. [CrossRef]
37. Sutherland, S.; Delph, L.F. On the importance of male fitness in plants: Patterns of fruit-set. *Ecology* **1984**, *65*, 1093–1104. [CrossRef]
38. Sutherland, S. Why hermaphroditic plants produce many more flowers than fruits: Experimental tests with *Agave mckelveyana*. *Evolution* **1987**, *41*, 750–759. [CrossRef]
39. Queller, D. Pollen removal, paternity, and the male function of flowers. *Am. Nat.* **1997**, *149*, 585–594. [CrossRef]
40. Queller, D. Sexual selection in a hermaphroditic plant. *Nature* **1983**, *305*, 706–707. [CrossRef]
41. Lloyd, D.G. Parental strategies of angiosperms. *N. Z. J. Bot.* **1979**, *17*, 595–606. [CrossRef]
42. Uriu, K. Periods of pistil abortion in the development of the olive flower. *Proc. Am. Soc. Hort. Sci.* **1959**, *73*, 194–202.
43. Cuevas, J.; Pinney, K.; Polito, V.S. Flower differentiation, pistil development and pistil abortion in olive. *Acta Hortic.* **1999**, *474*, 293–296. [CrossRef]
44. Rosati, A.; Caporali, S.; Paoletti, A.; Famiani, F. Pistil abortion is related to ovary mass in olive (*Olea europaea* L.). *Sci. Hortic.* **2011**, *127*, 515–519. [CrossRef]
45. Diggle, P.K. Architectural effects and the interpretation of patterns of fruit and seed development. *Annu. Rev. Ecol. Evol. Syst.* **1995**, *26*, 531–552. [CrossRef]
46. Wolfe, L.M.; Denton, W. Morphological constraints on fruit size in *Linaria canadensis*. *Int. J. Plant. Soil Sci.* **2002**, *162*, 1313–1316. [CrossRef]
47. Rosati, A.; Lodolini, E.M.; Famiani, F. From flower to fruit: Fruit growth and development in olive (*Olea europaea* L.)—A review. *Front. Plant Sci.* **2023**, *14*, 1276178. [CrossRef]
48. Kelly, D.; Sork, V.L. Mast seeding in perennial plants: Why, how, where? *Ann. Rev. Ecol. and System.* **2002**, *33*, 427–447. [CrossRef]
49. Pías, B.; Salvande, M.; Guitián, P. Variation in predispersal losses in reproductive potential in rowan (*Sorbus aucuparia* L. Rosaceae) in the NW Iberian Peninsula. *Plant Ecol.* **2007**, *188*, 191–203. [CrossRef]
50. Lavee, S.; Rallo, L.; Rapoport, H.F.; Troncoso, A. The floral biology of the olive: Effect of flower number, type and distribution on fruitset. *Sci. Hortic.* **1996**, *66*, 149–158. [CrossRef]
51. Suarez, M.P.; Fernández-Escobar, R.; Rallo, L. Competition among fruits in olive II. Influence of inflorescence or fruit thinning and cross-pollination on fruit set components and crop efficiency. *Acta Hortic.* **1984**, *149*, 131–143. [CrossRef]
52. Rosati, A.; Zipančič, M.; Caporali, S.; Padula, G. Fruit weight is related to ovary weight in olive (*Olea europaea* L.). *Sci. Hortic.* **2009**, *122*, 399–403. [CrossRef]
53. Rosati, A.; Zipančič, M.; Caporali, S.; Paoletti, A. Fruit set is inversely related to flower and fruit weight in olive (*Olea europaea* L.). *Sci. Hortic.* **2010**, *126*, 200–204. [CrossRef]
54. Cuevas, J.; Rallo, L. Respuesta a la polinización cruzada en olivo bajo diferentes temperaturas. *Actas Hortic.* **1988**, *1*, 203–208.
55. Janzen, D.H. Escape of *Cassia grandis* L. beans from predators in time and space. *Ecology* **1971**, *52*, 964–979. [CrossRef]
56. Ehrlen, J. Why do plants produce surplus flowers? A reserve-ovary model. *Am. Nat.* **1991**, *138*, 918–933. [CrossRef]
57. Brown, A.O.; McNeil, J.N. Fruit production in cranberry (*Vaccinium macrocarpon*): A bet-hedging strategy to optimize reproductive effort. *Am. J. Bot.* **2006**, *93*, 910–916. [CrossRef]
58. Guitián, J.; Guitián, P.; Navarro, L. Fruit set, fruit reduction, and fruiting strategy in *Cornus sanguinea* (Cornaceae). *Am. J. Bot.* **1996**, *83*, 744–748. [CrossRef]
59. Guitian, J. Selective fruit abortion in *Prunus mahaleb* (Rosaceae). *Am. J. Bot.* **1994**, *81*, 1555–1558. [CrossRef]
60. Mulas, M. Characterisation of olive wild ecotypes. *Acta Hortic.* **1999**, *474*, 121–124. [CrossRef]
61. Khadivi, A.; Mirheidari, F.; Saeidifar, A.; Moradi, Y. Morphological characterizations of *Olea europaea* subsp. *cuspidata*. *Genet. Resour. Crop Evol.* **2024**, *71*, 1837–1853. [CrossRef]
62. Darwin, C.R. *On the Origin of Species by Means of Natural Selection, or the Preservation of Favoured Races in the Struggle for Life*, 5th ed.; John Murray: London, UK, 1869.
63. Willson, M.F. Sexual selection in plants. *Am. Nat.* **1979**, *113*, 777–790. [CrossRef]
64. Stephenson, A.G.; Winsor, J.A. *Lotus corniculatus* regulates offspring quality through selective fruit abortion. *Evolution* **1986**, *40*, 453–458. [CrossRef] [PubMed]
65. Cuevas, J.; Oller, R. Olive seed set and its impact on seed and fruit weight. *Acta Hortic.* **2002**, *586*, 485–488. [CrossRef]
66. Charlesworth, D. Why do plants produce so many more ovules than seeds? *Nature* **1989**, *338*, 21–22. [CrossRef]
67. Lee, T.D. Patterns of fruit maturation: A gametophyte competition hypothesis. *Am. Nat.* **1984**, *123*, 427–432. [CrossRef]
68. Cuevas, J.; Rallo, L.; Rapoport, H.F. Pollen tube growth and ovule abortion in *Olea europaea* (Oleaceae): A case if ovule selection? In *Pollination Mechanisms, Ecology and Agricultural Advances*; Raskin, N.D., Vuturro, P.T., Eds.; Nova Science Publisher Inc.: New York, NY, USA, 2011; pp. 57–72.
69. Cuevas, J.; Rallo, L.; Rapoport, H.F. Staining procedure for the observation of olive pollen tube behaviour. *Acta Hortic.* **1994**, *356*, 264–267. [CrossRef]

70. Sauter, M.A. A guided tour: Pollen tube orientation in flowering plants. *Chin. Sci. Bull.* **2009**, *54*, 2376–2382. [CrossRef]
71. Kanaoka, M.M. Cell–cell communications and molecular mechanisms in plant sexual reproduction. *J. Plant Res.* **2018**, *131*, 37–47. [CrossRef]
72. Iwano, M.; Ngo, Q.A.; Entani, T.; Shiba, H.; Nagai, T.; Miyawaki, A.; Isogai, A.; Grossniklaus, U.; Takayama, S. Cytoplasmic Ca^{2+} changes dynamically during the interaction of the pollen tube with synergid cells. *Development* **2012**, *139*, 4202–4209. [CrossRef]
73. Takeuchi, H.; Higashiyama, T. Attraction of tip-growing pollen tubes by the female gametophyte. *Curr. Opin. Plant Biol.* **2011**, *14*, 614–621. [CrossRef]
74. Hiei, K.; Ohara, M. Variation in fruit- and seed set among and within inflorescences of *Melampyrum roseum* var. *japonicum* (Scrophulariaceae). *Plant Species Biol.* **2002**, *17*, 13–23. [CrossRef]
75. Medrano, M.; Guitián, P.; Guitián, J. Patterns of fruit and seed set within inflorescences of *Pancratium maritimum* (Amaryllidaceae): Nonuniform pollination, resource limitation, or architectural effects? *Am. J. Bot.* **2000**, *87*, 493–501. [CrossRef]
76. Hapuarachchi, N.S.; Kämper, W.; Hosseini Bai, S.; Ogbourne, S.M.; Nichols, J.; Wallace, H.M.; Trueman, S.J. Selective retention of cross-fertilised fruitlets during premature fruit drop of Hass avocado. *Horticulturae* **2024**, *10*, 591. [CrossRef]
77. Alcaraz, M.L.; Hormaza, J.I. Fruit set in avocado: Pollen limitation, pollen load size, and selective fruit abortion. *Agronomy* **2021**, *11*, 1603. [CrossRef]
78. Stutte, G.W.; Martin, G.C. Effect of killing the seed on return bloom of olive. *Sci. Hortic.* **1986**, *29*, 107–113. [CrossRef]
79. Stanton, M.L. Seed variation in wild radish: Effect of seed size on components of seedling and adult fitness. *Ecology* **1984**, *65*, 1105–1112. [CrossRef]
80. Nakamura, R. Seed abortion and seed variation within fruits of *Phaseolus vulgaris*: Pollen donor and resource limitation effects. *Am. J. Bot.* **1988**, *75*, 1003–1010. [CrossRef]
81. Zhang, J.; Maun, M.A. Seed size variation and its effects on seedling growth in *Agropyron psammophilum*. *Bot. Gaz.* **1990**, *151*, 106–113. [CrossRef]

Article

Role of Endogenous Hormones on Seed Hardness in Pomegranate Fruit Development

Haoxian Li [1,2], Lina Chen [1,3,4], Ruitao Liu [1,4] and Zhenhua Lu [1,2,3,4,*]

[1] Zhengzhou Fruit Research Institute, Chinese Academy of Agricultural Sciences, Zhengzhou 450000, China; lihaoxian@caas.cn (H.L.); chenlina@caas.cn (L.C.); liuruitao@caas.cn (R.L.)
[2] National Nanfan Research Institute (Sanya), Chinese Academy of Agricultural Sciences, Sanya 572000, China
[3] National Horticultural Germplasm Resources Center, Chinese Academy of Agricultural Sciences, Zhengzhou 450000, China
[4] Zhongyuan Research Center, Chinese Academy of Agricultural Sciences, Xinxiang 453000, China
* Correspondence: luzhenhua@caas.cn

Abstract: Seed hardness is a unique trait for edibility and an important breeding target for pomegranates. We compared changes in hormones during the development of soft- and hard-seeded varieties in order to identify key hormones and developmental stages that affect seed lignin synthesis and accumulation. During the development of pomegranate seeds, lignin accumulates significantly in the stereid layer, and the degree of lignification is higher in Shandazi than in Huazi cultivars. The results showed that the accumulation of lignin in the stereid layer of the outer pomegranate seed coat is the reason for the differences in seed hardness between the soft-seeded variety and the hard-seeded variety. The hardness of pomegranate seeds was positively correlated with endogenous indole-3-acetic acid (IAA) and jasmonic acid (JA), while it was negatively correlated with cytokinins (CTKs), abscisic acid (ABA), gibberellins (GAs), salicylic acid (SA), and strigolactones (SLs). The highest contents of IAA and JA were 8.615 ng·g^{-1} and 4.5869 ng·g^{-1}, respectively, in the hard-seeded variety. In the soft-seeded variety, the maximum values of dihydrozeatin (DZ), dihydrozeatin-7-glucoside (DHZ7G), ABA, gibberellin A1 (GA$_1$), SA, and 5-deoxystrigol (5-DS) were 281.82 ng·g^{-1}, 1542.889 ng·g^{-1}, 61.273 ng·g^{-1}, 5.2556 ng·g^{-1}, 21.15 ng·g^{-1}, and 0.4494 ng·g^{-1}, respectively. IAA, CTKs, ABA, GA$_1$, and SA play major roles in the formation of lignin in pomegranate seeds, collectively determining seed hardness.

Keywords: pomegranate; seed hardness; endogenous hormones

Academic Editor: Esmaeil Fallahi

Received: 29 November 2024
Revised: 25 December 2024
Accepted: 31 December 2024
Published: 3 January 2025

Citation: Li, H.; Chen, L.; Liu, R.; Lu, Z. Role of Endogenous Hormones on Seed Hardness in Pomegranate Fruit Development. *Horticulturae* **2025**, *11*, 38. https://doi.org/10.3390/horticulturae11010038

1. Introduction

Pomegranate (*Punica granatum* L.) is a perennial small tree or shrub of the order Myrtales, Lythraceae, *Punica* Linn. It is generally believed that pomegranate cultivation and domestication have a history dating back more than two thousand years in China [1]. Compared to other fruits like peach, pear, and apple, the edible part of the pomegranate is the outer seed coat. Pomegranate can be roughly categorized according to seed hardness into soft-seeded, semi soft-seeded (or semi hard-seeded), and hard-seeded varieties [2–4]. Soft-seeded pomegranates are preferred by most consumers as they can be chewed easily, while semi soft-seeded and hard-seeded pomegranates have a worse chewing experience. In view of this, seed hardness is one of the most important quality indicators for pomegranate fruit.

Pomegranate is a type of berry, and its seeds consist of three parts: the aril (outer seed coat), the seed coat (inner seed coat), and the embryo. From the perspective of plant development, the seed coat develops from the inner integument. Like most angiosperms, pomegranate has a double integument. The inner integument of plants such as *Arabidopsis* gradually degenerates during development as the outer seed coat accumulates more anthocyanins, eventually developing into a hard seed coat. The two layers of pomegranate integuments develop normally, forming two seed coat layers. The edible part, the aril, is rich in sugars, organic acids, and polyphenols and is developed from outer epidermal cells of the seed [5,6]. The seed coat layer arising from the integument contains high lignin, cellulose, and hemicellulose contents [7]. Observation of their microstructures reveals that the aril is composed of a parenchymatous cell layer and a palisade tissue layer, while the seed coat is composed of a cork layer, a stereid layer, and a parenchymatous cell layer [8]. The formation and aggregation of lignin monomers is well understood in model plants, such as *Arabidopsis*, tobacco, and aspen [9]. Lignin monomers are formed by the catalysis of a series of enzymes such as ammonia-lyases (Als) from phenylalanine or tyrosine, which are then polymerized into lignin under the action of polymerase [10,11]. In hawthorn (*Crataegus* spp.), the accumulation of lignin in the inner seed coat of soft-endocarp varieties is significantly lower than that of hard-endocarp varieties [12]. In the study of stone fruit trees such as peach, it was found that the accumulation and lignification of lignin in the endocarp are the main reasons for seed hardening [13].

Existing research on the gene linkage markers of soft seed traits in pomegranate suggests that seed hardness is controlled by major genes [14]. Multiple studies have found that the formation of cell walls in pomegranate seeds is mainly regulated by a hierarchical network of transcription factors, including *NAC*, *WARK*, *MYB*, and *MYC*, which participate in the synthesis of cellulose, hemicellulose, and lignin [15,16]. In many lignin-related transcription factor families, some *MYB* family genes are regulated by auxin and jasmonic acid [17]. The outer seed coat development marker gene INNER NO OUTER (*INO*) and its transcriptional regulator BELL1 (*BEL1*) have been positively selected in the process of genome evolution and specifically expressed in arils [18].

Many researchers have found that the regulation of endogenous hormones is closely related to lignin development, which not only requires the dynamic balance of multiple hormones but may also require the dominant role of a certain hormone at a specific period [19,20]. There is a concentration gradient of endogenous auxin in plant cambium differentiation and xylem development, which has position signaling and growth-promoting effects [21,22]. It was found that 50 mg/L and 100 mg/L indole-3-acetic acid (IAA) significantly promoted the accumulation of lignin in the inner skin of walnuts, while 150 mg/L and 200 mg/L IAA inhibited lignin synthesis [23]. Lignin formation for the primary phloem fibers and the secondary xylem in the stem of *Coleus blumei* Benth is controlled by IAA (0.1% w/w) and gibberellin A3 (GA$_3$) [24]. Exogenous gibberellin (GA$_3$) and cytokinin (6-benzylaminopurine) elevated lignin accumulation in secondary xylem development and the expression of genes involved in lignin biosynthesis in carrot taproot [25,26]. Foliar spraying with GA$_3$ (10.0 mg·L^{-1}) at the first blooming stage was found to significantly reduce seed hardness in pomegranate, but application of different concentrations of GA$_3$ (10.0, 20.0, 30.0 mg·L^{-1}) + 2, 4-D (10.0 mg·L^{-1}) to the flower stalk increased the seed hardness [27].

Due to the significant difference in seed hardness between pomegranate soft seed varieties and hard seed varieties, we speculate that there should also be differences and some correlations in the changes in endogenous hormones during the seed development process. Therefore, this study measured the endogenous hormones and seed hardness of two pomegranate varieties at 60, 90, and 120 days after blooming to determine their change patterns and correlations.

2. Materials and Methods

2.1. Plant Materials

Two cultivars, Huazi (HZ), a soft-seeded cultivar with 1.28 kg/cm² seed hardness, and Shandazi (SDZ), a hard-seeded cultivar with 8.13 kg/cm² seed hardness, were collected from the pomegranate variety nursery at the National Horticultural Germplasm Resources Center in Zhengzhou, China (E113°22′13.6″, N34°57′9.97″), and grown under the same growth conditions. Pomegranate fruits were gathered at 60, 90, and 120 days after blooming (DAB), respectively (Figure 1). Nine morphologically normal fruits were sampled and divided into three groups to achieve three biological replicates. The seeds were removed and immediately frozen in liquid nitrogen and stored at −80 °C in a freezer until further analysis. The seeds of 'Huazi' and 'Shandazi' cultivars are abbreviated as HZ and SDZ, respectively. TA-XT Texture Analyzer was used to determine seed hardness. Ten seeds were taken from each sample for testing, and experiments were repeated three times. Days after blooming (DAB) is calculated starting from the day the bisexual pomegranate flowers are slightly open.

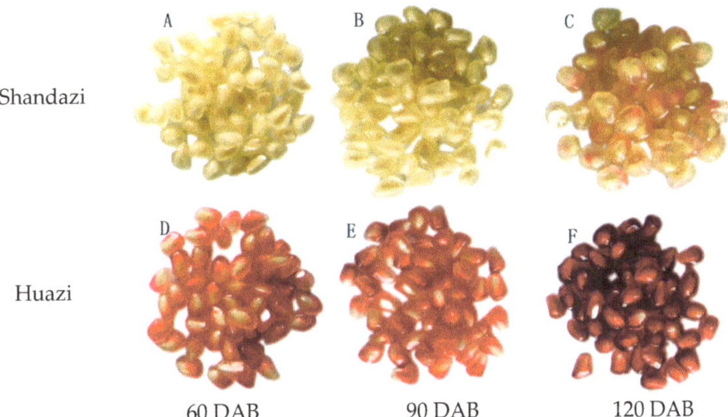

Figure 1. Morphology of two varieties of pomegranate seeds in three developmental stages. (**A–C**) represent the seeds of 'Shandazi' at 60, 90, and 120 DAB. (**D–F**) represent seeds of 'Huazi' at 60, 90, and 120 DAB.

2.2. Microscopy Observation of Lignin in Pomegranate Seeds

Pomegranate seeds were embedded in paraffin and then cut into 7 μm slices using a pathology slicer (RM2016, Shanghai Leica Instrument Co., Ltd., Shanghai, China). After the slices were dried, they were stained with safranin o-fast green staining for 2 h and sealed with neutral gum. They were observed under an upright optical microscope (ECLIPSE E100, NIKON, Tokyo, Japan) and imaging system (DS-U3, NIKON, Tokyo, Japan). The cytoderm, which has lignified, has a red color and the color of the cellulose cell wall is green.

2.3. Reagents and Standards

Chromatography-grade methanol and acetonitrile were sourced from Merck (Darmstadt, Germany). Chromatography-grade formic acid and acetic acid were purchased from Sigma-Aldrich (St. Louis, MO, USA). Standard products (purity greater than 99%) were purchased from Olchemim Ltd. (Olomouc, Czech Republic) and isoReag (Shanghai, China). The 1 mg/mL standard mother liquor was prepared using an 80% methanol solution. The standard solution was stored in a refrigerator at 4 °C away from light for future use. Ultrapure water (Millipore, Bradford, PA, USA) was used in all experiments.

2.4. Sample Pre-Processing

A certain mass of the sample was accurately weighed and ground into a homogenate by adding liquid nitrogen under low light. Then, 4 mL of 80% methanol extractant was added and the sample was sonicated for 1 h, followed by soaking overnight at 4 °C in the dark. The mixed solution was then centrifuged at 10,000 r/min for 10 min, and the supernatant was collected. The residue was dissolved, combined with the supernatant, and resuspend after concentrating by nitrogen-blowing. Next, a 50 μL sample was dissolved in 1 mL methanol/water/formic acid (15:4:1, $V/V/V$) solution, mixed well, and concentrated until dry. The sample was resuspended in 100 μL 80% methanol (V/V) and filtered through a 0.22 μm membrane filter for further LC-MS/MS analysis.

2.5. UPLC-MS/MS Conditions

Sample analysis and data collection were completed through UPLC (ExionLC™AD, AB Sciex, Framingham, MA, USA) and MS/MS system (Applied Biosystems 6500 Triple Quadrupole, AB Sciex, Framingham, MA, USA). The UPLC conditions were as follows. The chromatographic column (Waters, Milford, MA, USA) was Waters ACQUITY UPLC HSS T3 C18 (100 mm × 2.1 mm i.d., 1.8 μm). The mobile phase was composed of ultrapure water (A) and acetonitrile (B), with 0.04% acetic acid added to both. The gradient elution process was 5% from 0 to 1 min, then with 95% B from 1 to 9 min, and finally 5% B from 9.1 to 12 min. The flow rate, column temperature, and injection volume were set to 0.35 mL/min, 40 °C, and 2 μL, respectively. The MS/MS conditions were as follows. The interface of the electric spray ion source was Electrospray Ionization (ESI), and the temperature was set at 550 °C. The mass spectrometry voltages in positive ion mode and negative ion mode were 5500 V and −4500 V, respectively, and the curtain gas (CUR) was determined to be 35 psi. Each ion pair was scanned and detected based on optimized declustering potential and collision energy.

2.6. Statistical Analysis

Comparative analysis of seed hardness was performed using Microsoft Excel 2013 and SPSS (version 25.0). The data were calculated as mean ± standard deviation. One-Way ANOVA was used to detect differences between two cultivars. A p-value < 0.05 was considered significantly different. The MWDB database (Meatware, Wuhan, China) was structured based on standard samples for qualitative analysis of mass spectrometry data. Quantitative analysis of mass spectrometry data was conducted using the multiple reaction monitoring (MRM) mode of a triple quadrupole mass spectrometer. Data processing was carried out using Analyst 1.6.3 software (AB Sciex, USA) and Multiquant 3.0.3 software (AB Sciex, USA). Data on phytohormone content were processed using unit variance scaling (UV). Heat maps were drawn using R software (Version 3.3.5), and hierarchical clustering analysis (HCA) was performed on the accumulation patterns of phytohormones among different samples. Phytohormones were screened according to the requirement of fold change ≥ 2 and fold change ≤ 0.5 and were considered to have significant differences between each other. The detected phytohormones were annotated using KEGG database (http://www.kegg.jp/kegg/ (accessed on 24 May 2024)), and phytohormones with significant differences were classified and enriched.

3. Results

3.1. Seed Development Comparison for Morphology and Hardness

The seed hardness of two pomegranate varieties was compared at the three key growth stages of 60, 90, and 120 DAB. The results showed that the hardness of the seeds in the 'Huazi' (HZ) cultivar was lower than that in the 'Shandazi' (SDZ) cultivar. The seed hardness of 'Huazi' at 90 DAB was the lowest, measured at 0.970 kg·cm^{-2}, and the seed hardness of 'Shandazi' increased between 60 and 120 DAB. The maximum value of seed hardness was 8.124 kg·cm^{-2}. These findings show that there is a significant difference in seed hardness between the two varieties in three different stages, as shown in Table 1. The present study as well as previous research showed a positive correlation between seed hardness and lignin content [15,28].

Table 1. Seed hardness of 'Huazi' and 'Shandazi' cultivars at three stages.

Breeds	Seed Hardness/kg·cm^{-2}		
	60 DAB	90 DAB	120 DAB
Huazi	1.030 ± 0.325 [b]	0.970 ± 0.148 [b]	1.288 ± 0.724 [b]
Shandazi	6.896 ± 0.207 [a]	7.656 ± 0.211 [a]	8.124 ± 0.140 [a]

Note: Huazi is a soft-seeded variety, and Shandazi is a hard-seeded variety. Letters (a, b) represent significant difference, $p < 0.05$.

3.2. Microscopic Examination of Lignin in Pomegranate Seeds

By examining the seed hardness at 60, 90, and 120 DAB, the results (Figure 2) showed that the mature pomegranate seed coat can be roughly divided into five layers, forming a circular structure formed by cells of different sizes and densities. These layers consist of a palisade layer (PL), a parenchymatous cell (PC), a stereid layer (STL), a cork layer (CL), and a parenchymatous cell layer, from the outside to the inside. The outermost layer is an edible vesicle, which is composed of large, long columnar, highly lignified cells, with the long axis of the cells perpendicular to the surface of the seed coat and a thickness of about 100–400 μm. It is composed of one to multiple layers of cells without gaps between them. Near the palisade layer is the parenchymatous cell layer, usually consisting of multiple rows of cells with a thickness of 200–600 μm, an irregular shape, and large intercellular gaps. The stereid layer is a lignified layer of cells that gradually become denser from the outside to the inside. In the cork layer, the thickness the cells is about 50 μm, and the intracellular protoplasts have degenerated, leaving only cork and thickened lignified cell walls without intercellular spaces. The innermost layer is the parenchymatous cell layer, which is composed of a single layer of cells arranged in an orderly manner. The increase in hardness of pomegranate seeds is accompanied by the lignification of the outer cell wall of the seed coat.

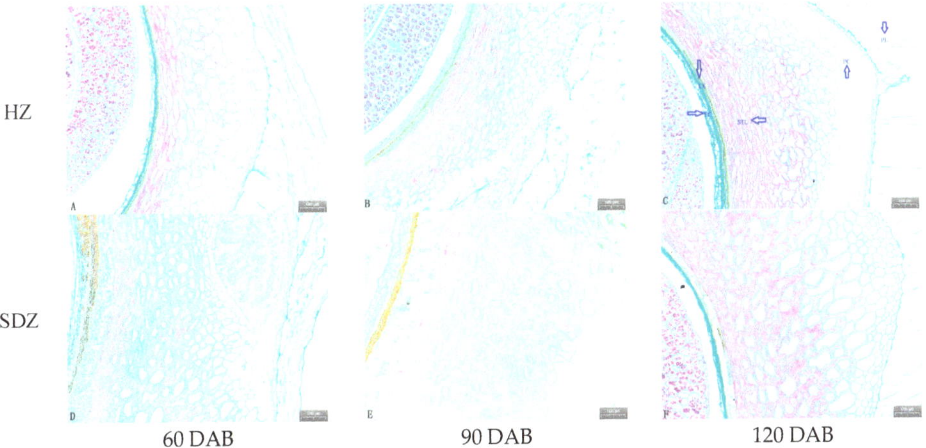

Figure 2. Safranin O-Fast green staining showed lignin deposition in cell walls of seeds. The lignified cell wall appears red, with darker colors indicating greater lignin deposition. The cellulose cell wall

appears green. (**A**) cell walls of 'HZ' (soft-seeded) at 60 DAB; (**B**) cell walls of 'HZ' at 90 DAB; (**C**) cell walls of 'HZ' at 120 DAB; (**D**) cell walls of 'SDZ' (hard-seeded) at 60 DAB; (**E**) cell walls of 'SDZ' at 90 DAB; (**F**) cell walls of 'SDZ' at 120 DAB. Note: PL: Palisade layer; PC: Parenchymatous cell; STL: Stereid layer; CK: Cork layer.

Through our observations, it was found that the stereid layer of the lignified seed coats of both pomegranate varieties presented a large red-colored area at 60 DAB, but the color was darker in 'SDZ'. At 90 and 120 DAB, the red area and depth of the lignified stereid layer in the 'SDZ' seed coat were greater than those of the 'HZ' seed coat. The red area and depth of the lignified stereid layer in the seed coat of both pomegranate varieties revealed an initial decrease and then an increase at 60, 90, and 120 DAB. In addition, the lignified stereid layer of the seed coat in 'SDZ' was denser than that in 'HZ'.

3.3. Cluster Analysis of Phytohormone Content in Pomegranate Seeds at Three Stages Between HZ and SDZ

We detected a total of fifty-one endogenous hormones from pomegranate seeds, including two abscisic acids and their derivatives, eleven auxins, twenty-one cytokinins, one ethylene, five GAs, eight jasmonic acids, two salicylic acids, and one SL. The clustering analysis results showed that the phytohormone content of the two varieties exhibited certain characteristics in each of the three stages (Figure 3). At 60 DAB, the contents of GA_3, IP, tZ, and oT were high in HZ pomegranate seeds, while mT, tZR, JA, JA-ILE, JA-Val, OPDA, TRA, IAA, and IAA-Glu were mainly detected in SDZ. At 90 DAB, IAA-Ala, MeJA, SA, SAG, oTR, tZOG, iP9G, DHZR, ACC, ABA, and 5-DS were extensively detected in HZ pomegranate seeds, while IAA-Gly and 2MeScZR were detected in SDZ. At 120 DAB, BAPR, IPR, ABA, MEIAA, IAA-Glc, IAA-Asp, TRP, GA_{15}, and GA_{20} were detected mainly in HZ pomegranate seeds, while H_2JA, cZROG, OPC-6, BAP7G, mT9G, and ABA-GE were detected in SDZ.

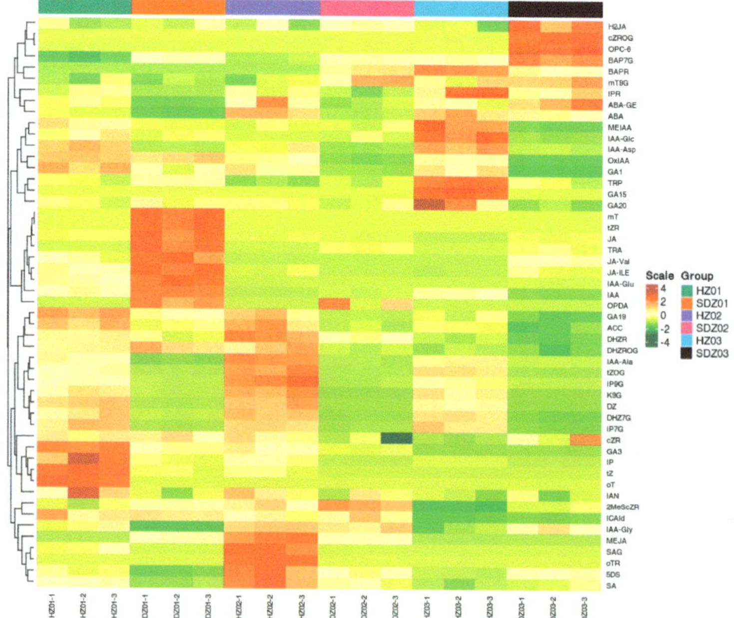

Figure 3. A cluster analysis heatmap of phytohormone content in pomegranate seeds at 60, 90, and 120 DAB between HZ and SDZ. Note: The horizontal axis represents the sample names from two varieties

in three periods, the vertical axis (right side) represents hormone information, and the vertical axis (left side) represents classification of hormones. Different colors represent the values obtained after standardized processing of relative content, with red representing high contents and green representing lower contents. H2JA, cZROG, OPC-6, BAP7G, BAPR, mT9G, IPR, ABA-GE, ABA, MEIAA, IAA-Glc, IAA-Asp, OxIAA, GA$_1$, TRP, GA$_{15}$, GA$_{20}$, mT, tZR, JA, TRA, JA-VAL, JA-ILE, IAA-Glu, IAA, OPDA, GA$_{19}$, ACC, DHZR, DHZROG, IAA-Ala, tZOG, iP9G, K9G, DZ, DHZ7G, iP7G, cZR, GA$_3$, IP, tZ, oT, IAN, 2MeScZR, ICAld, IAA-Gly, MEJA, SAG, oTR, 5DS, and SA refer to Dihydrojasmonic acid, *cis*-Zeatin-O-glucoside riboside, 3-oxo-2-(2-(Z)-Pentenyl)cyclopentane-1-hexanoic acid, N^6-Benzyladenine-7-glucoside, 6-Benzyladenosine, meta-Topolin-9-glucoside, N^6-isopentenyladenosine, ABA-glucosyl ester, Abscisic acid, Methyl indole-3-acetate, 1-O-indol-3-ylacetylglucose, Indole-3-acetyl-L-aspartic acid, 2-oxindole-3-acetic acid, Gibberellin A1, L-tryptophan, Gibberellin A15, Gibberellin A20, meta-Topolin, *trans*-Zeatin riboside, Jasmonic acid, Tryptamine, N-[(-)-Jasmonoyl]-(L)-valine, Jasmonoyl-L-isoleucine, Indole-3-acetyl glutamic acid, Indole-3-acetic acid, *cis*(+)-12-Oxophytodienoic acid, Gibberellin A19, 1-Aminocyclopropanecarboxylic acid, Dihydrozeatin ribonucleoside, Dihydrozeatin-O-glucoside riboside, N-(3-Indolylacetyl)-L-alanine, *trans*-Zeatin-O-glucoside, N^6-Isopentenyl-adenine-9-glucoside, Kinetin-9-glucoside, Dihydrozeatin, Dihydrozeatin-7-glucoside, N^6-Isopentenyl-adenine-7-glucoside, *cis*-Zeatin riboside, Gibberellin A3, N^6-isopentenyladenine, *trans*-Zeatin, ortho-Topolin, 3-Indoleacetonitrile, 2-Methylthio-cis-zeatin riboside, Indole-3-carboxaldehyde, Indole-3-acetyl glycine, Methyl jasmonate, Salicylic acid 2-O-β-glucoside, ortho-Topolin riboside, 5-Deoxystrigol, and Salicylic acid, respectively.

3.4. Dynamic Changes in Phytohormone Content in Pomegranate Seeds at Three Stages Between HZ and SDZ

Across the three growth stages, the content of IAA in HZ pomegranate seeds first decreased and then increased (Figure 4A), while the IAA content in SDZ seeds gradually decreased. At 60 and 90 DAB, the content of IAA in HZ was lower than that in SDZ. At 60 DAB, the IAA content in the SDZ and HZ pomegranate seeds was the highest, with values of 8.615 ng·g^{-1} and 4.5869 ng·g^{-1}, respectively. However, the content of IAA in HZ was higher than that in SDZ at 120 DAB. In the first two periods, the content of IAA in HZ pomegranate seeds was lower than that in SDZ seeds, but the IAA content in both was clearly opposite in the last period. Tryptophan (Trp), as a precursor for the biosynthesis of IAA, also underwent similar changes between HZ and SDZ. The TRP content in HZ pomegranate seeds was the highest at 120 DAB (Figure 4B), with a value of 8742.2 ng·g^{-1}. At 60 DAB, the TRP content in SDZ pomegranate seeds was the highest, with a value of 5481.3 ng·g^{-1}.

The trend of dihydrozeatin (DZ) content in HZ pomegranate seeds was first to increase and then to decrease, with the maximum value of 281.82 ng·g^{-1} at 90 DAB (Figure 4C). The content of DZ in SDZ pomegranate seeds gradually decreased, and the value at 60 DAB was 26.152 ng·g^{-1}. Dihydrozeatin-7-glucoside (DHZ7G), as a glucose conjugate of DZ, had no biological activity. The changes in the DHZ7G content in HZ and SDZ pomegranate seeds were consistent with the changes in DZ in both (Figure 4C,D). The contents of DZ and DHZ7G in HZ pomegranate seeds were significantly higher than those in SDZ at each stage. The N^6-isopentenyladenine (IP) content in HZ and SDZ pomegranate seeds rapidly decreased, with the maximum values occurring at 60 DAB, which were 2.1439 ng·g^{-1} and 0.38778 ng·g^{-1}, respectively (Figure 4E). The content of IP in HZ pomegranate seeds was significantly higher at each stage than in SDZ seeds. The *trans*-zeatin (tZ) content in HZ pomegranate seeds was sharply decreased in the first two periods, and the content was extremely low at 120DAB and could no longer be detected. At 60 DAB, the maximum content of tZ in HZ pomegranate seeds was 2.7271 ng·g^{-1}. The content of tZ in SDZ pomegranate seeds at 60 DAB was 0.2850 ng·g^{-1} and could not be detected in the last two stages. In the first two stages, the content of tZ in HZ pomegranate seeds was higher than that in SDZ, but still could not be detected in the last stage (Figure 4F).

The ABA content in HZ pomegranate seeds first increased and then slightly decreased, while the ABA content in SDZ gradually increased (Figure 4G). At 90 DAB, the ABA content in HZ pomegranate seeds was the highest, with a value of 61.273 ng·g^{-1}. The ABA content in SDZ pomegranate seeds was the highest at 120 DAB, with a value of 42.618 ng·g^{-1}. At the same time, the ABA content in HZ pomegranate seeds was consistently higher than that in SDZ during the three stages. 1-aminocyclopropane-1-carboxylic acid (ACC) is a direct precursor for ethylene biosynthesis, and the ACC content was positively correlated with the biosynthesis of ethylene. The content of ACC in HZ pomegranate seeds first remained high and then gradually decreased, while the content of ACC in SDZ pomegranate seeds gradually decreased (Figure 4H). At 90 DAB, the ACC content in HZ pomegranate seeds was the highest, with a value of 177.19 ng·g^{-1}. The ABA content in SDZ pomegranate seeds was the highest at 60 DAB, with a value of 119.47 ng·g^{-1}. The contents of ABA and ACC in HZ pomegranate seeds were notably higher than in SDZ seeds in each period.

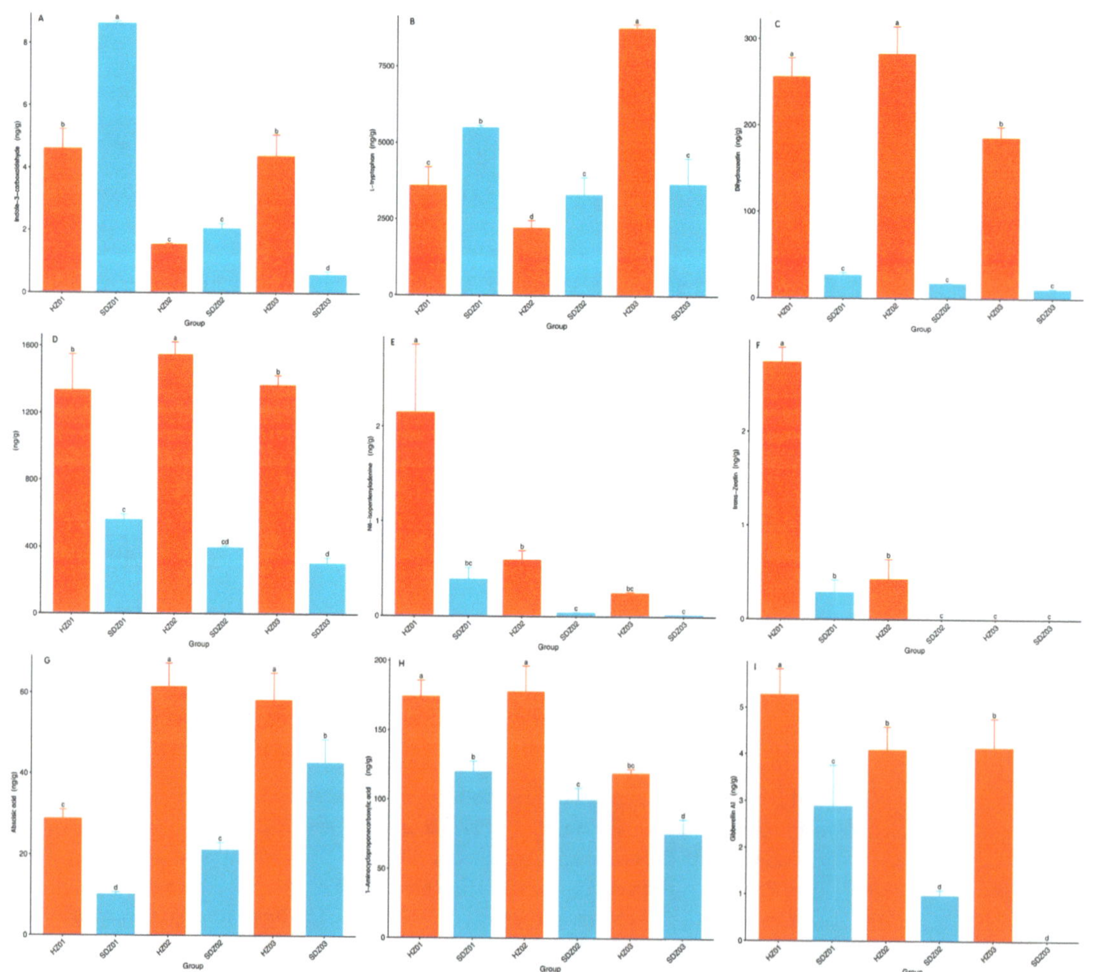

Figure 4. *Cont.*

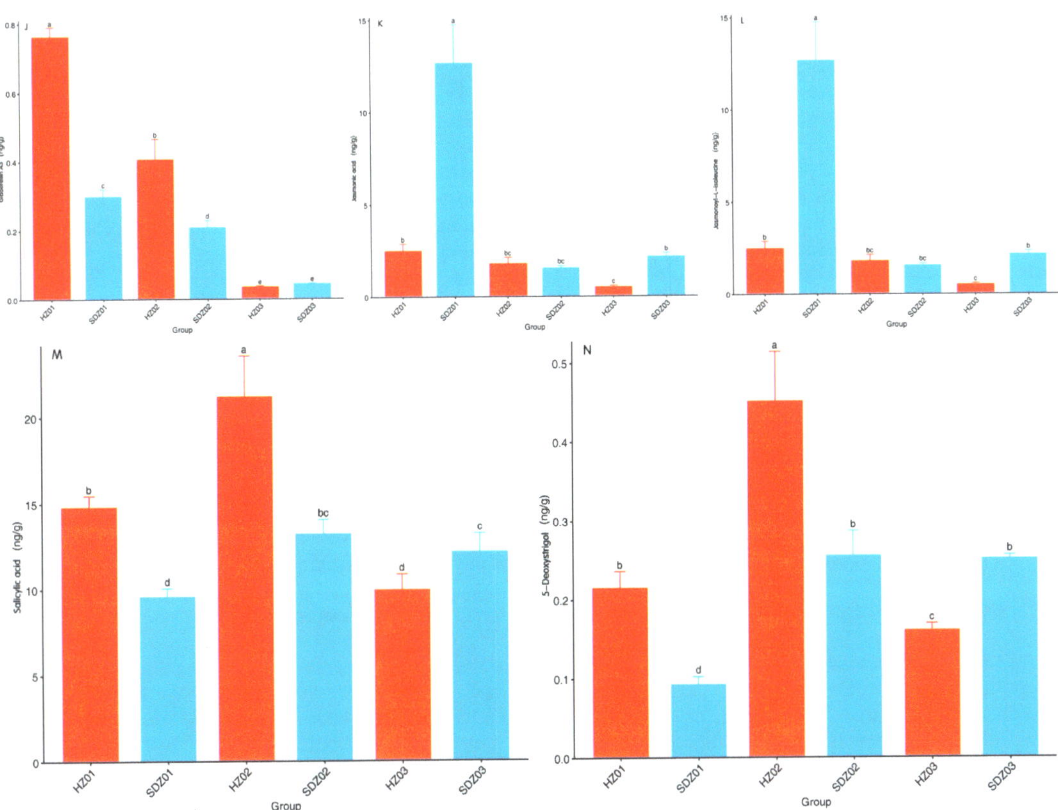

Figure 4. Changes in indole-3-acetic acid (IAA), L-tryptophan (TRP), dihydrozeatin (DZ), dihydrozeatin-7-glucoside (DHZ7G), N6-isopentenyladenine (IP), *trans*-Zeatin (tZ), abscisic acid (ABA), 1-aminocyclopropanecarboxylic acid (ACC), gibberellin A1 (GA$_1$), gibberellin A3 (GA$_3$), jasmonic acid (JA), jasmonoyl-L-isoleucine (JA-ILE), salicylic acid (SA), and 5-deoxystrigol (5-DS) contents in pomegranate seeds at three stages between HZ01, HZ02, HZ03 and SDZ01, SDZ02, SDZ03. Note: The determination of the phytohormone contents is based on ultra-high-performance liquid chromatography–tandem mass spectrometry (UPLC-MS/MS) technology. The horizontal axis represents the sample names from two varieties in three periods, and the vertical axis represents hormone contents. Letters (a–e) represent significant difference, $p < 0.05$. (**A**) represents the changes in IAA content of two varieties at three periods. (**B**) represents the changes in TRP content of two varieties at three periods. (**C**) represents the changes in DZ content of two varieties at three periods. (**D**) represents the changes in DHZ7G content of two varieties at three periods. (**E**) represents the changes in IP content of two varieties at three periods. (**F**) represents the changes in tZ content of two varieties at three periods. (**G**) represents the changes in ABA content of two varieties at three periods. (**H**) represents the changes in ACC content of two varieties at three periods. (**I**) represents the changes in GA$_1$ content of two varieties at three periods. (**J**) represents the changes in GA3 content of two varieties at three periods. (**K**) represents the changes in JA content of two varieties at three periods. (**L**) represents the changes in JA-ILE content of two varieties at three periods. (**M**) represents the changes in SA content of two varieties at three periods. (**N**) represents the changes in 5-DS content of two varieties at three periods.

The content of gibberellin A$_1$ (GA$_1$) in HZ pomegranate seeds first decreased and then remained stable, with a maximum value of 5.2556 ng·g^{-1}. The GA$_1$ content in SDZ pomegranate seeds rapidly decreased in the first two stages and could not be detected in the third stage (Figure 4I). At 60 DAB, the maximum content of tZ in SDZ pomegranate seeds was 2.8711 ng·g^{-1}. The content of GA$_1$ in HZ pomegranate seeds was higher than that in SDZ at each stage. Across all three stages, the content of gibberellin A$_3$ (GA$_3$) in HZ pomegranate seeds rapidly decreased, while the content in SDZ pomegranate seeds decreased more gradually (Figure 4J). The maximum value of GA$_3$ occurred at 60 DAB, with values of 0.7601 ng·g^{-1} and 0.2966 ng·g^{-1}, respectively. In the first two periods, the content of GA$_1$ in HZ pomegranate seeds was higher than that in SDZ seeds, but the GA$_1$content of both cultivars was basically the same in the last period.

Across the three stages, the content of jasmonic acid (JA) in HZ pomegranate seeds decreased gradually (Figure 4K), and JA-isoleucine (JA-ILE) showed a similar trend. The maximum values of the JA and JA-ILE content were 2.4451 ng·g^{-1} and 0.2479 ng·g^{-1}, respectively, at 60 DAB. The content of JA in SDZ pomegranate seeds first dropped sharply and then slowly rose over the three periods, and JA-ILE had a similar pattern (Figure 4L). At 60 DAB, the maximum values of JA and JA-ILE were 12.644 ng·g^{-1} and 0.7628 ng·g^{-1}, respectively.

Additionally, the content of salicylic acid (SA) in HZ pomegranate seeds increased first and then decreased, and the changes in the SA content in SDZ seeds were also similar (Figure 4M). The maximum value of the SA content in HZ and SDZ was 21.15 ng·g^{-1} and 13.149 ng·g^{-1}, respectively, at 90 DAB. The content of 5-deoxystrgol (5-DS) in HZ pomegranate seeds increased first and then decreased quickly, and the content of 5-DS in SDZ first increased rapidly and then maintained stability. The maximum value of 5-DS content in HZ and SDZ was 0.4494 ng·g^{-1} and 0.2540 ng·g^{-1}, respectively, at 90 DAB (Figure 4N).

3.5. Statistical Analysis of Differential Phytohormone Contents in Pomegranate Seeds at Three Stages Between HZ and SDZ

The results of the comparative analysis showed that there were 28 significantly different phytohormones in HZ01_vs._SDZ01, of which 18 were upregulated and 10 were downregulated (Table 2). TRA was upregulated in three combinations, namely HZ01_vs._SDZ01, HZ02_vs._SDZ02, HZ03_vs._SDZ03, while IAA-Ala and IAA-Asp were downregulated in these comparisons (Table 3). IAA was downregulated in the HZ03_vs._SDZ03 comparison. In addition, in the comparisons of HZ01_vs._SDZ01, HZ02_vs._SDZ02, HZ03_vs._SDZ03, DZ, IP, and tZ were downregulated. Also, ABA was downregulated in the comparisons HZ01_vs._SDZ01 and HZ02_vs._SDZ02, but was upregulated in HZ01_vs._HZ02 and SDZ01_vs._SDZ02. In addition, GA$_1$ and GA$_3$ were downregulated in the comparisons of HZ01_vs._SDZ01, HZ02_vs._SDZ02, and HZ03_vs._SDZ03. JA, JA-ILE, and MEJA were upregulated in the HZ01_vs._SDZ01 and HZ03_vs._SDZ03 comparisons, while JA-ILE and MEJA were downregulated in HZ02_vs._SDZ02. Finally, in the comparisons of HZ01_vs._HZ02 and SDZ01_vs._SDZ02, ABA and 5-DS were upregulated, while TRA, IAA-Glu, IAA, IP, tZ, JA-ILE, and JA-VAL were downregulated.

Table 2. A summary of statistics on the number of differential phytohormones.

Different Comparisons	All Significant Differences	Downregulated	Upregulated
HZ01_vs._SDZ01	28	18	10
HZ02_vs._SDZ02	21	18	3
HZ03_vs._SDZ03	26	17	9
HZ01_vs._HZ02	16	9	7
HZ02_vs._HZ03	22	15	7
SDZ01_vs._SDZ02	19	14	5
SDZ02_vs._SDZ03	17	10	7

Table 3. The list of differentially expressed phytohormones in pomegranate seeds at three stages between HZ and SDZ.

	Index	Compounds	Class	Type
HZ01_vs._SDZ01	ABA	Abscisic acid	ABA	down
	ABA-GE	ABA-glucosyl ester	ABA	down
	IAA-Ala	N-(3-Indolylacetyl)-L-alanine	Auxin	down
	IAA-Gly	Indole-3-acetyl glycine	Auxin	down
	TRA	Tryptamine	Auxin	up
	IAA-Glu	Indole-3-acetyl glutamic acid	Auxin	up
	IAA-Asp	Indole-3-acetyl-L-aspartic acid	Auxin	down
	IAN	3-Indoleacetonitrile	Auxin	down
	tZR	*trans*-Zeatin riboside	CTKs	up
	mT	meta-Topolin	CTKs	up
	iP7G	N^6-Isopentenyl-adenine-7-glucoside	CTKs	down
	iP9G	N^6-Isopentenyl-adenine-9-glucoside	CTKs	down
	tZOG	*trans*-Zeatin-O-glucoside	CTKs	down
	BAP7G	N^6-Benzyladenine-7-glucoside	CTKs	up
	DZ	Dihydrozeatin	CTKs	down
	IP	N^6-isopentenyladenine	CTKs	down
	K9G	Kinetin-9-glucoside	CTKs	down
	DHZ7G	Dihydrozeatin-7-glucoside	CTKs	down
	tZ	*trans*-Zeatin	CTKs	down
	oT	ortho-Topolin	CTKs	down
	GA_3	Gibberellin A3	GA	down
	JA-ILE	Jasmonoyl-L-isoleucine	JA	up
	JA-Val	N-[(-)-Jasmonoyl]-(L)-valine	JA	up
	MEJA	Methyl jasmonate	JA	up
	JA	Jasmonic acid	JA	up
	OPDA	*cis*(+)-12-Oxophytodienoic acid	JA	up
	SAG	Salicylic acid 2-O-β-glucoside	SA	down
HZ02_vs._SDZ02	ABA	Abscisic acid	ABA	down
	ABA-GE	ABA-glucosyl ester	ABA	down
	IAA-Ala	N-(3-Indolylacetyl)-L-alanine	Auxin	down
	TRA	Tryptamine	Auxin	up
	IAA-Asp	Indole-3-acetyl-L-aspartic acid	Auxin	down
	OxIAA	2-oxindole-3-acetic acid	Auxin	down
	oTR	ortho-Topolin riboside	CTKs	down
	iP7G	N^6-Isopentenyl-adenine-7-glucoside	CTKs	down
	iP9G	N^6-Isopentenyl-adenine-9-glucoside	CTKs	down
	tZOG	*trans*-Zeatin-O-glucoside	CTKs	down
	DZ	Dihydrozeatin	CTKs	down
	IP	N^6-isopentenyladenine	CTKs	down
	K9G	Kinetin-9-glucoside	CTKs	down
	DHZ7G	Dihydrozeatin-7-glucoside	CTKs	down
	tZ	*trans*-Zeatin	CTKs	down
	BAPR	6-Benzyladenosine	CTKs	up
	GA_1	Gibberellin A1	GA	down
	JA-ILE	Jasmonoyl-L-isoleucine	JA	down
	MEJA	Methyl jasmonate	JA	down
	OPDA	*cis*(+)-12-Oxophytodienoic acid	JA	up
	SAG	Salicylic acid 2-O-β-glucoside	SA	down

Table 3. *Cont.*

	Index	Compounds	Class	Type
	IAA-Ala	N-(3-Indolylacetyl)-L-alanine	Auxin	down
	IAA-Gly	Indole-3-acetyl glycine	Auxin	up
	TRP	L-tryptophan	Auxin	down
	TRA	Tryptamine	Auxin	up
	MEIAA	Methyl indole-3-acetate	Auxin	down
	IAA-Glu	Indole-3-acetyl glutamic acid	Auxin	up
	IAA-Asp	Indole-3-acetyl-L-aspartic acid	Auxin	down
	IAA-Glc	1-O-indol-3-ylacetylglucose	Auxin	down
	IAA	Indole-3-acetic acid	Auxin	down
	OxIAA	2-oxindole-3-acetic acid	Auxin	down
	2MeScZR	2-Methylthio-cis-zeatin riboside	CTKs	up
	cZROG	*cis*-Zeatin-O-glucoside riboside	CTKs	up
	iP7G	N^6-Isopentenyl-adenine-7-glucoside	CTKs	down
HZ03_vs._SDZ03	iP9G	N^6-Isopentenyl-adenine-9-glucoside	CTKs	down
	tZOG	*trans*-Zeatin-O-glucoside	CTKs	down
	DZ	Dihydrozeatin	CTKs	down
	IP	N^6-isopentenyladenine	CTKs	down
	K9G	Kinetin-9-glucoside	CTKs	down
	DHZ7G	Dihydrozeatin-7-glucoside	CTKs	down
	GA_{15}	Gibberellin A15	GA	down
	GA_{20}	Gibberellin A20	GA	down
	GA_1	Gibberellin A1	GA	down
	OPC-6	3-oxo-2-(2-(Z)-Pentenyl)cyclopentane-1-hexanoic acid	JA	up
	JA-ILE	Jasmonoyl-L-isoleucine	JA	up
	JA-Val	N-[(-)-Jasmonoyl]-(L)-valine	JA	up
	JA	Jasmonic acid	JA	up
	ABA	Abscisic acid	ABA	up
	TRA	Tryptamine	Auxin	down
	IAA-Glu	Indole-3-acetyl glutamic acid	Auxin	down
	IAA-Glc	1-O-indol-3-ylacetylglucose	Auxin	down
	IAA	Indole-3-acetic acid	Auxin	down
	oTR	ortho-Topolin riboside	CTKs	up
	iP9G	N^6-Isopentenyl-adenine-9-glucoside	CTKs	up
HZ01_vs._HZ02	tZOG	*trans*-Zeatin-O-glucoside	CTKs	up
	IP	N^6-isopentenyladenine	CTKs	down
	tZ	*trans*-Zeatin	CTKs	down
	oT	ortho-Topolin	CTKs	down
	JA-ILE	Jasmonoyl-L-isoleucine	JA	down
	JA-Val	N-[(-)-Jasmonoyl]-(L)-valine	JA	down
	MEJA	Methyl jasmonate	JA	up
	SAG	Salicylic acid 2-O-β-glucoside	SA	up
	5DS	5-Deoxystrigol	SL	up
	ABA	Abscisic acid	ABA	up
	ABA-GE	ABA-glucosyl ester	ABA	up
	IAA-Gly	Indole-3-acetyl glycine	Auxin	up
	TRA	Tryptamine	Auxin	down
	IAA-Glu	Indole-3-acetyl glutamic acid	Auxin	down
SDZ01_vs._SDZ02	IAA	Indole-3-acetic acid	Auxin	down
	OxIAA	2-oxindole-3-acetic acid	Auxin	down
	tZR	*trans*-Zeatin riboside	CTKs	down
	mT	meta-Topolin	CTKs	down
	cZR	*cis*-Zeatin riboside	CTKs	down
	IP	N^6-isopentenyladenine	CTKs	down
	K9G	Kinetin-9-glucoside	CTKs	down

Table 3. *Cont.*

Index	Compounds	Class	Type
tZ	*trans*-Zeatin	CTKs	down
BAPR	6-Benzyladenosine	CTKs	up
GA$_1$	Gibberellin A1	GA	down
JA-ILE	Jasmonoyl-L-isoleucine	JA	down
JA-Val	N-[(-)-Jasmonoyl]-(L)-valine	JA	down
JA	Jasmonic acid	JA	down
5DS	5-Deoxystrigol	SL	up

Note: ABA, CTKs, GA, JA, SA, and SL represent abscisic acid, cytokinins, gibberellin, jasmonic acid, salicylic acid, and strigolactone.

3.6. KEGG Analysis of Differential Metabolites in Pomegranate Seeds at Three Stages Between HZ and SDZ

In HZ01_vs._SDZ01, differential metabolites were enriched in nine metabolic pathways, and in particular the pathways of carotenoid biosynthesis and alpha-linolenic acid metabolism were significantly enriched (Figure 5A). The pathway of plant hormone signal transduction was significantly enriched in HZ02_vs._SDZ02 (Figure 5B). There were sixteen metabolic pathways involved in the enrichment of differential metabolites in HZ03_vs._SDZ03, among which the pathways of tryptophan metabolism, plant hormone signal transduction, and indole alkaloid biosynthesis were significantly enriched (Figure 5C).

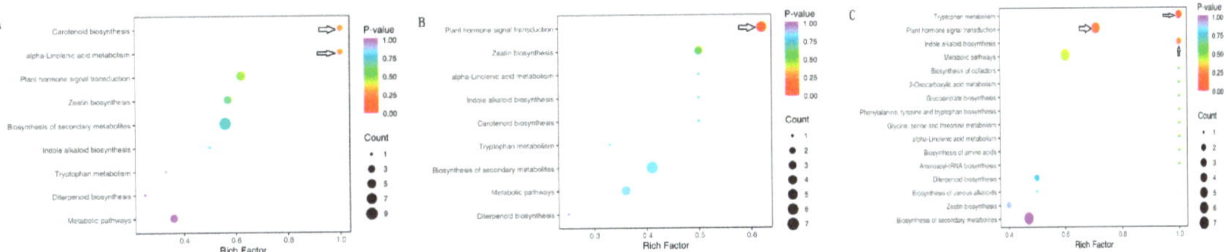

Figure 5. KEGG enrichment plot of differential metabolites in pomegranate seeds at three stages between HZ and SDZ. (**A**) HZ01_vs._SDZ01; (**B**) HZ02_vs._SDZ02; (**C**) HZ03_vs._SDZ03. Note: The horizontal axis represents the Rich factor corresponding to each pathway, the vertical axis represents the pathway name. The color of the dot corresponds to the *p*-Value, where the redder the dot, the more significant the enrichment level. The size of the dots represents the number of enriched differential metabolites. The Rich factor is the ratio of the number of differential metabolites in the corresponding pathway to the total number of metabolites annotated by that pathway. The magnitude of the value indicates the degree of enrichment. Arrows indicate significantly enriched pathways.

4. Discussion

4.1. The Relationship Between Lignin Formation and Seed Hardness in Pomegranate

A previous study found that the seed coat begins to develop 40 days after blooming in pomegranate [15]. The formation of seed hardness is the process of the continuous accumulation of lignin in the cells of the inner seed coat and the gradual thickening of the secondary cell wall, that is, the process of inner seed coat cell wall lignification. In addition, it was found through anatomical observation that only part of the inner seed coat of soft-seed varieties was lignified, while the entire inner seed coat of hard-seed varieties was lignified, except for the cells around the vascular bundles [29]. In the study of the development of ginkgo seed coats, it was found that closer fruit maturity, the cell wall of the middle seed coat significantly thickens and becomes lignified, forming a multi-layer structure mainly composed of a large number of lignified tracheids [30]. The results of our study indicate that the degree of lignification of the outer cell wall of the mature seed coat was very high in the hard-seed pomegranate variety, and there were significant differences in the thickness of the outer cell wall of the lignified seed coat between the two varieties studied. Comparing the hard-seeded and soft-seeded varieties, the soft-seeded variety had thinner lignified cell walls and larger inner spaces within the cell walls. Therefore, the accumulation of lignin in the hard-seeded variety was higher than that in the soft-seeded variety during seed development, which may be an important reason for the softness of the latter's seeds.

4.2. The Relationship Between the Content of IAA, CTKs, ABA, GAs, JA, SA, and SLs and the Hardness of Pomegranate Seeds

Phytohormone regulate the metabolic processes occurring in different tissues or organs of plants, even in the processes of cell aging and death, through the joint regulation of different hormones, thereby determining the quality of fruits [31,32].

Auxins are widely involved in various processes of plant growth and development, including embryo development; the morphogenesis of roots, stems, and leaves; fruit development; and polarity response [33,34]. During the hardening process of the inner skin of walnuts, the content of IAA consistently decreases during the incomplete hardening period and showed an upward trend in the later stage of hardening [35,36]. From the results of our study, it can be inferred that the accumulation and degree of development of seed lignin in the two pomegranate varieties were different. The AUX/IAA family of genes encoded a class of primary auxin-response-negative regulatory proteins and also a class of auxin-induced expression proteins, which regulated the expression of auxin response genes by interacting with specific auxin response factors (ARFs) [37]. An association analysis of genes and transcription factors related to lignin metabolism during the development of inner and outer seed coats found that AUX-IAA (*Pgr000815*) may be involved in the lignin metabolism, while AUX-IAA2 (*Pgr011491.1*), MYB (*Pgr022940.1*), and NAC66 (*Pgr017424.1*) may be involved in the metabolism of cellulose and hemicellulose [38].

Cytokinins play a role in the division and differentiation of plant vascular cells during cellular development and are essential for maintaining the structure and stability of procambial cells [19]. However, the function of cytokinins cannot be achieved without the cooperation of auxin [39]. Therefore, to maintain their structure in the same number of secondary wall cells in their seeds, soft-seeded pomegranate varieties need to maintain high concentrations of cytokinins, which is also the reason why their cytokinin content was always higher than that of hard-seeded varieties during the three fruit development stages in our analyses.

ABA is a very important plant hormone which can not only receive external environmental signals (including signals of drought, salt stress, low temperature, etc.) to participate in regulating plant growth and development, but which also plays an important role in fruit ripening [40]. This is consistent with the changes in endogenous ABA during the ripening process of peach (a concentration of ABA greater than 200 ng·g^{-1}), mango (a concentration of ABA greater than 1000 ng·g^{-1}), and cucumber fruits (a concentration of ABA more than 90 ng·g^{-1}) [41–43]. During the process of fruit ripening, there is a decrease in fruit hardness due to the massive accumulation of ABA, which increases the activity of enzymes such as cell wall-degrading enzymes, pectinesterase, and polygalacturonase that can alter the structure of fruit cell walls [44]. However, some studies suggest that exogenous ABA treatment (at concentrations of ABA of 50 mg·L^{-1}, 100 μM, and 25 mg·L^{-1}) can promote the synthesis and accumulation of lignin in plants [25,45,46]. It can be inferred that high concentrations of endogenous ABA inhibit lignin synthesis during the development of pomegranate seeds, while low concentrations promote it.

Gibberellins (GAs) can regulate stem elongation, secondary growth, and lignification of xylem in plants [47]. The application of exogenous GA$_3$ plays a positive promoting role in the lignification process of carrot (*Daucus carota* L.) roots, moso bamboo (*Phyllostachys edulis*), and sweet potato (*Ipomoea batatas*) vascular roots, as well as in increasing pear (*Pyrus ussuriensis* Maxim.) stone cell content [48–51]. Similarly, the content of endogenous hormones GA_1 and GA_3 in pomegranate fruits was higher in the soft-seeded variety than in the hard-seeded variety during the three developmental stages in our study. These findings confirm that gibberellins may participate in the regulation of plant xylem formation and development in multiple ways.

JA is an endogenous growth regulator that regulates the expression of downstream related genes in plants through the JA signaling pathway when stimulated by the environment or required for their own growth, thereby balancing plant growth [52]. JA participates in regulating lignin synthesis in *Arabidopsis* and melon stems in the form of active compounds when subjected to organ damage or external stress [53,54]. SA, as a small-molecule phenolic compound, is a plant growth regulator and important signaling molecule commonly found in plants [55]. SA can interfere with the signal transduction or response of gibberellin (GA), thereby inhibiting the expression of lignin synthesis-related enzyme genes in cells and affecting plant growth [56]. Strigolactones (SLs) are a novel class of plant hormones originated from the carotenoid pathway at the root and stem ends, closely related to lateral branch growth in plants [57]. 5-deoxystrigol (5-DS) is a precursor for the synthesis of SLs and is commonly found in many plants. The treatment of strawberry fruits during storage with 1 μM exogenous SL can significantly increase PAL and 4CL enzyme activity to promote lignin accumulation [58].

5. Conclusions

The results of this study indicate that the degree of cell wall lignification in pomegranate seeds increases with the development process, and there are differences in the thickness of stereid layer lignification among different pomegranate varieties. The seed hardness of pomegranate is mainly reflected in the accumulation of lignin in the stereid layer. The synthesis and accumulation of lignin in pomegranate seeds are regulated by multiple plant hormones, and there are also promotion and inhibition activities among hormones. A high level of IAA promotes the development of pomegranate seeds, and the transcription products of the IAA response factor accumulate in the seed coat. During the development of pomegranate seeds, the seed hardness is directly proportional to the content of endogenous IAA and JA and inversely proportional to the content of CTKs, ABA, GAs, SA, and SLs. In essence, the role of endogenous hormones is complex, and they may even have opposite effects at different concentration levels. IAA, CTKs, ABA, GA_1, and SA play a major role in the formation of lignin in pomegranate seeds, collectively determining the seed hardness. The results of this study help us to further understand the role of endogenous hormones in regulating pomegranate seed hardness.

Author Contributions: H.L. and Z.L. jointly proposed the experimental ideas; H.L. and L.C. designed specific experimental plans and operational steps. H.L. and R.L. used software to analyze data. H.L. completed the writing of the initial draft. H.L. and Z.L. confirmed the reliability of the data and improved the original draft. All authors have read and agreed to the published version of the manuscript.

Funding: This work was supported by the National Key Research and Development Program of China (no. 2021YFD1600801) from the China Rural Technology Development Center, and the Agricultural Science and Technology Innovation Program (no. CAAS-ASTIP-2024-ZFRI-05) from Chinese Academy of Agricultural Sciences.

Data Availability Statement: The data that support the findings of this study are available from the corresponding author upon reasonable request.

Conflicts of Interest: The authors declare that they have no competing financial interests or personal relationships that may have influenced the work reported in this study.

References

1. Cao, S.Y.; Hou, L.F. Introduction of Chapter one. In *China Fruit Monograph-Punica Granatum*, 1st ed.; China Forestry Publishing House: Beijing, China, 2013; pp. 1–2.
2. Jalikop, S.H.; Kumar, P.S. Use of soft, semi-soft and hardseeded types of pomegranate (*Punica granatum*) for improvement of firuit attributes. *Indian J. Agric. Sci.* **1998**, *68*, 87–91.
3. Melgarejo, P.; Sanchez, M.; Hernandez, Z.F.; Martnez, J.J.; Amoros, A. Parameters for determining the hardness and pleasantness of pomegranates (*Punica granatum* L.). *Options Méditerr. Sér. A Sémin. Méditerr.* **2000**, *42*, 225–230.
4. Lu, L.J.; Gong, X.M.; Zhu, L.W. Study on seed hardness of pomegranate cultivars in China. *J. Anhui Agric. Univ.* **2006**, *33*, 356–359.
5. Holland, D.; Hatib, K.; Bar-Ya'akoy, I. Pomegranate: Botany, horticulture, breeding. *Hortic. Rev.* **2009**, *35*, 127–191.
6. Hang, Z.Q.; Han, Q.B.; Xu, J.S. Components analysis of *Punica granatum* L. seed. *J. Anhui Agric. Sci.* **2010**, *38*, 18740–18741.
7. Dalimov, D.N.; Dalimova, G.N.; Bhatt, M. Chemical composition and lignins of tomato and pomegranate seeds. *Chem. Nat. Compd.* **2003**, *39*, 37–40. [CrossRef]
8. Qin, G.H.; Liu, C.Y.; Xu, Y.L.; Li, Y.L.; Qi, Y.J.; Pan, H.F.; Yi, X.K.; Gao, Z.H. Microstructure observation of seed coats and its development of pomegranate (*Punica granatum* L.). *Chin. J. Trop. Crops* **2018**, *39*, 489–493.
9. Boerjan, W.; Ralph, J.; Baucher, M. Lignin biosynthesis. *Annu. Rev. Plant Biol.* **2003**, *54*, 519–546. [CrossRef] [PubMed]
10. Humphreys, J.M.; Chapple, C. Rewriting the lignin roadmap. *Curr. Opin. Plant Biol.* **2002**, *5*, 224–229. [CrossRef] [PubMed]
11. Barros, J.; Shrestha, H.K.; Serrani-Yarce, J.C.; Engle, N.L.; Abraham, P.E.; Tschaplinski, T.J.; Hettich, R.L.; Dixon, R.A. Proteomic and metabolic disturbances in lignin-modifiedBrachypodium distachyon. *Plant Cell* **2022**, *34*, 3339–3363. [CrossRef] [PubMed]
12. Dai, H.Y.; Han, G.F.; Yan, Y.J.; Zhang, F.; Liu, Z.C.; Li, X.M.; Li, W.R.; Ma, Y.; Li, H.; Liu, Y.X.; et al. Transcript assembly and quantification by RNA-Seq reveals differentially expressed genes between soft-endocarp and hard-endocarp Hawthorns. *PLoS ONE* **2013**, *8*, e72910. [CrossRef]
13. Dardick, C.D.; Callahan, A.M.; Chiozzotto, R.; Schaaffer, R.J.; Piangnani, M.C.; Scorza, R. Stone formation in peach fruit exhibits spatial coordination of the lignin and flavonoid pathways and similarity to *Arabidopsis* dehiscence. *BMC Biol.* **2010**, *8*, 13–29. [CrossRef] [PubMed]
14. Sarkhosh, A.; Zamani, Z.; Fatahi, R.; Ranjbar, H. Evaluation of genetic diversity among Iranian soft-seed pomegranate accessions by fruit characteristics and RAPD markers. *Sci. Hortic.* **2009**, *121*, 313–319. [CrossRef]
15. Xue, H.; Cao, S.Y.; Li, H.X.; Zhang, J.; Niu, J.; Chen, L.N.; Zhang, F.H.; Zhao, D.G. De novo transcriptome assembly and quantification reveal differentially expressed genes between soft-seed and hard-seed pomegranate (*Punica granatum* L.). *PLoS ONE* **2017**, *12*, e0178809. [CrossRef]
16. Xia, X.C.; Li, H.X.; Cao, D.; Luo, X.; Yang, X.W.; Chen, L.N.; Liu, B.B.; Wang, Q.; Jing, D.; Cao, S.Y. Characterization of a NAC transcription factor involved in the regulation of pomegranate seed hardness (*Punica granatum* L.). *Plant Physiol. Biochem.* **2019**, *139*, 379–388. [CrossRef]
17. Demura, T.; Fukuda, H. Transcriptional regulation in wood formation. *Trends Plant Sci.* **2007**, *12*, 64–70. [CrossRef]
18. Qin, G.H.; Xu, C.Y.; Ming, R.; Tang, H.B.; Guyot, R.; Kramer, E.M.; Hu, Y.D.; Yi, X.K.; Qi, Y.J.; Xu, X.Y.; et al. The pomegranate (*Punica granatum* L.) genome and the genomics of punicalagin biosynthesis. *Plant J.* **2017**, *91*, 1108–1128. [CrossRef]
19. Fukuda, H. Signals that control plant vascular cell differentiation. *Nat. Rev. Mol. Cell Biol.* **2004**, *5*, 379–391. [CrossRef]
20. Xiao, Y.; Yi, F.; Ling, J.J.; Yang, G.J.; Lu, N.; Jia, Z.R.; Wang, J.C.; Zhao, K.; Wang, J.H.; Ma, W.J. Genome-wide analysis of lncRNA and mRNA expression and endogenous hormone regulation during tension wood formation in *Catalpa bungei*. *BMC Genom.* **2020**, *21*, 609. [CrossRef] [PubMed]
21. Uggla, C.; Sandberg, G.; Sundberg, B.; Moritz, T. Auxin as a positional signal in pattern formation in plants. *Proc. Natl. Acad. Sci. USA* **1996**, *93*, 9282–9286. [CrossRef] [PubMed]
22. Tuominen, H.; Puech, L.; Fink, S.; Sundberg, B. A radial concentration gradient of indole-3-acetic acid is related to sec-ondary xylem development in hybrid aspen. *Plant Physiol.* **1997**, *115*, 577–585. [CrossRef]
23. Li, Y.X.; Yu, S.Q.; Guo, Z.Z.; Fu, J.Z.; Lu, H.L.; Wang, H.X.; Zhang, R.; Mu, T.L.P. Effect of exogenous IAA on the growth and development of walnut endocarp. *J. Fruit Sci.* **2024**, *41*, 941–955.
24. Aloni, R.; Tollier, M.T.; Monties, B. The role of auxin and gibberellin in controlling lignin formation in primary phloem fibers and in xylem of coleus blumei stems. *Plant Physiol.* **1990**, *94*, 1743–1747. [CrossRef]
25. Jin, M.J.; Jiao, J.Q.; Zhao, Q.X.; Ban, Q.Y.; Gao, M.; Suo, J.T.; Zhu, Q.G.; Rao, J.P. Dose effect of exogenous abscisic acid on controlling lignification ofpostharvest kiwifruit (*Actinidia chinensis* cv. hongyang). *Food Control* **2021**, *124*, 107911. [CrossRef]
26. Khadr, H.; Wang, Y.H.; Zhang, R.R.; Wang, X.R.; Xu, Z.S.; Xiong, A.S. Cytokinin (6-benzylaminopurine) elevates lignification and the expression of genes involved in lignin biosynthesis of carrot. *Protoplasma* **2020**, *257*, 1507–1517. [CrossRef]

27. Chen, L.N.; Niu, J.; Liu, B.B.; Jing, D.; Luo, X.; Li, H.X.; Xia, X.C.; Yang, X.W.; Zhang, F.H.; Cao, D.; et al. Effects of foliar application of plant growth regulators at different flowering stages on fruit setting and quality in pomegranate. *J. Fruit Sci.* **2020**, *37*, 244–253.

28. Zhang, S.M.; Gong, L.Y.; Cao, D.Q.; Zhang, Y.J.; Yang, J. Total lignin content in pomegranate seed coat and cloning and expression analysis of *PgCOMT* gene. *J. Trop. Subtrop. Bot.* **2015**, *23*, 65–73.

29. Xie, X.B.; Huang, Y.; Tian, S.Y.; Li, G.L.; Cao, S.Y. Relationship of seed hardness development and microstructure of seed coat cell in soft seed pomegranate. *Acta Hortic. Sin.* **2017**, *44*, 1174–1180. [CrossRef]

30. Wang, L.; Qi, Y.M.; Lu, Y.; Jin, B. Morphogenesis and structural characteristics of mesotesta in *Ginkgo biloba* L. *J. Chin. Electron Microsc. Soc.* **2009**, *28*, 573–578.

31. Bowman, J.L.; Floyd, S.K. Patterning and Polarity in Seed Plant Shoots. *Annu. Rev. Plant Biol.* **2008**, *59*, 67–88. [CrossRef]

32. Du, W.; Ding, J.; Ruan, C.J. Dynamic changes of endogenous hormone levels during the development of sea-buckthorn fruit. *Chin. Bull. Bot.* **2018**, *53*, 219–226.

33. Pattison, R.J.; Csukasi, F.; Catalá, C. Mechanisms regulating auxin action during fruit development. *Physiol. Plant.* **2014**, *151*, 62–72. [CrossRef] [PubMed]

34. Zheng, L.W.; Gao, C.; Zhao, C.D.; Zhang, L.Z.; Han, M.Y.; An, N.; Ren, X.L. Effects of brassinosteroid associated with auxin and gibberellin on apple tree growth and gene expression patterns. *Hortic. Plant J.* **2019**, *5*, 93–108. [CrossRef]

35. Zhao, S.G.; Liu, K.; Wen, J.; Wang, H.X.; Zhang, Z.H.; Hua, L. Effect of GA_3 and 6-BA on the development of walnut shell. *Acta Hortic. Sin.* **2019**, *46*, 2119–2128.

36. Yu, S.Q.; Zhang, R.; Guo, Z.Z.; Song, Y.; Fu, J.Z.; Wu, P.Y.; Ma, Z.H. Dynamic changes of auxin and analysis of differentially expressed genes in walnut endocarp during hardening. *Acta Hortic. Sin.* **2021**, *48*, 487–504.

37. Li, Y.Y.; Qi, Y.H. Advances in biological functions of *Aux/IAA* gene family in plants. *Chin. Bull. Bot.* **2022**, *57*, 30–41.

38. Qin, G.H.; Liu, C.Y.; Li, J.Y.; Qi, Y.J.; Gao, Z.H.; Zhang, Z.L.; Yi, X.K.; Pan, H.F.; Ming, R.; Xu, Y.L. Diversity of metabolite accumulation patterns in inner and outer seed coats of pomegranate: Exploring their relationship with genetic mechanisms of seed coat development. *Hortic. Res.* **2020**, *7*, 10. [CrossRef] [PubMed]

39. Fukuda, H. Tracheary element differentiation. *Plant Cell* **1997**, *9*, 1147–1156. [CrossRef]

40. Leng, P.; Yuan, B.; Guo, Y.D. The role of abscisic acid in fruit ripening and responses to abiotic stress. *J. Exp. Bot.* **2014**, *65*, 4577–4588. [CrossRef]

41. Cao, Y.Q.; Leng, P.; Pan, X.; Yan, Z.G.; Ren, J. Role of abscisic acid in fruit ripening of peach. *Acta Hortic. Sin.* **2009**, *36*, 1037–1042.

42. Sakimin, S.Z.; Zora, S.; Gregory, M.S.; James, B.R. Role of brassinosteroids, ethylene, abscisic acid, and indole-3-acetic acid in mango fruit ripening. *J. Plant Growth Regul.* **2012**, *31*, 363–372.

43. Wang, Y.P.; Wu, Y.; Ji, K.; Dai, S.J.; Hu, Y.; Sun, L.; Li, Q.; Chen, P.; Sun, Y.F.; Duan, C.R.; et al. The role of abscisic acid in regulating cucumber fruit development and ripening and its transcriptional regulation. *Plant Physiol. Biochem.* **2013**, *64*, 70–79. [CrossRef]

44. Li, C.; Hou, B.Z. Molecular mechanism of abscisic acid in regulating fruit ripening. *J. Fruit Sci.* **2023**, *40*, 988–999.

45. Peter, G.M.; David, M.C. Suppression by ABA of salicylic acid and lignin accumulation and the expression of multiple genes, in *Arabidopsis* infected with *Pseudomonas syringae* pv. Tomato. *Funct. Integr. Genom.* **2006**, *7*, 181–191.

46. Zhang, Z.P.; Tan, Y.X.; Li, B.J.; Li, Y.C.; Bi, Y.; Li, S.Q.; Wang, X.J.; Zhang, Y.; Hu, D. Effects of exogenous abscisic acid treatment on periderm suberification of postharvest mini-tuber potato from aeroponic system and its possible mechanisms. *Sci. Agric. Sin.* **2023**, *56*, 1154–1167.

47. Mauriat, M.; Moritz, T. Analyses of *GA20ox-* and *GID1*-over-expressing aspen suggest that gibberellins play two distinct roles in wood formation. *Plant J.* **2009**, *58*, 989–1003. [CrossRef] [PubMed]

48. Wang, G.L.; Que, F.; Xu, Z.S.; Wang, F.; Xiong, A.S. Exogenous gibberellin enhances secondary xylem development and lignification in carrot taproot. *Protoplasma* **2017**, *254*, 839–848. [CrossRef] [PubMed]

49. Zhang, H.X.; Wang, H.H.; Zhu, Q.; Gao, Y.b.; Wang, H.Y.; Zhao, L.Z.; Wang, Y.S.; Xi, F.H.; Wang, W.F.; Yang, Y.Q.; et al. Transcriptome characterization of moso bamboo (*Phyllostachys edulis*) seedlings in response to exogenous gibberellin applications. *BMC Plant Biol.* **2018**, *18*, 125. [CrossRef]

50. Singh, V.; Sergeeva, L.; Ligterink, W.; Aloni, R.; Zemach, H.; Doron-Faigenboim, A.; Yang, J.; Zhang, P.; Shabtai, S.; Firon, N. Gibberellin promotes sweetpotato root vascular lignification and reduces storage-root formation. *Front. Plant Sci.* **2019**, *10*, 1320. [CrossRef]

51. Wang, X.Q.; Shang, Y.; Liu, W.C.; Liu, C.; Du, G.D.; Lv, D.G. Effect of growth regulators, Ca and B fertilizers on calyx abscission, fruit quality, and lignin metabolism in Nanguo pear. *J. Shenyang Agric. Univ.* **2019**, *50*, 399–405.

52. Lyons, R.; Manners, J.M.; Kazan, K. Jasmonate biosynthesis and signaling in monocots: A comparative overview. *Plant Cell Rep.* **2013**, *32*, 815–827. [CrossRef]

53. Denness, L.; McKenna, J.F.; Segonzac, C.; Wormit, A.; Madhou, P.; Bennett, M.; Mansfield, J.; Zipfel, C.; Hamann, T. Cell wall damage-induced lignin biosynthesis is gegulated by a reactive oxygen species- and jasmonic acid-dependent process in Arabidopsis. *Plant Physiol.* **2011**, *156*, 1364–1374. [CrossRef]

54. Liu, W.; Jiang, Y.; Jin, Y.Z.; Wang, C.H.; Yang, J.; Qi, H.Y. Drought-induced ABA, H_2O_2 and JA positively regulate *CmCAD* genes and lignin synthesis in melon stems. *BMC Plant Biol.* **2021**, *21*, 83. [CrossRef] [PubMed]

55. Chen, Z.; Zheng, Z.; Huang, J.; Lai, Z.; Fan, B. Biosynthesis of salicylic acid in plants. *Plant Signal. Behav.* **2009**, *4*, 493–496. [CrossRef]

56. Lina, G.G.; Luis, E.T.; Lisa, A.J.; Richard, A.D. Salicylic acid mediates the reduced growth of lignin down-regulated plants. *Proc. Natl. Acad. Sci. USA* **2011**, *108*, 20814–20819.

57. Gomez-Roldan, V.; Fermas, S.; Brewer, P.B.; Puech-Pagès, V.; Dun, E.A.; Pillot, J.P.; Letisse, F.; Matusova, R.; Danoun, S.; Portais, J.C.; et al. Strigolactone inhibition of shoot branching. *Nature* **2008**, *455*, 189–194. [CrossRef] [PubMed]

58. Huang, D.D.; Wang, Y.Y.; Zhang, D.C.; Dong, Y.F.; Meng, Q.X.; Zhu, S.H.; Zhang, L.L. Strigolactone maintains strawberry quality by regulating phenylpropanoid, NO, and H_2S metabolism during storage. *Postharvest Biol. Technol.* **2021**, *178*, 111546. [CrossRef]

MDPI AG
Grosspeteranlage 5
4052 Basel
Switzerland
Tel.: +41 61 683 77 34

Horticulturae Editorial Office
E-mail: horticulturae@mdpi.com
www.mdpi.com/journal/horticulturae

9 783725 837991